Basic and Applied Entomology
— *An Encyclopaedia* —

Basic and Applied Entomology
— *An Encyclopaedia* —

by
K.C. Kumawat
S.R. Kumawat

2013
Daya Publishing House®
A Division of
Astral International Pvt. Ltd.
New Delhi – 110 002

ISBN 9788170359623

Published by : **Daya Publishing House®**

A Division of
Astral International Pvt. Ltd.
ISO 9001:2008 Certified Company
4760-61/23, Ansari Road, Darya Ganj
New Delhi-110 002
Ph. 011-43549197, 23278134
E-mail: info@astralint.com
Website: www.astralint.com

Laser Typesetting : **Classic Computer Services**
Delhi - 110 035

Printed at : **Salasar Imaging Systems**
Delhi - 110 035

PRINTED IN INDIA

DEDICATED

TO

GOD TIRUPATI BALAJI

Preface

The discipline of entomology and pest management is very vast. The different branches of science like biology, ecology, physiology, toxicology, chemistry, biotechnology, microbiology, engineering, plant breeding, economics and many others with diverse background have contributed a lot to its development. As such the terminology used in the subject is vast and of diverse nature, not fixed and arise with the advancement of knowledge. We are engaged in collecting the different terms since our college education from the books of renowned writers only for the sake of convenience in our own lessons. This became a continuous quest during our occupancy as research scientist, as master trainer of officers of the State Department of Agriculture and as teaching professional of undergraduate and post graduate students. Many times, it was felt urgent to publish it as many students and colleagues approached us either before or sometimes after appearing in a competitive examination with problem of some new words; therefore, we find it now as the suitable and high time to take the mission in hand.

The present text *'Basic and Applied Entomology: An Encyclopaedia'* brought on paper largely encompasses the terminology and vocabulary frequently used in literature, lectures and discussions in class rooms and wherever the discipline is involved. The information contained herein is so essential for a vast group of people, students and experts who are increasingly becoming more enthusiastic to perceive the wordings in exact and concrete manner and cognizant for the thirst of knowledge; that it will provide better understanding of the subject. This type of information is hardly accessible at one place. Here, an attempt has been made to compile the fragmented and diffused information and to make available to the readers interested in the subject in a lucid manner. Hope, this will serve as the valuable source of information for the undergraduate and post graduate students of the agricultural universities, students of zoology having special paper of entomology in the universities of basic sciences, teachers, researchers and other cronies who are directly or indirectly interested in the field of entomology and pest management. This should not be understood that the information contained in this edition is self sufficient but it will be an incessant and uninterrupted pursuit of the authors to find new words and that of the inadvertently omitted ones. Any useful suggestions or criticisms from readers and book lovers are heartily welcome for further improvement of the manuscript.

The authors are highly thankful to Dr. A. K. Dahama, Hon'ble Vice-Chancellor, SK Rajasthan Agricultural University, Bikaner for his constant help, guidance and encouragement during writing of this book. We would like to thank Dr. G.L. Keshwa, Dean, SKN College of Agriculture, Jobner for going through the manuscript and imparting constructive suggestions for necessary improvement of the text. We express deep sense of gratitude to Dr. M.C. Bhargava, Associate Professor and Head; Dr. Ashok Sharma, Dr. V.K. Agrawal, Dr. M.C. Gupta and Dr. B.L. Jat, Associate Professors; venerated and renowned teachers and former faculty, Dr. S.C. Bhardwaj, Dr. K.N. Bhatnagar, Dr. P.M. Kanwat, Dr. J.K. Sharma and Dr. S.R. Dadhich, Department of Entomology, SKN College of Agriculture, Jobner; Dr. D.B. Ahuja, Principal Scientist, National Centre for Integrated Pest Management, New Delhi and Dr. R.P. Ghasolia, Asst. Professor (Pl., Pathology), Jobner for generous help and cooperation. Words fail to express our wholehearted gratefulness to Dr. R. C. Kumawat, Associate Professor and Head, Department of Agricultural Economics for vetting the manuscript from language point of view. We express sincere thanks to all the colleagues whose names could not occupy the space and regret for making unintentional omissions. Last but not the least, we wish to thank Mrs. Yamuna Kumawat (wife of author) and Mrs. Prabha Nehra (wife of coauthor) for their calm inspiration and amicable temperament all the time during the course of preparation of manuscript.

K.C. *Kumawat*

S.R. *Kumawat*

List of Standard Abbreviations (Used in the Encyclopaedia)

l	litre
ml	millilitre
mg	milligram
kg	kilogram
g	gram
ha	hectare
pl.	plural
Gr.	Greek
sing.	singular
e.g.	For example
i.e.	that is
L.	Latin
ppm	parts per million
ppb	parts per billion
µg	microgram
A°	Angstrom
λ	Lambda
abbr.	Abbreviation
cal.	Calorie
kcal.	kilocalorie

A

A: It is an abbreviation used for anal vein in the wings of insects.

Abamectin: It is an insecticide of avermectin group (isolated from soil microbe, *Streptomyces avermitilis*. It is effective against sucking insect pests, Dipterans, Psyllids, leaf miners and phytophagous mites, and marketed under trade names, Vertimec, Avid, Agrimec etc. See emamectin benzoate.

Abbott's formula: Mathematical formula given by Abbott which is used to correct the mortality in the untreated check, *viz.*, Corrected per cent mortality = per cent mortality in treatment - per cent mortality in check or control x 100/ (100- per cent mortality in check or control).

Abdomen: The most posterior (rear) of the three main body divisions (tagma) of insects (other two body divisions are head and thorax); it is generally of ten or eleven segmented (rarely twelve segmented) bearing no segmental paired appendages except cerci, styli and external genitalia mostly confined to the terminal segments. The abdominal segments are secondarily reduced in large number of insects. The eleventh segment bears a pair of cerci and the twelfth segment is modified into the paraprocts. The adult insects possess female genitalia on eighth and ninth and male genitalia on eighth (sometimes on eighth to tenth) segments. In Collembola, abdominal segments are reduced to six. The abdomen bears no functional legs in the adult stage. The fleshy abdominal legs (prolegs) found in larvae (Lepidoptera, Hymenoptera) bear a characteristic arrangement of crochets. In the honey bee, the abdomen encloses the honey stomach, true stomach, intestine, sting, and reproductive organs. In spiders, the abdomen is posterior of the two major divisions of body; the anterior one is cephalothorax.

Abdomen belly: Part of the body corresponding to the abdomen having digestive tract in vertebrates, lower orders of animals and insects.

Abdominal appendages, Collembola: The Collembolans do not bear external genitalia. They, however, bear three pregenital abdominal appendages;

collophore, tenaculum and forficula. They are situated medioventrally and are unpaired appendicular structures in the adult Collembola. (i) Collophore: The first abdominal segment bears a median tube or collophore. It is a long cylindrical tube terminating in a bilobed protrusible vesicle filled with the blood. It is a movable structure, articulated by a pair of retractor muscles and functions as an adhesive organ, (ii) Tenaculum: A minute organ present ventrally on the third abdominal segment. It is also called the retinaculum or hamula. It consists of basal corpus and distal paired processes called rami. The rami contain teeth on outer margins and internally an abductor muscle. It helps in springing process by holding the flexed forficula, (iii) Forficula: It is a springing or leaping organ situated ventrally on the fourth abdominal segment. It consists of a basal median manubarium and a pair of arms.

Abdominal gills: These occur as the lateral segmental integumentary outgrowths in the form of foliate or lamellate gills (Ephemeroptera nymphs) or filamentous gills (nymphs of Plecoptera, Zygoptera and Trichopteran larvae). These are richly supplied internally with tracheae. These occur primarily on first seven abdominal segments and form seven pairs of lateral abdominal gills in most of the aquatic forms (Ephemeroptera and Zygoptera) while secondarily these may develop in the form of finger shaped or copious tufts on various parts of the body of Plecopteran nymphs like, submentum, coxae, first two or three abdominal segments or on caudal abdominal region.

Abdominal legs: See prolegs.

Abdominal process: Intercoxal process of abdomen.

Abdominal segments: In the Pterygote insects, the abdominal segments are commonly referred to as the pregenital or visceral segments to those bearing no appendages and the genital segments to those possessing the genital aperture and the external genitalia. There is single genital segment, the ninth in the males, while two (eighth and ninth) in the females.

Abdominal sternite: Ventral sclerite of an abdominal segment (includes concealed as well as externally visible ones). Sternite number corresponds to that of true abdominal segment.

Abdominal style: Paired abdominal appendages.

Abdominal ventrite: Visible ventral abdominal sclerite. Ventrite number does not correspond to true sternite number except in rare cases where sternite one is visible.

Abduction: Refers to retraction or drawing back.

Abductor muscle: A muscle which extends in the appendage or draws it away from the body.

Abiotic: Non-living.

Abiotic damage: Abiotic damage to a crop includes damage by solar radiation, wind, temperature, humidity, pH etc.

Abiotic degradation: Degradation of a chemical or pesticide through physical or chemical mechanisms, *e.g.*, hydrolysis and photolysis.

Abiotic factors: Abiotic means inanimate, inorganic, non-living or physico-chemical factors. In agriculture, abiotic factors like, temperature, moisture, light, heat etc. play important role on the crop production.

Abomasum: Fourth and final region of stomach in *Ruminantia.*

Abrasion: A process which refers to wearing away by rubbing or grinding. It also includes the scrape or scratch that breaks or wears out the cuticle.

Abrasive: Property of any substance or compound which wears or grinds down an article or entity, *e.g.*, airborne dusts have abrasive action on the skin.

Abscission: Formation of a layer of cells in plants which is responsible for fall of leaves, fruits etc.

Absconding swarm: An entire colony of honey bees that abandons the hive because of disease, wax moth and some other maladies.

Absolute alcohol: Ethyl alcohol (ethanol, C_2H_5OH) containing not less than 99 per cent purity by weight is an important solvent and raw material for a number of chemicals.

Absorbent: A substance capable of absorbing other substances usually on the basis of its wetting property.

Absorption: (i) Transfer of a component from one phase to another or movement of a pesticide from the environment (*e.g.*, water, ingested food, leaf surface etc.) across a biological membrane into an organism. (ii) In the cockroach, in which the food is digested largely in the crop, a small amount of absorption, notably of fats, may occur in that segment of the foregut, but the greater part of absorption undoubtedly takes place in the mid gut. The cells seem never to take up solid particles; the dissolved food stuffs like the digestive enzymes diffuse through the peritrophic membrane, when such is present and are absorbed through the epithelial cells. Absorption in the midgut of the cockroach or the locust seems to be largely a matter of facilitated diffusion; for example, as soon as glucose or other sugars enter the circulating blood, they are converted into trehalose. A steep gradient of concentration of the sugar in question as between the gut contents and haemolymph is thus maintained and facilitates uptake by diffusion.

Absorptive clay: A special kind of clay which has ability to absorb the insecticidal spills.

Acalyptrata: The muscid flies in which the calypters are absent or vestigial.

Acanaloniidae: A family of superfamily Fulgoroidea of suborder Homoptera and order Hemiptera. Spines are wanting in their hind tibiae except at apex. Forewings very broad and held vertically at sides of the body at rest.

***Acanthiophilus helianthi* Rossi:** Belongs to order Diptera and family Tephritidae. It is an insect pest of safflower, commonly known as the safflower bud fly.

Acanthus (pl., acanthae): A spine or spur on the insect body.

Acari: An order in class Arachnida of the phylum Arthropoda which includes the mites and ticks. They are characterized by an oval and a minute to small body size.

Acaricidal carbinol: This is one of the most important group of acaricides and derivatives of 1, 1-di (chlorophenyl) methanol; 1, 1-bis (4-chlorophenyl) ethanol (chlorofenthol); 1,1-bis (4-chlorophenyl) -cycloprophlmethanol (procholnol); 1,1-bis (4-chlorophenyl)-2, 2, 2-trichloroethanol (dicofol), 1, 1-bis (4-chlorophenyl)-1 –ethoxycarboxy methanol (chlorobenzilate) etc.

Acaricide (Miticide): Toxic material to acarids (ticks and mites) which are used in their control or management.

Acarine disease: It is a disease of adult honeybee caused by parasitic mite, *Acarapis woodi* (Rannie). The mites enter the tracheal system, suck body fluid, lay eggs and block the tracheae.

Acarinum: Small invagination in the abdomen of old world carpenter bees providing protection for symbiotic mites.

Acarology: Refers to the study of mites.

Acceptable daily intake (ADI): The daily ingested intake of a pesticide (mg kg^{-1} body weight per day) that over the entire life time of a human being (a standard man= 60 kg body weight) appears to be without appreciable health risk on the basis of all known facts at a specified time. It is usually expressed in mg kg^{-1} of body weight per day.

Accessory cells: The cells involved in secreting the cuticular structures which make receptor, *e.g.*, trichogen and tormogen cells.

Accessory genital appendages (Odonata): The genital appendages on venter of the second abdominal segment in Odonata.

Accessory genitalia: It is seen only in Odonata on sternites of second and third abdominal segments.

Accessory hearts: Contractile organs which function to move haemolymph of appendages. Accessory pumps (hearts) are often present in the thorax, these aspirate blood from the thoracic cavity through certain wing veins and return it through connecting vessel either to the aorta itself (*Dytiscus*, *Sphinx*) or to the body cavity again (*Tabanus*, *Vespa*, *Apis*). Accessory pumping organs irrigate the antennae (*Periplaneta*, Hymenoptera etc.), the legs (Hemiptera), and the wings (Diptera). Circulation through the appendages is affected by muscular movements and pressure changes in the abdomen brought about by the respiratory movements.

Accessory lobes (Apicale of tegmen): Paired articulated lobes arising from the apicale of tegmen. Referred to by Spilman (1952) as lateral lobes but the latter term usually synonymous with parameres.

Accessory mesocoxal articulation: A ball and socket joint formed mesally between the mesocoxa and mesosternal or metasternal process. It may consist of a lateral process at the apex of either process and a coxal impression or, occasionally, a mesocoxal knob and mesosternal impression.

Accessory procoxal articulation: A ball and socket joint formed mesally between the procoxa and prosternal process. It may consist of a prosternal knob and coxal impression or occasionally a coxal knob and prosternal impression.

Accessory pulsatile organ: See accessory hearts and pulsatory organs.

Accessory pumps: See accessory hearts.

Accessory reproductive glands: Secretory cells or tissues associated with the internal reproductive organs. See accessory sex glands.

Accessory sex glands: These glands occur in one to four pairs and are either tubular or vesicular. The male accessory glands are of two types: (i) Ectadenia: They are ectodermal in origin and open into the ejaculatory duct in Coleoptera and some other orders, (ii) Mesadenia: They are mesodermal in origin and open into the vasa deferentia, seminal vesicles or even in the ejaculatory duct. These occur in Orthoptera and other insects. In beetles, both ectadenia and mesadenia may occur together. The accessory glands are absent in Apterygota, Odonata and some Diptera. In female cockroaches, a pair of glands open at the venter of ninth abdominal segment secreting an adhesive material for the formation of a case known as ootheca for the eggs; in males, the mushroom shaped glands open in the ejaculatory duct and secrete substances which mixed in sperms and facilitate sperm transmission to female.

Accessory vein: An extra branch of a longitudinal vein or a secondary longitudinal vein in an insect wing.

Accretive release: A method of periodic introduction of biotic agents (natural enemies of insects) in which annual early season liberations of biotic agents or natural enemies against fairly abundant pest populations increase naturally in response to rising pest densities as the season progresses.

Accumulative pesticides: The pesticides which accumulate in the animal system or in the environment.

Acephate: See Orthene.

Acerentomidae: A family of order Protura. Tracheae and spiracles absent. Third abdominal appendage one-segmented. Two transverse rows of setae on abdominal tergite, with one two grooves and laterotergites. Combs are usually present on the eighth abdominal segment. For example, *Acerentomon* sp., *Gracilentulus* sp.

***Aceria* sp. (*Eriophyes* sp.):** Belongs to order Acarina and family Eriophyidae. It is a pest of fruit and ornamental plants, commonly known as the leaf curl mite.

***Aceria jasmini* Chanana:** Belongs to order Acarina and family Eriophyidae. It is a pest of ornamental plants, commonly known as the eriophyid mite.

***Aceria mangiferae* Sayed:** Belongs to order Acarina and family Eriophyidae. It is a mite pest of mango, commonly known as mango bud mite.

Acetamiprid: An insecticide of neonicotinoid group. Chemically, it is (E)-N^1 (6-chloro-3-pyridylmethyl)-N^2- cyano- N^1–methylacetamidine. It is marketed under trade name Pride 20 SP. Mode of action is same as imidacloprid and effective against insect pests of vegetables, fruit trees, tea etc.

Acetyl cholinesterase: An enzyme that catalyzes the breakdown of acetylcholine. See choline acetyl transferase also.

Acetylcholine (ACh): A chemical which forms at the axonal terminals and stored there in the synaptic vesicles (located at axonal terminals) to conduct the nerve impulse over the gap between the nerves and muscles or glands, thus acts as a synaptic neurotransmitter. In normal nerve activity, it is hydrolyzed by the enzyme acetyl cholinesterase after each impulse. Some pesticides work by blocking the function of acetyl cholinesterase enzyme.

Acetylcholine (ACh) receptors: These are specific receptor sites on the surface of post-synaptic membrane which allow interaction of the Ach molecules. These receptors are made up of two functional units- one is the recognition site binding ACh and another is the ionic channel which lets the ions move through the membrane down their electrochemical gradient.

***Achaea janata* Linn.:** Belongs to order Lepidoptera and family Noctuidae. It is an insect pest of castor, commonly known as the castor semilooper.

***Acherontia styx* Westwood:** Belongs to order Lepidoptera and family Sphingidae. It is an insect pest of sesame, commonly known as the til hawk moth.

Acicular: Pointed, needle shaped.

Acid: A compound whose H atom is replaceable by positive ions or radicals to form salts.

Acid rain: Burning of fossil fuels like coal etc. increases oxides of nitrogen and sulphur in atmosphere; they interact with rain water and converted into strong acidic compounds such as sulphuric acid and nitric acid and come down as acid rain.

Acidic: Acid in reaction resulting from excess of H^+ ions over hydroxyl ions ($OH\text{-}^-$) in solution.

Acidopore: Flexible setae fringed nozzle in formicine ants.

***Acigona steniellus* Hamps:** Belongs to order Lepidoptera and family Crambidae. It is an insect pest of sugarcane, commonly known as the Gurdaspur borer.

Acinar: Resembling a cluster.

Acinus: Resembling a single grape in a cluster.

Aclerdidae: A family of superfamily Coccoidea, suborder Homoptera and order Hemiptera.

Acone eye: A type of compound eye in which the ommatidia lack both crystalline cone and corneal exocone lenses, in place of which there is a group of elongated transparent cells. See also ommatidium, types.

Acoustical communication: Communication by sound and hearing.

Acre: An area equivalent to 43,560 $feet^2$, used in measurement of land.

Acrididae: A family of order Orthoptera, suborder Caelifera and super family Acridoidea. Pronotum laterally compressed, largest group of grasshoppers and locusts.

Acridoidea: A super family of order Orthoptera and suborder Caelifera. Tarsi three segmented, *e.g.*, locusts and short horned grasshoppers.

***Acridotheres tristis* (Linn.):** Belongs to order Passeriformes and family Saturnidae (class Aves). It is a bird predator of insect pests, commonly known as the common myna.

Acron: Anterior, preoral body cap of segmented animals (arthropods), the portion of a sternum anterior to the antecostal suture.

Acrosternite: The narrow marginal projection anterior to the antecosta of a definitive sternal plate that includes the preceding primary intersegmental sclerotization, characteristic of abdominal sterna of insects but it is absent on thoracic sterna.

Acrostichal bristles: Two rows of bristles on middle of the dorsum of mesothorax in Diptera.

Acrotergite: The portion of a tergum anterior to the antecostal suture.

Acrotrophic egg tube: Egg tube in which the trophic cells are apical.

Acrylonitrile: A highly flammable and toxic fumigant. Acrylonitrile (66 per cent) is mixed with carbon tetrachloride (34 per cent) before use to reduce the risk of fire and explosion. This mixture is a good fumigant, effective against nearly all insects and is safe to apply to food processing units as does not affect viability of seeds and does not leave any residue, however, it must be handled carefully with safety and the person applying it must wear gas mask due to its high mammalian toxicity.

Actaletidae: A family of order Collembola. Third abdominal tergite much reduced and abdominal segments 4 - 6 fused; hypognathous head and have tracheae. For example, *Actaletes* sp.

Action potential: A brief electrical depolarization that propagates along a fibre.

Activated charcoal: Finely ground high quality charcoal which absorbs liquids and gases easily.

Activation centre: A region in a fertilized egg where cleavage cells induce- further growth and development.

Activator: A substance added to a pesticide that increases its toxicity resulting in more effective control.

Active immunity: Resistance to a disease acquired by an animal as the result of antibody production in response to antigens.

Active ingredient (A.I.): Pesticide present in a formulation as described by the common name or the part of a pesticide formulation from which the biological effect is obtained.

Active principle: Part of a substance which has the biological effect, not the inert carrier or other substance added to it.

Active space: The space within which the concentration of a pheromone or other behaviourally active substance is concentrated enough to generate the required response, remembering that like light and sound, pheromones become more dilute, further they radiate out from their source.

Active ventilation: Ventilation aided by movements such as muscular pumping.

Actual dosage: The amount of active ingredient (not formulated product) which is applied to an area or other target.

Aculea (pl., aculeae): Fixed hair of the cuticle.

Aculeate: Pertaining to the stinging Hymenoptera (Aculeata) which includes bees, ants and many other wasps.

Acuminate: Tapering to a long point or ending in a point.

Acute: (i) Refers to short duration, characterised by sharpness or severity. Opposite of chronic. (ii) Pointed, forming an angle of less than 90°.

Acute dermal toxicity: Resultant poisonous effect of a single dose of chemical or pesticide taken through skin.

Acute inhalation toxicity: Resultant poisonous effect of a single dose of chemical or pesticide taken through inhalation in lungs.

Acute oral toxicity: Resultant poisonous effect of a single dose of chemical or pesticide taken through mouth.

Acute poisoning: See acute toxicity.

Acute toxicity: The toxicity that causes damage or death from a single dose of chemical or exposure. The toxicity of the chemical is determined at the end of 24 hours.

Adaptability: Capacity of evolutionary change.

Adaptable: Capable to adapt due to inheritable (and/ or non-inheritable), structural or functional changes.

Adaptation: Any morphological, physiological or behavioural change in an organism which enables it to take advantage of its environment and enhances its chance of survival.

Adaptation importation: Importation involving the introduction of those exotic natural enemies of insect pests which have fortuitously adapted to a native pest at other places where it has established.

Adaptive colour change: A rapid change in body colour of an organism in response to a change in colour characteristics of the surrounding environment.

Adaptive radiation: Evolutionary variance in members of a single phyletic line in a series of rather different niches or adaptive zones.

Additive: Any material which is added to a pesticide, *e.g.*, wetting agent, surfactant etc.

Adductor muscle: Any muscle that draws an appendage into the body.

Adecticous: A type of pupa in which the mandibles are immovable and non-functional.

Adelgidae: A family of superfamily Aphidoidea, suborder Homoptera and order Hemiptera. Sexual and parthenogenetic females oviparous; Rs not separate on forewings; Cu arises separately from A on forewings; wings held roof like over abdomen when at rest. Apterous parthenogenetic females covered with flocculent wax. Infest conifers.

Adelphoparasitism: See autoparasitism.

Adenotrophic viviparity: See viviparous.

Adephaga: A suborder of order Coleoptera. This includes a single superfamily Caraboidea. These are mostly predatory in habit and a few are phytophagus.

Adephagan aedeagus: Aedeagus which lacks a phallobase so that parameres are articulated directly with the base of penis and sometimes to each other. This type occurs in the suborders Archostemata, Myxophaga and Adephaga.

Adfrontal area: The area between the adfrontal sutures and the ecdysial lines on the lepidopterous larval head.

Adherence: The ability of a material to stick to a particular surface.

Adhesive: See sticker.

Adhesive agent: Material added to the formulated pesticide to increase its retention power on the treated surface or plant.

Adhesive organs: Many insects can walk on smooth leaves and even glass surface which is possible due to adhesive organs which are commonly the expansions of tarsal surface or of special pulvilli. The usual mechanism seems to be the provision of 'tenant hairs' moistened with greasy secretions which have soft delicate extremities that can be brought into such intimate contact with the surface that seizure or adhesion takes place and the insect is held by surface molecular forces.

Adipocytes: These are blood cells laden with fat globules and charged with wax. They are found in certain Hemiptera and their relation with other blood cells is uncertain.

Adipokinetic hormone (AKH): Holwarda *et al.* (1977) have analyzed the intrinsic hormone from the glandular lobe of corpus cardiacum of locust and found that it is the octapeptide of molecular weight, 850 Daltons. Each corpus cardiacum was found to contain about 0.2 ng hormone. In locust, the same substance acts as the adipokinetic as well as the hyperglycemic hormone.

Adjuvant: A spray additive substance that improves the properties of a pesticide formulation, *e.g.*, wetting agents, spreaders, emulsifiers, dispersing agents, foam suppressants and penetrants.

Adjuvant oils: These are vegetable and mineral oils used to improve the effectiveness of pesticide sprays. Vegetable oils are used as sole pesticide carriers in the control droplet application sprayer, whereas, the mineral oils are used exclusively as adjuvant in water based sprays. The oils help to keep the pesticide diluted and reduce the rate of evaporation on treated surface leading to adsorption.

Adoption substance: A substance secreted by the social parasite which is accepted by the host insect as a member of the colony.

***Adoretus nitidus* Arrow:** Belongs to order Coleoptera and family Rutelidae. It is an insect pest of ber, commonly known as ber beetle.

***Adoretus pallens* Arrow:** Belongs to order Coleoptera and family Rutelidae. It is an insect pest of ber, commonly known as ber beetle.

***Adoretus* sp.:** Belongs to order Coleoptera and family Rutelidae. It is commonly known as defoliating beetle and is a minor pest of ber and grapevine.

Adrenergic blocking agents: Agents that block the nerve fibres which liberate sympathin, a substance like adrenaline.

Adsorption: The phenomenon by which materials are held or bound firmly to the surface in such a manner that chemical is available slowly. In many cases, clay soils and soils that contain high amount of organic matter tend to adsorb pesticides.

Adult: The last or mature stage in the life cycle of an insect which attained puberty and ready to reproduce to begin offspring.

Adult transport: The carrying or dragging of an adult social insect by a nest mate, generally during the colony emigrations. It is very common and frequent in ant which is stereotyped behaviour.

Adulterate: To reduce the purity of a material below the standards it is supposed to represent. For example, a pesticide that does not conform to the professed standard or quality as documented on its label.

Adulterated: When any foreign substance is added in the chemical or any part of it is subtracted, the purity of chemical is deteriorated and becomes sub-standard.

Adulterated honey: Any product labeled 'Honey' or 'Pure Honey' that contains ingredients other than honey but does not show these on the label. (Suspected mislabeling should be reported to the Food and Drug Administration or Quality Control Administration).

Adulticide: Chemical substance that kills the adult stage of arthropod pests (insects, mites).

Adultoid reproductive: In the higher termites, a supplementary reproductive individual which is fully developed imago and is morphologically indistinguishable from the primary reproductive individual it replaces.

Adventitious vein: A secondary vein which is neither accessory nor intercalary but usually the result of a number of cross veins lined up to form a continuous vein.

Adverse effect: Change in morphology, physiology, growth and development or life span of an organism which results in impairment of functional capacity or which increases susceptibility to ions under which molecular oxygen serves as the terminal electron acceptor in respiration and the harmful effects of other environmental influences.

Aedeagal apodeme: An apodeme of the aedeagus.

Aedeagus: The external genital apparatus in insects, including a phallobase, parameres (sometimes absent or fused together) and a penis containing an invaginated endophallus, but excluding the genital capsule.

***Aedes*:** A genus of mosquito.

Aeolothripoidea: Superfamily of suborder Terebrantia, order Thysanoptera. Ovipositor curved upwards; forewings broad and rounded at apex, with two longitudinal veins, front margin without the fringe of long hairs. Antennae nine segmented. This super family is represented by a single family Aeolothripidae, *e.g., Mymarothrips, Franklinothrips, Erythrothrips,* etc.

Aerate: To expose the grain receptacle to fresh air in order to remove the toxic gas of a fumigant chemical.

Aerial: Atmospheric or belonging to the air.

Aerial application: Pesticides applied with the use of helicopter or aeroplane when the incidence of a pest or disease occurs over a wide area. Such pesticides are applied either without dilution of pesticide with water or with very little quantity of water.

Aerial dusting: Application of dust formulation of pesticide either with helicopter or aeroplane in case of pest problem over a wide area.

Aerobic: Conditions under which molecular oxygen serves as the terminal electron acceptor in respiration or in metabolic oxygenation.

Aerobic respiration: It is the oxidative process of liberating energy within the cells.

Aeropyle: The openings in the chorion (egg shell) through which air enters.

Aerosol: A system of fine solid or liquid particles (0.1-5.0 μm diameter) dispersed from a gas aerosol can using an inert compressed propellant. It leaves very small droplets of pesticides suspended in the air. It is a common mean of dispensing insecticides for domestic use.

Aeshnidae: A family of order Odonata. The largest and most powerful group of dragonflies with compound eyes in contact for a considerable area on dorsum of head. Examples, *Anax, Aeshma, Epiaeschna*.

Aeshnoidea: A super family of order Odonata. Antenodals incomplete, triangular area of fore and hindwings not different in size and position,

Aesthetic pest: A pest whose mere presence is objectionable for psychological reasons.

Aestivate: To pass the summer in a dormant state.

Aestivation: Dormancy during a warm or dry season which allows insects to survive during periods of extreme heat or drought.

Aeterogonic growth: Growth patterns in which some parts of the insect body develop at different rates than others.

Afferent neurone: A nerve that conducts from the periphery towards a nerve center, the axon of a sensory neurone. See sensory neurone also.

Affinity: Likeness, especially in relationship.

Afforestation: Replanting of trees in the areas of deforestation.

AFID: An abbreviation stands for alkali flame ionization detector. It is used in gas chromatography.

Aflatoxin: Toxic organic compound (metabolite) which is produced by the fungus, *Aspergillus flavus*. This fungus grows on groundnut kernels, maize grains etc. and produce aflatoxin responsible to cause jaundice. This toxin is also carcinogenic.

African sleeping sickness: A disease caused by infection with *Trypanosoma* sp., protozoans and transmitted by the bite of the tsetse fly.

After swarm: A small swarm, usually headed by a virgin queen which may leave the hive after the first or prime swarm has departed.

Agamic reproduction: Reproduction without union of gametes. Reproducing parthenogenetically is an example of agamic reproduction.

Agamospecies: An asexual species whose members are of common origin.

Agar gel: A gel formed of agar-agar which is dried mucilaginous substance obtained from certain oriental seaweeds.

Agaristidae (Phalaenoididae): Family of super family Noctuoidea, suborder Ditrysia and order Lepidoptera. Mostly tropical insects, diurnal in habit and their coloration and antennal structure differentiate them from Noctuidae. The common Indian species are *Aegocera venulia, A. bimacula* and *Exsula victrix.*

Age distribution: The proportions of individuals in different age groups of a population at any given time.

Age polyethism: The regular changing of labour roles by colony members as they grow older.

Aged residue: Pesticide and degradation products present in environmental system after application and following a period long enough to allow transport, adsorption, metabolism and dissipation process to alter the distribution and chemical nature of the applied pesticide.

Aggregate sample: Sample made up of set proportion of other samples, typically an average.

Aggregation: Assembling at a place by members of a species in response to an attractive stimulus.

Aggregation pheromone: An intraspecific chemical messenger that causes insect congregation at food sites, reproductive habitats or hibernation sites.

Aging: All changes in structure and function that occur from the beginning to the termination of life.

Agitate: To keep the pesticide materials mixed up or to keep it from separating or settling in the spray tank.

Agitator: A mechanical device in the spray tank which remain in motion during spray operation to ensure uniform distribution of toxicant and to prevent sedimentation by keeping pesticide in the suspension.

Aglycon: Non-sugar part of glycoside or glucuronide conjugate derived from pesticide.

Agricultural chemical: Chemicals used to enhance agricultural production and to protect crops from insect pests and diseases (plant hormones, chemical fertilizers, pesticides etc.).

Agricultural Chemicals Approval Scheme: A government scheme under which proprietary brands of crop protection chemicals can be officially approved by the Agricultural Chemicals Approval Organization for specific uses in U.K. conditions if label and advertising claims are fulfilled. The scheme enables users to select and advisors to recommend efficient and appropriate chemicals and to discourage the use of unsatisfactory ones. A list of approved chemicals is published annually under this scheme.

Agricultural commodity: Any plant or part of a plant, animal or animal product which is to be bought or sold.

Agrioidea: A super family of order Odonata. Arculus nearer to wing base than to nodus; antenodals complete, five or more distinct.

Agrionidae: A family of order Odonata. Large broad winged damselflies with base of wings devoid of stalks, *e.g.*, *Coenagrion* sp. and *Ischnura* sp.

Agrochemical: Agricultural chemical used in crop and food production including pesticide, feed additive, veterinary drug and related compounds.

Agroecology: It is the study of ecology in relation to agricultural systems.

Agroecosystem: The ecosystem composed of cultivated land and surrounding or intermixed uncultivated areas, the plants contained or grown thereon and associated flora and fauna. This is a man-made crop ecosystem to maximize the production of an agricultural commodity and differs from natural ecosystems.

Agromyzidae: A family of order Diptera.

***Agrotis flammatra* Schiffer Mueller:** Belongs to order Lepidoptera and family Noctuidae. It is a species of cutworm or surface caterpillar, commonly known as the gram cutworm.

***Agrotis ypsilon* (Rott.):** Belongs to order Lepidoptera and family Noctuidae. It is a species of cutworm or surface caterpillar, commonly known as the greasy cutworm.

Air blast nozzle: See nozzle types.

Air blast sprayer: A spray machine adapted for high and low volume spray. It is used for spraying the gardens, orchards, shade trees and vegetables to control insect pests.

Air mediated transfer of pesticide: It includes carrier transport of pesticides in water vapour or dust in addition to evaporation of pesticides through volatilization. The deposition of the particles to earth is mainly through the rain and to lesser extent through settling.

Air pollution: Air pollution can be defined as any atmospheric condition in which certain substances are present in such conditions that they can produce undesirable effects on man and his environment. The air pollutants include gaseous pollutants and particulate pollutants. The primary gaseous pollutants include CO, CO_2, SO_2, NO, hydrocarbons and most suspended particulate matter (SPM) and secondary gaseous pollutants include NO_2, SO_3, HNO_3, H_2SO_4, O_3, most NO_3 and SO_4 salts, peroxyacyl nitrate (PAN), aldehydes, ketones etc. The particulate atmospheric pollutants are not gases but may be solid particles, suspended droplets or mixture of both. These include aerosols, smoke and grit, smog, heavy metals (like Pb, Mn, Ni etc.), pesticides and radioactive elements etc.

Air quality standards: Specifications of air pollutants that are considered tolerable for human beings.

Air sac: A dilated portion of a trachea or pouch like expansion of a trachea which is thin walled and lacks taenidia in the walls. Treviranus suggested the main function of these collapsible tracheae or air sacs as the ventilation of respiratory system, likewise lungs in vertebrates. This aids in the flight of insects due to decreasing the specific gravity.

Air sac mite (Pulmonary mite): *Cytodites nudus*, a mite pest of fowls present in respiratory tract and certain internal organs.

Ak grass hopper: *Poekilocerus pictus* Fab. is a pest of papaya which also causes damage in brinjal, fig, castor and tomato. The nymphs and adults feed on foliage causing defoliation. The pest voraciously feeds on ak, *Calotropis procera*, a perennial weed.

Akinesis: The state of immobilization.

Alar squama: The middle of three flap-like outgrowths at base of the wing in various flies.

Alarm defense system: Defensive behaviour which works as alarm in the ants.

Alarm pheromone: An intraspecific chemical messenger produced in the exterior environment to elicit aggressive or defensive behaviour. The honey bee produces 2-heptanone in the mandibular glands and a series of other compounds in the sting apparatus. Isoamyl acetate, the principal alarm pheromone of the sting apparatus is used to signal other workers to sting at the same place, thus amplifying the defense capabilities of the group.

Alarm recruitment system: A communication system that enables the nest-mates to come together at a particular place to aid the defense of colony. The odour trail system in the species of lower termites is prevalent due to which the colony members are recruited to fight against the intruders.

Alary muscle: A wing shaped muscle that connects the heart to the lateral portions of terga, also called aliform muscle or wing muscles of the heart.

Alary polymorphism: Morphism due to intraspecific diversity in wing, *e.g.* elate, brachypterous, cropterous etc., also called pterygopolymorphism.

Alatae: Winged forms.

Alcohol: The ethyl alcohol (C_2H_5OH) is commonly referred to as alcohol, the term applied to hydroxyl organic compounds beginning from one carbon compound methanol or methyl alcohol (CH_3OH). The methyl and ethyl alcohol are common solvents used in the pesticide formulations.

Aldicarb: A highly toxic systemic oxime carbamate insecticide, acaricide and nematicide, chiefly used as soil treatment against soil insect pests. It is chemically, 2-methyl-2-(methylthio) propionaldehyde O-(methylcarbamoyl) oxime ($C_7H_{14}N_2O_2S$). It is white crystalline solid, slightly soluble in water (0.6 per cent or 600 mg l^{-1}) but more soluble in organic solvents. It is highly toxic to mammals. In plants, it is oxidized to sulfoxide which is further oxidized to less toxic sulfone. It is available commercially as Temik 10 G having systemic action.

Aldicarb, metabolism: The metabolism of aldicarb is different than other carbamate insecticides because it has an extra hydrolysable bond. Since, it is also a thioether, oxidation is followed by detoxification. It is rapidly oxidized to sulfoxide but much more slowly to sulfone. In all cases, the hydrolysed products (oximes) become propionaldehyde analogues which are further oxidized to acids. In the rats, aldicarb is metabolized and extracted in the urine principally as the aldicarb sulfoxide and oxime sulfoxide. Nitrite fragments are also found in animal and plants in addition to the metabolites mentioned above.

Aldrin ($C_{12}H_8Cl_6$): A synthetic insecticide; a chlorinated hydrocarbon of not less than 95 per cent 1, 2, 3, 4, 10, 10-hexachloro-1, 4, 4a, 5, 8, 8a- hexahydro-1, 4-endo exo-5, 8-dimethano- naphthalene; moderately toxic to mammals, acute oral $LD\text{-}_{50}$ for rats is 44 mg kg^{-1}; phytotoxicity is none when properly formulated but some crops are sensitive to solvents in certain formulations. It is a contact and stomach insecticide. In 1949, Kearns and his associates discovered the toxic properties of aldrin. It was named after a German chemist Kurt Alder who got Nobel Prize in chemistry in 1950 along with his co-worker Otto Diels for the work on diene synthesis. Their Diel-Alder reaction is very popular and is used in the formulation of cyclodienes. The technical material is a tan to brown solid containing about 78 per cent active aldrin. The pure material is white crystalline with melting point 104 °C. The impurities in technical material slowly form HCl in the storage; hence epichlorohydrin which inhibits acid is added as stabilizer. It is oxidized to its epoxide dieldrin in soil, insect body and liver of animals.

Aldrin, metabolism: Aldrin is converted to its epoxide analogue dieldrin by mammals, soil microorganisms, plants and insects. Dieldrin is further metabolized into hydrophilic metabolites (2-ketodieldrin in urine of rat and sheep, 9-hydroxydieldrin in faeces of rat and sheep, *cis*-aldrindiol in cockroach and *trans*-aldrindiol in rabbit). The active insecticide is excreted either as the insecticide itself or as a metabolic product and is not stored in the animal tissue.

Aleukia: The situation in which number of leukocytes (white blood corpuscles) are decreased drastically in the blood.

***Aleurocanthus woglumi* Ashby:** Belongs to order Hemiptera and family Aleyrodidae. Commonly known as citrus blackfly and is an insect pest of citrus, avocado, grapevine, mango, guava, pear, plum etc.

***Aleurolobus barodensis* Maskell:** Belongs to order Hemiptera and family Aleyrodidae. It is an insect pest of sugarcane, commonly known as the sugarcane whitefly.

***Aleurothrips fasciatipennis*:** A predatory thrips of citrus whitefly, *Dialeurodes citri* (Ashmead).

Aleyrodidae: A family of superfamily Aleyrodoidea of suborder Homoptera and order Hemiptera. These are small inconspicuous insects, primarily restricted to tropics and subtropics but a few are found in the cooler regions. All species are found on the leaves of plants, mostly on the undersurface. Examples, sugarcane whitefly, *Aleurolobus barodensis*; sugarcane spotted aleyrodid, *Neomaskellia bergii*; castor whitefly, *Trialeurodes barodensis*; cotton whitefly, *Bemisia tabaci*; citrus whitefly, *Dialeurodes citri*; citrus blackfly, *Aleurocanthus woglumi* etc.

Aleyrodoidea: A superfamily of suborder Homoptera and order Hemiptera.The aleyrodoidea consists of a unique small family Aleyrodidae commonly called whiteflies or aleyrodids.

Alga (pl., algae): Primitive aquatic green plants ranging from unicellular species to giant sea weeds. Algae are heterogeneous group of chlorophyll bearing vascular thallophytes without true roots, stems and leaves in their thalloid plant body.

Algicide: A chemical substance used to control algae.

Algology: (L., *alga*= sea weed, *logos*= study) or phycology (Gr. *phycos*= sea weed). The study of algae is called algology.

Algorithm: A precisely defined sequence of rules stating how to produce specific outputs from specific inputs through a given number of steps.

Alienicolae: Wingless female aphids that reproduce parthenogentically in enormous numbers on herbaceous plants (secondary hosts) throughout the spring and the summer; eventually they give rise to winged forms that fly to another plant (primary host) in the autumn and reproduce by a normal sexual process.

Aliform muscles: Muscle fibres.

Alighting board: A small projection or platform at the entrance of the hive.

Alimentary canal: The tube of the digestive system, extending from the mouth to the anus, broadly dividing into foregut, mid gut and hind gut. It is merely a simple tube or completely degenerated as the nutrients are absorbed directly through general body surface exceptionally in parasitic Hymenoptera. It is short in length in the primitive insects, such as Aperygota, Ephemeroptera, Odonata, Dermaptera, Phasmida and the larvae of butterflies, moths, bees, wasps, mosquitoes etc, while it is enormously elongated in some insects, particularly, in Homoptera and Cyclorrhapha (Diptera). The alimentary canal is often longer in the phytophagous insects than in carnivorous insects, though, it cannot be applied strictly as a general rule for all groups of insects.

Alinotum: Wing bearing notal plate of the dorsum of mesothorax or metathorax of Pterygote insects.

Aliphatic: A term applied to the open chain or fatty series of hydrocarbons.

Aliphatic compounds: Compounds possessing open-chain structures. Generally, these are considered less toxic to plants than aromatic compounds.

Aliquot: Known fractional portion of a homogeneous material. The term is usually applied to volumetric sub sampling fluids.

Alitrunk: That part of the thorax to which the wings are attached; in the higher Hymenoptera this includes also the first abdominal segment (propodeum).

Alkali: Inorganic compound, any soluble hydroxide (OH) of the alkali metals: lithium, sodium, potassium, rubidium, and cesium. More broadly, ammonium hydroxide (*see* ammonia) and soluble hydroxides of the alkaline earth metals are also called alkalies. Strong bases that turn litmus paper blue, they react with acids to yield salts, are caustic, and in concentrated form corrode tissues. Sodium hydroxide (caustic soda) and potassium hydroxide (caustic potash) are very important industrial chemicals, used in the manufacture of soaps, glass, and many other products. The mineral trona, a compound of sodium carbonate and bicarbonate, is one naturally occurring alkali. It may be mined or recovered from dry lake beds. It is the opposite of an acid. It is usually dangerous in concentrated form.

Alkaline: Having the reaction of an alkali resulting from an excess of hydroxyl ions (OH) over hydrogen (H^+) ions in solution and pH more than 7.0.

Alkaline soil: Soil containing alkali salts, usually in amounts that are harmful to crop growth. The pH above 8.5 is detrimental to most of the crops.

Alkaloids: Many substances found in plants having powerful pharmacologic action, and characterized by content of nitrogen and the property of combining with acids to form salts.

Allantoine: Many insects will break down their uric acid wholly or in part to allantoine or allantoic acid. Allantoine is plentiful in the excreta of blow fly maggots. This substance was isolated from the secretions of certain blue bottle flies to heal infected wounds of soldiers in first and Second World War. The enzyme uricase converts uric acid to allantoine is particularly active in many Heteroptera (*e.g.*, *Dysdercus*) which excrete most of their nitrogen in this form.

Allastostatin (Allastostatic hormone, ASH): It is secreted by the cerebral neurosecretory cells, mostly the lateral neurosecretory cells, as found in locusts. It inhibits juvenile hormone synthesis in corpora allata during vitellogenic cycle of egg maturation of virgin female cockroach, *Diploptera punctata* through phosphatidyl inositide pathway.

Allatectomy: Removal of the corpora allata.

Allatotropin (Allatotropic hormone, ATH): The insect peptide neurohormone which stimulates juvenile hormone synthesis. It is secreted by the cerebral medial neurosecretory cells (MNC), mostly during development and reproductive period.

Allele: A Greek word which means belonging to one another and used to refer to the individual members of a gene pair.

Allelochemics (Allelochemicals): Substance emitted by an organism that mediates inter-specific interactions among organisms and modifies the be-havioural response in species other than the emitter. Depending on the type of the response induced, these chemicals have been classified into four groups, *viz.*, allomones, apneumones, kairomones and synomones.

Allelomorph: Contrasting pair of traits for expression of contrasting feature of a character are called allelomorph or allele. For example, factors for length of pea plant T and t are allelomorph or allele.

Allergens: Insect secretions, usually proteins whose effects depend on the physiology.

Allergy: Refers to hypersensitivity of the body to a physical or chemical agent.

Allethrin: A synthetic pyrethroid insecticide having quick knock-down and insect flushing action. Allethrin is highly effective against some insects, but less effective against others. Allethrin has low toxicity and is used selectively since it is highly effective against some insects but less effective against others. It is especially useful as a quick knock-down insecticide for almost any type of insect or arachnid pest. It is related to the plant derived pyrethrins; relatively harmless to warm blooded animals with acute oral LD_{50} for rats, 920 mg kg^{-1}.

Allochronic species: Species not occurring at the same time interval; not contemporary.

Allometabola: See neometabola.

Allometric growth: Abnormal development of one part of the body which is also known as heterogenic growth. The disproportion generally follows the allometric law, *i.e.*, the logarithm of dimension of the part is proportional to the logarithm of the dimension of the whole: $y = Kx^a$, where, x = dimension of the whole; y = dimension of the part; a = growth coefficient, and K = another constant, each part increasing geometrically at the rate peculiar to itself. See allometry also.

Allometry: The study and measurement of the relative growth of a part of an organism in combination of the whole. Any size relation between two body parts that can be expressed by $y = bx^a$, where a and *b* are fitted constants. In the special case of isometry, a = 1, and the relative proportions of the body parts, therefore, remain constant with change in total body size. In all other cases, the relative proportions change as the total body size is varied.

Allomone: A compound released by one organism which evokes a behavioural or physiological reaction in an individual of a different species (receiver) and that is favourable to the emitter but not to the receiver, *e.g.*, (i) Plant allomones which deter feeding; and (ii) Alarm pheromones in ants. Thus, it is an interspecific chemical messenger that benefits the releaser but not receiver organisms. See kairomone also.

Alloparental: When individuals other than the parent assist in the caring for that parents offspring.

Allopatric: Populations or species which are mutually exclusive, but usually inhabit adjacent geographical regions. These are geographically isolated. Thus, two or more forms of species have essentially separate distributions. See sympatric also.

Allopatric hybridization: The crossing between individuals belonging to two allopatric populations along a well defined contact zone.

Allopatric speciation: Formation of species during geographical isolation.

Allotype: A paratype of opposite sex to the holotype.

Ally: Any organism akin to another by structure etc.

Allyl acetate: Allyl acetate is a colourless liquid fumigant with acrid odour having molecular weight 100.1 and boiling point 103.5 °C, insoluble in water, highly inflammable and has low vapour pressure. It is toxic to human by ingestion, inhalation and contact, however, not carcinogenic. The rust red flour beetle is tolerant to allyl acetate, whereas, all developmental stages of lesser grain borer are so susceptible that cent per cent mortality is achieved at the dosage of 150 g per m^3. A dosage of 200 g per m^3 with exposure period of 72 hours effectively controlled the stored grain insect pests of wheat.

Almond moth: See *Cadra cautella.*

Alphataxonomy: A taxonomic level concerned with the characterization and naming of species.

Alpine: In mountains the region above the tree line.

***Alternaria*:** A saprophytic fungus which is responsible for causing early blight and rot diseases in many plants.

Alternating generations: When two generations are produced within a life cycle each producing individuals of only one sex, either male first and then female or *vice-versa.*

Alternative host: Host plant other than the main host on which an organism (bacterium, fungus, insect, virus etc.) can develop to complete its life cycle. Weeds and wild plants are often alternative hosts of certain insect pests and disease causing microorganisms.

Alticidae (formerly Halictidae) (flea beetles): Family of superfamily Chrysomeloidea, suborder Polyphaga and order Coleoptera. These are small jumping beetles, black, blue, greenish or brownish in color. Some have light markings. Though it closely resembles Chrysomelidae this group is characterized by closely approximated antennae, globose forecoxae, deeply bilobed third tarsal segment, grooved tibiae and swollen hindfemora. A number of flea beetles are known as important pests. *Longitarsus belgaumensis* on sunnhemp, *L. nigripennis* on pepper, *Phyllotreta downsei* on cruciferous vegetables and *Chaetocnema pusaensis* on millets are well-known pests in India. The steel blue beetle *Altica cyanea,* often found in large numbers on rice feed on the weed plant *Ammonia* in rice fields. It also breeds on the foliage of singhara (*Trapa bispinosa*) crop in Delhi.

Altosid (Methoprene): A low toxicity compound. Altosid is a slow release larvicide that is used in mosquito larvae control in water before they mature to adults. Altosid can be used to kill mosquito and other water breeding insects around the house.

Altruistic: Self destructive or potentially self destructive behaviour performed for the benefit of others.

Alula: In insects (not birds), the outermost of the three flap-like outgrowths at the base of wing in various flies, really a part of the wing membrane.

Aluminium phosphide: It was introduced as a fumigant pesticide by A.G. Degesch in 1955. In India, the aluminium phosphide tablets of 3 g and 1 g, capable of liberating 1/3 in weight as phosphine gas (PH_3) are available under the trade names of Celphos, Quickphos, Phostoxin, Phosfume etc. It produces carbide or garlic type odour. It is heavier than the air and has low solubility in water. It is highly toxic to all stages of insects having good penetration power and also highly toxic to mammals. It is formulated as tablets, pellets or small sachets of powder in trilaminated pouch with additional materials such as ammonium carbamate (40 per cent) and aluminium oxide (5 per cent) to regulate the release of fumigant and to suppress flammability. It affords effective control of stored grain insect pests infesting cereals, millets, pulses, milled products, spices, oil seeds, oil cakes, bales and other packages. It is banned for general public use and used by public agencies only. There is no chemical reaction of the PH_3 with the treated material at recommended dosage but the food stuffs at high moisture content will increase the total phosphorus content. No ill effect on the organoleptic qualities of cooked food is observed. The MRL in India on food grains and milled food grains are 0.05 and 0.01 ppm, respectively.

Alutaceous: Covered with minute cracks like the human skin.

Alveolus: A hair socket.

Amatidae (Syntomidae): Family of super family Noctuoidea, suborder Ditrysia and order Lepidoptera. These are tropical insects, small to medium sized, day-flying and usually inactive. Many are wasp like in appearance resembling sawflies and other Hymenoptera and brightly colored. The larva is short with verrucae bearing numerous setae. It pupates in a cocoon of felted hairs and silk.

Amaze: A highly toxic insecticide. Amaze is used principally as a soil insecticide against the corn rootworm.

Amber fossilized gum: It is a hard, sometimes clear, yellowish-brown material, often containing fossilized insects.

Amblycera: A suborder of Mallophaga. Antennae four segmented, capitate, concealed, mandibles more horizontal, maxillary palpi four segmented. Meso and metathorax usually separate.

Ambrosia: The fungus cultivated by wood boring scolytoid beetles. The term is also used to designate the part of fungus that grows out into the burrows and is eaten by the beetles.

Ambrosia beetle: A wood boring scolytoid beetle that cultivates fungus (ambrosia) for food.

Ambulatorial leg: Walking type leg or ambulatorial leg is typical of Dermaptera, ants and some beetles. These are characterized by having flattened femora, long tibia and a well defined femorotibial joint. The pretarsi are provided with the long claws and flattened arolia to enable the insects to move perfectly on even the rough surface.

Ambush: See permethrin.

Amensalism: A type of association in which one population is inhibited and the other is not affected.

Americian foulbrood: A brood disease of honey bees caused by the spore forming bacterium, *Bacillus larvae*.

Ametabola (No metamorphosis): Insects which do not undergo a metamorphosis or development without change of form, *i.e.*, without metamorphosis. After hatching from eggs looks exactly except size and different in spines and setae. It is a primitive condition and the adult closely resembles the young stage. Examples, Thysanura (silverfish), Collembola, Protura and Diplura.

Ametabolous: See ametabola.

Amide: Compound derived from carboxylic acid by replacing the hydroxyl of the -COOH by the amino group, $-NH_2$.

Amine: An organic compound containing nitrogen, derived from ammonia (NH_3) by replacing one or more hydrogen atoms by as many hydrocarbon radicals.

Amino acid: A group of nitrogenous organic compounds containing an amino group (NH_2) and a carboxylic group (COOH). Amino acids are building blocks of proteins. Some amino acids, such as leucine, histidine, histamine, arginine etc. occur predominantly; while tourine, glycine valine phenylalanine and alanine are present in trace amount in the urine of various insects, particularly, *Rhodnius*, *Melanoplus*, and *Bombyx*. In the carpet beetle, *Attagenus* and the wool moth, *Tineola*, cystine is the main sulphur containing excretory product, about 55 per cent of the total water soluble sulphur. Other nitrogenous excretory compounds include: xanthine and hypoxanthine, *e.g.*, *Galleria*, *Melophagus*, *Pieris*; allantoine, *e.g.*, Collembolan, *Gryllus*, *Carausius*, *Bombyx*, *Aglais*, blow fly maggots; and allantoic acid, *e.g.*, larval and adult Lepidoptera and larval Hymenoptera.

Aminocarb: An insecticide of carbamate group (phenyl carbamate). It is the common name of the product Metacil having 3- methyl-4-dimethylaminophenyl-N-monomethyl carbamate ($C_{11}H_{16}O_2N_2$). It is a white crystalline odourless substance, melting point 93-94 °C. It is slightly soluble in water, moderately soluble in aromatic solvents and readily soluble in organic solvents. It is unstable in strongly alkaline medium; effective against biting insects with particular reference to

Lepidopterous larvae and also some acaricidal properties; acute oral and dermal LD_{50} values to rat are 30-50 and 275-1000 mg kg^{-1}, respectively. It is, generally, formulated as 5 per cent dust and 50 per cent wettable powder.

Aminoniotelic: Aquatic insects excreting ammonia as nitrogenous waste.

Ammonia (NH_3)**:** A pungent, colourless, alkaline and water soluble gas. The larvae of *Calliphora* and *Lucilia* which are meat feeders, excrete large amount of ammonia as the excretory end product. It is also excreted by the aquatic insects as the main nitrogenous component of the urine. It is a very toxic substance and its elimination requires a large quantity of water in the urine. In some terrestrial insects also, it occurs in very small quantity in the dried excreta. In *Melanoplus*, 0.6-0.7 mg of ammonia is excreted along every 10.6-15.6 mg uric acid.

Amnion: It is the cellular, membranous covering of the embryo formed from folds of the extra-embryonic blastoderm or specifically the inner layer of each amniotic fold or of the completed covering.

Amnion cavity: A space or cavity in the developing egg formed by the amniotic folds containing the germ band.

Amniotic cavity and extra embryonic membrane formation: Soon after formation of the germ band, the serosa along with little quantity of yolk from either side extends laterally until both the extensions meet and fuse in the ventral mid line. As a result, formation of small cavity on the ventral surface of germ band take place, called the amnion. The amnion thus is derived from the serosa which surrounds externally the yolk. The amnion and serosa may be connected or separated mid ventrally, *i.e.*, at the site of fusion.

Amoebic disease: Protozoans, *Malpighamoeba mellificae* attack Malpighian tubules and midgut epithelium and are excreted along with faecal matter and infect other bees.

Amphibian: Animals of the Phylum Vertebrata those are intermediate between fish and reptiles. They are cold blooded animals and have moist skin.

Amphientomidae: A family of order Psocoptera and suborder Troctomorpha. Forefemora with a row of saw like teeth. Wings scaled. Forewings with two anal veins.

Amphipneustic: A type of insect respiration in which first thoracic and post-abdominal spiracles are functional only, *e.g.*, some Dipterous larvae.

Amphitoky: Production of both males and females parthenogentically.

Amphoteric: A compound with the property of behaving either as an acid or base. An amphoteric surface active compound is capable of anionic or cationic behaviour depending on whether it is in an acidic or basic system.

Amplexiform: A type of wing coupling in which an enlarged humeral area of the hindwing is broadly overlapped by the forewing.

Ampulla: A small dilatation in a canal or duct.

***Amrasca biguttula biguttula* (Ishida):** Belongs to order Hemiptera and family Cicadellidae. It is an insect pest of fibre crops, commonly known as the cotton jassid.

***Amritodus atkinsoni* (Lethierry):** Belongs to order Hemiptera and family Cicadellidae. It is same as *Idiocerus atkinsoni* commonly known as mango hopper.

***Amsacta moorei* Butler:** Belongs to order Lepidoptera and family Arctiidae. It is an insect pest of Kharif crops, commonly known as the red hairy caterpillar.

Amylase: An enzyme that catalyzes the breakdown of starch into maltose units.

Anabasine: The alkaloid other than nicotine and nornicotine found in the tobacco. Anabasine which is also called neonicotine is also obtained from new twigs of *Anabasis aphylla* and nornicotine from *Dubosia hopwoodii*.

***Anacanthotermes macrocephalus* (Desneux):** It is a species of termite found in Eastern Afghanistan, Pakistan and arid zone of Western India.

Anaemia: Any condition in which the number of red blood cells and the amount of haemoglobin are less than normal. It is frequently manifested by pallor of the skin and mucous membranes.

Anaerobic metabolism: Insects are remarkably resistant to lack of oxygen. Down to a low level of oxygen tension, the precise level varying in different insects, the oxygen uptake remains unchanged. Below this level it falls off; but many species can survive for long periods in the complete absence of air. Adult *Musca* can recover from 12-15 hours without oxygen. The extra metabolic activity very quickly ceases, and the insect becomes quiescent; the basal metabolism continues anaerobically. The type of chemical change does not alter; the unoxidised metabolites simply accumulate. Thus, by the time the insect is again restored to the air, it is greatly in arrears for oxygen; and in order to pay this debt and oxidize the lactic acid and other substances that have accumulated its rate of oxygen consumption remains above the normal level for a long time. During anaerobic period, the blood naturally becomes more acidic, the pH falls, carbon dioxide capacity is reduced; consequently, at very low tensions of oxygen, there may be liberation of combined carbon dioxide.

Anaerobic: Living or functioning in the absence of air or free oxygen or condition under which reductive conditions prevail.

Anaerobic respiration: Refers to liberation of energy within the cell without oxygen.

Anal: Pertaining to last abdominal segment which bears the anus.

Anal angle: The small apical area enclosed by the inner and outer margins of the hindwing.

Anal appendages: Movable appendages at the end of abdomen; the external genitalia of Odonata.

Anal area of wing: The posterior portion of the wing which usually includes veins.

Anal cell: A cell in the anal area of wing.

Anal fold: A fold in the inner margin of hindwing.

Anal hooks: Hooked or clubbed setae at the posterior end of pupa, used to attach the pupa to the cocoon or a pad of silk.

Anal lobe of hind wing: The lobe formed by a posterior incision near the wing base between veins AA and AP and often at the end of anal fold; the anal area when it is delimited apically by an incision.

Anal margin: The wing margin of insects near the anus.

Anal organ: In aquatic Brachyceran and Cyclorrhaphan larvae, the anal organs are present on either side of the anus (Stoffolano, 1970). The wall is composed of epithelial cells showing all cytological and histochemical characteristics of ion transporting epithelia, particularly, the abundance of mitochondria and apical plasma membrane infoldings. The organs are main site of Na and K uptake into the haemolymph from the external medium (Komnick, 1977).

Anal papillae: Besides columnar epithelium layer of hind gut in larvae of mosquito, *Aedes* sp., the anal papillae also absorb the traces of chloride; sodium and potassium found in fresh water and so maintain actively the normal ionic composition of blood. Hindgut of Lamellicorn larvae has special areas of cells which are believed to be concerned with absorption. Water, inorganic salts and amino acids are certainly absorbed in rectum of many insects.

Anal sac: In firebrat, *Thermobia* sp. (Thysanura), the hindgut can be divided into three regions, ileum, rectum and anal sac. The anal sac is filled with air and functions as not an air bladder but a site for water vapour uptake. The wall of anal sac is highly convoluted and made up of a single layer of epithelial cells. It is internally lined with thin cuticle while externally enclosed in muscle coat. In between the epithelium and cuticle, lies a large subcuticular space. The epithelial cells are provided with apical membrane infoldings or leaflets containing filamentous mitochondria. Due to such modifications of end part of hindgut, the firebrat is capable of absorbing water from the atmosphere at even 50 per cent relative humidity.

Anal segment: Segment pertaining to the last abdominal segment which bears the anus.

Anal shield: A sclerotized mid-dorsal plate on 10th abdominal segment of the larva.

Anal tube: The membranous posterior end of the intestinal tract.

Anal veins (A): All the longitudinal veins in the anal region. According to the Comstock-Needham system, the veins here called postcubitus and vannals. The anterior most anal vein lying just behind the cubitus is termed as post-cubitus (PCu) by some workers. These vary from 1-12 in different groups and are generally convex.

Analgestic: Pain relieving.

Analogue: A compound analogous in both structure and formula to another compound.

Analysis: This refers to the estimation of pesticide residue. The methods of residue analysis are divided into chemical assay (where chemical reactions are involved in estimation) and microbioassay (where living organism is involved for estimation). This also refers to the statistical analysis of the data.

Analytical range: Measurement range of test method where the performance has been validated and quality standards such as action limit has been developed.

Analytical standard pesticide: Pesticide reference material of high and defined purity (more than 95 per cent) for preparation of calibration standards.

Anamorphosis or Anamorphic development: The addition of abdominal segments to the body by intercalary development progress as in Protura.

Anaphothrips flavicinctus: A species of thrips feeding on grasses and cereals.

Anaphylactic shock: Constriction of the muscles surrounding the bronchial tubes of a human, caused by hypersensitivity to venom and resulting in sudden death unless immediate medical attention is received.

Anaphylaxis: Hypersensitivity to a foreign proteinaceous substance or poison (such as a venom) marked by a systemic reaction.

Anaplasmosis: Infection with *Anaplasma*, a genus of Sporozoa in red blood cells.

Anapleurite: The dorsal supracoxal sclerotization of a generalized thoracic pleuron.

Anar butterfly: See *Virachola isocrates* Fab.

Anasa wilt: A wilt disease of cucurbits caused solely by the feeding of squash bug, no parasitic microorganism is involved.

Anatomy: (Gr., *Ana*=up, *temno*= cut). A branch of biology which deals with the structure of organisms as revealed by the dissection.

Anautogenous: Requiring an exogenous source of protein for vitellogenesis.

Ancestor: An early form or type from which a latter organism is developed.

***Andraca bipunctata* Walker:** Belongs to order Lepidoptera and family Notodontidae. It is an insect pest of tea, commonly known as the bunch caterpillar.

Andrenidae (mining bees): Family of super family Apoidea, suborder Apocrita and order Hymenoptera. These are medium sized solitary bees and gather pollen and nectar from flowers of a few species of plants. On the head below each antennal socket two subantennal sutures are present. These insects nest in the ground. The nest consists of a vertical tunnel with lateral braches ending in a single cell. Often they nest close together. In rare cases common entrance is used by several females. Many are parasitized by *Nomada* (Apidae) and Strepsiptera.

Androconia (sing., Androconium): In male butterflies, specialized wing scales (often called scent scales) possessing special glands which produce a chemical attractive to females.

Androconial scales: See androconia.

Anemic: Deficient in blood quantity or quality.

Anemophily: Transfer of pollens by wind, a type of pollination.

Anemotaxis: Directed response relative to an air current, commonly the insect is excited by the odour in the air stream and moves up or down wind, often using tactile or visual cues. See klinotaxis, tropotaxis and telotaxis also.

Anepimeron: The portion of the anapleurite anterior to the pleural suture.

Anepisternum: The portion of anapleurite posterior to the pleural suture.

Aneuploidy: Addition or loss of one chromosome. In human beings, aneuploidy occurs when homologues pair of two chromosomes do not separate out during cell division (In human being, chromosomal numbers 2n= 46). See polyploidy also.

Angoumois grain moth: The pest (*Sitotroga cerealella*) infests stored grains like wheat, rice, maize and millets. Infestation passes on from field through grain and straw. Larva bores into grains, eats kernels inside and renders grain hollow; damage restricted to surface layer.

Angstrom: One hundred millionth part of a centimeter, a unit used in measuring the length of light waves.

***Anguina tritici*:** Belongs to order Tylenchida and family Tylenchidae. It is a pest of wheat, commonly known as the wheat gall nematode.

Animal sign: Evidence of an animal's presence in an area.

Animal: All vertebrate and invertebrate species.

Anionic: An ion having a negative charge. When the surface active portion of surfactant molecule possesses a negative charge, it is termed as anionic surface active agent. Cationic is the antonym.

Anionic surfactant: (i) A surface active additive to a pesticide having a negative surface charge. These perform better in cold, soft water. Most wetting agents are of this class. (ii) Salt of an organic acid, the structure of which determines its surface activity.

Anisoptera: Without petiolate fore and hind wings, the later more broadened at base. The discoidal cell of forewing differentiated in triangular and supratriangular areas. Eyes large mostly contiguous. Wings depressed in repose. Larvae with rectal gills and with 4-8 folds in gizzard. Penis jointed.

Anisozygoptera: A suborder of order Odonata. Mostly fossil species and living species includes only two species. Adult has a Zygopteran venation but the general form of the body is anisopterous and the nymphs resemble Anisoptera.

***Anjeerodiplosis peshawarensis* Mani:** Belongs to order Diptera and family Cecidomyiidae. It is an insect pest of fig, commonly known as the fig midge.

Annoying insects: Those causing discomfort to human beings and animals by their sounds, bad odours, tastes of their secretions, crawling over one's body, irritation in the eyes and giving frightening look, *e.g.*, house flies, house crickets, cockroaches, mole crickets, stink bugs, hairy and slug caterpillars, eye flies, cattle flies etc.

Annual plant: One that completes life cycle (seed to seed) in one year or one growing season.

Annular: Ring shaped.

Annulate: Formed in ring like segments or with ring like markings.

Annulated: With ring like segments or divisions (usually referring to the antennae).

Annulus: A ring of membrane, sclerite or pigment surrounding a joint or segment. (pl., annuli), *e.g.*, The Tortricinae has two annuli of scales per flagellar segment.

Anobiidae (death-watches): Family of superfamily Bostrychoidea, suborder Polyphaga and order Coleoptera. The beetles are small and cylindrical with hypognathous head concealed by prothorax. The three apical segments of the serrate, clavate or pectinate antennae are free and large.

***Anomala dimidiata* (Hope):** Belongs to order Coleoptera and family Scarabaeidae. It is commonly known as defoliating beetle and is a minor pest of grapevine.

***Anomis flava* (Fab.):** Belongs to order Lepidoptera and family Noctuidae. It is an insect pest of fibre crops, commonly known as the green semi-looper.

***Anomis sabulifera* Guenee:** Belongs to order Lepidoptera and family Noctuidae. It is an insect pest of jute, commonly known as the jute semi-looper.

Anopheles: A genera of mosquito.

Anoplura: Gr., *Anopl*= unarmed, *ura*= tail. An insect order which includes the sucking lice which are small wingless, tough skinned, dorsoventrally flattened, often dark coloured insects, ectoparasites on all kinds of domestic and wild mammals and man. These suck the blood of hosts. The head is prognathous, small, freely movable, without ocelli with vestigial or moderately developed compound eyes and short, thick antenna composed of three to five segments. The mouth parts are adapted for piercing and sucking with a short proboscis or rostrum armed inside with minute denticles. The rostrum which is the piercing organ is exerted while feeding to have a grip or hold on the host. Thorax is short without distinct segments and sclerites undifferentiated, legs short, stout and adapted for clasping or grasping with the aid of a long, claw at the end of one segmented tarsi which work against tibial processes. Abdomen distinctly segmented and cerci absent.

Anopluran piercing and suking mouth parts: Three superimposed stylets in these mouth parts are (i) maxillae forming dorsal stylet, (ii) hypopharynx, the middle stylet and (iii) labium, the ventral stylet. The dorsal and ventral stylets are forked at the base. All parts are entirely withdrawn into a long, slender pocket or sac in the head concealed beneath the pharynx during non-feeding period. The proboscis or rostrum or mouth cone is snout like eversible sac formed from the labrum. It is armed internally with small curved hooklets that serve as anchorage when reverted. It opens into a preoral cavity which in turn opens into the dorsal sucking pump and ventral stylet sac. The sucking pump is a two chambered structure; anterior one is the buccal cavity, while the posterior one is pharynx.

Ant: Any of various social insects of the family Formicidae (order, Hymenoptera). Only the males and fertile females have wings. They are living in colonies that have a complex social organization.

Antagonism: An ecological association between organisms in which one or more of the participants are harmed or has its activities limited.

Antagonist: (i) Substance that lowers down the biological or phar-macodynamic efficacy of a toxicant or an agent or substance that counteracts the action of another. (ii) An organism that depresses the activity of another organism.

Ante: Before, in front of.

Anteapical cell: A cell in the distal part of the wing.

Anteapical: Just proximal of the apex.

Anteclypeus: A transverse sclerite on the anterior border of the frontoclypeus in the larva, with which the labrum articulates.

Antecosta (pl., antecostae): An internal ridge on the anterior portion of a tergal or sternal plate which serves as the site of attachment of the longitudinal muscles.

Antecoxal sclerite: A metasternal sclerite in front of the hind coxae. It can be seen in between the trochantin and the episternum and articulates directly with coxa by performing a ventral articulation. Sometimes, the second antecoxal piece may be present along with the first one.

Antenna (pl., antennae): A pair of jointed appendages (feeler) located on the head in procephalic region of an insect above the mouthparts, slight anterior and ventral to the compound eye, generally used for smelling (olfactory) and feeling or tactile (touching), communicating organ, knowing humidity changes, wind velocity, wind direction, detecting danger, locating their food, finding their mates, and in some cases for hearing (auditory). The base of each antenna is movable, fixed into a small antennal socket. All insects except proturans have a pair of antennae. They are movable jointed appendages of the head segment. The parts of the antenna are (i) Scape: It is the first and basal segment of the antennae and usually longer than any of the succeeding segments, (ii) Pedicel: It is the second segment and contains sensory apparatus known as Johnston's organ, and (iii) Flagellum (Clavoke): It forms the remainder part of antenna. The antenna in different insects is extremely different in shapes: Bipectinate: fringed (pectinate) on both sides (Lymantriidae), Capitate: ending in a capitulumor knob (*Khapra* beettes), Clavate: club- shaped, gradually clubbed (Cur-culionidae), Elbowed **(**clubbed geniculate**)** bent at union of scape and pedicel (Chalcidoidea), Filiform Thread like (Gryllidae), Genniculate abruptly bent (Formicoidea), Knobbed: swollen-tipped, abruptly (Papilionidae), Lamelatte expanded in flattened plates (Melolonthidae), Moniliform: like a string of beads (termites-Isoptera), Pectinate: toothed like a comb (male moths), Plumose: feather-like (mosquito).Serrate: notched like a saw (Elateridae), Setaceous: bristle-like (locust, grasshopper).

Antenna cleaner: A concavity lined with hairs and/or stout setae and located near the apex of the protibia or mesotibia. Occurs in taxa with more or less filiform antennae and used to clean the antenna by sliding it over the hairs.

Antennal cavity: A prothoracic cavity for housing the whole antenna or a portion of the antenna.

Antennal club: It refers to enlarged portion of the antennal apex, consisting of a variable number of antennomeres (often 3). In an incrassate antenna, the antennomeres gradually enlarge towards the apex, but if there is an abrupt change in length or width at some point, then the antennomeres beyond this are considered to be part of the club. In butterfly, it is enlarged distal segments of a clubbed antenna.

Antennal fossa: A cavity or depression in which the antennae are located.

Antennal groove: A groove in the head capsule into which the basal segment of the antenna fits.

Antennal insertion: Point of attachment for the antennae, consisting of an opening in the head capsule, sometimes with a reinforced sclerotized ring.

Antennal lobes: Lobes in the insect brain that receive both the sensory and the motor axons.

Antennal suture: A suture that surrounds the antennal base or antennal socket. Each suture gives internally the submarginal ridge. The submarginal ridge in large number of insects bear a pivot like process known as antennifer. The antennifer provides articulation to the antenna.

Antennation: Refers to touching with the antennae. The movement can serve as a sensory probe or as a tactile signal to another insect.

Antennectomy: Removal of a part of antenna or entire antenna.

Antennifer: Articular point of antenna located in the antennal socket.

Antennomere: Antennal segment; including scape, pedicel and flagellomeres.

Antennule: The first antennae of Crustacea.

Antenodal: Preceding or before the nodus.

Antenodal cross veins: Cross veins along the costal border of wing, between the base of the wings and the nodus, extending from the costa to the radius.

Antenodal veins: Small cross-veins at the front of the dragonfly or damselfly wing, between the wing base and the nodus.

Antepenultimate: The second one from the last.

Anterior: Concerning or facing the front, towards the head.

Anterior angles: The fore corners of the pronotal disc. These may be rounded; angulate or produced forward forming rounded or acute processes.

Anterior cross vein: The r-m cross vein.

Anterior imaginal ring: Ring of embryonic cells located at the posterior end of larval foregut.

Anterior intestine: The portion of hind gut between the Malpighian tubules and the rectum.

Anterior mesenteron rudiment: The anterior group of cells of ventral endoderm remnant that regenerates the mesenteron.

Anterior notal process: Anterior wing articulation point on the notum.

Anterior pharyngeal part: The precerebral part of pharynx in insects having a pharyngeal section of the stomodaeum behind the brain.

Anterior process of metendosternite: An extension of the apical portion of the metendosternite which usually bears the anterior tendons.

Anterior tendons of metendosternite: A pair of tendons borne on the anterior process or the lateral arms of metendosternite.

Anterolateral callosity: More or less flattened or callus like area at or very near to the anterolateral corner of pronotum, and often containing a gland opening.

Anther: Part of stamen that contains the pollen.

Anthesis: The period during which the flower remains open.

Anthocoridae: Family of superfamily Cimicoidea, suborder Heteroptera and order Hemiptera which includes flower bugs and minute pirate bugs. Small predaceous insects, elongate oval and flattened. Possess ocelli, three segmented rostrum, distinct cuneus and embolium, scent gland on meta thorax and three segmented tarsi. *Piezostethus sp.* is predacious on sugarcane mealy bugs; *Orius tantillus* (anthocorid) on aphid, *Aphis gossypii*, nymphs of mites on crops and thrips on sorghum ears and *O. indicus* on the thrips, *Taeniothrips distalis* infesting flowers of *Cajanus cajan*.

Anthomyiidae: Family of suborder Cyclorrhapha, order Diptera. Two subfamilies are recognized in this family. The Anthomyiinae are dark colored and possess on the undersurface of the scutellum fine erect hairs and more than one sternopleural bristle. They are mostly phytophagous, some are scanvengers and othe rare parasitic. The scatomyzinae are devoid of such erect fine hair on the undersurface of scutellum and only sternopleural bristle is present. They bred in cow dung. Different species of *Atherigona* are known to infest a wide variety of cereals and are often serious pests.

Anthoxanthins: Whitish or yellow pigments found in insects.

Anthracnose: The disease having characteristic lesions, necrosis and hypoplasia on stems, leaves and fruits. This disease is usually caused by certain imperfect fungi, *e.g.*, *Colletotrichum*, *Gloeosporium* etc.

Anthraquinone: Anthraquinone derivatives form the red colour of cochineal insect (carminic acid), remarkable pigment of aphids (aphins), lac insect (laccainic acid) etc.

Anthribidae (fungus weevils): Family of superfamily Curculionoidea, suborder Polyphaga and order Coleoptera. These insects are numerous in Indo-Malaysian region. They are chiefly found in dead branches, beneath loose bark and in fungi; *Brachytarsus* is predaceous on scales. The larvae live on fungi, some on seeds and a few bore in dead wood. The insects are small to minute, 0.5 to 11 mm in length, possess short and broad beak and non-geniculate antennae which are short with a three-segmented club and in some long and slender. The tarsi are four with stalked compound eyes.

Anthropogenic: Brought about or influenced by human beings.

Anthropomorphism: Interpreting behaviour patterns observed in animals in terms of human characteristics.

Antibiosis: (i) It is an association between two or more organisms that is detrimental to one or more of them. (ii) It is also used in the sense of three fold mechanism of host plant resistance (Painter, 1951) for those preventative, injurious or destructive effects on the insect life history which result from the use of a resistant host. A mechanism of plant resistance to pests whereby harmful effects occur to the insect due to certain biochemical in a particular plant.

Antibody: A proteinaceous substance found in serum which is produced in the body of animal in response to entering or introduced antigens.

Anticoagulant: Any substance that inhibits or delays or suppresses blood clotting.

Anticoagulin: A substance which is antagonistic to the coagulation of blood. In those insects which suck the blood of living animals, the saliva usually contains an anticogulin, probably antithrombin. It has been shown that if the salivary glands are removed from the living tsetse fly, although it continues to take food for some time, ultimately its crop and proboscis become blocked with coagulated blood. See anticoagulant also.

Antidiuretic hormone: The insect neurohormone which increases water retention.

Antidote: An immediate treatment or remedial measures taken to counteract poisonous effect of a toxicant.

Anti-drift agent: A substance added to pesticides to reduce the number of fine droplets produced at the spray nozzle.

Antifeedant: A substance that deters or inhibits feeding by an insect but does not necessarily kill it. Antifeedants are not repellent in the true sense of the word, since insects are not driven away or kept away. Such chemicals inhibit feeding of insects and thus protect the crop or commodity and eventually starve to death without eating the leaves. There are following types of compounds which show antifeedant property: Triazene (compound 24055), organotins (Breston, Brestenol), carbamate (Baygon) and botanical extracts (pyrethrins, *Neem* seed etc.).

Antifrothing agent: Material added to prevent frothing of liquid in the spray tank.

Antifungal: A chemical substance that kills or inhibits fungi.

***Antigastra catalaunalis* Duponchel:** Belongs to order Lepidoptera and family Pyralididae. It is an insect pest of sesame, commonly known as the sesame leaf and capsule caterpillar.

Antigen: Any foreign substance (usually protein) which is introduced or gain entrance into the blood of host and stimulates the production of antibodies.

Antigenic: Having the properties of an antigen.

Anti-insect agents: Chemicals that kill, control or deter the growth of pests, *e.g.*, insecticides, acaricides, nematicides, antifeedants, insect chemosterilants, growth regulators, antimetabolites and hormones used for insect control.

Antimicrobial: A chemical substance that kills or inhibits micro-organisms.

Antipyretic: Fever reducing.

Antiserum: Serum containing antibodies.

Anti-siphoning device: A device attached to a filling hose to prevent water in the spray tank from draining into the water source.

Anti-transpirant: A chemical applied directly to the plants which reduces the rate of transpiration or water loss by the plants.

Antiviral: A chemical substance that inhibits or kills viruses.

Antixenosis: Refers to non preference of the host by insect pests.

Antlion: A common name for insects of the family Myrmeleontidae (Order Neuroptera). They dig cone or funnel shaped pitfalls with its large jaws in the sand and wait for stray ants to fall in and serve as their prey.

Antu ($C_{11}H_{10}N_2S$): Chemical name, alpha-napthylthiourea. A rodenticide that can develop tolerance by ingestion of sublethal doses and bait shyness occurs that may last up to six months. Besides, dogs are almost as susceptible as rat and even secondary poisoning is possible. Antu is used at the rate of 1-3 per cent in bait for brown rats, but is not recommended for any other rodent species. It takes 16-30 hrs or even 8 days to kill an animal. It was very popular poison in USA but now days disappeared from the market.

Anus: Posterior opening of the alimentary canal system.

***Aonidiella aurantii* (Maskell):** Belongs to order Hemiptera and family Diaspididae. It is commonly known as citrus red scale and is a pest of citrus, fig, grape, rose, *Acacia*, *Eucalyptus*, willow etc.

***Aonidiella orientalis* Newstead:** Belongs to order Hemiptera and family Diaspididae. It is commonly known as scale insect and is a minor pest of black plum (jamun).

Aorta: The anterior prolongation which is non-pulsatile portion of the dorsal blood vessel. It is principal artery of the blood. Its junction with the heart is frequently marked by the presence of aortic valve. The aorta extends forwards through the thorax to terminate in the brain.

***Apanteles glomeratus* (Linn.):** Larval parasitoid of cabbage caterpillar.

Apex (pl., apices): The tip of a root or shoot.

Aphicide: An insecticide especially effective against aphids.

Aphid: An insect belonging to the family Aphididae in the order Hemiptera, suborder Homoptera. Sometimes, it is also referred to as 'plant louse'.

Aphididae: A family of superfamily Aphidoidea, suborder Homoptera and order Hemiptera. Sexual forms oviparous, parthenogenetic females viviparous; Rs separate on forewings, if winged. A large group of small soft bodies, delicate pear shaped insects that infest the stem or leaves of a variety of plants in large numbers and suck up the sap causing serious damage. Examples are: cotton aphid, *Aphis gossypii*; mustard aphid, *Lipaphis erysimi*; *Aphis craccivora* etc.

Aphidoidea: A superfamily of suborder Homoptera and order Hemiptera. The three families included in this superfamily are Aphididae, Adelgidae and Phylloxeridae.

Aphins: Red, orange and yellow fat soluble pigments found in aphids.

***Aphis craccivora* Koch:** Belongs to order Hemiptera and family Aphididae. It is an insect pest of cotton and ornamental plants, commonly known as the groundnut aphid.

***Aphis gossypii* Glover:** Belongs to order Hemiptera and family Aphididae. It is an insect pest of ornamental plants, commonly known as the cotton aphid.

***Aphis nerii* Roy:** Aphid species of ornamental plants.

***Aphis solanella* Theobald:** Aphid species of ornamental plants.

Aphrodisiac: A chemical substance that stimulates sexual desire.

Apiary: Colonies, hives, and other equipments assembled in one location for beekeeping operations; bee yard.

Apical: Around the apex (the tip or apicad). It denotes position near or movement towards the apex of a body part. The apex of the head or pronotum is at the anterior end while that of the abdomen or an elytron is at the posterior end; on the legs or antennae, apical and distal are synonymous.

Apical area (of hind wing): A rather vaguely defined area of the hindwing, apical to the ends of radial and medial bars and last r-m cross vein.

Apical area (of the forewing): The area just inside of and contiguous with the apex.

Apical cell: A cell near the wing tip.

Apical cross vein: A cross vein near the apex of wing.

Apical margin: The outer edge of an insect wing.

Apical piece: See apicale.

Apicale of tegmen: The apical portion of the sheath like tenebrionoid tegmen which is divided transversely. This appears to be formed by the fusion of two parameres. It is also called the apical piece or paramere.

Apices: See apex.

Apiculture: Apiculture or bee keeping is the science and art of raising honey bees to get honey and bee wax.

Apidae: Family of super family Apoidea, suborder Apocrita and order Hymenoptera. A large number of solitary and parasitic bees and the social honeybees are included under this family. These insects are characterized by usually three submarginal cells in forewings. If two submarginal cells are present, the second cell will be much shorter. Rarely, it may be single. The scopa is present, usually on the hindlegs and rarely on gastral sternite also.

Apiology: Scientific study of bees.

***Apion corchori* Marshall:** Belongs to order Coleoptera and family Apionidae. It is an insect pest of jute, commonly known as the jute stem weevil.

Apionidae: Family of superfamily Curculionoidea, suborder Polyphaga and order Coleoptera. This is a small group of beetles with non-geniculate antennae. The insects develop in seeds and stems or roots of plants. The common genera included are: *Apion* and *Cylas.* The important species met with in India are the sweet potato weevil, *Cylas formicarius, Apion ampulum* breeding in green gram, cashew etc. and *Apion corchori* infesting jute in Northern India.

***Apis mellifera*:** Scientific name of honey bee found in the Europe and United States of America, now also being raised in India and other parts of the world.

Apneustic: The closed type of ventilatory system in which none of the spiracles are functional. In Collembola, Protura, parasitic Diptera and immature stages of endoparasitic Hymenoptera and aquatic insects, air enters through the general body surface or specialized extensions of it such as gills.

Apocrita (Clistogastra): Suborder of Hymenoptera. These insects are characterized by a deep constriction between the propodeum (first abdominal segment) and the second abdominal segment. The larvae are apodous. The Apocrita have highly specialized habits and are often social. The adults generally feed on sap, flowers and other plant materials and occasionally the parasitic insects may feed on the body fluid of the host. They possess a well developed ovipositor.

Apodeme: Invagination of the body wall that strengthens the exoskeleton for muscle attachment. See also apophyses.

Apodous larva: Larva without legs. These are characterized by the absence of trunk appendages, presence of three pairs of sensory papillae in place of thoracic legs and are derived mostly from oligopod type. These are of three subtypes: (i) eucephalous larva is with sclerotized head capsule and little reduced cephalic appendages, *e.g.*, Nematocera, Buprestidae, Cerambycidae and aculeate Hymenoptera. (ii) hemicephalous larva possesses reduced head capsule, *e.g.*, Tipulidae and Brachycera (Diptera). (iii) acephalous larva is characterized by absence of head capsule but buccopharyngeal apparatus and mouth hooks, *e.g.*, Cyclorrhaphous Diptera.

Apoidea: Super family of Suborder Apocrita, order Hymenoptera. This superfamily includes the solitary and social bees, the latter being confined to the families Apidae and Halictidae. These are commonly found visiting flowers and aid in cross-pollination of crops by carrying pollens which adhere to their body hairs. The antennae are 13-segmented in male and 12-segmented in female.

Apolysis: The separation of the old cuticle from the epidermis is known as apolysis. This is followed by shedding of old cuticle (ecdysis). Immediately after ecdysis, the tanning of cuticle (sclerotization) takes place. The apolysis starts with increase in the number and size of the epidermal (hypodermal) cells due to repeated mitotic divisions. The epidermal cells become tall, columnar in shape and closely packed. They exert marked tension at the cuticular surface and as a result separate themselves from the cuticle. Due to separation of hypodermis from the cuticle formation of a subcuticular space takes place. The epidermal cells or modified moulting glands discharge the moulting fluid (containing enzymes, proteinase and chitinase) in subcuticular space. This fluid digests the endocuticle of old cuticle. Initially, the moulting fluid is inactive but as soon as the cuticulin layer is formed (secreted by oenocytes), this became active. The cuticulin layer (impermeable to moulting fluid) moves forward with the synthesis of wax and polyphenol layers. After formation of the epicuticle, successive layers of procuticle are deposited beneath and subcuticular space is gradually occupied by newly synthesized cuticle. Some indigested material persists in the form of ecdysial membrane below the old cuticle. The other epicuticular layers, the wax, polyphenol and cement layers are formed from the substances transported by epidermal cells through the pore canals. The cement layer in some insects is secreted by the dermal glands. It is deposited over the wax layer at the last, *i.e.*, just before the ecdysis.

Apomictic: Parthenogenetic population.

Apophysis (pl., apophyses) and apodeme (pl., apodemes): There are two types of cuticular processes. The hollow invaginations are called apodemes and are the sites of muscle attachment. The solid invaginations unlike apodemes are called apophyses. These are internal or external tubercular or elongated projections arising from body wall. These structures form the endophragmal skeleton inside the body of insects that provide mechanical support to various organs and the whole body. They form the distinct skeletal structures such as tentorium in the head and endothorax in the thorax.

Aporous: Without a pore.

Aposematic colouration: Warning colouration displayed by many species.

Appendage: An external limb or similar structure or organ attached to the body or any other main structure. The appendages of cuticle include setae (microtrichia), spines and spurs.

Appendicular ovipositor: The true ovipositor formed from appendages.

Appendix: In insects, a short vein, especially a short continuation after the main vein has changed direction.

Apposition image (eye): These contain a very long rhabdon extending from the lens to the basement membrane. According to Exner's elaboration of mosaic theory, there are some insect eyes (in Orthoptera, Hymenoptera, Diptera) in which the light received in one facet illuminates a single retinal element to produce an apposition image (See superposition image also).

Apposition mechanism: It is a mechanism of image formation in the bright day light in most of the diurnal insects, Odonata, Orthoptera, Hymenoptera, Diptera, butterflies and various Coleoptera. The pigment cells envelop the central point of dioptric apparatus. The other peculiarity of ommatidia of these insects is that the focal length of apparatus is equal to the total length of cones. The formation of reverse image, therefore, takes place by virtue of the light entering from comparatively smaller part of the visual field and that too at the retinula where it comes in intermediate contact with the cone. The apposition of all such individual reverse images of different ommatidia finally produces a single complete, erect image of an object. See superposition image also.

Approved common name: It is the generic name of a pesticide.

Approved product: It is the proprietary brand of pesticide officially approved by the government.

***Apsylla cistella*:** See mango shoot gall psyllid.

Aptera (Diplura): Apterygote insect order.

Apterae: Wingless forms of insects.

Apterous: Without wings or wingless.

Apterygota: Subclass of the wingless insects including most primitive orders with no metamorphosis. Except for the size, all nymphal stages closely resemble the adults (which are wingless).

Apterygote: Any of the primitive insects which have never developed wings during their history, *e.g.*, bristletails.

Apterygote orders: The subclass apterygota consists of four orders, *viz.*, Thysanura, Diplura, Protura and Collembola. They are widely diversified from each other representing heterogenous assemblage having polyphyletic origin and some abnormal insect characters. Thysanura, moreover, show their close relationship with the main evolutionary stalk of Pterygota.

Aquatic: Living in the water. See terrestrial.

Aquatic respiratory system: Aquatic respiratory system are evolved variably: abdominal gills (nymphs of Plecoptera, Zygoptera and Trichopteran larvae), caudal gills (nymphs of Zygoptera), rectal gills (nymphs of Anisoptera), spiracular gills (Tipulid and Simulid larvae), blood gills (Trichopteran and *Chironomus* larvae), physical gills (aquatic bugs and aquatic beetles) and plastron (aquatic beetles, bugs, eggs etc.). These are species specific in their structure and are of taxonomic significance.

***Arachis hypogaea* L.:** Scientific name of groundnut.

Arachnida: A class of phylum Arthropoda that includes the spiders, scorpions, ticks and mites; have a cephalothorax (head and thorax fused together) and an abdomen; six pairs of limbs, four pairs for walking and first pair as jaws and devoid of simple eyes and antennae.

Arachnologist: Person who studies arachnids (spiders, scorpions, ticks, mites, etc.).

Arachnophobia: Fear of spiders, ticks, mites and scorpions.

Aradidae: Family of superfamily Aradoidea of division Gymnocerata, suborder Heteroptera and order Hemiptera. These insects are broad flattened bugs living among fungi, beneath loose bark or in crevices of dead or decaying wood, *e.g.*, flat bugs or fungus bugs, *Aradus* sp.

Aradoidea: Superfamily of division Gymnocerata, suborder Heteroptera and order Hemiptera. This includes the families Aradidae, Termatophylidae, Salididae, Mesovellidae, Thaumastotheriidae, Hebridae and Hydrometridae.

Araneae: An order belonging to the class Arachnida, *e.g.*, spiders. They are characterized by two body parts (cephalothorax and abdomen), four pairs of legs, and a pair of fangs (chelicerae).

Araneology: Study of spiders.

Arboreal: Living on trees and other large plants.

Arboricide: Chemicals for the destruction of undesirable arborial and bush vegetation.

***Arboridia viniferata*:** It is a species of jassid which is minor pest of grapevine.

Arbovirus: Term applied to a virus that can replicate in both arthropods and vertebrates. It transmitted by mosquitoes or other blood sucking insects.

Arch: Primitive; original.

Archedictyon: A fine primitive network of cuticular ridges or of veins on the wings of insects.

Archeognatha: Insect order (Gr., *Archeo*= ancient; *gnath*= jaw), *e.g.*, jumping Microcoryphia.

Archicerebrum: The primitive suprastomodaeal nerve mass of the prostimium.

Archiciphalon: The primitive annelid and arthropod head or prostomium.

Archimetabola: See heterometabola.

Archostemata: A suborder of order Coleoptera. A single superfamily Cupesoidea is recognized. This includes two families, *viz.*, Cupesidae (reticulated beetles) and Micromalthidae (micromalthid beetles) and these beetles are not known to occur in India so far.

Arctic region: The region of the world found North on the globe. North Pole region.

Arctiidae [Lithosiidae] (tiger moths, footman moth): Family of super family Noctuoidea, suborder Ditrysia and order Lepidoptera. This includes mostly tropical insects with stout body and moderately broad brightly colored wings with spots or bands on them. They are noctural and species of several genera are capable of production of sound. The larvae of Arctiinae are densely clothed with long hairs and feed on herbaceous plants. Their cocoons are made up of larval hairs and silk. The larvae of Lithosiinae are sparsely hairy and feed on lichens. The larvae of Nycteolinae are not prominently hairy and pupate in boat shaped cocoons. They feed on leaves and fruiting bodies of plants. *Diacrisia obliqua*: Commonly known as the Bihar hairy caterpillar, this is a pest of pea, cotton, jute, sunnhemp, *Sesamum indicum*, Chrysanthemum, mulberry, castor, cauliflower, cabbage, lettuce, Lucerne, *Phaseolus radiatus, Pennisetum typhoideum,* linseed, maize, groundnut, sunflower, *Abutilon indicom,* tobacco, turmeric, mustard, soyabean, *Eleusine corcana,* lantana, sweet potato, lablab, cowpea etc. It is a notorious pest of jute in Bengal, Bihar, Orissa and Assam. *Amsacta albistriga* and *A. moorei*: The red hairy caterpillar; the insects are distributed all over India.

Arcuate: Bent like a bow, or arched.

Arcuate vein: See veina arcuata.

Arculus: (i) A basal cross vein between the radius and the cubitus. (ii) Taxonomically significant wing vein distally enclosing the median space in Odonata.

Arecanut mirid bug: See *Carvalhoia arecae.*

Areola: A small opening or space or a small area between things. An accessory cell (See also basal areole).

Areolet: A small cell in the wing in the family Ichneumonidae (Hymenoptera), the small sub-marginal cell opposite to the second recurrent vein.

***Argas persicus* (Oken):** Belongs to order Acarina and family Argasidae. It is a pest of farm animals, commonly known as the fowl tick.

Argidae: Family of super family Tenthredinoidea, suborder Symphyta (Chalastogastra), order Hymenoptera. Small to medium sized sawflies with characteristics antennal segment is U-shaped and Y-shaped in few cases.

Arista: A large bristle located on the dorsal side of last or apical antennal segment in the order Diptera. When stouter and more rigid, it is called a style.

Aristate: Bristle-like three segmented antenna in which the last segment bears heavy bristle or arista, *e.g.*, house fly, *Musca nebulo* Wiedemann; syrphid fly, *Syrphus balteatus* and tsetse fly.

Arixeniidae: Family of order Dermaptera and suborder Arixeniina. Mandibles flattened with dense hair on inner margin and toothed only at apex. Forceps very weak and setose, apterous, viviparous, ectoparasitic mostly on bats, eyes vestigial, *e.g.*, *Arixenina* sp.

Arixeniina: A suborder of Dermaptera.

Army ant: An ant species that shows nomadic and group predatory behavior. The nest site is changed relatively at frequent interval, in some cases daily and workers collect in groups (same as the legionary ant).

Army worm: See *Mythimna separata.*

Arolium (pl., arolia): A pad like structure at the apex of last tarsal segment between the claws (in Caelifera, Orthoptera and Apoidea, Hymenoptera) or a pad like structure at the base of each tarsal claw in order Hemiptera and many other insects.

Aromatic compound: Organic compound derived from benzene with aroma (aroma means odour).

Aromatic solvent: Chemically, those solvents that have benzenoid molecular structure. They are generally more phytotoxic than aliphatic materials.

Aromatics: Oil or oil like materials that kill plants and animal tissue similar to burning. Chemically, these are compounds having a closed ring structure.

Arrested growth: See diapause.

Arrhenotoky: In parasitic Hymenoptera, the fertilized eggs are diploid giving rise to females, whereas, the unfertilized eggs are haploid giving rise to males, *e.g.*, many Hymenoptera like *Trichogramma exiguum, T. minutum, Telenomus remus.*

ARS: Agricultural Research Service of the US Department of Agriculture and also of ICAR in India.

Arsenate: An arsenical compound with a hydrocarbon group connected to the arsenic atom; hence, an organic arsenical.

Arsenic trioxide (As_2O_3): Highly toxic to man and animals, also known as white arsenic and arsenous oxide; arsenic trioxide is used chiefly as a rodenticide. Fairly well accepted by rats as lack odour and a moderate taste are not objectionable to them. Although, it kills slowly, it is effective against rats and mice. The rats develop tolerance at sublethal doses, however, it must be used with extreme caution around children and pets.

Arsenic: An element present in pesticides; may occur in soils; a trace element.

Arsenicals or arsenic compounds: One of the most important groups of early insecticides comprised principally of the arsenates and arsenites. The killing power of these materials is directly related to the per cent of metallic arsenic contained in these materials. The examples are lead arsenate, calcium arsenate, sodium arsenate and Paris green.

Arterogamy: Alternation of bisexual with parthenogenetic reproduction.

Arthrodial membrane: Soft and stretchable cuticle between the segments.

Arthropleona: A suborder of order Collembola. Body more or less elongate; distinct segmentation.

Arthropod: An invertebrate animal with jointed appendages, a member of the phylum Arthropoda.

Arthropoda: A phylum within the animal kingdom. It includes Crayfish, Millipedes, Centipedes, Spiders, and Insects. Arthropoda have a segmented body with a hard outer skeleton (exoskeleton). Usually the segments are grouped in two or three body regions. The body has a bilateral symmetry and it has paired segmented appendages.

Arthropodin: A group of ordinary proteins of the cuticle that tanned to form sclerotin are called arthropodins. It is a mixture of various proteins and is devoid of Sulphur containing aminoacids. It resembles sericin or silk gelatin in its general properties. It is water soluble and constitutes primarily the pro and endocuticle representing the untanned protein.

Articulation: It is a joint between two segments of insect body.

Artificial barrier: For preventing the crop from migratory insect pests such as army worms, termites, locusts etc., artificial barriers can be employed successfully. These are (i) Sticky bands or oil cloth bands which are often employed around the tree trunk against the insects climbing or crawling over the trunk and attacking the flowers and fruits. Application of grease or coal tar bands on the trunk of mango tree may prevent the attack of mango mealy bug, *Drosicha mangiferae* or red ants. (ii) Trenching and fencing can be employed against chinch bug, army worm and hairy caterpillars in order to check the entry of these pests from infested to the non-infested or healthy areas. (iii) Screening is a regular method which is applied on the windows and doors against mosquitoes.

Artificial control: Refers to applied control.

Artificial diet: An unfamiliar food that has been formulated, synthesized, processed and/or concocted on which an insect in captivity can develop through all or part of its life cycle.

Ascalaphidae: Family of superfamily Myrmeleontoidea, suborder Planipennia and order Neuroptera. It is closely related to the antlions (Myrmeleontidae).

Aschelminthes: A phylum of invertebrate animals that include roundworms, rotifers and several other classes.

***Asclepias syriaca*:** A plant species which is a source of nicotine.

Ascorbic acid: Vitamin C; a white crystalline substance, stable when dry but very easily oxidized in solution, especially neutral or alkaline solutions.

Aseptic: Free of disease causing organisms.

Asexual reproduction: This refers to production of young ones without sexual reproduction. In such organisms, the reproductive organs are incompletely developed and produce eggs or young ones either by cell budding or parthenogenesis.

Asilidae (robberflies): Family of suborder Brachycera, order Diptera. Asilidae are quite commonly found in allshorts of places and are moderate to large sized bristly flies with horny piercing type of mouthparts and powerful, prehensile legs adapted for catching the prey and piercing its body. A tuft of hair and protruding eyes are very characteristics of these robberflies. The leg ends in a bristle-like empodium and large pulvilli. Some of the species noticed in India are *Allocotosia aurata, Hyperechia xylocopilormis* and *Philodicus femoralis.*

***Asphondylia sesami* Felt.:** Belongs to order Diptera and family Cecidomyidae. It is an insect pest of sesame, commonly known as the sesame gall fly.

***Aspidiotus destructor* Signoret:** Belongs to order Hemiptera and family Diaspididae. It is an insect pest of coconut, mango, date palm and tamarind, commonly known as coconut scale or mango hard scale or date palm scale.

***Aspidiotus tamarindi*:** Belongs to order Hemiptera and family Diaspididae. It is an insect pest of tamarind, commonly known as hard scale of tamarind.

Aspirator: A simple device with which insects can be picked up by suction.

Assembling scents: Chemicals (pheromones) that when released, result in the formation of a group of individuals of the same species.

Association or internuncial neurone: See interneurone.

Association: A grouping of communities defined primarily by the presence of a common set of characters in species.

Associative learning: Learning based on the presence of reward or punishment; includes classical conditioning, instrumental learning and shock avoidance conditioning.

Astelocyttarus: Pertaining to nests, normally those of social wasps in which the comb is attached directly to the support.

Aster yellow: A viral disease of many kinds of plants transmitted by the six spotted leaf hopper and characterized by stunting of plants, sterility, and chlorosis in foliage.

Asterolecaniidae: A family of superfamily Coccoidea, suborder Homoptera and order Hemiptera.

Astronomy: In this branch of science, planets, stars, constellations, galaxies etc. found in the space are studied.

Asymmetrical: Organs or body parts not alike on either side of a dividing line or plane.

Asymptote: The point in the growth of a population at which numerical stability is reached.

Asynchronous muscle: A rapidly contracting muscle in which the individual contractions are not initiated by a neuronal impulse; compare synchronous or neurogenic muscle; also called the resonating muscle.

ATBI: All Taxa Biodiversity Inventory in Costa Rica.

***Athalia lugens proxima* (Klug):** Belongs to order Hymenoptera and family Tenthredinidae. It is an insect pest of mustard, commonly known as the mustard sawfly.

***Atherigona soccata* Rond.:** Belongs to order Diptera and family Anthomyiidae. It is an insect pest of sorghum, commonly known as the sorghum shootfly.

Atom: Atom is the fundamental constituent unit of a substance. The molecules are formed by the combination of two or more atoms in a constant ratio. Normally, atoms cannot exist independently.

Atomic weight: The atomic weight of any element is that number by which it is determined that how many times one atom of that element is heavier than weight of 1.008th part of one atom of hydrogen or weight of sixteenth part of one atom of oxygen 16 or weight of twelfth part of one atom of carbon.

Atomizer: Device for breaking up a liquid stream into very fine droplets by a stream of air.

ATP: Adenosine tri-phosphate; a molecule having high energy bonding through a phosphate group.

Atrial aperture: The outside opening of a chambered or atriate spiracle.

Atrium (pl., atria): Spiracular chamber housing the opening to a trachea, a chamber just inside a body opening.

Atrophied: Reduced in size, rudimentary or vestigial.

Atrophy: Decrease in size of a tissue, organ or part after full development has been obtained; a condition in which the affected cells undergo degenerative and autolytic changes and become smaller and lesser functional.

Atropine: A poisonous, crystalline alkaloid used in medicine; a specific antidote used in case of poisoning by organophosphate insecticides.

Attapulgite: A magnesium silicate clay, $Mg_3(AlSi)_4O_{10}(OH)_2$ that is used as a carrier for the insecticide in the dust formulations. The powder is white in colour. It is also known as fuller's earth.

Attenuated: Very slender and gradually tapering distally.

Attini: Foliage cutting ants which feed on fungi grown on cut foliage, subfamily Myrmicinae.

Attractants: Substances which elicit a positive directional response; chemicals having positive attraction for animals such as insects, usually in low concentration and at considerable distance. Several proprietor lures are manufactured.

Auditory organs: Any structure that functions as an ear; specialized structures covered by a membrane on the anterior tibia or the base of abdomen in Orthoptera.

***Aulacaspis tegalensis* (Zehn.):** Belongs to order Hemiptera and family Diaspididae. It is an insect pest of sugarcane, commonly known as the sugarcane scale.

Aulacidae: Family of super family Evanioidea, suborder Apocrita and order Hymenoptera. These have the antennae inserted just above the clypeus. On forewing cross-veins 2m-cu and 2r-m are present. Anal lobe is absent in hindwing. These insects are chiefly parasites of coleopterous insects. Indian species include *Aulacus bituberculatus.*

Auricle: A small lobe or ear like structure.

Austroperlidae: A family of order Plecoptera. Mandibles normal and the third segment is not longer than first and second together; characterized by more than six anal veins in hindwings, *e.g.*, *Austroperla* sp. and *Tasmanoperla* sp.

Autecology: The branch of ecology which deals with the interrelationships of the individual organism and its environment. Also see synecology.

Autoamputation: Capacity to detach a limb grasped by a predator, usually at the level of trochanter as seen in nymphs of walking sticks.

Autocidal control: The autocidal control involves the employment of an organism of one or both sex to destroy the members of same species and differs from chemical control in which the poisonous chemicals are used or from the biological control in which the organisms of other groups or species are used. The autocidal control is based on the principle of decreasing or suppressing the population of rapidly reproducing individuals by reducing their birth rate rather than by increasing death rates as the chemical and biological methods are initiating. This method constitutes the sterile male technique, genetic technique, pheromonal technique, synthetic hormones, antihormones and insect growth regulators as the potent sterility agents.

Autocide: The control of a pest by the sterile-male technique, genetic technique, pheromonal technique, synthetic hormones, antihormones and insect growth regulators.

Autogenous: Able to carry out vitellogenesis without an exogenous source of protein; compared with anautogenous.

Autoinfection: Infection of host by a microorganism or virus produced within or upon the body of the same individual host.

Autolysis: See histolysis.

Automatic uncapper: Automated device that removes the cappings from honey combs, usually by moving heated knives and metal teeth.

Autoparasitism (Adelphoparasitism): A special type of parasitism in which the female develops as a primary parasitoid, but the male is a secondary parasitoid through females of its own species (Flanders, 1937). For example, males of *Coccophagoides utilis* Doutt develop as a hyperparasite from unfertilized egg laid by parent female of C. *utilis,* a primary parasitoid of purple citrus scale, *Parlatoria oleae.*

Autotomictic parthenogenesis: A normal reduction division occurs followed by the fusion of two nuclei as seen in phasmids, coccids etc.

Autotomy: Same as autoamputation.

Autotoxicity: The property of a plant or organism to produce chemicals that deter its growth or kill its own species.

Auxillae: Small plates beneath the bases of the pretarsal claws, bearing the pulvilli when the latter are present.

Auxillary spray material: Any agent (surfactant, solvent, carrier or diluent etc.) added to a pesticide spray to improve its physical properties or performance.

Auxillary vein: The subcosta.

Auxin: A generic term for compounds characterized by their capacity to induce elongation in shoot cells. They resemble indole-3- acetic acid in physiological action.

Available name: A scientific name published in accordance with the requirements specified in articles 8-20 of the International Code of Zoological Nomenclature (ICZN).

Average cost: The cost per unit at a given level of production; it is equal to the total costs divided by production.

Avermectines: These are macrocyclic lactones, natural products isolated from soil microorganism mycelia of *Streptomyces avermitilis* and closely related to milbemycines. Avermectines affect insects and mites by binding to ligand gated chloride channels (Glutamate and GABA), thereby, chloride channels opens into neurons which results in loss of cell function and disruption of nerve impulses.

Avicide: A substance to control the bird (Avian) pests.

Axenic: Rearing of one or more individuals of a single known species in or on a non-living medium.

Axiidae: Family of super family Geometroidea, suborder Ditrysia and order Lepidoptera. Members of this family possess the tympanal organs on the seventh abdominal segment whereas in the others they are seen on the first or second abdominal segment. It includes a few Mediterranean species.

Axil: The upper (usually acute) angle between a leaf stalk and the stem from which its branches emerge. Buds of new vegetative growth or flowers also emerge from the axils.

Axilla: A triangular or rounded sclerite laterad of the scutellum.

Axillary area: Of wing. A topographical area of the wing, basal, and which contains the articular sclerites.

Axillary cell: A cell in the anal area of the wing.

Axillary cord: The thickened, corrugated posterior edge of the articular membrane of wing base, continuous with the posterior marginal fold of alinotum.

Axillary sclerites: The small sclerites at base of wings in the wing flexing insects.

Axon: A fibre or process of a nerve cell that transmits impulses from the cell body or soma. The neurone or nerve cell consists of a prominent nucleated cell body or soma and a long cytoplasmic filamentous extension, the axon. The cell body of one neurone communicates through a transitional gap (synapse) with the axonus terminus of the other neurone in a linear fashion so that a continuous chain is formed along which the electrical disturbances (impulses) can be transmitted from one part of the body to the other.

***Azadirachta indica* A. Juss:** A tree species commonly known as neem and a member of mahogany family, Meliaceae. The tree is known for having the pesticidal alkaloids, like azadirachtin, meliantriol, salannin, nimbin, nimbidin etc.

Azadirachtin: A chemically active ingredient of the neem tree used in pesticides which acts as powerful insect antifeedant, repellent and growth regulator. The demonstration of its ability to prevent locusts chewing on crops was the first scientific proof for neem's traditional use for insect control on Indian crops.

Azinphos ethyl (Ethyl guthion): A synthetic insecticide and acaricide; an organophosphate, O, O-diethyl S-4-oxo-1, 2, 3-benzotriazin-3 (4H)-yl- methyl phosphorodithioate; highly toxic to mammals; acute oral LD_{50} for rats 7 to 18 mg kg^{-1}; used on cotton and potatoes. Highly toxic Guthion is a general usage, broad spectrum insecticide that is effective against wide variety of insects on fruits, vegetables, flowers and ornamental plants.

Azobenzenc: A synthetic acaricide, $C_6H_5N= NC_6H_5$, used by volatilizing from hot water pipes of green house, or by ignition in a pyrotechnic mixture, or as a spray, either emulsion or suspension; chronic toxicity is 1000 ppm in diet which kills rats in a few days; phytotoxicity has injured roses and *Asparagus* fern.

Azodrin: Trade name of monocrotophos.

B

Bacillidae: A family of order Phasmida, includes stick or walking insects. The stick insects possess elongate and cylindrical body and are often apterous. The first abdominal segment fused into the metathorax.

Bacilluria: Having bacteria in the urine.

***Bacillus*:** A genus of rod shaped bacterium.

***Bacillus cereus*:** It is a non-crystalliferous spore forming bacterium. It produces septicemia in the host, like other spore forming bacteria. It also produces phospholipase-C in quantities sufficient to damage gut cells of the susceptible species thus aiding the entry of bacterium in the haemocoel. It is effective against Coleoptera, Hymenoptera and Lepidoptera. Its host range is limited by pH of the gut, as highly alkaline gut content inactivates the toxic phospholipase C.

***Bacillus larvae*:** The bacterium responsible to cause American foulbrood disease.

***Bacillus popilliae* Dutky:** It is an obligate spore forming bacterium widely employed for the control of Japanese beetle, *Popillia japonica* Newman. It is a fastidious organism and cannot be propagated on artificial media. It is prepared commercially by injecting or feeding the host with the bacterium. The bacterium multiplies in great numbers in the body of host. They are recovered by grinding the host with talc. The standardized powder contains 100 million spores per gram. Upon ingestion by the host, it produces lethal septicemia.

***Bacillus thuringiensis* (B.t.):** It is a crystalliferous, facultative, spore forming bacterium which is widely used microbial insecticide to control certain insect pests especially caterpillars. It was isolated by Berliner (1915) in Germany from the diseased larvae of Mediterranean flour moth, *Ephestia kuhniella*. It is most effective against the insects having the alkaline gut content (pH 9.0-10.5) and also against those which have enzymes to dissolve the crystal. It is marketed under the trade names, Sporeine, Thuricide, Bakthane, Dipel, Biotrol, Kajrab, Agritol, Dandrobacillin etc. The first commercial product is Sporeine from France in 1938. *B. t. kurstaki* is a potential strain, isolated from France in 1938. *B.t.* strains are highly specific in their activity are *B.t.* var. *kurstaki*, *B.t.* var. *aizwai*, *B.t.* var. *israelensis* and *B.t.* var. *tenebrionis* which are effceive against caterpillars, wax moth larvae, Dipterans (mosquitoes, black fly and fungus gnat larvae) and beetles (Colorado beetle grub), respectively. The various toxic crystalliferous proteins isolated from these bacteria are *Cry-I*, *Cry-II*, *Cry-III*, *Cry-IV* and *Cry-V* which are effective against Lepidoptera, Lepidoptera and Diptera, Coleoptera, Diptera, and Lepidoptera and Coleoptera, respectively.

Bacteraemia: Bacteria in the blood.

Bacteria (sing., bacterium): Unicellular organism, some of which cause diseases in plants and animals. They are too small that they cannot be seen without a microscope.

Bactericide: A substance that is toxic to bacteria.

Bacteriophage vector: Bacteriophages contain linear DNA molecules that break into two fragments by cleavage at single site. The insertion of desired DNA in between two fragments can lead to the formation of recombinant DNA. Bacteriophages infect the bacterial cells and follow the lytic cycle and thus multiple copies of the recombinant DNA can be generated. The non-essential genes of bacteriophage DNA are deleted prior to insertion of desired DNA. Foreign DNA of 20-25 kb can then be inserted into the phage DNA. Among the several types of bacteriophages used as vector, the most common are lambda and M_{13} phages.

Bacteriophages (sing., bacteriophage): Gr. *phages*= eaters. Viruses which infect, grow on and kill bacteria are known as bacteriophages. This term was coined by d' Herelle (1917).

Bacteriostatic: Preventing the growth of bacteria.

***Bactrocera (=Dacus) cucurbitae* (Coquillett):** Belongs to order Diptera and family Tephritidae. It is an insect pest of brinjal, commonly known as the melon fruit fly.

***Bactrocera (=Dacus) dorsalis* (Hendel):** Belongs to order Diptera and family Tephritidae. It is an insect pest of mango, loquat, fig etc., commonly known as the fruit fly.

Baculoviruses: The group of insect viruses with rod shaped DNA and with occlusion in a proteinaceous crystal. These belong to the family Baculoviridae.

Bacunculidae: A family of order Phasmida, includes stick or walking insects with long antennae and forelegs. The stick insects possess elongate and cylindrical body and are often apterous. First abdominal segment free from thorax.

Baetoidea: A super family of Ephemerida. M_{1+2} and Cu_1 of forewing are parallel at their bases; four movable tarsi on hind legs and the fifth segment fused with the tibia, if present. This includes the families Oligoneuriidae, Baetidae, Caenidae, Leptophlebiidae and Ephemerellidae.

***Bagrada cruciferarum* (L.):** Belongs to order Hemiptera and family Pentatomidae. It is an insect pest of oilseed crops, commonly known as the painted bug.

Bait: A formulation including attractants, toxicants, and/or mechanical devices used as a lure for killing pests. Generally, poisoned foodstuff used to attract and kill insects or rodents.

Bait shyness: The tendency for rodents, birds or other pests to avoid poisoned baits or to discard the poisons by animals after the initial mortality of their fellows.

Bait stations: Containers that are used for placing bait for feeding of the rats.

Balostomatidae: This is a family of division Cryptocerata, suborder Heteroptera and order Hemiptera. The largest members of the Heteroptera are found in this family. The bug, *Lethocerus grandis* exceeds 190 mm in length. These are brownish oval and somewhat flattened and possess four segmented antennae, forelegs are prehensile, middle and hindlegs have tibiae flattened with fringes of hairs and adapted for swimming. Two retractile apical appendages are present. Air is stored beneath the wings in the reservoir formed by convcave dorsum of the abdomen. These bugs are predaceous on fishes, tadpoles, snails and aquatic insects, *e.g.*, giant water bug, *Belostoma* sp.

Banana scale moth: See *Nacoleia octasema.*

Banana stem weevil: *Odioporus longicollis* Ol., a weevil pest of banana; reddish to dark brown grub with long curved snout, bores inside stem killing the plant.

Banana weevil: *Cosmopolites sordidus* (Germar) is an insect pest of banana. Grub bores into rhizome, tunneling the root stock and travelling up the stem.

Band: A transverse marking broader rather than a line.

Band application: Treatment of a band of soil in row crops, usually covering plant rows, with either sprays or granules.

Banded blister beetle: See *Mylabris phalerata.*

Bandicoot: *Bandicota indica,* a biggest rat species found in fields and slums; terrific burrower; damages crops in the fields, weakens foundation of buildings by making tunnels; omnivorous in habit, feeding on grains, vegetables, fruits; feeds on and destroys potato and peas in the field.

Bantonite, $(AlMg)_2Si_4O_{10}(OH)_2$: It is a hydrated alumina silicate; montmorillonite type mineral; slightly grey in colour. It is used as inert material in the insecticide dust formulations.

Barium carbonate ($BaCO_3$): A rodenticide which is also known as witherite, very toxic to mammals, less and slow acting poison, develop rapid bait shyness, therefore, it is not popular. The baits are prepared by addition of 16 parts of barium carbonate in 83 parts of food bait and 1 part vegetable oil. Now-a-days, it has been replaced by newer rodenticides.

Bark eating caterpillar: *Indarbela quadrinotata* Wlk., an insect pest of guava; caterpillar feeds on bark and destroys tissues; also attacks citrus, loquat, pomegranate, mango and litchi. Badly damaged plants stop growing and soon die off.

Barrier application: The use of pesticide or another agent to check the pests from entering a container, area, field or building.

Barthrin: A synthetic analogue of pyrethrum.

Basal: Proximal, at the base or near the point of attachment (opposite to apical).

Basal anal cell: An anal cell near the wing base or a cell at the base of wing between 1A and 2A.

Basal apodeme: An apodeme of the phallobase.

Basal areole: A small cell at the base of wing.

Basal cell: A cell near the base of wing, bordered at least in part by the unbranched longitudinal veins.

Basal fold: See plica basalis.

Basal lamina: One of two layers that form the basement membrane, an amorphous sheet which underlies epithelial cells.

Basal piece: See basale and phallobase.

Basal plates: Sclerites of the phallobase.

Basal treatment: A treatment applied to the stem or trunks of plants at or just above the ground line.

Basal vein: A vein in middle of the forewing extending from the median vein to the subcostal vein.

Basalare (Basalar sclerite): An epipleurite located anterior to the pleura.

Basale of tegmen: The basal portion of a sheath like tegmen which is divided transversely. This appears to be homologous to the phallobase. It is also called basal piece.

Base oils: These are solvents or diluents that are used in many oil based pesticides. Base oil diluents are usually high grade (highly refined) kerosenes that are essentially colourless and odourless and that tend to dry leaving very little if any oil film on treated surface is there.

Basement membrance: The noncellular, undifferentiated innermost layer (0.5 μm thick) of the insect integument on which epidermis rests.

Basic nephridial system: In a large number of insects, the distal ends of Malpighian tubules absorbs various waste substances, salt and water from haemolymph and in *Carausius, Dacus* and *Drosophila* bear distinct dilations at their distal ends. All tubules often open independently but in some insects, such as Gryllids, all tubules open proximally into a common ureter and the latter finally opens into the gut. Similarly in some beetles, *Haltica* and *Donacia*, a group of four tubules open into a common ampulla, while remaining two open independently. In the case of red cotton bug, *Dysdercus,* all tubules of one side open into a common ampulla or bladder which ultimately opens into the gut. In Hemiptera, the tubules are differentiated into the distal clear part filled with fluid and proximal opaque half, containing solid or granular substances.

Basic product: Specifically the actual product.

Basiconic sensilla: Sense organs of which the external part has the form of a minute cone or peg projecting above the general body surface of insect, probably an olfactory structure.

Basicosta: The proximal sub marginal ridge of the inner wall of a leg segment.

Basicostal suture: The external groove of a leg segment forming the basicosta.

Basicoxite: Usually narrow basal rim of the coxa proximal to the basicostal suture and its internal ridge.

Basipodite: The basal segment of the telopodite or second segment of a generalized appendage; in Crustacea bearing the exopodite (first trochanter).

Basisternum: The part of the sternum anterior to the sternacostal suture.

Basitarsus: The first segment of the tarsus- usually the largest.

Basivalvulae: Small sclerites, sometimes occurring at the bases of the first valvulae, often confused with the first valvifers.

Batch: Quantity of material or product which is known or assumed to be produced under uniform conditions.

Batesian mimicry: A palatable, edible or harmless species that gains protection by virtue of its close resemblance (in the body colouration) to an unpalatable or harmful species.

***Batocera rubus* Linnaeus:** Belongs to order Coleoptera and family Cerambycidae. It is an insect pest of mango, commonly known as mango stem borer.

***Batocera rufomaculata* Dejean:** Belongs to order Coleoptera and family Cerambycidae. It is an insect pest of mango, commonly known as mango stem borer.

Batumen: A protective layer of propolis or hard cerumen (sometimes vegetable matter, mud or various mixtures) that encloses the nest cavity of stingless bees.

Bauxite ($Al_2O_3.2H_2O$): It is high alumina clay compound essentially of aluminium hydrated mineral; light greyish in colour; used as inert material in the insecticide dust formulations.

Bawd: Towards the base.

Baygon: Trade name of propoxur.

Baytex: Trade name of fenthion.

Bead: A fine raised margin.

Beak: Term commonly used to denote elongate mouthparts of sucking insects proboscis or rostrum.

Beak proboscis: The long, protruding mouthpart structures of an insect with piercing-sucking mouthparts. This type of mouthparts can be found in the suborders Heteroptera (true bugs) and Homoptera (cicadas and others).

Bean fly: See *Ophiomyia phaseoli*.

***Beauveria bassiana*:** A species of entomopathogenic fungi causing white muscardine disease in many insect species. It is marketed under trade names, Larvocel, Boverin, FBB, Dispel, Racer BB (1.15 WP) etc.

Bed: (i) A narrow flat-topped ridge on which crops are grown with a furrow on each side for drainage of excess water. (ii) An area in which seedlings or sprouts are grown before transplanting.

Bedbug: See *Cimex lectularius.*

Bed up: To build up beds or ridges with tillage implement.

Bee: Any of several insects of the super family Apoidea (order Hymenoptera). Bees are winged, hairy-bodied, usually stinging insects, including both solitary and social species. They are characterized by chewing and lapping type mouthparts for gathering nectar and pollen.

Bee blower: An engine with attached blower used to dislodge bees from combs in a honey super by creating a high velocity, high volume wind.

Bee bread: A mixture of collected pollen and nectar or honey, deposited in the cells of a comb to be used as food by the bees.

Bee brush: A brush or whisk broom used to remove bees from combs.

Bee candy: Artificial food for bees made from sugar and substances like honey.

Bee escape: A device used to remove bees from honey supers and buildings by permitting bees to pass one way but preventing their return.

Bee farming: See apiculture.

Bee fold: An enclosure for beehive.

Bee hive: A box or receptacle with movable frames, used for housing a colony.

Bee keeping: See apiculture or bee farming.

Bee metamorphosis: The three stages through which a bee passes before reaching maturity, *viz.*, egg, larva and pupa.

Bee pasturage: Also bee forage; plants yielding nectar and pollen, and within flying distance of a hive.

Bee smoker: A device consisting of a funnel, firepot and bellows for blowing smoke into a beehive to force out the bee colony.

Bee space: 1/4 to 3/8 inch space between combs and hive parts in which bees build no comb or deposit only a small amount of propolis.

Bee tree: A tree with one of more hollows occupied by a colony of bees.

Bee veil: A cloth or wire netting for protecting the beekeeper's head and neck from bee stinging.

Bee venom: The poison secreted by special glands attached to the sting of bee.

Bee wax: A complex mixture of organic compounds secreted by special glands on the last four visible segments on the ventral side of the worker bee's abdomen and used for building comb. Its melting point is from 143.6 to 147.2 °F. Chemically, it is a mixture of esters and paraffin hydrocarbons.

Beetle: Any of numerous insects of the order Coleoptera. Beetles have biting mouth parts, their forewings are modified to form a strong covering (known as elytra) that protects the underlying membranous hindwings when at rest.

Behavioral resistance: A mechanism in insects of resistance to insecticides which involves a behavioral change whereby contact with insecticides is avoided.

Behaviour: The whole complex of observable, recordable or measurable activities of a living animal.

Bellow type duster: See duster, bellow type.

***Bemisia tabaci* (Genn.):** Belongs to order Hemiptera and family Aleyrodidae. It is an insect pest of ornamental plants, cotton, chillies, coriander, cumin, pulses etc., commonly known as the cotton whitefly.

Benchmarks for pesticide behaviour in environment: Physicochemical properties and characteristics as a tool to predict the behavior of pesticides in environment. These include generally, the lipophilicity which determines accumulation in fat; volatility, hydrolysability and oxidative behavior which determine persistence.

Beneficial insects: Insects that serve the interest of man. For example, insect pest predators and parasitoids that helps to keep pest populations under control. The bees and other pollinating insects, silkworm, lac insect etc. are beneficial insects.

Benthic: Pertaining to the bottom of seas, lakes and ponds.

Benthos: Non-planktonic animals (not suspended in water) associated with fresh water substrata (upper layer of the sediment in rivers and ponds) at the sediment water interface.

Bentonite: A clay composed mainly of silica and aluminium silicate; mined in Mississippi, Wyoming and elsewhere; used as dust diluent and for lining ponds to hold water.

Benzaldehyde: A volatile, almond-smelling chemical used to drive bees out of honey supers.

Benzene hexachloride: Chemical name of BHC (common name). A synthetic insecticide, chlorinated hydrocarbon, 1, 2, 3, 4, 5, 6-hexachlorocyclohexane of mixed isomers; slightly more toxic to mammals than DDT, acute oral LD_{50} for rats is about 200 mg kg^{-1}; phytotoxicity: more toxic than DDT, interferes with germination, suppresses growth and reduces yields except at low concentration; certain crop plants, as potato absorb crude BHC with consequent tainting of tubers.

Benzenoid chemical: One having the benzene ring form of molecular structure.

Ber beetle: See *Adoretus pallens.*

Ber fruit fly: *Carpomyia vesuviana* Costa, an insect pest of jujube, the maggot feeds on fruit pulp, deform them and ultimately the fruits fall off.

Berlese funnel: A sampling device that uses heat to drive insects from a sample of soil or litter.

Berlese-Jeschikov's theory: Berlese proposed the theory of metamorphosis in 1913. According to this theory, the metamorphosis arose by shifting of hatching of larva to an earlier stage of embryonic development so that the completion of morphogenesis of the imaginal structures must necessarily takes place during post embryonic development. According to this theory, the hemimetabolous insects hatch at a latter stage of morphogenesis, the oligopod stage with only thoracic appendages; while the holometabolous insects hatch at the earlier or so called polypod stage, characterized by embryonic appendages on the abdominal segments. The above concept was latter advanced by Jeschikov (1929-41). He proposed that the hemimetabola eggs hatch at the late stage (postoligopod) due to its large size and relatively large yolk content whilst the holometabola eggs hatch at the earlier stage as it is small and contains much less yolk or the poorest quality yolk as found in parasitic Hymenoptera which undergo hypermetamorphosis. Because of this reason, the juvenile forms of hemi and holometabola differ widely.

Beta taxonomy: A taxonomic level concerned with the arrangement of species into a natural system of lower and higher taxa.

Betelvine bug: *Disphinctus politus* W., an insect pest of betelvine which is reddish brown bug and its nymph sucks juice from tender leaves in between veins, making them to develop black spots.

Bethylidae: Family of super family Bethyloidea, suborder Apocrita and order Hymenoptera. Bethylidae are ant-like, small to medium sized dark colored insects parasitic on larvae of Lepidoptera and Coleoptera. A few species are capable of stinging men. Many species are apterous. In a few, winged and wingless forms occur in both sexes.

Bethyoidea: Super family of Suborder Apocrita, order Hymenoptera. Bethyloidea are parasitic on other insects. These insects are characterized by absence or poor development of a trochantellus. An anal lobe is usually present in hindwings. The ovipositor is short.

BHC ($C_6H_6Cl_6$): It is a common name of a synthetic chlorinated hydrocarbon insecticide, 1, 2, 3, 4, 5, 6-hexachlorocylohexane, mixed isomers and a specified percentage of gamma; slightly more toxic to mammals than DDT, acute oral LD_{50} for rats is about 200 mg kg^1. Phytotoxicity more than DDT, interferes with germination, suppresses growth and reduces yields except at low concentration; certain crop plants such as potato absorb crude BHC with consequent tainting of tubers. The compound was first synthesized by Faraday in 1825. Its insecticidal properties were discovered by British and French workers independently from 1941-1942. Its active principle, gamma isomer was isolated by Van der Linden (1943). It contains 10-80 per cent active gamma isomer together with four other isomers of varying proportions (alpha, beta, delta and epsilon). The crude product is greyish or brownish amorphous solid with a characteristic musty odour having melting point 65 °C.

Bibionidae: Family of suborder Nematocera, order Diptera. This family includes generally robust flies with antennae 8 to 16-segmented, with antennal segments bead-like and closely apposed. Wings large and with anterior veins strikingly prominent. Males with eyes holoptic. Adults occur in grassy vegetation or decaying vegetation and larvae are gregarious. The larvae of *Bibio* are the most primitive of dipteran larvae. *Bibio obscuripennis* occurs in India.

Bidrin: Trade name of dicrotophos.

Biennial: Occurring in cycles of two years.

Bifid: Forked or divided into two parts.

***Bifiditermes beesoni* (Gardner):** A species of termites which occurs in Pakistan and North-western India and attacks dead and dying wood.

Biforous spiracles: Spiracles of coleopterous larvae having two pouches of the atrium originally supposed to open separately to the exterior.

Bihar hairy caterpillar: *Diacrisia obliqua* W., an orange coloured hairy caterpillar that feeds on the leaves of sweet potato and a number of other crops.

Bilateral symmetry: Similarity in the form of one side with the other.

Biliverdin: These are blue green granules derived from the break down of traces of haemoglobin that have been absorbed unchanged from the gut.

Bilobate aedeagus: Aedeagus in which the phallobase and parameres are well developed but the penis is membranous or sclerotized only at base, where there may be a pair of anterior struts. This type is characteristic of many Scarabaeoidea.

Bilobed: Divided into two lobes.

Bimenthrin: A synthetic pyrethroid marketed under trade name Talstar (10 EC).

Binapacryl ($C_{15}H_{18}N_2O_6$): A synthetic acaricide; a nitrophenol derivative, 2-(1-methyl-n-propyl)-4, 6-dinitrophenyl 3-methyl crotonate; non-systemic acaricide, very effective against red spider mite with fungicidal action also; moderate mammalian toxicity; acute oral LD_{50} for rats is 136 to 225 mg kg^{-1}; controls mites on tree fruit and nuts also. It is marketed under trade name Morocide by Hoechst and formulated as 40 EC or 50 WP.

Bink: A wasps' or bees' nest.

Binocular vision: This vision helps in knowing the distance of object; the eye is divided into two parts, one of which is apposition type and the other is superposition type, *e.g.*, preying mantis.

Binomen: The designation of scientific name of a species consisting of generic and specific parts.

Binomial nomenclature: The system of classification of organisms as advanced by Carolus Linnaeus by naming scientific names of species having both generic and specific components and adopted by the International Congress of Zoology in International Code of Zoological Nomenclature (ICZN).

Binomial: Having two names, as in a scientific name, composed of a genus name and a trivial name of specific epithet, *e.g.*, *Rhynocoris kumarii.*

Bioaccumulation of pesticide: Accumulation of the toxicant inside dynamic ecological system depending on (i) physicochemical characteristics of insecticide, (ii) concentration vs. dilution or degradation in the system, (iii) rate of food consumption and body size, and (iv) dynamics of accumulation, *i.e.*, balance of uptake and elimination.

Bioactivation: Transformation of a pesticide within system of an organism into a more biochemically active metabolite.

Bioassay: It is the combination of two words, *bios* means life and *assay* means determination. Thus bioassay stands for determination of relative toxicity of insecticides by studying and examining their effects on living organisms. It is the estimation of amount of toxicant in a sample by measurement of its effect on test organisms; usually estimation of amount necessary in a standard type of application to produce 50 per cent kills.

Biochemistry: The branch of chemistry that deals with chemical aspects in plants, animals and their life processes.

Biocide: A chemical substance obtained from plant or animal which is toxic to a wide range of organisms (*e.g.*, insects, snails, birds and people).

Bioconcentration: Process leading to higher concentration of a pesticide in an organism than in environmental media to which it is exposed.

Biodegradable: Capable of being broken down by micro-organisms. It usually refers to biological processes in soil, water and sewage. It can also refer to man-made organic compounds such as pesticides.

Biodiversity: Biodiversity means presence of different kinds of living organisms in a particular ecosystem. It includes complete spectrum of living organisms from microscopic level to macroscopic level. There are four types of biodiversity: (i) Genetic biodiversity, (ii) Species biodiversity, (iii) Community biodiversity, and (iv) Ecosystem biodiversity.

Biogas: The biogas is obtained from the biomass which occupies an important place in the non-conventional energy sources. In the rural areas, the biogas is better source of fuel than wood. The constituents of biogas are methane (65 per cent), carbon dioxide (30 per cent), hydrogen (1.2 per cent), hydrogen sulphide (0.8 per cent), oxygen (1.2 per cent), nitrogen (1.2 per cent) and carbon monoxide (0.8 per cent).

Biogeochemical cycles (Inorganic-organic cycles or nutrient cycles): The chemical elements including all the essential elements of the protoplasm (nutrient material) tend to circulate in the biosphere in characteristic paths from environment to the organism and back to the environment. These more or less circular paths are known as biogeochemical cycles or inorganic- organic cycles. The various nutrient cycles are (i) hydrological cycle (water cycle), (ii) carbon cycle, (iii) nitrogen cycle, (iv) phosphorus cycle, and (v) sulphur cycle.

Biological assessment of exposure: Assessment of exposure of a living organism to pesticides using biological specimen (blood, urine etc.) taken in the environment (work place, field etc.) with analysis either directly by chemical determination of parent or metabolite, or indirectly by measurement of a relevant biochemical parameter (*e.g.*, plasma cholinesterase activity for organophosphorus compounds).

Biological classification: The arrangement of animals into taxa on the basis of inferences of their genetic relationship.

Biological control agent: Any biological agent that adversely affects pests.

Biological control: The control of pests by employing predators, parasites, or diseases; the natural enemies are encouraged and disseminated by man. The use of natural enemies such as predators, parasites, and pathogenic micro-organisms or antagonists is done to control pest populations, diseases and weeds. This can be achieved either through conservation and stimulation of indigenous natural enemies or by the importation and mass introduction of exotic natural enemies. Smith, 1919 used this term to signify the use of natural enemies for control of insect pests. With the advancements, many techniques such as resistant varieties, hormonal, genetic and sterile male techniques were clubbed together. However, widening the subject makes it ambiguous; therefore, it is desirable to use classical connotation only.

Biological diversity: The number of different kinds of organisms per unit area or volume; the richness of species in a given area.

Biological insect pest suppression: The use or encouragement, by man of living organisms or their products for the population reduction of insect pests; Also see biological control.

Biological magnification (Biomagnification): When a small organism feeds on several micro-organism containing residues of a pesticide, the pesticide may get deposited in its body. A predatory animal which feeds on several such small organisms; and the human beings at the top of food chain consume quantity of the pesticide much larger than that was present in the micro-organisms. This increase in chemical concentration as it passes through the food chain is called biological magnification, bioconcentration or biomagnification.

Biological oxygen demand: The quantity of oxygen used by micro-organisms in the biochemical oxidation of organic matter and oxidisable inorganic matter by aerobic biological action.

Biological pesticide: A pesticide, the active ingredient of which consists of a living organism or its by-product.

Biological race: Strains of species which are alike morphologically but differ in some physiological way, such as a parasite and saprophyte with a particular host requirements or a free living organism with a food or habitat preference.

Biological species: A group of interbreeding organisms reproductively isolated from other closely similar groups.

Biological species concept: A concept stressing reproductive isolation and the possession of a genetic programme affecting such isolation (for true species).

Biological transmission: The insects play a part in the life of pathogenic organism. It may be of three types: (i) Propagative transmission: The pathogen undergoes no cyclic changes but simply multiplies inside the body of insect as in an artificial culture tube. Most of the bacterial and viral multiplication occurs in the same way, *e.g.*, bubonic plague. The *Bacillus pestis* germs are swallowed by a flea during feeding the blood from an infected rat. They multiply in a sort of filtering chamber of the gut and then they are transmitted to a man in the next biting. (ii) Cyclopropagative transmission: The pathogen undergoes cyclic changes and multiplies in the body of insect, *e.g.*, malaria parasite, *Plasmodium*, which is spread by mosquitoes, *Anopheles*. They receive the pathogen from infected man through the blood they suck. These germs undergo reproductive cycle and multiply in the body of insect, which in second bite are transmitted to the other indiduals. (iii) cyclodevelopmental transmission: The pathogen undergoes cyclic developmental changes inside the insect body but does not multiply, *e.g.*, filariasis worm, *Wuchereria*. It is a tropical disease. The thread like nematode, *Wuchereria bancrofti* are known as filaria and they cause cutaneoous lesions and may also occlude the lymph vessels, leading to elephantiasis. The filarias are transmitted by mosquitoes, *Culex* and *Aedes*; black flies, *Simulium*; biting midges, *Culicoides* and mites, *Liponyssus*. In these vectors, the pathogens grow in size without undergoing multiplication.

Biology: (Gr., *Bios*= life). The science which includes natural history of plants (Botany) and animals (Zoology) is termed as biology.

Bioluminescence: The ability of certain organisms to produce light, *e.g.*, in some insects of order, Collembola, Diptera and Coleoptera.

Biomarker: Indicator (molecular, biochemical, cellular or organism) signaling an event or condition in a biological system or sample and giving a measure or exposure to a xenobiotic.

Biomass: The total living mass in a defined segment of an ecosystem expressed as the living weight per unit area or mass. Soil microbial biomass is often used as an indication of potential microbial activity level in the soil. Agriculture remains, animal excreta, municipal waste and inflammable substances produced from plants come under the category of biomass. It is an unlimited source of non-conventional energy.

Biome: Major communities of animals and plants having a similar life form.

Biometer flask: Experimental apparatus commonly used in the laboratory studies of pesticide degradation in soil. Contains separate compartments for aerobic incubation of soil and for media to trap carbon dioxide and volatile products.

Bionomics: A branch of biology that deals with the habits, breeding and adaptation of living organisms to their environment.

Biorational pesticides: Pesticides based on bacteria, viruses, fungi or protozoa; includes pest control agents and chemical analogues of naturally occurring biochemicals (*e.g.*, pheromones, insect growth regulators etc.).

Biordinal: Of two alternating sizes.

Bioresmethrin: Insecticide with low toxicity and wide spectrum activity that is especially useful against stored food insects.

Biosphere: The land, water and atmospheric portions of our planet in which life exists.

Biosynthesis: The method of synthesis of complex molecules within the living organism.

Biosystematics: A branch of biology incorporating ecological, biological, morphological, cytological, biochemical aspects in the identification and classification of organisms, especially within an evolutionary framework.

Biota: The fauna and flora of a given habitat.

Biotechnology: Any technique that uses living organisms or substances from those organisms to make or modify a product, to improve plants or animals or to develop micro-organisms for specific uses. As evident from the word itself, this science is made up by combination of biology and technology.

Biotic: Relating to life.

Biotic community: The plants and animals of a given habitat.

Biotic factors: Biological factors include competition among insects, natural enemies, diseases, quality and quantity of food, plant physiological factors like tolerance, vigour, rapid growth and distasteful characters etc. that influence the growth and population density.

Biotic pesticides: Formulations containing micro-organisms, known as insect pathogens and their by-products or products obtained from other organisms. These are applied in the same manner as conventional pesticides to control pest species; also known as microbial pesticides or biopesticides.

Biotic potential (Bp): The measure of innate ability (inherent capacity) of an individual to reproduce and survive if left to itself and isolated from its natural enemies, diseases, or any other inhibiting factors.

Biotop: A small area with uniform environment occupied by unified community of organism.

Biotransformation pathway: Sequence of the changes occurring in the structure of pesticide when it is introduced into a specific biological test system.

Biotransformation: Conversion of chemical structure of a pesticide either *in vitro* or *in vivo.* This process is catalyzed by enzymes.

Biotype: A biological strain of an organism, genetically and morphologically indistinguishable from other members of its species, but exhibiting distinctive physiological characteristics particularly with regards to its ability to successfully utilize pest resistant host organisms or to act as an effective beneficial species.

Bipectinate: It is a type of antenna which differs from pectinate antenna in having the long projections on both sides, *e.g.*, silk moth.

Bipolar neurone: A type of neurone possessing the axon at one pole and dendrite (dendron) at the other.-

Biramous: Double appendages or with two branches.

Bird damage: Top fruit buds and ripe soft fruit are often taken by birds. Farm crops suffer especially in winter and spring when they may be the only food source. *Brassicas*, clovers and cereals are destroyed during winter and newly sown seeds of cereals, peas and beans may be eaten, particularly by pigeons in the spring. Grain from lodged crops and grain shed during harvest is also eaten, and sugar beet can be damaged in summer when the fleshy parts of leaves are eaten. Controls include shooting, bird scarers and narcotic baits.

Bird repellent: A substance which drives away birds or discourages them from roosting.

Bird scarer: Device to repel birds from crops (*e.g.*, scarecrow).

Biserial: In two series, often concentric.

Bisexual: Having two sexes distinct and separate; *i.e.*, a species with males and females.

Biting and chewing mouthparts: Basic or primitive type of mouthparts adapted for biting and chewing the food with well developed mandibles, *e.g.*, cockroach. These are also called cutting and chewing type mouth parts.

Biting lice: Lice which infest mammals and birds; belong to order Mallophaga. Include the common dog louse (*Trichodectes canis*), *Heterodoxus spiniger* also affecting dogs, *Bovicola bovis* (*Damalinia bovis*) parasitic on cattle, *B. equi* on horses, *B. ovis* on sheep and *B. caprae* on goats, body louse or vent louse (*Menacanthus stramineus*) affecting the vent, breast and thighs of poultry birds, shaft louse (*Menopon gallinae)* on its body feathers, head louse (*Lipeurus heterographus*) found on the skin and feathers of the birds' head and neck, wing louse *(L. variabillis or L. caponis)* found on the underside of large wing feathers, and the fluff louse (*Goniocotes gallinae*) found in the fluff at the base of feathers; cause lousiness.

Biting response: The onset of feeding which may or may not be followed by complete feeding.

Bittacidae: Family of suborder Eumecoptera, order Mecoptera. Legs very long and slender with only one claw, raptorial tarsi. Abdomen of males not scorpion like.

Bituberculate: With two tubercles or swellings.

Bivalved: With two valves.

Bivoltine: Having two generations per year.

Bivouac: The mass of army ant workers within which the queen and brood live while the colony is not on the move.

Black ant: See *Monomorium indicum.*

Black flies: *Simulium* sp. (Simulidae, Diptera) are the pests of medical importance and also called buffalo gnats. They are cosmopolitan, abundant in warm countries, causing great annoyance and irritation. Swarms of black flies are very troublesome. These are mostly active during morning and evening hours, and females are biting flies.

Black fly transmitted diseases: The onchocerciasis and elephantiasis are the diseases transmitted by the black flies. See details.

Black headed caterpillar: See *Nephantis serinopa.*

Black palm beetle: See *Oryctes rhinoceros.*

Blank material (Blank sample): Laboratory simulated test material known to be free of the pesticide being analyzed. A portion of blank material is used to test the method, apparatus and reagents for interferences or contamination.

Blanket application: The application of spray or dust over an entire area rather than on rows of beds.

Blasticotomidae: Family of super family Tenthredinoidea, suborder Symphyta (Chalastogastra), order Hymenoptera. The larvae are devoid of abdominal legs. The stems of ferns are bored by the larvae.

Blastocoele: The cavity of blastula (segmentation cavity).

Blastoderm: The one-cell thick layer of cells that surrounds the yolk of an egg early in embryological development. The energids may undergo one or more mitotic divisions and retain the distinct cell walls and subsequently form a continuous layer of cells, called the blastoderm. The blastoderm in true sense, is the primary germinal epithelium. It lies just beneath the vitelline membrane.

Blastogenesis: The origin of different castes within a species from the egg by means other than genetic.

Blastokinesis: All displacements, rotations or revolutions of the embryo in the egg.

Blastomeres: The cleavage cells, or cells produced by the division of the egg or its nucleus that form the blastoderm.

Blastopore: The mouth of the gastrulation cavity.

Blastula: The early stage of embryo in which the only cell layer is blastoderm.

Blender: A device for blending soft or liquid materials.

Blending: (i) Mixing of ingredients for proper formulation. (ii) The production of a condition intermediate between two hereditary characters.

Blepharoceidae: Family of suborder Nematocera, order Diptera. These are elongate flies with long legs, confined to hilly areas and breeding in the clear waters of swift running hill streams. The larvae hold on to submerged stones. The females are predaceous on small Diptera. The larval body consists of head, thorax and the first two abdominal segments fused together, the remaining segments being incised laterally. A longitudinal row of median ventral suckers are present on the abdomen.

Blight: Common name for a number of different diseases on plants. Blight is characterized by the rapid death of plant tissue (*e.g.*, leaf blight, blossom blight, shoot blight).

Blister beetles: Beetles belonging to the family Meloidae which are pests of pearlmillet, cucurbits etc. These have acids and cantharides in their bodies.

Blood: The body cavity liquid and its contents.

Blood brain barrier: Thin layer of cells rich in mitochondria surrounding the nerve cells and serve as a storage and transport site of nutrients and metabolites and thus acting as a blood brain barrier.

Blood cells: The cellular liquid and its contents.

Blood gills: In many aquatic insect larvae, tubular or digitiform outgrowths from the body surface which may contain blood only, or blood with a rather sparse supply of tracheae. These structures are called blood gills. These structures are located at anal extremity in 4-6 numbers in the larvae of Trichoptera and two pairs on penultimate segments and two pairs on anal end in *Chironomus* larvae (Diptera). They are more osmoregulatoy than the respiratory structures.

Blood sucking lice: Lice belonging to the order Siphunculata, and parasitic on mammals. Comprise *Haematopinus eurysternus*, parasitic on cattle, *H. tuberculatus* on buffaloes *H. suis* on pigs, *H. asini* on horses, *Linognathus vituli* on calves, *L. stenosis* on goats, *L. pedalis* on sheep, *L. setosus* on dogs, and *Solenoptes capillatus* parasitic on cattle, cause lousiness in host.

Bloom period: The period during which flowers are opened.

Blossom stage: Crop stage from the time the first blossoms open until the petals fall.

Blotch: A plant disease characterised by large, necrotic and irregular in shape, spots or blots on leaves, shoots and stems.

Blow flies: See *Calliphora erythrocephala.*

Blow off: The removal by high winds of a material (*e.g.*, toxicant) as solid particles from the surface of the treated plant, foliage or soil.

Blue butterfly: *Cosmolyce (Polymmatus) boeticus* L., an insect pest of pulses. Short pale green caterpillar feeding on leaves, flowers and pods.

Blue rock pigeon: See *Columba livia.*

BNHS: Bombay Natural History Society of India; an organization popular in conservation of nature.

Boardman feeder: A device for feeding bees in warm weather, consisting of an inverted jar with an attachment allowing access to the hive entrance.

Body cavity: The definitive cavity of the body and appendages, not strictly equivalent in all animals.

Body wall: The integument of the body, formed of the ectoderm consisting of epidermis, cuticula and basement membrane.

Bolls (sing., boll): The fruiting bodies in which the lint and seeds of cotton develop.

Bombycidae: Family of super family Bombycoidea of suborder Ditrysia and order Lepidoptera. These insects are mainly Ethiopian and Oriental. Both sexes have pectinate antennae. *Eupterote* and its allies have a frenulum and their larvae have tufted long hairs and numerous secondary setae but verrucae are absent.

Bombycoidea: Super family of suborder Ditrysia and order Lepidoptera. Members of this superfamily do not possess maxillary palpi, tympanal organs and chaetosema. Antennae are pectinate, especially in males. Frenulum is vestigial or atrophied and proboscis may be rarely developed.

Bombycol: Female sex pheromone of the silkworm moth.

Bombyliidae (beeflies): Family of suborder Brachycera, order Diptera. Popularly called the beeflies, with robust densely pubescent bodies. Legs very long and slender; proboscis generally very long, often projecting forwards. Males with holoptic eyes. Wings show often reduced venation, with well developed squamae. Larvae are usually parasitic or predaceous on grasshopper eggs and beetles, or on smaller insects. Adults feed on nectar of flowers. *Bombylius* is a very common cosmopolitan form.

***Bombyx mori*:** Scientific name of silkworm, a beneficial insect.

Booklouse: See *Liposcelis transvaalensis* Enderlein.

Book lung: A respiratory cavity containing a series of leaf like folds in spiders.

Boom: Horizontal or vertical light frame carrying several spray nozzles.

Boom sprayer: A long pipe to which several nozzles are attached to spray a wide area.

Boopiidae: A family of suborder Amblycera of order Mallophaga. Ectoparasites on kangaroos. Legs long and slender and body clothed with stiff slender spines. Example, *Boopia and Heterodoxus*.

Booting stage: Growth stage of grain crops when the leaf sheath swells up due to growth of developing spike or panicle.

Boreidae: Family of suborder Eumecoptera, order Mecoptera. Wings rudimentary; legs normal and tarsi two segmented; labial palpi short; ovipositor conspicuous in females, *e.g.*, *Boreus* sp.

Bostrichoid aedeagus: Aedeagus in which the phallobase broadly overlaps the parameres, and the latter are fused basally to each other and to the base of penis; a secondary bridge may also unite the parameres. This type of aedeagus is found in most Bostrichoidea (Coleoptera).

Bostrychidae: Family of superfamily Bostrychoidea, suborder Polyphaga and order Coleoptera. The beetles are small or large, black or reddish brown, elongate, cylindrical, with hood-like pronotum concealing the hypognathous head, clubbed antennae and smooth or variously sculptured elytra sloping and toothed posteriorly. The coxae of forelegs are large and continguous. The larva is characteristic with reduced head and much enlarged thorax.

Bostrychoidea: Superfamily of suborder Polyphaga, order Coleoptera. Bostrychoidea are characterized by a hood like pronotum. Certain species, often associated with timbers are quite destructive. This includes the following four families: Anobiidae, Ptinidae, Bostrychidae and Lyctidae.

Bot: The larva of certain flies that is parasitic in the body of mammals.

Botanical pesticide: A pesticide obtained from plants, *e.g.*, pyrethrum, nicotine, azadirachtin, rotenone etc.

Botany: Branch of life science which deals with the internal and external structure of plants, mechanism of various plant organs and their all life activities.

Bottom board: The floor of a beehive.

Bound residue: Chemical species in soil, plant or animal tissue originating from a pesticide (generally radio labeled) that are unextractable by a standard method such as Soxhlet solvent extraction which does not significantly change the chemical nature of the residues. These unextractable residues are considered to exclude small fragments recycled through metabolic pathways into natural products.

Box hive: A bee hive in which the combs are not removable.

Brace comb: A bit of comb built between two combs to fasten them together, between a comb and adjacent wood, or between two wooden parts such as top bars.

Brachycera: Suborder of Diptera, in which the antennae are shorter than the thorax and are generally three segmented, the last being elongate. The arista or style is terminal when present. The palpi are one or two segmented and porrect. The larvae possess an incomplete usually retractile head and vertically biting mandibles. The pupa is free except in Stratiomyidae.

***Brachymeria excarinata* Cab.:** It is a pupal parasitoid (Chalcididae, Hymenoptera) of diamondback moth.

***Brachymeria* sp.:** Larval parasitoid (Chalcididae, Hymenoptera) associated with citrus caterpillar, *Papilio demoleus* Linn.

Brachypterous: Having very short wings that do not cover the abdomen.

***Bracon greeni* Ashmead:** Larval parasitoid (Braconidae, Hymenoptera) associated with pink bollworm of cotton.

Braconid: Insect belonging to the family Braconidae (Hymenoptera). This is a family of parasitoid wasps. Many braconid species are important in biological control of pests.

Braconidae (braconids): Family of super family Ichneumonoidea, suborder Apocrita and order Hymenoptera. These are very small stout bodied insects found among flowers and at light. They are distinguished from the Ichneumonidae in the cross vein 2m- Cu being absent in the forewings and therefore, there is only a single recurrent vein as against two recurrent veins in the ichneumonids. Further, the first sector of M+Rs which is present in the Braconidae is absent in the Ichneumonidae.

Bract: Small leaf like structure at the base of flower.

Brahmaeidae: Family of super family Bombycoidea of suborder Ditrysia and order Lepidoptera. A very small group of moths found in the tropics. They are characterized by the presence of proboscis and large upturned labial palpi. *Brahmaea wallichii* and *B.hearseyi* occur in India in hilly forest areas.

Brain: The supraoesophageal ganglion or brain is a combination of the presegmental and two anterior head segments. The three portions of the brain corresponding to these three ganglia are known as protocerebrum, deutocerebrum and tritocerebrum. In addition, there is a suboesophageal ganglion which is connected with the brain by paraoesophageal connectives. It is formed by the fusion of ganglion of mandibular, maxillary and labial segments.

Brain hormone: (AH, activation hormone or PTTH, prothoracicotropic hormone) secreted by neurosecretory cells of insect brain and passes through the axons that end blindly in the corpus cardiacum lying just behind the brain. The active material is believed to be liberated from the corpus cardiacum; it is probably a protein or polypeptide that activates prothoracic gland or thoracic gland, *i.e.*, the source of hormone, ecdysone which induces moulting.

Brain sheath: See cerebral neurolemma.

Branchiae: See processes of body wall.

Brand name or trade name: Name given to a product sold by a company to distinguish it from similar products made by other companies.

Brand: A name, term, sign, symbol or design or a combination of them intended to identify the goods or services of one seller and to differentiate from those of competitors.

Branding: Endowing products and services with the power of a brand.

***Braula caeca* Nitzsch:** The scientific name of a wingless fly (Braulidae, Diptera) commonly known as bee louse (1.2-1.5 mm in length) found in various parts of the world in beehives and feeds on the pollens and nectar at the bee's mouth.

Breaking: The separation of the phases from emulsion.

Breeding season: A particular period of year in which a particular animal species breeds.

***Brevipalpus phoenicis* (Geijskes):** Belongs to order Acarina and family Tenuipalpidae. It is a pest of tea, commonly known as the red crevice tea mite.

Brinjal lacewing bug: See *Urentius sentis.*

Brinjal leaf eating beetle: *Epilachna vigintioctopunctata*, an insect pest of brinjal. Young ones and adults feed on leaves and other tender parts.

Brinjal shoot and fruit borer: *Leucinodes orbonalis* Guen., a serious insect pest of brinjal, attacking top shoot and fruits; pale white caterpillar with violet markings bores into shoot, eats away tissues, causing attacked part to droop and wither away; cuts young fruits near calyx.

Brinjal stem borer: *Euzophera perticella* Rag., an insect pest of brinjal; pale white caterpillar which attacks stem.

Bristle: A stiff hair.

Broad leaved plants: Plants that are not mosses and not grasses. These have broad leaf lamina.

Broad spectrum pesticide: A non-selective pesticide that has activity against a wide range of pests and thus it kills a wide range of insects.

Broadcast application: Application over an entire area rather than only on rows, beds or middles (also referred to as blanket application), *e.g.*, broadcast application of a pesticide or a fertilizer.

Brodifacoum ($C_{31}H_{23}BrO_3$): It is anticoagulant rodenticide, chemically 3-[3-(4′ bromo[1,1′-biphenyl]-4-yl)-1,2,3,4-tetrahydro-1-naphalenyl]-4-hydroxy-2H-1-benzophyran-2-one. Brodifacoum is the most potent rodenticide currently used for the commensal rodents. It is available in 0.005 per cent pellet formulation and in wax blocks. Because of its acute toxicity, its lethal dose can be obtained in a single feeding, although death is delayed for 4 or 5 days. It is marketed under trade name Talon, Klerat and Havoc.

Bromadiolone ($C_{30}H_{23}BrO_4$): It is an anticoagulant rodenticide marketed under trade names Maki, Contrac, Supercaid and LM 635. Chemically, it is 3-[3-(4 bromo[1,1 biphenyl]-4-yl)-3-hydroxy-1-phenylpropyl]-4-hydroxy-2H-1-benzopyran-2-one. It is not quite as toxic as brodifacoum but can result in the same level of control. It is available in 0.005 per cent pellet formulation and in wax blocks.

Bromethalin: A non-anticoagulant rodenticide (marketed under trade name Assault, Vengeance, Trounce); chemically, N-methyl-2, 4-dinitro-N- (2, 4, 6-tribromophenyl)- 6- (trifluoromethyl) benzenamine. Bromethalin is a single dose rodenticide that causes central nervous system depression and paralysis leading to death in 2-4 days. Bait should be renewed at interval of several days. Continuous bait availability (as with anticoagulants) is not required but bait needs to be presented long enough to allow all animals in the area to feed. Bait shyness has not been reported.

Brood: In insects, a group of individuals of a given species which have hatched into young or which have become adult at approximately the same time and which live together in a defined area. It often refers to the immature stages of ants, bees, wasps and other insects.

Brood capping: A porous substance with which honey bees seal the brood cells in a bee hive.

Brood care: Parental care of offspring.

Brood cell: A special chamber or pocket built to house immature stages.

Brood chamber: The part of hive in which the brood is reared; may include one or more hive bodies and the combs within.

Brood comb: A wooden frame enclosing a wax sheet from which bees built the cells of honeycomb, placed in the brood chamber of a hive.

Brood disease: Various bacteria, fungi, viruses attack the brood of honey bee causing heavy toll of larvae. American and European brood diseases are caused by *Bacillus pluton*, *B. alvei* and *Streptococcus apis*, while the chalk brood disease by a fungus, *Aspergillus flavus* and the sac brood disease by a virus.

Brown Plant Hopper (BPH): Rice pest in South and South-East Asia. See *Nilaparvata lugens.*

Bruchidae (Lariidae) (seed beetles): Family of superfamily Chrysomeloidea, suborder Polyphaga and order Coleoptera. Small, short, stout-bodied, beetles covered by setae or scale and are usually dull grayish or brownish in color. The elytra are short and do not cover the tip of abdomen. The head is hypognathous, and produced into a broad short snout anteriorly. Antennae are usually clavate, serrate or pectinate. Body is somewhat narrowed anteriorly. The elytra are smooth, striate and covered with scales. The tarsi are five-segmented ending in claws hooked at base, the femora often swollen and dentate.

***Brumus suturalis* (Fab.):** Belongs to order Coleoptera and family Coccinellidae. It is commonly known as lady bird beetle and is a predator of aphids, coccids, psyllids etc.

Brush: Economically less uselful woody plants, including a range in size from small shrubs to large trees; woody weeds.

Brush control: Control of woody plants.

Bubonic plague: A bacterial disease of rodents and man caused by *Pasteurella pestis* and transmitted chiefly by the oriental rat flea; marked by chills, fever, and inflammatory swelling of lymphatic glands.

Buccal: Pertaining to the mouth.

Bud moth: See *Phycita infusella.*

Bud spray: A pesticide application to trees any time between the time the first colour appears in the ends until the buds begin to open.

Budding: Colony fission; the creation of new colonies by the departure of one or more reproductive females accompanied by a group of workers specifically to establish a new colony.

Budworm: Common name in the USA for various Tortricid larvae.

Buff comb: A bit of wax built upon a comb or upon a wooden part in a hive but not connected to any other part.

Buffalo fly: The common buffalo fly, *Lyperosia exigua* is a serious pest of buffalo in Asia belonging to family Muscidae (Diptera). It is known to transmit *Trypanosoma evansi* causing surra disease in buffalo. In other animals, it causes great irritation due to repeated blood feeding. Animals loose body weight and milk yield. See *Siphona (Haematobia) exigua.*

Buffer zone: Distance for environmental protection between the edge of an area where pesticide application is permitted and a sensitive non-target area, *e.g.*, water interface.

Bug: True bugs are insects of suborder Heteroptera (order Hemiptera). These are wingless or four winged with mouthparts adapted for piercing and sucking. The term bug is sometimes used to refer to any insect or similar organisms such as centipedes and mites.

Bulb: Rounded expansion of an organ.

Bulla (pl., bullae): A blister or blister like structure.

Bulletin board services: When it is required to give any information to many internet users, then it is displayed on the bulletin board.

Bunch caterpillar: See *Andraca bipunctata.*

Buprestidae (metallic wood-borers): Family of superfamily Buprestoidea, suborder Polyphaga and order Coleoptera. These are mostly tropical insects and inhibit moist forests. The beetles are minute or large sized, somewhat flattened or cylindrical and the elytra with a scutellar stria. They are golden, green, bronzy or blue metallic colored beetles and in view of the splendour of their metallic lusture they are used in embroidery and in jewellery in India and other countries. The beetles possess 11-segmented, short, serrate antennae. The hindangles of the prothorax are not produced backwards and the prosternal spine is received in a mesosternal cavity. The metasternum has a well-marked suture. The thorax and abdomen are firmly united. The tarsi possess membranous expansions. In India *Trachys dasi* and *T.pacifica* on jute, *T.bicolor* on *Butea frondosa, T.virescens* on cotton and lady's finger *T.ipomoea* on sweet potato, *T.manseuta* on *Sida* sp. on *Zizphus jujuba* and *Barleria cristata* are some leaf miner pests. *Sphenoptera gossypii* on cotton, *S. perotetti* on groundnut, *Psiolptera fastuosa* on teak and *Belinota prasina* on mango are other buprestid pests known.

Buprestoid aedeagus: Aedeagus in which the phallobase is more or less reduced and often fused to the parameres, which form a tube surrounding the penis; the entire structure is usually somewhat flattened. This type occurs only in Buprestidae.

Buprestoidea: Superfamily of suborder Polyphaga, order Coleoptera. This includes a single family Buprestidae.

Bursa communis: In some insects (Odonata), the bursa copulatrix opens into the vagina through a well defined tubular connection called bursa communis.

Bursa copulatrix: In some insects, there is a dilation or invagination of the genital chamber or vagina of female to form the accessory organ or separate pouch known as bursa copulatrix in which sperms are received during copulation. Such insects have two reproductive openings in female, *i.e.*, one vulva for receiving the sperms; other is ovipore for discharging the eggs, *e.g.*, water beetle and Ditrysian Lepidoptera.

Bursicon: A polypeptide hormone (molecular weight, 40,000 Daltons approximately) secreted by certain cells of the nervous system (neurosecretory cells of pars intercerebralis and of the ventral ganglia mostly the terminal abdominal ganglion as found in *Periplaneta americana*) that control cuticular hardening and darkening, also called the tanning.

Butterfly: Any of various insects of the order Lepidoptera. Butterflies have slender bodies, knobbed antennae, and four broad, usually colourful wings. Most butterflies are day flying insects (See also, moth).

By-products: Substances obtained incidentally during the manufacture of some other substances. By-products are sometimes as important as the manufactured substances itself (*e.g.*, the by-products of coal gas manufacture include ammonia, coaltar and code); but in most cases the by-products are unwanted (*e.g.*, 1, 2-dichlorobenzene in manufacture of chlorobenzene; inactive hexachloro-cyclohexane isomers in the manufacture of lindane etc.).

Byrrhoidea: Superfamily of suborder Polyphaga, order Coleoptera. This is small group of insects characterized by deflexed head with no visible clypeus, the coax of foreleg is large with trochantin exposed and a prosternal spine which fits into a mesosternal cavity. Two families, Byrrhidae and Georyssidae are recognized.

Byssoid: To consist of fine threads.

C

Cabbage borer: *Hellula undalis* is an insect pest of cabbage.

Cabbage butterfly: See *Pieris brassicae* Linn.

Cabbage flea beetles: See *Phyllotreta cruciferae.*

Cabbage semilooper: See *Thysanoplusia nigrisigna* (Walker) and *Plusia orichalcea* (Fab.).

***Cadmilos retiarius* Dist.:** Belongs to order Hemiptera and family Tingidae. It is an insect pest of ornamental plants, commonly known as the sunflower lacewing bug.

***Cadra cautella* (Wlk.):** Belongs to order Lepidoptera and family Phyctidae. It is an insect pest of stored grains and other products, commonly known as the almond moth.

Caecum (pl., caeca): A sac like or tube like structure, open at only one end. A pouch like structure in which some animals house bacterial population, essentially involved in the digestion of cellulose.

Caelifera: A suborder of Orthoptera. Antennae shorter with less than 30 segments, tympanal organ one on either side of the first abdominal segment.

Calcar: Modified tibial spur, often with comb of teeth.

Calcareous: Referring to soils or rocks, possessing those elements which result in alkaline or basic reactions.

Calcaria: Movable spurs at the apex of tibia.

Calciferol: (i) Calciferol or vitamin D is not necessary for insects. (ii) The Chemical, 3-(2-p-diphenyl-1, 2, 3, 4-tetrahydronapth-1-yl)-4-hydroxycoumarin is marketed under trade name Calciferol (empirical formula, $C_{28}H_{44}O$) is an effective anticoagulant rodenticide. In bait, it is used at 0.1 per cent concentration with or without 0.025 per cent warfarin.

Calcite ($CaCO_3$): Most common and widely distributed mineral of the earth and occurs as original constituent of many rocks like lime stone. The powder is white in colour used as inert carrier in the insecticide dust formulations.

Calcium arsenate: It is an inorganic insecticide, white powder (25 per cent dust). Commercial formulation consists of tricalcium arsenate, $Ca(AsO_4)_2$ and calcium arsenate $CaHAsO_4$, stomach poison; applied as dust or spray usually for controlling leaf eating insects. It was used as an insecticide as early as in 1907 but the compound contain so much water soluble arsenic that it burnt the foliage badly.

Calcium cyanide: Extremely toxic, like other forms of cyanide (a deadly poison). Calcium cyanide should be used only by the certified pest control operators. This is a powdered or granular material that reacts with moist air to produce hydrocyanic acid gas, a deadly fumigant poison effective against all pests.

Calibrate: To determine the amount of pesticide that will be applied to the target area.

Callidulidae: Family of super family Calliduloidea of suborder Ditrysia and order Lepidoptera. Moths resemble small butterflies and are seen in the Orient. They are day-flying insects and inhabit the hills.

Calliduloidea: Super family of suborder Ditrysia and order Lepidoptera. Moths without tympanal organs and maxillary palpi and with chaetosma, proboscis, simple or pectinate antennae and a small frenulam.

Calling: A virgin female moth releasing sex pheromone to attract males for mating.

Callipharixenidae: Family of order Strepsiptera. They are parasites of Heteroptera. The male is characterized by five-segmented antennae in which the third and fourth segments are laterally flabellate. In the female cephalothorax may be greatly elongated with two pairs of spiracles or the head and thorax may be distinct. The genital tubes may vary from four to five.

***Calliphora erythrocephala* (Meigen):** Belongs to order Diptera and family Calliphoridae. It is an insect pest of farm animals, commonly known as the blow fly.

Calliphoridae (blowflies): Family of suborder Cyclorrhapha, order Diptera. Blue bottles or green bottles or blowflies are large or medium sized flies of metallic shades of green or blue-green. Majority of them breed on carrion and some are serious pests causing myiasis in animals and man. The larvae are 12 segmented and three straight slits are present on each of the spiracles. Several species are pests of sheep and the important genera are *Calliphora, Lucilia* and *Chrysomyia.*

Calliphorin: A protein produced in the fat body and stored in the haemolymph, *e.g.*, family Calliphoridae, Diptera.

***Callosobruchus analis* (Fab.):** Belongs to order Coleoptera and family Bruchidae. It is an insect pest of stored grains and other products, commonly known as the mung dhora (pulse beetle).

Callosobruchus chinensis **Linn.:** Belongs to order Coleoptera and family Bruchidae. It is an insect pest of stored grains and other products, commonly known as the gram dhora (pulse beetle).

Callow: Newly enclosed workers in social insect colonies whose exoskeletons are still soft and whose colour has not fully matured.

Callus: (i) A rounded swelling: applied especially to swollen regions at the front or back of the thorax in various flies. (ii) A rounded swelling or tissue overgrowth around a wound or canker in plant.

Calocoris angustatus **Leth.:** Belongs to order Hemiptera and family Miridae. It is an insect pest of sorghum, commonly known as the sorghum earhead bug.

Calorie (abbr., cal.): A unit of heat or energy. One calorie is the amount of heat required to raise the temperature of one gram of water by 1° C. The kilocalorie (abbreviated as kcal.), equal to 1000 calorie is also used. As both calorie and joule refers to the same quantity, they are related to each other. 1 cal.= 4.186 J; 1 kcal.= 1000 cal.= 4186 J= 4.186 kJ.

Calypter (pl., calypters): One or two small lobes at base of the wings (above the halteres) in order Diptera.

Calyptodomous: Of the nests of wasps, referring to those which are surrounded by an envelope.

Calyx: The bulbous junction of the pedicels of ovarioles.

Camera lucida: A device when attached to the eye piece of a microscope enables accurate drawings of objects seen through a microscope.

Camouflage: Crypsis, in which an organism is indistinguishable from its environmental objects in colour, shape and structure, *e.g.*, some butterflies, mango looper, surface grasshopper, some species of mantids, nymphs of certain assassin bugs etc.

Campaniform sensilla (*sensilla campaniformia*): Type of mechanoreceptor consisting of a pit covered with a cuticular dome.

Campanulate: Bell shaped.

Campodeidae: A family of order Aptera (Diplura). Styli on the first abdominal segment absent but lobe-like appendages present; cerci are multi-segmented and not perforated apically. For example, *Campodea* sp., *Lepidocampa* sp., *Indocampa* sp.

Campodeiform larva: Campodeiform or oligopodous (without abdominal prolegs) is a larva having the form of the genus *Campodea* (order Diplura or Aptera). It is elongate and flattened, and has well developed legs and antennae. Usually they are active, *e.g.*, Neuropteran, Trichoptera and Coleoptera.

Campoletis chlorideae **Uchida:** Larval parasitoid of *Helicoverpa armigera* (Hub.).

Canker: A necrotic, sometimes sunken or cracked area surrounded by living tissue on a stem. Canker is a necrotic symptom of disease in woody plant parts. The necrosis is restricted to an area surrounded by callus.

Cannibalism: The phenomenon of eating of an individual's own body or those of other individuals of the same species usually because of mineral or vitamin deficiency, a lack of natural food, overcrowding etc., *e.g., Helicoverpa* larva, assassin bugs etc.

Cannibalistic: Feeding on other individuals of the same species.

Cantharidae (soldier beetles): Family of superfamily Cantharoidea, suborder Polyphaga and order Coleoptera. These are elongate, soft-bodied, yellowish, black or brown beetles found feeding on petals, pollen and nectar in flowers; some feed on mealy bugs, aphids and other small insects. The head is not concealed by pronotum. Males are winged and females apterous. They are mostly predaceous in their adult and larval stages. Phosphorescent organs are absent in these insects. The genera included under this family are *Cantharis, Chauliognathus* and *Podabrus.*

Cantharidin: The product which is obtained from the blood and accessory glands of the male genitalia of several species of Meloid beetles is a poison known as cantharidin. This poison, although, does not harm many species of insects but brings about appreciable mortality in several insect species such as fern beetle, *Phyllopertha horticola*; silkworm, *Bombyx mori*; stick insect, *Carausius morosus* etc.

Cantharoidea: Superfamily of suborder Polyphaga, order Coleoptera. Cantharoidea are soft bodied insects with filiform or clavate antennae. They are predaceous in their larval stages and suck the body juice of their victims by means of the internally channeled or perforated mandibles. This superfamily comprises of eight families, *viz.* Brachypsectridae, Homalisidae, Karumiidae, Drilidae, Phengodidae, Lampyridae, Cantharidae and Lycidae.

Canthus (pl., canthi): A sclerotized bar encroaching on the eye.

Cap cell: The distal or peripheral cell of a sense organ, or one of the component units of the organ, probably corresponding to the tormogen, or socket forming cell of a seta.

Capillarity: The rising or moving of a liquid in minute tubes by surface tension; capillary action.

Capitate: Terminal segment forms large knob in this type of antenna, *e.g.*, in Nitidulid beetles.

Capitulum: Head like structure of ticks which bears the feeding organs.

Capniidae: A family of order Plecoptera. Cerci long and multi-segmented, *Capnia* sp.

Capped blossom: Apple blossom in which the petals are unable to open and form a brown cap over the buds as a result of eating the heart of blossom by apple blossom weevil.

Capped brood: Pupae whose cells have been sealed with a porous cover by mature bees to isolate them during their non-feeding pupal period; also called sealed brood.

Capping melter: Melter used to liquefy the wax from cappings as they are removed from the honey combs.

Cappings: The thin wax covering of cells full of honey; the cell coverings after they are sliced from the surface of a honey filled comb.

Capsid bug: One of a large family of greenish or brownish plant bugs, slender or elongate, about 6 mm long. The common green capsid (*Lygocoris pabulinus*) is a pest of fruits and vegetables. The apple capsid (*Plesiocoris rugicollis*) feeds on the stem tips, fruit and foliage of apple trees.

Capsule: The granular inclusion body which is characteristic of a granulosis viral infection; the proteinaceous covering of a granulosis virus rod produced in the infected tissue cell.

Captan: A protective fungicide, particularly for foliage application; chemically, N-trichloro-methylmercapto-4-cyclohexane-1, 2-dicarboximide; low acute toxicity to warm blooded animals, acute oral LD_{50} for rats is greater than 15,000 mg kg^{-1}; no evidence of phytotoxicity.

Carabidae (ground beetles): A family of superfamily Caraboidea, suborder Adephaga, order Coleoptera. Tropical insects which are found under stones, in the soil, in moss and rotting wood, under bark, etc. They are oval, broad and somewhat flattened beetles exhibiting variation in size, shape and color. The antennae are conspicuous and project anteriorly. The legs are adapted for running or borrowing, in the former they are long and in the latter they are short and thickened. The elytra in some species are soldered together and the hind membranous wings are wanting; such species cannot fly. Mostly species are active only at night feeding on other insects and a few are attracted to light. A few belonging to the genera *Ophonus, Zabrus, Omophron* and *Amara* are known to feed on cereals, and seeds of plants, though majority of carabidae are carnivorous in their larval and adult stages. Beetles belonging to tribe Brachinina secrete from the anal end of the body a foul-smelling defensive fluid, in some the fluid volatilizes into a gas, the discharge being accompanied by a distinctly audible sound. The fluid has caustic properties. Some of the more common species found in India belong to the genera *Anthia, Agacetus, Calosoma, Macrochilus, Dioryche, Rombus, Amara, Parena, Pheropsophus* etc.

Carabiform larva: A larva shaped like the larva of a carabid beetle, that is elongate, flattened and with well developed legs; filaments lacking on end of abdomen.

Caraboid: Resembling a carabid larva.

Caraboidea: A superfamily of suborder Adephaga, order Coleoptera. This includes families, *viz.* Rhysodidae, Cicindelidea, Carabidae, Paussidae, Amphizoidae, Hygrobiidae, Dytiscidae and Gyrinidae.

Carapace: The hard upper plate of the cephalothorax of a spider.

Carbamates: Carbamate insecticides (a chemical class of insecticides) are synthetic derivatives of physostigmine, commonly known as eserine. It is the principal alkaloid of the plant calabar bean, (*Physostigma venenosum*). A number of synthetic analogues of physostigmine such as prostigmine and NU683 were developed which were effective inhibitors of cholinesterase but since they are water soluble they could not get importance as insecticides. Hence, the modern carbamates were modified by eliminating the polar moiety of physostigmine so that they can easily penetrate the insect cuticle and nerve sheath. Carbamates are derived from carbamic acid and dithiocarbamic acid, and have anticholinesterase activity. They have an OCON= group in the molecule.

Carbamates, metabolism: The major characteristic of the carbamates is the oxidative degradation through the mixed function oxidase system. As a result of this degradation, less polar metabolites like carbamic esters, napthols and phenols are produced.

Carbaryl ($C_{12}H_{11}O_2N$): Highly useful, moderate toxicity, general purpose carbamate pesticide; white crystalline solid; the technical material is 95 per cent pure, practically insoluble in water but soluble in most of the organic solvents; contact and stomach poison with long residual effect. It is hydrolysed to 4-hydroxy and 5-hydroxynaphthyl methyl carbamate which further break down with rupture of aromatic ring. It is effective against many insects and arachnid pests (ticks, mites, spiders etc.) both indoors and outdoors. It is marketed under trade name Sevin. Sevin is also effective against many insect species that are resistant to other poisons. However, it is not effective against houseflies, carpet beetles and termites. Incidentally, Sevin is highly toxic to honey bees. Generally, it can be used with effectiveness against all types of ticks in yards, houses and dog pens. It is also effective against evergreen bagworms. Chemically, it is 1-naphthyl N-methyl carbamate; moderately toxic to mammals, acute oral and dermal LD_{50} for rats is 500-700 mg and 7500 mg kg^{-1}, respectively; no evidence of phytotoxicity in use at normal rates, a product of the then Union Carbide Corporation, New York. Now, it is manufactured by Rhonepoulenc. It is formulated as 5 and 10 per cent dusts and 50 per cent wettable powder. See Sevimol and Sevidol also.

Carbaryl, metabolism: Carbaryl is metabolized through MFO system in plants and animals to produce four major metabolites: (i) N-hydroxymethyl derivative, (ii) 4-hydroxy derivative (iii) 5-hydroxy derivative, (iv) 5-6 dihydroxy derivative. Depending on the organism, there is some hydrolysis of carbaryl which leads to formation of 1-naphthol, CO_2 and methylamine. In cotton plant, carbaryl is catalysed by esterase to 1-naphthol and N-methyl carbamic acid. The latter undergoes spontaneous degradation to give up CO_2. Carbaryl is hydrolysed to 1-naphthol in soil mostly by chemical hydrolysis but it is also degraded by a variety of soil organism. It is degraded by the fungus, *Gliocladium roseum* to N-hydroxy methyl derivative, 4- and 5-hydroxy derivatives. *Aspergillus terreus* hydrolises carbaryl to N-hydroxy derivative which is further hydrolysed to 1-naphthol.

Carbofuran: A synthetic carbamate insecticide with nematicidal action, plant systemic applied to soil; chemically 2, 3-dihydro-2, 2-dimethyl-7-benzofuranyl methyl carbamate ($C_{12}H_{15}NO_3$); highly toxic to mammals, acute oral LD_{50} for rats is 5 mg kg^{-1}; used to control corn rootworms by incorporating into soil at planting time. It is white crystalline solid, soluble in water (250-700 mg l^{-1}). It is a product of

Chemagro Corporation, Kansas City, and Niagara Chemical Division, Middlepoint, New York. Highly toxic to man and animals, like most soil applied systemic insecticides. It is used effectively against white grubs, corn rootworms, European corn borers, sugar beet root maggots, alfalfa weevils, grasshoppers, shoot fly and potato insects. It is marketed under trade name Furadan.

Carbohydrate: A group of organic compounds composed of carbon, hydrogen and oxygen only, *e.g*, sugar, starch, cellulose etc.

Carbon bisulphide or carbon disulphide (CS_2): This is one of the first fumigants employed on a large scale, first used against grain infesting insects in France (1858). Its use in France in 1869 against the grape phylloxera, *Icerya vitifoliae* is a land mark in the history of applied entomology. It is also called carbon disulphide, extremely absorptive and penetrative, a clear, highly inflammable liquid, used as fumigant for stored grain insect pests, mice and rodents. It is also used for fumigating insect galleries or holes in the trees, as soil fumigant for Japanese and Asiatic beetles and to rid horses from bots and roundworms; highly toxic to mammals and toxic to plants and certain seeds; moderately toxic rodenticide. Carbon disulphide displays a stringent, irritating odour, therefore, never be used in, under, or near residential buildings. Rather, it should be used only in rat or mole burrows and grain godowns that are out in the open and located away from combustible things. For stored grain insect pests, the dosage is 160 g per m^3 of space with exposure period of 36-72 hrs. The fumigant is highly phytotoxic, but neither affects germination of seeds nor impairs the milling quality of grain.

Carbon cycle: Carbon is the basic building block for all organic matter. Three main sources of carbon are (i) CO_2 in air and dissolved water, (ii) rocks containing carbonates, and (iii) fossil fuels like coal and petroleum. There are three main ways to utilize atmospheric or aquatic CO_2 for various purposes. Firstly, CO_2 is converted into organic matter like carbohydrate and other complex organic substances by photosynthesis and then transferred to producers or plants. From producers, it goes to consumers, *i.e.*, herbivores and carnivores through food chain. Second way is biochemical fixation through which it goes to consumers. Third is chemical combination by which it converts into coal, peat, oil and carbonate rocks. This form of carbon is also obtained by the death of producers and consumers. This accumulated carbon dioxide in various forms is again released in the atmosphere by three different ways and then complete the cycle: (i) respiration of consumers (herbivores and carnivores), (ii) decomposition of consumers by the microbes, and (iii) combustion, weathering and volcanic activities occurring in coal, peat, other fuels and deep in earth which convert them into free CO_2.

Carbon dioxide (CO_2): Heavy colourless gas that does not support combustion.

Carbon monoxide (CO): Highly toxic, this very toxic gas is a primary component resulting from the combustion (burning) of petroleum (carbon) products. Carbon monoxide (CO) is particularly prevalent in automobile exhaust emission and in the exhaust from heaters and other burning materials. In some cases, the CO emitted in automobile exhaust can be used safely, effectively and cheaply to fumigate certain pests. For example, a hose to a car's exhaust pipe can be connected and then open end of the hose can be placed into the opening of underground burrows to fumigate rats, mice, snakes, moles, gophers and many insect pests. This method is particularly effective and cheap and even spares the task of disposing of the dead carcasses which are already buried. Carbon monoxide can also be used to fumigate bats in attics and between the walls of houses. However, this method is very dangerous. It must be employed only with the very strict supervision and caution. All human beings and pets must be removed from the building before it is fumigated. CO is heavier than air and will settle downward from the attic, replacing the air and permitting all rooms within the house. Therefore, following the fumigation, all doors and windows should be opened and the house or building should be ventilated thoroughly occupying it again. The CO is colourless, odourless and tasteless, therefore, one cannot detect its presence. Symptoms of carbon monoxide poisoning are frontal headache, drowsiness, confusion etc.

Carbon tetrachloride (CCl_4): It was first employed as fumigant for the control of stored grain pests by A.P. Morse in 1908. It is a colourless liquid having boiling point 76.8°C, molecular weight 153.84 with specific gravity 1.595 at 20°C temperature and vapour pressure 114.5 mm Hg at 20°C temperature. It is insoluble in water but soluble in organic solvents. The gas has specific gravity of 5.31, is not inflammable or explosive and possesses pungent chloroform like odour, comparatively less toxic to insects and slow acting and has no effect on viability of seeds. The recommended dosage is 2.5 l per m^3 space with exposure period of 14 days. The fat containing foods may absorb large quantities and residues remain without chemical reaction. The chemical gets desorbed on aeration, movement and heating or cooking.

Carbonyl sulphide: Carbonyl sulphide gas is an effective and safe alternative to phosphine. This has got significant potential as a new fumigant for use as an alternative to phosphine. The fumigant properties of carbonyl sulphide, an industrial gas has been elaborately studied in Australia and has been patented in 1993. It is moderately toxic and can achieve satisfactory control in a period of 6 hrs to 7 days. The fumigant does not affect the quality of treated wheat and germination. It is likely to be registered in Australia as a grain fumigant under the trade name Cosmic.

Carbophenothion: A synthetic insecticide and acaricide; an organic phosphate, S-(p-chlorophenylthio) methyl O, O-diethyl phosphorodithioate, marketed under trade name Trithion; highly toxic to mammals, acute oral LD_{50} for rats is 30 mg kg^{-1}; slightly injurious to fruit and foliage of certain apple varieties. It is especially useful against insect pests of fruits, vegetables, flowering plants etc.

Carcinogen: A substance that can cause the initiation of tumour formation (cancer).

Carcinogenic: The term used to describe the cancer producing property of a substance or agent.

Cardamom hairy caterpillar: See *Lenodera vittata.*

Cardamom thrips: *Sciothrips cardamomi* is an insect pest of cardamom; attack fruits which shrink and show corky encrustation on skin.

Cardia: The anterior part of the ventriculus; in many Diptera taking the form of a small spherical sac, often mistaken for a proventriculus.

Cardiac cycle: See mechanism of circulation.

Cardiac sinus: The channel of the haemocoel dorsal to the alimentary canal.

Cardiac valve: A valve at the junction of proventriculus (gizzard) and mid gut (ventriculus). It prevents regurgitation of food from the foregut. It is also known as proventricular or stomodeal valve.

Cardioaccelaratory hormone (CAH): Gersch *et al.* 1960 separated the neurohormone D, accelerating the insect heart beats, from the extracts of corpora cardiaca and homologous heart. Accelerating hormone was also analyzed from the terminal abdominal ganglion of the American cockroach as a peptide of molecular weight ranging from 3500- 7500 Daltons along with another substance of molecular weight about 150 Daltons, representing mostly acetylcholine.

Cardioblasts: The dorsal strands of mesodermal cells that form the dorsal blood vessel.

Cardo (pl., cardines): The basal segment or division of a maxilla.

Carina (pl., carinae): A pronounced ridge.

Carinula: A small carina, a small ridge.

Carnivore: Flesh eating. An organism (usually an animal) which feeds on animals (See also herbivore and omnivore).

Carnivorous: Feeding on the flesh of other animals (See also, herbivorous).

Carpel: Structural unit of the pistil in which the seeds develop.

***Carpomyia vesuviana* Costa:** Belongs to order Diptera and family Tephritidae. It is an insect pest of ber, commonly known as ber fruit fly.

Carpopodite: The fifth segment of a generalized appendage (tibia).

Carposinidae: Family of super family Tortricoidea of suborder Ditrysia and order Lepidoptera. In handwings M_1 and often M_2 wanting. The well known pests of importance are *Meridarchis scyrodes* and *M.reprobata* with the larve feeding on fruit of *Zizyphus jujuba* and *Syzygium jambolanum* respectively.

Carrier: (i) Inert material serving as diluent and vehicle for a chemical compound (active ingredient) to facilitate its application. For example in pesticides that are formulated as a dust, (ii) Organism harbouring a parasitic organism without itself showing symptoms of the disease spread by that organism.

Carrions' disease: Sand fly, *Lutzomiya verrucarum* spreads carrions' disease (bartonellosis) in Peru, Equador and Southern Columbia by transmitting the bacteria, *Bartonella bacilliformis*. The visceral form is highly fatal, called as oraya fever.

Carry over: (i) Analytical carry over: Unintended contamination of a sample undergoing analysis with material from a previous sample. (ii) Field carry over: Persistence of pesticide residues in soil after use in one crop such that uptake is observed in a succeeding, possibly more sensitive crop.

Cartap hydrochloride: A synthesized related compound of nereitoxin available commercially. Cartap, 13-bis (carbamoylthio)-2-(N,N-dimethylamino) propane hydrochloride, is an effective insecticide formulated as 2 per cent dust, 50 per cent WP and 2, 4, 5 and 10 per cent granules. It is marketed under trade name Padan and Caldan. It is very effective against rice stem borer and cabbage diamond back moth.

Cartridge: The part of a respirator which adsorbs fumes and vapours from the air before the applicator breathes there in.

Carumen: A brown mixture of wax and propolis used for nest construction by social bees.

***Carvalhoia arecae* Miller and China:** Belongs to order Hemiptera and family Miridae. It is an insect pest of arecanut, commonly known as the arecanut mirid bug.

Cashew tree borer: See *Plocaederus ferrugineus.*

Cashewnut leaf eating caterpillar: *Cricula trifesnestrata* H. is an insect pest of cashewnut; hairy, dark brown caterpillar with reddish white strips on the sides; defoliates tree.

Cashewnut leaf thrip: *Selenothrips rubrocinctus* G. is an insect pest of cashewnut; damages leaves of young plants by suking sap, making them turn pale brown and crinkled.

Cassididae (tortoise beetles): Family of superfamily Chrysomeloidea, suborder Polyphaga and order Coleoptera. Broadly oval or circular beetles with head concealed under prothorax and body and elytra broadly widened. The mouth is inferior in position. Antennae are short and not inserted on frontal prominences. The beetles are 5 to 6 mm in length, brightly colored with markings. The somewhat flattened elongate oval larvae are spiny and bear at the posterior end a forked process which is usually bent upward and forward over the body. The cast skins and excrement are attached to this process which forms a shield-like covering over the body. Both larvae and adults are phytophagous scraping and biting holes on leaves of a wide variety of plants. *Aspidomorpha miliaris, Chirida sexnotata* and *Metriona circumdata* infest sweet potato and other *Ipomoea* sp.

Caste determination: (i) Determination of different caste systems in social insects such as termites, honey bees etc. Castes in honey bees are determined partly by fertilization (unfertilized eggs become drones or males) and partly by the type of food provided to the larvae (if royal jelly is given only for two days to female larvae from fertilized eggs they become worker bees). (ii) The cast differences in termites are controlled solely by two pheromones; the pheromone gives off by the reproductive caste may be the juvenile hormone itself; while a juvenile hormone inhibitor or a corpus allatum inhibitor may be produced by the soldiers. Juvenile hormone is concerned also in the development of queens in the honey bee.

Caste: Morphologically and functionally different forms within a colony of social insects, as in Isoptera and some Hymenoptera. A form or type of adult in a social insect colony, *e.g.*, the soldiers, workers, queen and drones in a colony of termites. The three types of bees that comprise the adult population of a honey bee colony: workers, drones, and queen.

Castnioidea: Super family of suborder Ditrysia and order Lepidoptera. This includes a single family castniidae comprising of brightly colored day-flying moths. Found in Tropical America and the Indo-Malaysian and Australian regions.

Castor butterfly: *Ergolis merione* C. is an insect pest of castor, a green spiny caterpillar of a snuff coloured butterfly that feeds on leaves.

Castor capsule borer: *Dichocrocis punctiferalis* Guen., an insect pest of castor; also attacks fruits of guava, stem of turmeric, rhizomes and stem of ginger.

Castor hairy caterpillar: *Euproctis lunata* Wlk. is an insect pest of castor; dark reddish brown hairy caterpillar that feed on leaves.

Castor semi-looper: *Achaea janata* L., a serious insect pest of castor; caterpillar feeds on leaves and make holes; badly damaged leaves dry up and fall off. Sometimes attacks stem when serious, defoliates plant completely.

Castor whitefly: See *Trialeurodes ricini.*

Catabolism: Oxidative biodegradation of a pesticide to provide chemically available energy and generate metabolic intermediates.

Catalyst: A substance which by its mere presence in small quantity, alters (accelerates or retards) the rate of a chemical reaction without itself undergoing any chemical change is known as catalyst or catalytic agents. The phenomenon is known as catalysis.

Catapleurite: The lower and inner of the two incomplete subcoxal rings that form the thoracic pleurite, also called the catepleurite, katepleurite and coxopleurite.

Catastrophe: A sudden disaster; a sudden and violent upheaval in some part of the surface of Earth.

Catchment: Land and water confined within a single drainage basin.

Category: In taxonomy, it designates rank or level in a hierarchic classification; or a class, the members of which are all taxa assigned a given rank.

Caterpillar: An eruciform larva of an insect belonging to the order Lepidoptera (butterflies and moths). Caterpillars usually have a cylindrical body, a distinct head capsule, chewing mouthparts, thoracic legs, and abdominal prolegs. The name caterpillar is also used for larvae of scorpionflies and sawflies.

Catfacing: The injury caused by the feeding of such insects as plant bugs and stink bugs on developing fruit which results in uneven growth and a deformed mature fruit.

Catharophily: Transfer of pollens by beetles, a type of pollination.

Cationic: An ion having a positive charge is a cation. When the surface active portion of a surfactant molecule possesses a positive charge it is termed a cationic surfactant.

Cationic surfactant: Material in which surface activity is determined by the basic part of a compound.

Cattle biting louse: *Damalinia (Bevicola or Trichodectes) bovis* (Trichodectidae, Mallophaga) is the common cattle biting louse or red louse of the cattle which is a pest of veterinary importance. The disease transmitted by this insect is cutaneous myiasis.

Cattle blood sucking louse: *Linognathus (Haematopinus) vituli* and *H. eurysternus* (Haematopinidae, Anoplura) are long nosed and short nosed sucking cattle lice, respectively and pests of veterinary importance.

Cattle ticks: See *Hyalomma anatolicum.*

Cattle warble fly: *Hypoderma bovis* (Oestridae, Diptera) is a pest of veterinary imortance. It causes internal myiasis which damages skin, meat and loss of life of cattle.

Cauda: The pointed end of the abdomen in aphids. This is terminal protrusion, such as that on aphid for manipulating honeydew.

Caudal: Pertaining to the tail or posterior end of the body.

Caudal filament: Long, thread like appendage at the posterior end of the abdomen.

Caudal ganglion: The posterior most abdominal ganglion; innervates the genitalia.

Caudal gill: Most of the Zygopteran nymphs bear the median gill representing an appendix dorsalis and two lateral, the cerci. The caudal gills are filamentous and hairy triquetral structures. They are modified into swollen saccoid gills or flattened lamellate gills in the higher nymphs. The gills are internally highly tracheated and entire internal cavity is filled with tracheolar system, nerves, blood channels and the alveolar tissue.

Caudal sympathetic nervous system: The nerves arising from the last abdominal ganglion which innervates reproductive systems and the posterior end of the gut.

Caudal visceral system: Part of the visceral nervous system associated with segments of the abdomen.

Causal agent: That which is capable of causing disease.

Causal organism: The pathogen that produces a given disease.

Cauterization: Destruction of tissue by burning usually with an electrical current.

Caution: A warning to the user of pesticide chemicals used on labels of pesticide containers having slightly toxic pesticide in a slightly category as defined by the Federal Insecticide, Fungicide and Rodenticide Act. Also see signal words.

***Cavelerius excavatus* (Distant):** Belongs to order Hemiptera and family Lygaeidae. It is an insect pest of sugarcane, commonly known as the sugarcane black bug.

Cavernicolous: Found in caves, cave dwelling, also called the xeromorphs.

Cecidomyiidae: Family of suborder Nematocera, order Diptera. These are minute delicate insects with relatively long antennae and legs. The antennae are of monoliform type adorned with conspicuous whorls of hair. Ocelli are present or wanting. The venation is greatly reduced to a few longitudinal veins which are unbranched for the most part. Cross-veins are absent. Coxae are not elongated and tibial spurs are absent. The larvae is white, yellow, orange or bright red and occasionally brown in color, peripneustic with a reduced head.

Cecropia C18-JH: See juvenile hormone.

Celsius and Fahrenheit, relation: The temperature of a body has different values on the Celsius and Fahrenheit scales. If C and F denotes the temperature of a given body on Celsius and Fahrenheit scales, the two are related according to the following formula: $F-32/9= C/5$.

Cell: (i) A functional unit of living matter bounded by a membrane and containing nuclei or nuclear materials; (ii) the hexagonal compartment of a honey comb, (iii) an area of the wing bounded by a number of veins. A cell is closed (closed cell) if it is completely surrounded by veins and opens (open cell) if it is bounded partly by the wing margin.

Cell bar: A wooden strip on which queen cups are placed for rearing queen bees.

Cell body: The swollen portion of a neurone that contains the nucleus and is metabolic and maintenance functions of the neurone, also called the perika.

Cell culture: The growing of cells in vitro.

Cell cup: Base of an artificial queen cell, made of bee wax or plastic and used for rearing queen bees.

Cellulase: An enzyme that catalyzes the breakdown of cellulose. In the vast majority of insect species, the cellulose is quite unaffected by passage through the intestinal canal: those plant cells that are not ruptured by jaws of the caterpillar pass through the gut with their walls intact, although their contents may be digested; the excrements of many wood boring beetles contain all the lignin and cellulose that have been ingested with the food; the very starch grains may be protected from digestion by their pectin covering. Many insects can digest cellulose; sometimes by means of cellulase enzyme they secrete themselves, sometimes with the help of symbiotic microorganisms; and hemicellulose (lichenin etc.) is digested by most phytophagous insects.

Cellule: Any part of the wing included between veins.

Cellulose: A carbohydrate composed of repeating units of the sugar, glucose etc. Cellulose is a main component of the cell walls of plants and algae. It is an inert carbohydrate, the chief component of the solid framework or woody part of many plants.

Cement layer: A thin layer on the surface of insect cuticle formed by the hardened secretion of dermal glands. This is outer most layer of the epicuticle which is also known as tectocuticle and is protective to the wax layer (lipid epicuticle or cuticulin). It is secreted by dermal glands or so called Verson's gland in Lepidoptera. The nature of the cement layer is not fully known, but there is evidence that in cockroach and other insects, it contains resinous material resembling shellac, secreted by the lac insect, *Laccifer lacca*.

Centigrade: A thermometer scale in which freezes at 0 °C and boils at 100 °C. To change to degrees Fahrenheit, multiply degrees centigrade by nine fifth and add 32. Also, Celsius (C) is the preferred and approved by the Ninth General Conference on Weights and Measures in 1948.

Centipede: Animal belonging to the class Chilopoda. Some characteristics of centipedes are an elongate, flattened, worm like shape, a hard external skeleton, two body sections (head and many-segmented trunk), 15 to 30 pairs of legs (one pair per trunk segment), a pair of antennae with 14 or more segments, and simple eyes.

Central body: Neuropilar mass connecting the two lobes of the protocerebrum.

Central nervous system: The chain of paired segmental ganglia connected by the location of the integrative activities of the insect nervous system.

Central place foragers: Insects such as ants and bees that return to their nests after foraging.

Cephalad: Toward the head or anterior end. Cephalic or attached to the head, anterior.

Cephalic: Pertaining to the head.

Cephalic appendages: There are two types of cephalic appendages in the insects- the preoral appendages are the antennae while the postoral appendages are the mandibles, maxillae and labium.

Cephalic lobes: The head lobes of the embryo comprising the region of the prostomium and usually that of the tritocerebarl somite.

Cephalopharyngeal skeleton: Reduced, invaginated mouthparts that provide articular points for the mouthparts of higher Dipteran larvae.

***Cephalopina titillator*:** Belongs to order Diptera and family Oestridae. It is an insect pest of farm animals, commonly known as the camel nasal bot fly.

Cephalothorax: The body region made up of the head and thorax in Arachnida and Crustacea joined together, *e.g.*, the anterior (front) of the two major divisions of the body of a spider.

Cephoidea: Family of super family Megalodontoidea of suborder Symphyta (Chalastogastra), order Hymenoptera. This includes a single family Cephidae, members of which are called stem sawflies. The insects are slender, compressed sawflies, mostly black or darkly colored with or without narrow, yellow bands. The prothorax is large and articulates movably with the following segment. The apodous larvae bore into stems of grasses, berries etc. The larvae sometimes may have three pairs of reduced tubercle like thoracic legs. *Cephus cinctus* and *C. pygmaeus* are the wheat stem sawflies in the USA.

Cerambycidae (long horned wood boring beetles): Family of superfamily Chrysomeloidea, suborder Polyphaga and order Coleoptera. This is a large family of phytophagous species. The adult beetle is large, cylindrical with long antennae usually at least half as long as body or often much longer. They are often brightly colored and others exhibit striking cases of cryptic coloration. In most species the inner margin of eyes are notched and antennae may arise within this notch or just opposite to it. The tibial spurs are well developed and the tarsi five segmented, the third segment being bilobed conceals the further segments. Many cerambycids stridulate either by working the hindmargin of prothorax against a specialized striated area at the base of the scutellum or by the hindfemora rubbing against the edges of the elytra. Both types of stridulatory organs may occur in the same insect as in the Hawaiian *Plagithmysus.* This family is divided into subfamilies, like Lamiinae, Prioninae and Cerambycinae. These beetles are also known as longicorn beetles or long horned beetles.

Cerambycinae: A subfamily of Cerambycidae, Coleoptera. It is a large subfamily comprising adults of varied sizes and colourations. They are usually large, somewhat flattened and bear very long antennae. The larvae attack all parts of plants. The genera included under this subfamily are: *Chelidonium, Clytocera, Aglaophis, Caloclytus, Demonax, Gnatholea, Merionaeda, Neocerambyx, Plocaederus, Stromatium, Xylotrechus* etc.

Ceratopogonidae (biting midges): Family of suborder Nematocera, order Diptera. These are the bitting midges and include many bloodsucking forms. The head is not concealed by the thorax and the antennae are pilose in the females and plmose in the males and are 12 to 15 segmented. The mouthparts are adapted for piercing. Ceratopogonids are predatory as adults and suck the blood of vertebrates, including cattle. Of the Indian genera, *Ceratopogon, Culicoides, Leptoconops and Brachypogon,* the first two are most common.

Cerci, types: The cerci are simple in orthoptera, annulated in Dictyoptera and filamentous in Thysanura, Ephemeroptera and Plecoptera. In some insects, they represent sexual dimorphism: (i) in *Calliptamus* (Orthoptera) females have simple, conical cerci, while males bear long, flattened, lobulated structures, (ii) in Embioptera, female possesses simple cerci, while male bears asymmetrical structures in which left circus modifies into a clasping organ, and (iii) in Dermaptera, the cerci are straight and smooth in the female but they become curved and toothed structures in male. They are primarily tectile sense organs and are provided with the trichoid hairs but may function as the sound receiver, or defensive devices (Dermaptera) or clasping organs (Embioptera), feeding organs (raptorial) as in Japygidae, or may modify into lateral gills performing respiration and ionic exchange (Zygoptera) or swimming organs (Ephemeropteran nymphs).

Cercopidae: A family of superfamily Cicadoidea, suborder Homoptera and order Hemiptera. It is characterized by two ocelli; one or two stout spines present on hind tibiae and usually a series or circlets of spines at the apex. They are measuring rarely over 13 mm long. Hind coax short and conical. Example, frog hoppers or spittle bugs or cuckoo spits. They feed on herbaceous plants.

Cercus (pl., cerci): One of a pair of appendages at the posterior end of abdomen (XI segment) found in Apterygota, all Hemimetabola except Hemiptera, and Hymenoptera and Mecoptera in Holometabola. This is located at the base of the supra-anal plate and paraprocts, variable sized and shaped, segmented or not.

Cereal: A grass grown for its edible seed or grain, such as maize, wheat, rice, barley and millet.

Cerebral ganglion: Supra- oesophageal ganglion. The brain of an insect lies between the supporting apodemes of the tentorium, just above the oesophagus.

Cerebral neurolemma: It is the brain sheath composed of two layers, the outer neural lamella and the inner perineurium. The neural lamella is non-cellular while the perineurium is a cellular layer. The neural lamella is a thick layer composed of the protein, collagen like fibres having repeated banding pattern. The fibres are 130 angstrom in diameter and randomly oriented, though often arranged parallel to each other. The fibres are embedded in the matrix of amorphous granular substance containing mostly the mucopolysaccharides. The perineurium is a thin cellular layer composed of two types of cells lying beneath the neural lamella as its inner layer. The cells of one type are rich in mitochondria, fat droplets and glycogen granules. The adjacent cells are communicated each other by a series of tortuous channels or complex interdigitations or comb shaped septate desmosomes. These cells are considered as the nutritive cells providing various nutrients to the cortical neurones.

***Ceroplastes rubens* Mask.:** Belongs to order Hemiptera and family Coccidae. It is an insect pest of jack fruit, commonly known as the pink waxy scale.

Certification: Means any recognition by a certifying agency that a person is competent and thus authorized to use or supervise the use of restricted pesticides.

Certified applicator: Means any individual who is certified to use or supervise the use of any restricted use pesticide covered by his certification.

Certified reference material: Reference material, accompanied by a certificate, whose pesticide concentrations are certified by procedures which establish their traceability and for which each certified concentration is accompanied by an uncertainty at a stated level of confidence. Storage conditions and period for which the certification remains valid may also be included for unstable materials.

Cerumen: A mixture of wax and propolis used by social bees in nest construction.

Cervical sclerite: A sclerite located in the lateral part of the cervix between the prothorax.

Cervical: Concerning the neck region, just behind the head.

Cervix: In most of the orders, the intersegmental region or the neck region is membranous which is known as cervix. The cervix connects the head with the thorax and possibly includes the posterior unsclerotized part of the labial segment and the anterior part of the prothorax.

Cetoniidae (flower beetles): Family of superfamily scarabaeoidea, suborder Polyphaga and order Coleoptera. The Cetoniidae are principally pollen feeders and are found under loose bark or in debris and a few are myrmecophilous. The larvae develop in organic matter in the soil and a few infest roots of plants. These are medium-sized to large beetles and the well known goliath beetle, *Goliathus goliathus* measures 12.5 cm long. They are brilliantly colored and dorsally flattened with a large scutellum. The mandibles are not expanded and are invisible from above. The labrum is concealed. *Anatona stillata, Anthracophora crucifera, Chiloloba acuta, Oxycetonia versicolor, Protaetia alboguttata* and *P.aurichalcea* are some common Indian species.

CFTRI: Central Food Technology Research Institute is located at Mysore and was established in 1950 with a mandate to train personnel and evolve technology regarding food processing and preservation.

Chaetae (sing., chaeta): Stiff hairs or bristles.

Chaetosema, chaetosemata: A group of setae, usually divergent sensilla present on the head of butterflies and some moths.

Chaetotaxy: The study of arrangement of the bristles or hairs or chaetae on an insect: especially important in the classification of the Diptera, Collembla and several other groups. Often of taxonomic value.

***Chaetothyrium mangiferae*:** It is a sooty mould which grows on the honeydew secreted by mango hopper, *Amritodus atkinsoni.*

Chafers: Various beetles, the grubs of which are pests in old grassland, usually those which are in slightly dry situations.

Chagas' disease: A parasite, *Trypanosoma cruzi* causes this major health problem in millions of Latin Americans. It lives and reproduces inside nerve and muscle cells, particularly, those of the heart and drains its victims of all their energy. The kissing bugs (one of the species is *Rhodnius prolixus*) lives in the cracks and crevices of the houses and during their bite in the night period, besides sucking the blood, they live their droppings or excrements along with infective stage of the parasite on human skin (the parasite develops and multiplies in the bug's hind gut) and when the victim wakes up and scratches the itchy bump, the excrement together with the parasite's infective stage is rubbed into the wound and enters the blood stream.

Chain transport: The relying of food from another in the course of transporting it back to the nest.

Chain: In chemistry, a series of atoms connected by bonds, forming the skeleton of a number of compounds.

Chalastogastra: A suborder of the Hymenoptera.

Chalaza: A sclerotized, conical area of the larval integument, bearing a single plumose seta or up to three simple setae.

Chalcidfly: Any small insect of the super family Chalcidoidea which in its early stage exists in or on other larvae or pupae, thus destroying many undesirable pests.

Chalcidoidea: Super family of Suborder Apocrita, order Hymenoptera. This is a very large group comprising some of the smallest members of Insecta. They are generally found on flowers and foliage.

Chalcididae: Family of super family Chalcidoidea, suborder Apocrita and order Hymenoptera. Chalcididae are characterized in possessing rows of short teeth on hindfemur which is greatly swollen. When at rest the wings are not folded longitudinally. Prepectus is wanting. The ovipositor is straight and short.

Chamaemyiidae (Ochthiphilidae): Family of suborder Cyclorrhapha, order Diptera. The flies are minute, grey, sometimes black spotted and are predaceous on aphids and coccids. In India, *Leucopis luteicornis* is predaceous on the cotton coccids *Pseudococcus* sp. The aphids *Aphis gossypii, Macrosiphum granarium* and *Myzus persicae* upon by *Leucopis griseola.*

Channel: On internet, chating between more than one people is possible. Before starting chating, selection of group of persons concerned with a specific topic from the available topics is termed as channel.

Character: A trait.

Character correlation: The degree of association between two or more variable character.

***Charops* sp.:** Larval parasitoid (Ichneumonidae, Hymenoptera) associated with citrus caterpillar, *Papilio demoleus* Linn.

Chat: In chating, a computer user can talk with any other computer user in the world. This system may be used to make friends any where in the world and helpful in discussion on a topic.

Cheek: The lateral part of the head between the compound eye and the mouth.

Chelicera: (pl., chelicerae): The anterior pair of appendages in arachnids, the fangs. The first pair of appendages of adult Chelicerata, innervated from the tritocerebral ganglia of the brain; equivalent to the second antennae of Crustacea.

Chelisochidae: Family of order Dermaptera, suborder Forficulina and superfamily Forficuloidea. Aedeagus unpaired, tergum of tenth segment well developed, eleventh tergum reduced or absent, second tarsal segment with a narrow lobe extending beneath the third segment, *e.g., Chelisoches morio.*

***Chelonus pectinophorae* Cushman:** Larval parasitoid (Braconidae, Hymenoptera) of pink bollworm of cotton.

Chemical control: Control of pests with synthetic pesticides. The chemical control is the quick, effective and most popular method of pest control. It provides a long term control of the pests. It is a technical method which requires the proper selection of chemical agents, pesticides and proper application by using a suitable formulation and equipment at a critical period of the pest development and multiplication.

Chemical name: Scientific name of a pesticide following the recommendations of IUPAC for naming of chemical compounds or other accepted naming convention (Chemical abstracts).

Chemical reaction: When two or more substances are combined and as a result, undergo a complete change to make new substances or materials. The whole chemical reaction is represented by symbols and formulae.

Chemically inactive: Does not easily react with any other chemical or material.

Chemistry: In this branch, structure of matter, composition, behaviour, characteristics and their laws of mutual combination are studied. All the above properties are studied in various branches such as, organic, inorganic, physical, and analytical and biochemistry.

Chemoreception: The perception of atoms and molecules as stimuli.

Chemoreceptor: Receptor activated by chemicals. In insects, chemoreceptors for smell (olfactory) are usually situated on the antennae. Receptors for taste (gustatory) may be found on all parts of the insect's body, but they are located mainly on mouthparts and their tarsi. Some insects, including bees and wasps, have taste organs on their antennae. According to Dethier and Chadwier (1948), the chemoreceptors can be classified into three principal categories: (i) Common chemoreceptors: Providing common chemical sense or contact with chemicals of high concentration, (ii) Gustatory chemoreceptors: Providing sense of taste that is mediated by chemical stimuli acting with liquids at high concentration (gustatory chemoreception), and (iii) Olfactory chemoreception: Providing sense of smell (olfactory sense) that is mediated by chemical stimuli in a gaseous state at low concentration. There are, however, some receptors which appear to mediate responses to chemicals in both gaseous or aqueous phases in some aquatic beetles. On the basis of morphology, the chemoreceptors are often grouped into four types: (i) Sensilla trichoidea olfactoria, (ii) Sensilla basiconica, (iii) Sensilla coeloconica, (iv) Sensilla placodea.

Chemosterilant: Chemical used to render an insect sterile without killing it.

Chemotherapy: The treatment of a desired plant or animal with chemicals to destroy or control a pathogen without seriously harming the plant.

Chewing and lapping type mouth parts: It is a type of combination between chewing, and piercing and sucking type mouth parts. These have long lapping tongue like glossa, and lateral labial and maxillary palps and galeae, *e.g.*, honey bees.

Chewing mouth parts: This type of mouth parts represents most primitive complex of the feeding appendages comprising a labrum, epipharynx, hypopharynx, a pair of mandibles, a pair of maxillae and a labium. Example, Apterygota, Plecoptera, Dictyoptera, Orthoptera, Coleopteran and other insects feeding on solid food material. Various types of chewing mouth parts are: (i) Grasping mandibular mouth parts, (ii) Grasping sucking mouth parts, and (iii) Mandibulosuctorial mouth parts.

Chigger: The parasitic larva of trombiculid mites.

Chill coma: The cockroach, *Blatta* goes into a state of chill coma when temperature is lowered to 9.5 °C; but if it is not kept for more than 24 hrs. at 15 °C, it does not go into chill coma until the temperature has fallen to 2 °C.

Chilled brood: Immature bees that have died from exposure to cold; commonly caused by mismanagement.

Chilli thrips: *Scirtothrips dorsalis* Hood., an insect pest of chilli; lacerates leaf tissues, feeds on sap of foliage, curling up leaves.

***Chilo auricilius* Dudgeon:** Belongs to order Lepidoptera and family Crambidae. It is an insect pest of sugarcane, commonly known as the stalk borer.

***Chilo infuscatellus* Sn.:** Belongs to order Lepidoptera and family Crambidae. It is an insect pest of sugarcane, commonly known as the sugarcane shoot-borer.

***Chilo partellus* (Swinhoe):** Belongs to order Lepidoptera and family Crambidae. It is an insect pest of maize, commonly known as the maize borer.

***Chilo polychrysa* (Meyr.):** Belongs to order Lepidoptera and family Crambidae. It is an insect pest of cereals and millets, commonly known as the dark-headed stripped borer.

***Chilo sacchariphagus* (Boyer):** Belongs to order Lepidoptera and family Crambidae. It is an insect pest of sugarcane.

***Chilo suppressalis* (Wlk.):** Belongs to order Lepidoptera and family Crambidae. It is an insect pest of cereals and millets, commonly known as the pale-headed stripped borer.

***Chilomenes sexmaculata* Fab.:** Belongs to order Coleoptera and family Coccinellidae, commonly known as lady bird beetle and is a predator of aphids, coccids, psyllids etc.

Chilopoda: A class of arthropods (See centipede).

China clay ($Al_3O_32SiO_22H_2O$): It belongs to kaolinite group of clays. The powder is pinkish cream in colour.

Chironomid: An insect of the order Diptera whose young ones are blood red and worm like are invariably found at the bottom of ponds living in tubes.

Chironomidae (midges): Family of suborder Nematocera, order Diptera. Small mosquito-like delicate flies commonly called gnats or midges, with antennae plumose in males and pilose in females, 6 to 15 segmented. Wings not scaled. Mouthparts poorly developed; legs long and slender; head small, often concealed by the thorax. The adults are abundant near lakes, ponds and streams and appear before sunset in swarms. The larvae are aquatic and with two pairs of elongate blood gills on segment 11 and two pairs of papillate anal gills. *Chironomus vicarious, C. cubicularum* and *C. socius* are common Indian species.

Chitin: It is a nitrogenous polysaccharide named by Odier (1823) occurring in the cuticle of insects and other arthropods. It is made up of long chains of acetylated glucosamine residues, N-acetyl glucosamine $(C_8H_{13}O_5N)_x$.

Chlorbenside: This is a low toxicity chemical is effective against mites on fruits, flowers and plants. It is marketed under trade name Mitox.

Chlordane ($C_{10}H_6Cl_8$): It is an insecticide having low toxicity, like most other chlorinated hydrocarbons of the DDT family, chlordane usage has been restricted by the EPA because of its chemical stability and long term residual power that make it threat to man and animals through the food and water chains. It was discovered by Hyman but its insecticidal properties were described by Kearns and associates in 1945. Chlordane is used mainly to control subterranean insects (termites) beneath buildings. Chemically, it is 2, 3, 4, 5, 6, 7, 8, 8-octachloro-2, 3, 3a, 4, 7, 7a hexahydro-4, 7-methaniondene. Chlorinated terpene with chlorine content 64 to 67 per cent remains longer than the BHC but not as long as DDT, has multiple effect on insects, acting as stomach and contact poison, has slight fumigant effect; toxic to mammals and fish. It is used against large number of soil pests and foliage feeders. Moderately toxic to mammals, acute oral and dermal LD_{50} for rats is about 250 mg and 1600 mg kg^{-1}, respectively; high concentrations injurious to some vegetables and residues in soil depress germination. Highly toxic to bees.

Chlordecone: Highly toxic insecticide, very effective in poison bait form to control cockroaches and ants.

Chlordimeform: Highly toxic, this poison is especially useful against the eggs and larvae of several moths that attacks crops, as well as against most stages of ticks and mites and many other insect species. It is marketed under trade names, Galecron and Fundal.

Chlorfenvinphos: Highly toxic organophosphate insecticide; chemically, 2-chloro-1-(2, 4 dichlorophenyl) vinyl diethyl phosphate ($C_{12}H_{14}O_4Cl_3P$); compound 4072, Sapecorn and Birlane are the trade names of chlorfenvinphos. Its insecticidal properties were discovered by Chamberlain and his associates in 1962. It is an amber coloured oily liquid, slightly soluble in water (145 mg l^{-1} at 20 °C) but readily soluble in organic solvents. It is a contact insecticide, used principally to kill flies and does not leave harmful residues on the treated surface.

Chloride cells: Komnick (1977) reported the integumental chloride cells are found in fresh water insects belonging to order Ephemeroptera, Plecoptera and Heteroptera. In Ephemeropteran larvae, they occur throughout the body surface but often concentrated on the tracheal gills and lateral parts of abdominal sternites and tergites. In Plecopteran larvae, they are concentrated on the ventero-lateral parts of tracheal gills. All chloride cells are recognized in histochemical preparations and show all ultrastructural characteristics of ion transporting functions. They play osmoregulatory function by absorbing the electrolytes from the external medium.

Chlormephos: Highly toxic soil insecticide is very effective against wireworms, millipedes, symphilids and many other pests.

Chlorobenzilate: A synthetic acaricide, chlorinated hydrocarbon and relative of DDT; chemically, ethyl 4, 4-dichlorobenzilate, relatively non toxic to mammals, acute oral LD_{50} for rats is about 1900 to 3200 mg kg^{-1}. It is marketed under trade name Acaraben.

Chlorophacinone ($C_{23}H_{15}ClO_3$): Chemical name is 2-[(p-chlorophenyl) phenylacetyl-1, 3-indenedione. Low toxicity, trade names are Rozol, Liphadione, Caid, Delta, Quick, Ramucide, Topitox, Ravicac, Afnor etc. It is an anticoagulant cumulative poison that causes slow internal bleeding in rodents. It is used against Norway rats and house mice. Like most other anti- coagulant baits, Rozol requires repeated feedings over a period of days for good results. Rozol can be absorbed through the skin and thus should be handled with rubber gloves. The 0.25 per cent concentrate when mixed in the ratio of 1: 49 with bait provides 0.005 per cent chlorophacinone in bait.

Chlorophyll: Green pigment of most plants.

Chloropicrin: Insecticidal fumigant for treatment of stored grain and cereal products and for soil treatment to control insects, nematodes, weeds and fungi; empirical formula CCl_3NO_2; 0.05 oz./1000 cubic feet lethal in 10 minutes to mammals; very toxic to plants when injected into soil. It was first prepared by Steinhouse and insecticidal use was reported in 1907 by Fabricwerke. Fairly toxic tear gas and should be handled with caution. It is a good spot fumigant against a variety of pests and is used as warning gas for other more toxic, but less detectable fumigant. It is clear to yellow, volatile soil fumigant; a nematicide, rodenticide and herbicide, poisonous to human. Its dosage is 2 kg per 1000 q grain with exposure period of 48 hours. The fumigated grains should be thoroughly aerated as residues may affect milling and baking properties of the grain.

Chloropidae: Family of suborder Cyclorrhapha, order Diptera. These are small light colored insects without the anterior orbital bristles, the bare arista and the small oral fossa. The larvae are mostly phytophagous and a few are predaceous.

Chloroplasts: The organelles inside plant leaf cells to which the chlorophyll is confined and in which the reactions of photosynthesis are carried out.

Chloropropylate (acaralate): A synthetic acaricide related to DDT; chemically, isopropyl 4, 4-dichlorobenzilate; acute oral LD_{50} for rats is about 5000 to 34600 mg kg^{-1}.

Chlorosis: Yellowing or mottling of normally green tissue as a result of lack of chlorophyll, through its failure to develop, or destruction by chemical effects, disease etc.

Chlorphenamidine: A synthetic insecticide and acaricide; N-(4-chloro-o-tolyl)-N, N-dimethyl-formamidine, hydrochlorine; moderately toxic to mammals; acute oral LD_{50} for rats is about 295 to 335 mg kg^{1}; kills mites and their eggs on apple trees but its use is restricted to nonbearing trees or trees in dormancy. It is marketed under trade name Fundal S.P.

Chlorpyriphos ($C_9H_{11}Cl_{13}NO_3PS$): An organophosphate insecticide, chemically dimethyl 3,5,6 trichloro-2 pyridyl phosphorothionate a white crystalline solid with melting point 41-43 °C; highly soluble in organic solvents almost insoluble in water (2 mg l^{-1} at 20 °C) and readily soluble in organic solvents; relatively stable to hydrolysis and oxidation; moderately toxic; marketed under trade names, Dursban, Danusban, Lethal, Chlortox, Lorsban etc. Although, it acts somewhat slowly but provides good residual action, particularly on smooth and non- porous surfaces such as stainless steel, linoleum, tile etc. Though somewhat toxic, it is especially effective against cockroaches, termites and many other insects as well, both indoors and outdoors. It is formulated as 20 EC.

Chlorthion: An organophosphate insecticide having contact and stomach action with low mammalian toxicity, chemically dimethyl 3-chloro 4-nitrophenyl phosphorothionate ($C_8H_9NO_5PSCl$). It is yellow, crystalline solid with melting point 21 °C. It is highly soluble in aromatic hydrocarbons but practically insoluble in water (40 mg l^1 at 20 °C). Other properties are similar to methyl parathion.

Cholecalciferol: The non-anticoagulant rodenticide marketed under trade name vitamin D_3, Quintox, Rampage etc.). Chemically, it is 9, 10-Seocholesta-5,7,10 (19)-trein-3 betaol. It is a single dose or multiple dose rodenticide that causes mobilization of calcium from the bone matrix to plasma and death from hypercalcemia in 3-4 days after ingestion of a lethal dose. As the toxicant is slow acting, bait shyness apparently does not occur. As with bromethalin, once a rodent consumes a lethal dose, all food intake ceases.

Choline acetyltransferase: An enzyme detected from the nervous system of various insects which is responsible for catalyzing the synthesis of acetylcholine (ACh) from acetate and choline.

Cholinesterase inhibitor: A substance which inhibits the enzyme chlorinesterase. It prevents transmission of nerve impulses from one nerve cell to another or to a muscle.

Chordata: A phylum containing fishes, amphibians, reptiles, birds and mammals.

Chordotonal organ: See Johnston's organ.

Chorion: The oocyte covering is commonly referred to as the egg shell or chorion. It is composed of inner vitelline membrane and outer chorion. It serves various functions. It protects oocyte from mechanical stress, predators, environmental hazards like desiccation, allow sperm entry, provide elasticity for easy oviposition, ensure adequate O_2 supply, discharge of CO_2 and liberation of larva.

Chromatography: Method of chemical analysis developed from the fact that if a liquid mixture is allowed to trickle through a column of adsorbing material (*e.g.*, chalk) the components of the mixture may be adsorbed in separate layers in the column.

Chromosomes: At cell division the dark stained, rod-shaped structures which contain the hereditary units called genes.

Chronic: Of long duration, *e.g.*, chronic disease or infection, opposite of acute.

Chronic effect: A slow and continuous effect.

Chronic exposure: Continued exposures occurring over an extended period of time, or a significant fraction of the life time of the exposed individuals or test species.

Chronic symptoms: Symptoms that appear over a long period of time.

Chronic toxicity: Capacity of a pesticide or chemical to produce injury or death following prolonged and repeated exposure. There are three objectives of studying chronic toxicity effect on the organs and tissues, secondary effects such as carcinogenic (liver carcinoma), teratogenic and mutagenic (chromosomal aberrations) and no effect.

Chronological: Pertaining to arrangement of data in order of time of appearance.

***Chrotogonus trachypterus* Blanchard:** Belongs to order Orthoptera and family Acrididae. It is a polyphagous insect pest, commonly known as the surface grasshopper.

Chrysalis: The pupa of a butterfly.

Chrysididae (cuckoo or ruby-tailed wasps): Family of super family Bethyoidea, suborder Apocrita and order Hymenoptera. These are small green, green and ruby or blue colored insects with usually coarsely sculptured body. The venation on forewing is fairly complete and hindwing without closed cells. The gaster is ventrally flat or concave and only three or four segments are visibly dorsally. The ovipositor is concealed and retractable. The insect when disturbed curls up in a ball. They are ectoparasites of larvae of bees and wasps. A species of *Chrysis* is parasitic on the slug caterpillar *Spatulicraspeda castaneiceps* in South India.

***Chrysocoris stollii* Wolff.:** Belongs to order Hemiptera and family Pentatomidae. It is an insect pest of litchi and tamarind, commonly known as the litchi bug.

Chrysomelidae: Family of superfamily Chrysomeloidea, suborder Polyphaga and order Coleoptera. The beetles are oval, convex, brightly colored and measure 3.5 to 12 mm long. The antennae are widely separated at base, head not produced, eyes not prominent, prothorax laterally margined, prosternum broad and pygidium not exposed beyond elytra. The third tarsal segment is not bilobed and the forecoxae are transverse.

Chrysomeloid aedeagus: A modification of the cucujiform aedeagus in which the parameres are highly reduced or absent, the anterior tegminal strut often laterally compressed and the upper portion of the ring sometimes reduced to membrane so that the tegmen becomes fork like. This type occurs in most derived groups of Chrysomelidae.

Chrysomeloidea: Superfamily of suborder Polyphaga, order Coleoptera.

***Chrysopa scelestes* Banks:** It belongs to family Chrysopidae, order Neuroptera and is a natural enemy of sucking insect pests and many other insect pests.

***Chrysoperla carnea* (Steph.):** Belongs to order Neuroptera and family Chrysopidae. Commonly known as green lacewing and is a predator of aphids, coccids, psyllids, aleyrodids, larvae of many insects etc.

Chrysopidae: Family of superfamily Hemerobioidea, suborder Planipennia and order Neuroptera. With bright green bodies and wing veins and iridescent eyes; antennae filiform. R_3 arising from main separating from R_1 and does not fuse distally with Sc. Eggs with long pedicel.

Chunk honey: Honey cut from frames and placed in jars along with liquid honey.

Cibarium: The food packet of the extra oral or pre-oral mouth cavity between the base of hypopharynx and the under surfaces of the clypeus.

CIBC: Commonwealth Institute of Biological Control (a unit of the Commonwealth Agricultural Bureaux). It is an international organization headquartered in Trinidad (West Indies) which maintains insectaries and cultures of beneficial organisms in various parts of the world and supplies them, or collect other desired species on a contract basis to supply to needful governments, organizations or individual publisher of a catalogue of the parasites and predators of the terrestrial Arthropods, periodical technical bulletins and books on biological control, and provider of other useful reviews and documents on various aspects of the subject on request.

Cibicidae: Family of super family Tenthredinoidea, suborder Symphyta (Chalastogastra), order Hymenoptera. The insects are large, stout, bumble bee-like in appearance and possess clubbed antennae. The larvae are characteristics in remaining partly curled up and covered with waxy powder.

Cicadellidae (Jassidae): A family of superfamily Cicadoidea, suborder Homoptera and order Hemiptera. This family includes a large number of variously sized insects possessing different forms and colour; rarely exceeds 13 mm in length. They are generally small and elongate, tapering posteriorly. There are two ocelli. Antennae possess sensorial, apical part being long and setaceous. Hind tibiae with double rows of spines extending all over its length. Movable tibial spur absent. There are six nymphal instars. The wing rudiments being prominent from the third instar onwards. They have the habit of running side ways. Examples, green rice leaf hopper, *Nephotettix nigropicta*, *N. virescens*; sesamum cicadellid, *Orosius albicinctus*; mango hopper, *Amritodus atkinsoni*; castor leaf hopper, *Amrasca flavescens*; cotton leaf hopper, *Amrasca biguttula biguttula* (*Empasca devastans*).

Cicadidae: A family of superfamily Cicadoidea, suborder Homoptera and order Hemiptera. Three ocelli placed near together and thickened anterior femora which are spined beneath are characteristic; empodia absent. In some cases males have sound producing organs at base of the abdomen. Examples are Cicadas which are large insects ranging from 25-50 mm in length and known as locusts in the U.S.A. which is a misnomer, however, these are usually called as cicadas.

Cicadoidea: A superfamily of suborder Homoptera and order Hemiptera. Characterized by absence of sensilla on antennal pedicel and presence of sensory grooves on flagellum. Tegolae absent; middle coxae short, hind coxae mobile. This includes families, Cicadidae, Membracidae and Cercopidae.

Cicindelidae (tiger beetles): A family of superfamily Caraboidea, suborder Adephaga, order Coleoptera. Large active insects, measuring 10 to 25mm long, brightly colored, eyes markedly prominent, mandibles large and acutely toothed, long legs adapted for running rapidly and abdomen ventrally with six visible segments in female and seven in males. They are tropical and subtropical and are predaceous on a variety of small insects. The tiger beetles are extensively found in India and most of them belong to the genus *Cicindela*, *Cicindela cancellata*, *C. calligramma*, *C. cardoni*, *C. dives*, *C. duponti*,*C. sexpunctata*, *C. undulata*, *Collyris emarginatus*, *C. crassicornis*,*C. fonellii*, *Prothyma paradoxa* etc. are some species fond in India.

Cigarette beetle: *Lasioderma sericorne* Fab. is a serious insect pest of stored tobacco; also attacks coriander, mustard, cumin, dried ginger. Grub feeds on tobacco leaves, making large galleries; in severe attack, turns tobacco into powder; adult also cuts holes into tobacco leaves. It is also called tobacco borer beetle.

Ciliated: Bearing minute setae, hairs, cilia in many non-insects.

***Cimex lectularius* Linn.:** Belongs to order Hemiptera and family Cimicidae. It is an insect pest of household, commonly known as the bedbug.

Cimicidae: Family of superfamily Cimicoidea, suborder Heteroptera and order Hemiptera which includes bed bugs. These are blood sucking ectoparasites on man and animals. They are broadly oval, flattened insects having very short hemelytra. Ocelli are wanting and rostrum lies in a ventral groove. Legs are normal with three jointed tarsi. The bed bugs are nocturnal and hide during day time in crevices found on walls, floors or furniture in houses. The saliva introduced by its bite causes itching, burning and swelling. The bed bug is known to transmit Kala azar. The two common species of bed bugs come under the genus *Cimex*. *Cimex lectularius* is cosmopolitan and found through-out Europe and North America. *Cimex hemipterus* (*C. roundatus*) is distributed in Africa and Southern Asia.

Cimicoidea: Superfamily of division Gymnocerata, suborder Heteroptera and order Hemiptera.

Circadian rhythm: Physiological rhythm with an approximate 24 hour cycle.

Circulation, mechanism: The haemolymph circulation always takes place in postero-anterior direction. The dorsal vessel is the sole circulatory organ. It is a contractile organ and peristaltic wave often originates at the posterior most end and runs to the anterior end including pumping of the blood in similar direction. The heart undergoes uniform expansion and contraction comprising diastole and systole phases of a cardiac cycle, respectively. The diastole is initiated through myocardium and associated connective tissues but in some insects due to contraction of alary muscles (*Agrion*, *Hydrophilus* etc.). The systole is brought up by tetanic contraction of muscle fibres of the wall of dorsal vessel. The diastole phase is often longer than the systole except in *Lepisma*. Sometimes, there occurs an intermediate or resting phase, the diastasis in between the diastole and systole appears a presystolic notch. The incurrent ostia open during diastole allowing haemolymph inside the heart while remain closed during systole in order to prevent outward as well as backward flow of the haemolymph. The haemolymph is, therefore, drawn into the successive heart chambers through the ostia during systole. During its forward passage, some amount of blood is discharged out through the excurrent ostia and lateral segmental vessels.

Circulative virus: A virus that circulates through and passes some part of its life cycle within its vector.

Circulatory system: In animals, two types of blood circulatory systems are found: (i) Open type: In which the blood flows in the body cavity or haemocoel, *e.g.*, in insects. The alary muscles cause pulsation of heart by contraction and expansion by which circulation of blood takes place. (ii) Close type: In which blood flows in the veins and arteries.

Circumoesophageal connectives: The connectives between the brain and the ventral nerve cord embracing the stomodaeum. Primitively, the connectives from the archicerebrum to the first ventral ganglia; in insects the connectives between the tritocerebral ganglia and the mandibular ganglia.

Cirrospiloideus phyllocnistioides **(Narayan):** Larval parasite of citrus leaf miner, *Phyllocnistis citrella* Stainton.

Citrus aphids: *Toxoptera citricidus* Fab. and *T. aurantii* Fab. are dark aphids living on tender leaves and shoots of citrus trees and sucking cell sap out of them.

Citrus bark eating caterpillar: *Indarbela quadrinotata* Walk. is an insect pest of citrus trees, damages by eating bark in caterpillar stage.

Citrus blackfly: See *Aleurocanthus woglumi* Ashby.

Citrus blossom midge: See *Dasineura citri.*

Citrus caterpillar: See *Papilio demoleus.*

Citrus fluted scale: *Icerya purchasi* Mask. is also known as cottony cushion scale; yellow, red or brown insect coated with whitish or yellowish mealy substance; a pest on citrus trees. Adult females and nymphs suck cell sap from trees.

Citrus fruit sucking moths: *Othreis fullonia* Cramer. *O. ancillia* Cramer., *O. materna* Cramer, *Achaea janata* L. *Calpe emarginata* Fab. Destructive insect pest of citrus fruit; moth pierces fruits and suck juice, attacking both ripe and unripe fruits; causes premature fall of fruits. Also attacks grape fruit, pumello, malta and orange; *A. janata* also attacks guava.

Citrus green bug: *Rhynchocoris humeralis* Thunberg, an insect pest of citrus trees; nymphs and adults suck juice from fruits causing premature shedding.

Citrus leaf miner: *Phyllocnistis citrella* Stainton a major insect pest of citrus; greenish yellow caterpillar mines into leaf tissue and makes shining galleries on leaves; causes serious set-back in growth of trees. Also attacks orange, lime, pummelo and all varieties of grape fruit and malta.

Citrus mealy bug: See *Nipaecoccus viridis* and *Pseudococcus filamentosus* Cockerell.

Citrus mite: See *Paratetranychus citri.*

Citrus nematode: See *Tylenchulus semipenetrans.*

Citrus psylla: *Diaphorina citri* Kuw. An insect pest of citrus; nymphs and adults suck cell sap from tender parts of plants such as shoots, buds and flowers. Clusters round shoots, secretes honeydew; causes immature shedding of leaves and non-setting of fruits. Also attacks grape-fruit, pummello, malta, sweet lime, orange and lemon, especially *Kagzi* lime.

Citrus red scale: See *Aonidiella aurantii* (Maskell).

Citrus white fly: *Dialeurodes citri* Ashm. Is an insect pest of citrus; nymphs and adults suck sap from leaves, secrete honeydew leading to sooty mould. In severe attack, leaves turn yellow and fall prematurely.

Cixiidae: A family of superfamily Fulgoroidea of suborder Homoptera and order Hemiptera. No prolongation of head in front or moderately so; carinae may be absent and if present a median carina evident on frons, distinct claval suture evident. Widely distributed but mostly tropical. Example, *Oliarus hodgarti* on grasses in India.

Clade: A delimitable monophyletic unit formed by cladogenesis (a mode of evolution which consists in the splitting of an evolutionary line, such as a species, genera, families and other higher categories).

Cladism: When the organisms are ordered entirely on the basis of recency of common descent.

Cladistic: Based on the principle of cladism.

Cladogram: A branching from a common ancestor; a branching diagram showing the development of clade. A diagram showing nothing more than the sequence in which groups of organisms are interpreted to have originated and diverged in the course of evolution.

Clarifying: Removing visible foreign material from honey or wax to increase its purity.

Clasper: One of a pair of appendages at the tail end of a caterpillar by which it clings to leaves, twigs and other surfaces.

Clasporial legs: Clinging or clasping legs found in mammalian lice, Siphunculata and Mallophaga, a single segmented tarsus bears a long hook like curved claw with ventral pad. The claw opposes the tibial process so that the body hair can be firmly held as an adaptation of ectoparasitic mode of life. The forelegs of some aquatic beetles are provided with the suckers on the metatarsi and stumpy tarsomeres in males for holding the females during mating, *e.g.*, the male beetles of the family, Dytiscidae, Hydrophilidae, Carabidae, Cincindelidae and Meloidae.

Class: A division of the animal kingdom lower than a phylum and higher than an order, for example, the Insecta.

Classes of pesticides: There are four broad classes based on their evolution: (i) First generation- Inorganics:metal (arsenic trioxide), Bordeaux mixture (copper sulphate, lime, and water), sulphuric acid, Paris green (copper arsenite), calcium arsenite; (ii) Second generation- organochlorines (DDT, toxaphene, dieldren):a. broad spectrum and persistent, b. eliminated due to environmental consequences bioaccumulation, c. don't know the mechanisms of action for many; (iii) Third generation- organophosphates and carbamates: a. Typically less persistent but more acutely toxic than the organochlorines, b. primarily nerve toxins; (iv) Fourth generation life-cycle disrupters: a. Target a part of the insect life-cycle to reduce pest population, b. May be endocrine disrupters for non-target organisms.

Classical biological control: It involves deliberate introduction and establishment of natural enemies into areas where they did not previously occur and employed largely against pests of exotic origin, *e.g.*, *Rodolia cardinalis* (Muls.), an Australian ladybird was used against *Icerya purchasi* Mask in India, which displayed spectacular control.

Classical theory: Goodrich (1897) proposed on the basis of embryonic evidences that the procephalon is formed from the anterior most three somites along the unsegmented apical acron. These somites or metameres of procephalon are commonly called as the pre-antennal, antennal and intercalary segments, respectively.

Classification: The animal kingdom is divided in phyla, for example the phylum Arthropoda. Each phylum is subdivided in classes, for example the class Hexapoda (insects). Classes are subdivided into orders, for example the order Coleoptera (beetles). Orders are divided into families, families into genera (sing., genus), and genera are divided into species.

Claustral Foundation: A way of setting up of a new colony by a queen, or king and queen in the termites, which involves her/them being sealing her/themselves a way in a small chamber and raising the first group of workers entirely (or almost so) on stored body reserves (fat and often the flight muscles).

Claval fold: Lying just behind cup, it separates the remigium from the anal fold.

Clavate: It is also known as club like antenna in which the segments enlarges gradually from base to apex and last segment finally ending in the round cone or club, *e.g.*, in butterfly and some beetles.

Clavicornia: Section of superfamily Cucujoidea, order Coleoptera. This section includes about 20 families and following are of some interest.

***Clavigralla gibbosa* Spinola:** Belongs to order Hemiptera and family Coreidae. It is an insect pest of pulse crop, commonly known as the tur pod bug.

Clavoke: See flagellum.

Clavus: The sharply pointed anal area of the hemelytra of Heteroptera. In Noctuidae, a process from the dorsal margin of the sacculus.

Claw: A generally paired sharp structure at the end of the tarsal segments of the insect leg.

Clay: Type of soil with mineral soil particles that are less than 0.002 mm in diameter.

Clean up: Isolation of toxicant from interfering substances or solvent is called clean up.

Cleavage centre: It is located in between prospective head region and the differentiation centre in the prospective region of thorax. In other insects, positions of both centres may change but the differentiation centre is always found behind the cleavage centre.

Cleavage: The ova of most of the insects are centrolecithal and meroblastic and thus the cleavage is often superficial, though there are some exceptions in which the ova are isolecithal and holoblastic which under go total cleavage (Collembola). After fusion of egg and the sperm nuclei at periphery of egg, the diploid fused nucleus migrates to a centre of the egg. During its course of migration, it divides repeatedly and produces large number of daughter nuclei. Each daughter nucleus is surrounded by a hollow mass of cytoplasm forming an independent cell like nucleo-cytoplasmic unit called the energid or a cleavage cell.

Cleg: One of various species of blood sucking, two winged, tabanid flies (*Haematopota* sp.). They are ash grey with finely mottled wings and large bright eyes and are troublesome to horses, also called horse flies.

Cleptobiosis: The relation in which one species robs the food stores or scavenges in the refuse piles of another species, but does not nest in close association with it.

Cleptoparasitic: 'Cuckoo' bees do not make their own nests, instead oviposit in the cells of other bees, and their larva feed on the provisions supplied by their hosts.

Cleptoparasitism: (i) Where one female uses the resources and nest of another individual (usually of different species) to provide for her young thus usurping the owners efforts and preventing her from using them. (ii) A type of parasitism in which the adult parasitoid preferentially appropriates for its own progeny the previously paralyzed host of another parasitoid (Arthur, 1961). It is similar to inquilism but the host is killed in this type of parasitism. Destruction of *Rhyssella curvipes* larva by *Pseudorhyssa alpestris* larva appears and the parasite enjoys sole possession of a paralyzed wood wasp larva. The parasite steals the prey of the host and it is, therefore, called the cleptoparasite.

Cleridae (checkered beetles): Family of superfamily Cleroidea, suborder Polyphaga and order Coleoptera. Mostly tropical insects, beautifully colored and covered with a dense pubescence. They inhabit usually tree-trunks or are found on plants; a few occur in carcasses and skins. The tarsi are five-segmented and in many the first or the fourth segment is short. The adults and larvae are mostly predaceous on bark boring Coleoptera; some are saprophagus. In South India the clerid beetles *Dasyceroelerus eriaceus* and *Ommadius nilgiriensis* are predaceous on scolytid beetles.

Cleroid aedeagus: See Cucujiform aedeagus.

Cleroidea: Superfamily of suborder Polyphaga, order Coleoptera. The members of this superfamily are characterized by five segmented tarsi, lack of femoral plate on hindcoxa and the prosternal process not being received by the mesosternum. This comprises the families Ostomatidae, Chaetosomatidae, Cleridae, Melyridae and Phloiophilidae.

Click beetle: A beetle of the family Elateridae, characterized by their ability to leap into the air when they fall on their backs, making a clicking sound as they try to land right side up. Their larvae are wireworms, also known as skipiacks.

Click mechanism: The oscillatory contractions of the flight muscles in *Callipria* sp. (Diptera).

Climate: The total long-term characteristics of weather for any region. It includes rainfall, temperature, humidity, wind direction and wind velocity.

Climatograph: A polygonal diagram resulting from plotting temperature means against relative humidity.

Cline: A progressive, usually continuous change in one or more characters of a species over a geographic or altitudinal range or a series of form changes; gradient of biotypes; character gradient.

Clods: Hard lumps of soil formed by ploughing or cultivating the soil.

Clone: A population of individuals all derived asexually from the same single parent, *e.g.*, a group of plants originating by vegetative propagation from a single plant.

Closed cell: A cell bounded at all sides by the veins.

Closing apparatus of a spiracle: The closing mechanism and opening, formed either of the lips of the atrium or by a valve at the inner end of the atrium.

Closing band: The movable valvular fold of the inner closing mechanism of a spiracle.

Closing bow: The rigid but elastic lip of the inner closing mechanism of a spiracle opposite the valve.

Clothing hair: These are macrotrichia (cuticular modifications). The clothing hair cover the entire body and appendages as found in honeybees.

Clotrophic (ovarioles): Meroistic ovarioles in which trophocytes are connected to oocytes by nutritive cords.

Club: The thickened terminal (farthest from the head) end of the antennae.

Clubbed: With the distal part expanded, enlarged, or swollen, *e.g.*, a clubbed antenna.

Cluster: (i) A large group of bees hanging together, one upon another, (ii) Group of related or similar species into species group or higher taxa.

Clypeofrontal or epistomal suture: See epistomal suture.

Clypeolabral suture: This suture represents the lower margin of clypeus. The labrum is hanging down from this suture. The ridge possesses an apodeme at the centre which provides an insertion to the anterior retractor muscles of the labrum.

Clypeolabrum: Sclerite formed by the complete fusion of the clypeus and labrum, so that there is no suture present.

Clypeus: Lowest part of the insect face, just above the labrum. A large region or sclerite on the lower part of the face between the frons and labrum. It is divided into ante-clypeus and post-clypeus. Both parts of clypeus can be seen in the crickets. The post-clypeus modifies into a convex process termed as the ginglimus of the mandible (a sclerite of the head on which cibarial dilator muscles arise). The clypeus is also called epistoma.

***Cnaphalocrocis medinalis* (Linn.):** Belongs to order Lepidoptera and family Pyralididae. It is an insect pest of cereals and millets, commonly known as the rice leaf folder (leaf roller).

Coagulation: coagulation of the blood is brought about by plasma or cellular inclusion.

Coarctate pupa: A pupa enclosed inside a hardened shell formed by the previous larval skin which is typical of cyclorrhaphous Diptera. The pupa remains permanently pharate within a puparium formed from the larval skin.

Coccidae (Lecaniidae): A family of superfamily Coccoidea, suborder Homoptera and order Hemiptera.

***Coccinella rependa*:** Belongs to order Coleoptera and family Coccinellidae. Commonly known as lady bird beetle and is a predator of aphids, coccids, psyllids etc.

***Coccinella septempunctata* Linn.:** Belongs to order Coleoptera and family Coccinellidae, commonly known as lady bird beetle and is a predator of aphids, coccids, psyllids etc.

Coccinellidae (lady bird beetles): Family of superfamily Cucujoidea, suborder Polyphaga and order Coleoptera. This is a well known and widely distributed group of insects. The adults are small or medium-sized, oval or round, convex, often brightly colored red, black, yellow or brown and variously spotted and the head is partly concealed by pronotum. The tarsi are 4-4-4, the deeply bilobed second segment. Wide range of variations in color and patterns within the same species occur notably in *Adalia* and *Menochilus sexmaculatus.*

Coccoid: Sphere shaped.

Coccoidea: A superfamily of suborder Homoptera and order Hemiptera. It includes families, like Margarodidae, Ortheziidae, Lacciferidae, Pseudococcidae, Dactylopiidae, Coccidae (Lecaniidae), Aclerdidae, Asterolecaniidae and Diaspididae.

***Coccus*:** A spherical bacterium.

***Coccus* (=*Lecanium*) *virdis*:** Belongs to order Hemiptera and family Coccidae. It is an insect pest of coffee, commonly known as the soft green scale.

Cochineal insect (*Dactylopius coccus*): A coccid of the Mexican *Opuntia* (prickly pear) from which a red dye is obtained.

Cochlidiidae (Limacodidae or Eucleidae): Family of super family Psycoidea of suborder Ditrysia and order Lepidoptera. Slug caterpillars: moth is medium sized with stout body and wings heavily and loosely scaled. Sexual dimorphism is evident in some cases such as *Spatulicraspeda castaneiceps.* Larva has a thick short fleshy body, a small retractile head and minute thoracic legs. Larvae mostly possess poisonous spines on the body. Prolegs are wanting. Pupa closed in a hard round or oval cocoon being often irritating to touch. The cocoon may be found on the plants or in some cases inside the soil as in *Thosea* sp.

Cockchafer: One of the species of large beetle (*Melolontha* sp.), the larvae of which are destructive pests of crops. It is also called maybug or white grub.

Cockle: A small gall in wheat formed in place of wheat seed due to the attack from a nematode, *Anguina tritici*. See wheat ear cockle.

Cockroaches: See *Periplaneta americana.*

Coconut black headed caterpillar: *Nephantis serinopa* Meyr., an insect pest of palms; caterpillar feeds on leaves, making leaf dry up.

Coconut red palm weevil: *Rhyncophorus ferrugineus* Fab., is an insect pest of coconut affecting 4 to 15 year old trees; weevil breeds inside palm; larvae feed on plant tissue near crown, causing drying up of central shoot.

Coconut scale: See *Aspidiotus destructor.*

Coconut white grub: See *Leucopholis coneophora.*

Cocoon: (i) A case, made partly or completely of silk, which protects the pupa in many insects, especially the moths. The cocoon is made by the larva before it pupates. (ii) A silken case in which a pupa is formed. In case of silkworm, the pupa develops in a silk bound case. The silk like protective covering or case is spun by the same species prior to quiescence, diapause or pupation.

Cocoonase: An enzyme secreted by some advanced moths and used to soften or break down cocoon silk to allow escape of the adult.

Coddle: To denature (fix) the protein substances of a host insect by brief immersion in hot water, thus preparing it for more efficient utilization in parasitoid mass rearing programmes.

Code: The International Code of Zoological Nomenclature, abbreviated as ICZN.

Codling moth: A small moth (*Cydia pomonella*), the caterpillars of which eat the inside of growing apples and cause them to fall premature.

Coeloconica: Sensilla with their terminal bristle like process of sense cells sunk in shallow pits.

Coelome: A body cavity formed of the coelomic sacs only.

Coelomic sacs (coel): Small cavities in the mesoderm of a developing insect.

Coenagrioidea: A super family of order Odonata. Less than five antenodals present. Postnodals more or less completely in a line with the cross veins behind them. This includes family Agrionidae and Lestidae.

Coenzymes: In some cases, the non-protein groups attached to the enzymes to make the process active called prosthetic groups, these proteins are called coenzymes.

Coevolved: Closely associated species that have acted as agents of natural selection on one another.

Coffee berry borer: See *Hypothenemus hampei.*

Coffee green scale: *Coccus virdis* Gr. is a small, green scale insect pest of coffee. Clusters around stems and leaves, sucks sap; also exudes honeydew inviting mouldy growth.

Coffee red borer: *Zeuzera coffeae* Niet. Caterpillar bores into and damages coffee stem.

Coffee shot-hole borer: *Xyleborus morstatti* Hag. And *X. compactus* Hag. Major insect pest of robusta coffee; beetle bores into twigs of young plants and tunnels inside; young plants die off.

Coffee stem borer: It is also known as coffee white stem borer, *Xylotrechus quadripes* Chev. is an important insect pest on arabica coffee; larva attacks tips of branches of young and old trees and tunnel down into trunk. Borer affected plant shows yellowing and withering of leaves and drying of branches.

***Cofta spectra* (Distant):** Belongs to order Hemiptera and family Cicadellidae. It is an insect pest of cereals and millets, commonly known as the white rice leafhopper.

Cohesion: Mutual attraction of atoms or molecules; for example, the cohesion of water molecules.

Cohort: An indefinite taxonomic group used in different ways by different authorities, such as a group in rank above a super order, group between class and order, or group of related families.

Cold resistance: Resistance in insects against cold. It is usual for insects to reduce their water content before going into hibernation and this reduction may be one factor in the increased resistance of insects to cold during the winter. But the chief factor in cold resistance is a lowering of freezing point and undercooling point (the lower limit of super cooling) of the whole insect. One cause of this is an accumulation of glycerol in the haemolymph during diapause, where it may reach a level as high as 9 per cent (occasionally as high as 40-50 per cent). At the same time, there is a change of unknown nature in the haemolymph which leads to the elimination of the undefined nucleating agents needed for the initiation of freezing.

***Colemania sphenarioides*:** See Deccan wingless grasshoper.

Coleoptera: (*Coleo*= sheath, *ptera*= wings). An insect order containing the beetles. Beetles are characterized by hard front wings that meet in a straight line in the middle of the back, chewing mouthparts, and a complete metamorphosis, *e.g.*, beetles, weevils etc. The Coleoptera in view of their extremely varied habits, wide distribution, and countless adaptation, have occupied a dominant position as the largest group in the animal kingdom. In size they range from the minutest (0.25 mm) to the largest (cerambycids 150 mm long). They occur in soil, humus, rotten wood and other decaying organic matter, in timber, furniture and museum specimens; some are aquatic, and some littoral and other infest stored products. Some like the chrysomelids are carnivores and predatory. A few are cavernicolous and some are subterranean with eyes either absent or reduced.

Collar: A broad and often convex margin at the anterior or posterior end of the prothorax, set off by a distinct transverse groove.

Collaterial glands: Glands that secrete adhesive materials. In cockroach, these secrete substances to form ootheca. The Hymenopterans like ants, bees etc. produce poisons for paralyzing the prey for defense purposes.

Collembola: *Coll*= glue, *embol*= a wedge. A primitive insect order containing the springtails. They are characterized by wingless bodies with a spring mechanism, chewing mouthparts and no metamorphosis, *e.g.* springtails.

Colletidae (plasterer bees and yellow-faced bees): Family of super family Apoidea, suborder Apocrita and order Hymenoptera. Colletids are primitive apoids which nest in the soil or in cavities in stem or wood.

Colliculum: A sclerotized plate or thickening near the posterior end of the ductus bursae in the female genitalia.

Colloidal formulation: Solution in which the particle size is less than 6 µm in diameter, and the particles stay indefinitely dispersed.

Colloids: Liquids, and some pastes.

Collophore: A non-genital extendable tube like structure (adhesive organ) on the ventral side of first abdominal segment of the body of springtails (order Collembola).

Collum: The neck or collar; a thin connection between head and thorax in the Hymenoptera and the Diptera; or the posterior narrow part of the head as in the Coleoptera.

Colon: The proctodaeum or hind gut has three portions; the ileum, a short tube which tapers posteriorly and on the inside has folds to increase the absorptive capacity; the colon is not easily differentiated externally and the rectum is the terminal swollen portion. The colon also known as large intestine is thus situated between the ileum and the rectum.

Colonization: It is an attempted establishment of a community of organisms in a new locality. It includes the field release of imported species and such manipulation as is necessary for favouring their general increase and spread.

Colony: The aggregate of worker bees, drones, queen and developing brood living together as a family unit in a hive or other dwelling. A group of individuals of the same species of insects living together, often in a nest.

Colony fission: The multiplication of social insect colonies by the departure of one or more reproductive forms, accompanied by groups of workers, from the parental nest, leaving behind comparable units to perpetuate the parental colony. This mode is referred to occasionally as hesmosis in ant literature and sociotomy in termite literature. Swarming in honeybees can be regarded as a special form of colony fission.

Colony odour: The odour found on the bodies of social insects which is peculiar to a given colony. By smelling the colony odour of another member of the same species, an insect is able to determine whether it is a nestmate (See nest odour and species odour).

Colorado beetle: An American beetle (*Leptinotarsa decemlineata*), yellow with black stripes, a pest of potato crop, feeding on the leaves.

Colour hue, castaneus: Chestnut colour, bright reddish-brown.

Colouration: Certain white or colourless pesticides must be treated with a colouring agent which will produce a uniformly coloured product not subject to change in colour beyond the minimum requirements during ordinary conditions of marketing storage and use. Seed treatment materials are examples of pesticides which have been coloured to indicate upon visual inspection what seed has been treated.

***Columba livia*:** Belongs to order Columbiformes and family Columbidae. It is a pest of farm animals, commonly known as the blue rock pigeon.

Comb: (i) A group of spines on the leg of an insect specifically used for cleaning other parts of the insects body. (ii) A mass of six-sided cells made by honey bees in which brood is reared and honey and pollen are stored; composed of two layers united at their bases.

Comb foundation: A commercially made structure consisting of thin sheets of bee wax with the cell bases of worker cells embossed on both sides in the same manner as they are produced naturally by honey bees.

Comb honey: Honey produced and sold in the comb, in either thin wooden sections (4 x 4 inches or 4 x 5 inches) or circular plastic frames.

Combination cleavage: Total cleavage of the egg followed by peripheral cleavage.

Combined nitrogen: Nitrogen in a form in which it is bonded to other elements, *e.g.*, ammonia (NH_3), nitrate (NO_3), and nitrite (NO_2).

Combustible: Able to catch fire and burn.

Co-metabolism: Microbial metabolism of a pesticide where the derived energy is not used to support microbial growth.

Commensal: Two species living in association, in which one species is benefitted and the other is neither helped nor harmed.

Commensalism: (i) An association in which one individual is benefitted and the other is neither helped nor harmed. (ii) Any species that feeds on the surplus food supply of another without destroying the owner of the food supply.

Commission: The International Commission of Zoological Nomenclature.

Commissure: A bridge connecting any two bodies or structures on a body.

Commodity product: An expression used in the trade to distinguish a patent free substance from one over which a company exercises exclusive rights.

Common exposure route: Means a likely way (oral, dermal and respiratory) by which a pesticide may reach and enter an organism's body.

Common field cricket: *Brachytrypes achatinus* Stall. is a major pest of maize, pigeonpea and sunnhemp; attacks young seedlings, cutting them from base.

Common field rat: *Gerbillus indicus*, a species of rat found in sorghum and pearlmillet fields; make holes in the ground with numerous entrances; galleries terminate in chambers. Does not hoard food, but comes out every evening and morning for it. A voracious feeder of grains in fields and stores.

Common moiety: Molecular sub-unit which is common to the structures of several pesticides or metabolites.

Common myna: See *Acridotheres tristis*.

Common name: (i) Simple name assigned to a pesticide active ingredient by the International Organisation for Standardization (ISO) to be used as a generic or non-proprietary name. Many pesticides are known by a number of trade or brand names but have only one recognized common name. (ii) The name of an insect that is used only in a particular region or country (See also: Scientific name).

Common weaver bird: See *Ploceus philippinus*.

Communal: Applied to the condition or to the group showing it in which members of the same generation cooperate in nest building but not in brood care.

Communication: Action on the part of one organism (or cell) that alters the probability pattern of behavior in another organism in an adaptive fashion.

Community: Assembly of populations of different species of living organism (quite often interdependent on and interacting with each other) within a specified location in space and time.

Community biodiversity: It is the variation in biological communities in which species are living. There are three types of community diversity, *viz.*, alpha, beta and gamma diversity. Alpha diversity indicates the variation in the same ecological community and different organisms share the same habitat. Beta diversity indicates the interaction between two communities or two habitats and species migrate from one habitat to other due to change in environment. Gamma diversity shows variation in habitats over the total land or total geographical area.

Compartment: Part of an organism or ecosystem considered as an independent system for purpose of assessment of uptake, distribution and dissipation of a pesticide.

Compatibility: The ability to mix different pesticides without physical or chemical interactions which would lead to reduction in biological efficiency or increase in phytotoxicity.

Compatible: (i) In chemistry, two compounds are compatible when they can be mixed without undesirably affecting each other's properties. (ii) In biology, it refers to a pathogen being able to form a parasitic relationship with a host plant.

Compensation: The ability of plants or plant parts to make up for damage caused to other parts of the plant. For example, a rice plant that lost a tiller because of attack by a stem borer will produce new tillers to compensate for this.

Competition: Competition occurs when two or more organisms or populations interfere with or inhibit one another as they strive to secure a resource that is in limited supply. For example, weeds compete with crop plants for nutrients, moisture, light and other essential growth factors. Competition can also occur between individuals of the same species. Competition is of two types: (i) Interference type: Both interacting species or population actively inhibit each other in the mode of direct interference. (ii) Resource type: Each population adversely affects the other in the struggle for food or resources.

Competitive displacement: An ecological concept, based on Gause's law which suggests that one species of insect or other organism may be capable of replacing another species over a substantial area through competition for the same niche.

Complete metamorphosis or Complex metamorphosis: Metamorphosis in which the insect develops through four distinct stages, *e.g.*, ova or egg, larva, pupa and adult or imago; the wings (when present) develop internally during the larval stage (endopterygota). The insects involved in this type of metamorphosis are categorized as Holometabola.

Complex: Used for a number of related taxonomic units, especially those in which taxonomy is difficult or confusing.

Compliance (Residue): Meeting of official maximum residue limit (MRL) standards by residue levels in food consignments or combined samples and tested by approved methods.

Composite sample: Combined increment samples, or combined replicate samples or combined samples from replicate trials. This is preferred term for bulk sample which is ambiguous.

Compost: Mixture of organic materials used to improve soil structure and fertility. Compost is usually prepared by microbial degradation of plant materials.

Compound eye: An eye consisting of many individual elements or ommatidia each of which is represented externally by a facet. The compound eyes are chief visual organs of Machilidae (Thysanura) and majority of Pterygotes. The compound eyes are replaced by stemmata in the larval Holometabola and in some adult insects- Lepismatidae (Thysanura), Collembola, parasitic insects such as Siphunculata and Sterpsiptera. They are generally reduced in female coccids, cave dwelling insects and sterile forms of social termites and ants. Located dorsolaterally on the head, the surface of each eye is divided into a number of circular or hexagonal areas called facet. Each facet have lens of a single eye unit called ommatidium. The lenses fit together like the cells of a honey comb. Some insects can see sharp images and different colors. Compound eyes differ from the ocelli or stemmata in division of cornea into large number of facets often equal to that of ommatidia. The ommatidia in a compound eye may vary from one in worker ant, *Ponera punctatissima* to 10,000-28,000 or even more in dragonflies.

Compound nest: A nest containing colonies of two or more species of social insects, up to the point where the galleries of the nests anastomose and the adults sometimes intermingle but where the broods of the species are still kept separate (See mixed nest).

Compressed air sprayer: Sprayer in which pressure is created inside tank by hand pump which pressurizes liquid from top and forces it to go outside as spray; operated by carrying on back or shoulder of operator; largely used for spraying field crops; covers about half to one acre a day.

Compressed natural gas: In compressed natural gas (CNG), 90 per cent methane is present. It is obtained from the oil wells and used as fuel in the automobiles.

Compression sprayer: The sprayer which is fitted with an air pump is known as compression sprayer or pneumatic sprayer. The air pressure is employed in the sprayer for forcing a liquid through the nozzle for forming a spray.

Computer science: The study of the working and composition of computer.

Concentrate: Refers to a commercial pesticide preparation before dilution for use.

Concentrate spraying: Direct application of a pesticide concentrate without dilution.

Concentrated solution: Commercial pesticide preparation before dilution for use.

Concentration: Proportion of active ingredient in a pesticide preparation, before or after dilution. The concentration can be shown, for example, as g or ml per litre or as a percentage by weight or volume.

Concentration effect relationship: Association between the exposures concentration of the pesticide and the magnitude of the resultant continuously graded change either in an individual organism or in a population.

Condemnation: The removal of a crop or product which does not meet the legal standards for tolerance on foods and is therefore not to be sold.

Condensation: Chemical change in which one or more molecules react with the elimination of water.

Conductivity: The property of nervous or other protoplasmic tissue by which changes in metabolic activity are propagated through it.

Condyle: A process which articulates the mandible to the head.

Cone or semper cell: See ommatidium, structure.

Cone or swirl nozzle: See nozzle types.

Confined areas: The rooms, buildings and green houses with limited or inadequate ventilation.

Confluent: Flowing or coming together.

Confusant: The term is used for a pheromone or pheromone like analogue when it is used in a program of disruption of communication for insect pest suppression.

Congeneric: A species agreeing in all characters of generic value with others compared with it.

Conglobation: A type of body compaction in which the head and prothorax are capable of folding against the hind body, forming a sphere.

Coniopterygoidea: Superfamily of suborder Planipennia, order Neuroptera. Small forms with wings fragile and covered with a waxy powder. R_3 with two branches and venation reduced and and veins not bifurcate at margin. Antennae filiform.

Conjugation: Biosynthetic reaction in which a pesticide or its metabolite is linked to an endogenous compound.

Connate ventrites: Ventrites which are immovably united so that they can not slide over one another as they can when joined by membrane. This may be used as a synonym with fused ventrites, but they are always departed by a groove or line, while fusion sometimes involves the disappearance of any joining line.

Connective: A longitudinal cord of nerve fibers connecting successive ganglia.

Conopidae (thick-headed flies): Family of suborder Cyclorrhapha, order Diptera. Conopidae are medium brownish flies with long antennae, a very long and slender proboscis, and slender elongate abdomen. Many resemble solitary wasps. The larvae of conopidae are parasitic, especially on adult bees, wasps, grasshoppers etc. The flies usually chase and lay eggs on their hosts during flight. Some species that occur in India are *Conopsnubeculosus*, *C. erythrocephala*, *Physocephala bicolor*, *P. refescens*, *P.tenella*, *Myopa nigriventris* and *M.testacea.*

Conspecific: Individuals of the same species.

Contact herbicide: Herbicide that kills those plant parts with which it comes into contact.

Contact insecticide: Insecticide that kills insects by penetrating through the cuticle or through the spiracles. The pest may absorb the poison while walking on the treated surface, while flying through a mist or fine droplets or when they are hit directly during spraying or dusting of the insecticide.

Contaminant: (i) Minor impurity in a substance, (ii) Extraneous material added to a sample prior to or during chemical or biological analysis, (iii) Unintended pesticide residue in an agricultural commodity or environmental compartment (*e.g.*, ground water), (iv) The presence of unwanted organisms or micro-organisms in food and air.

***Contarinia sorghicola* (Coq.):** It is an insect pest of sorghum, commonly known as sorghum midge (Cecidomyiidae, Diptera).

Contemporaneous species: Species occurring at the same time period.

Contiguous: Touching, usually applied to eyes (see also Holoptic).

Construction gland: A gland of wasps producing a substance which enables them to make paper out of wood-pulp.

Copularium: The first chamber built by a newly mated pair of sexual termites.

Continental: A species or population inhabiting large and extensive land masses.

Control: (i) Untreated subjects used for comparison with those given a particular crop protection treatment. (ii) To reduce damage or pest density to a level below the economic threshold. The different techniques are: (a) Legislative: The use of legislation to control the importation and to prevent any spread of a pest within a country. (b) Physical: The use of mechanical (hand picking etc.) and physical methods (heat, cold, radiation etc.) of controlling pests. (c) Cultural: Regular farm operations designed to destroy pests. (d) Chemical: The use of chemical pesticides as smokes, gas, dusts, and sprays to poison pests. (e) Biological: The use of natural predators, parasites and disease organisms to reduce pest populations. (f) Integrated: The very carefully reasoned use of several different methods of pest control in conjunction with each other to control pests with a minimum disturbance to the natural situation.

Control sample (Field): Sample from a field test plot to which no pesticide was applied (a zero rate sample).

Control schedule: The time schedule for the application of control measures irrespective of presence or absence of the pests.

Controlled atmosphere: The controlled atmosphere refers to altering the normal atmosphere by introducing inert gases in sealed storage. The controlled atmosphere system protects the grain quality and controls insect and mite pests. The combination of inert gases with different fumigants needs to be tested with the objective to reduce the dosages of fumigants and to avoid various problems related with excessive use of pesticides. Complete mortality of all stages of resistant strains of *Tribolium castaneum*, *Oryzaephilus surinamensis*, *Ephestia cautella* and psocids was achieved in rice stacks following CO_2 treatment and there is no change in colour and odour of rice and wheat. It requires a period of 10 days and above for disinfestation of food grains.

Controlled droplet application: See sprayers.

Conventional: Everyday customary.

Convergence: Development of similarities between animals of different groups resulting from adaptation to similar habitats.

Coordinated biological insect pest suppression: A pest population reduction programme involving the directed use of available biological suppression techniques in compatible combination with the effects of beneficial organisms.

Copper arsenate: This compound has been developed comparatively recently as a safe and stable arsenical for foliage application. It had advantage of being quite insoluble and resistant to hydrolysis by water and is only slightly affected by exposure to carbon dioxide. It is as toxic as lead arsenate to a variety of insects.

Coprophagus: Feeding on animal dung.

Copularium: The first chamber built by a colony founding couple of termites.

Copulation: The union or mating of the sexes for reproduction.

Corbicula (sing., corbiculum): A concave, smooth space, edged by a fringe of hairs arising from the margins of the hind tibiae of bees; this is the pollen basket and its function is to hold the collected pollen.

Corbiculate legs: Pollen collecting legs, *e.g.*, hindlegs of worker honey bees. These legs are also evident in the dragonflies and same as forageal legs. The foretibia bears a terminal spur known as fibula, the calcar which can be accommodated into a proximal cavity of the tarsi. The antennae are inserted through this cavity so that the attached pollens on the antennae can be removed by the fine hairs of the cavity. These pollen grains are collected and carried out in the basket or corbicula formed by the combs (scopa) of the metatarsal setae of both the hindlegs. The legs of dragonflies are often directed forward as a result of unique modifications of the thorax. The legs can cover the mouth parts and pre-oral cavity from the ventral side like a spoon. The legs are filamentous, useless for walking but the femurs, tibia and tarsi are provided with long spines and bristles. These bristles or spines of the legs of one side combine with those of other side and form a sort of finely netted basket for trapping, collecting and transporting the mosquitoes etc.

Corcyra cephalonica **(Stainton):** Belongs to order Lepidoptera and family Galleriidae. It is an insect pest of stored grains and other products, commonly known as the rice moth.

Cordulegasteridae: A family of order Odonata.

Corduliidae: A family of order Odonata. Green eyed skimmers with metallic colours, *e.g.*, *Cordulaia* sp.

Coreidae: Family of superfamily Coreoidea, suborder Heteroptera and order Hemiptera. This includes the leaf footed bugs, medium to large insects with antennae inserted well up on sides of head and a four segmented rostrum. In some, the antennae show extra-ordinary dilations. Metathoracic gland openings are present and membrane of wings with numerous veins. Many species with flattened leaf like legs, tarsi are three segmented and pulvilli present. These insects are mostly dull coloured and phytophagous, few are predaceous. These bugs produce penetrating nauseous odour. Examples are *Leptocorisa acuta* on rice, *Paradasynus* sp. on cashew, *Clavigralla horrens* on pulses etc.

Coremata: Thin walled eversible organs of male moths used for the dissemination of pheromones. In Pyralids, these are expandable pencils of specialised scales.

Coreoidea: Superfamily of division Gymnocerata, suborder Heteroptera and order Hemiptera. This includes families, *viz.*, Coreidae (leaf footed bugs), Corizidae and Coriscidae. Coreoidea are important pests of crops.

Corethrogyne: Specialised scales near the tip of the abdomen in some female moths, usually used to cover the egg mass.

Coriaceous: Thick hard and somewhat rigid; like leather.

Corimelaenidae (Thyreocoridae): Family of superfamily Scutelleroidea of division Gymnocerata, suborder Heteroptera and order Hemiptera. Small, broadly oval and strongly convex with the scutellum covering most part of the abdomen and wings. These are phytophagous and commonly known as negro bugs, *e.g.*, *Coptosoma cribraria* on pulses.

Corium: The elongate, usually thickened, basal part of the front wing in members of the suborder Heteroptera (true bugs).

Corixidae: This is a family of division Cryptocerata, suborder Heteroptera and order Hemiptera. Aquatic bugs with dorsally flattened body which inhabit common fresh water pond and lakes and occasionally streams. The head is not inserted in the prothorax. The one or two segmented rostrum is concealed, antennae three or four segmented, stridulatory organs are present, *e.g.*, water boatmen, *Arctocorixa abdominalis*.

Corm: The swollen fleshy underground stem that grows to a considerable size, *e.g.*, elephant's foot.

Cormel: The secondary corm produced from an old corm.

Cornea: See ommatidium, structure.

Corneagen layer: See ommatidium, structure.

Corneagen/ corneagenous cell: See ommatidium, structure.

Cornicle: One of a pair of tubular structures extending from the posterior part of the abdomen of aphids (family: Aphididae).

Corolla: The inner floral circle usually consisting of petals.

Coronal suture: The stem of the Y-shaped suture on the vertex and extending forward.

Corpora allata: Paired endocrine glands located just behind the brain and responsible for secreting the juvenile hormone (Wigglesworth, 1934). They are spherical or oval bodies in most of the insects except Hemiptera, Dermaptera and adult Dipteran flies where both the corpora allata are fused together and form a single median corpus allatum lying below (Hemiptera) or above (Diptera) the aorta.

Corpora cardiaca (sing., corpora cardiacum): Neurohaemal organs where some brain hormone is stored and released. They may be oval, spherical or fusiform and are often fused medially and even with the corpora allata posteriorly. They are, however, situated on either side of the hypocerebral ganglion and sometimes connected with latter by the lateral nerves, the nervi cardiostomatogastrici.

Corpora optica: These are optic lobes and do not represent the ganglia. Each optic lobe contains a group of three neuropile masses: the outer, lamina ganglionaris or lamina; the middle, medulla externa or medulla and the inner medulla interna or lobula; and are connected to each other.

Corpora pedunculata: These are some mushroom shaped bodies each on either side of the pars intercerebralis. Each corpus pedunculatum is separated into two parts, an anterior flattened cap or calyx and the posterior stalk or peduncle running downwards ventrally. The stalk is distally divided into two branches or roots, the alpha-lobe (also called the outer stalk or posterior root) and beta-lobe (also called the median root or inner stalk). Corpora pedunculata are associated with the most complex sensory input in the brain and serve as a site of summation, translation of complex sensory information into the complex behavioural patterns and thus it also represents a seat of intelligence.

Corpora ventralia: A pair of ventral bodies or accessory lobes lies in the ventro-lateral region of the protocerebrum and both ventral bodies are connected with each other by a protocerebral transverse commissure. Each ventral body is divided into two halves, an oral and an ocular. Each ventral body is connected with the dorsal protocerebrum *via* median fascile and with the ellipsoid body by ellipsoidal fibres. Some axonal fibres from the pars intercerebralis also terminate into the ventral bodies.

Corporotentorium: Body of tentorium formed by the fusion of anterior and posterior arms. See tentorium also.

Corpus centrale: It is a median mass of neuropile at the centre of protocerebrum. It is an association centre from which the pre-motor outflow from the brain to the ventral nerve cord is conducted. It is connected to the various parts of the brain by a large number of axonal fibres, with the pars intercerebralis, deutocerebral antennal lobes, calyces etc.

Corpus luteum: The mass of degenerating follicle cells left in an egg chamber after the discharge of the egg.

Corrosive: Having the power to eat away slowly. Some pesticides eat or wear away rubber hoses, nozzles and other parts of spray machinery.

Cortex: The outer or superficial layer or peripheral region of insect brain and other ganglia. It is filled with the monopolar, motor neurones and glial cells. Some of the neurones are modified into the neurosecretory cells and terminate into the neurohaemal organs.

Cortical cytoplasm: The peripheral layer of cytoplasm in the egg.

***Corvus splendens* Vieillot:** Belongs to order Passeriformes and family Corvidae. It is a predator of insect pests of farm animals, commonly known as the house crow. It is also an important pest of crops.

Corydalidae: Family of superfamily Sialoidea, suborder Megaloptera, order Neuroptera. With ocelli, eight pairs of gills, *e.g.*, dobsonflies.

Cosmid vector: The plasmids containing the cos sites of lambda phage (between which the DNA segments are inserted) are called cosmids. These cos sites facilitates the packaging of plasmid DNA within the lambda particle. Similar to plasmids, the cosmids get stabilized in the bacterial cell but lack the lytic genes. Large DNA fragments (40 kbp= 40,000 bp) of eukaryotic genome can be cloned with the help of cosmids.

Cosmopolitan: A species occurring very widely throughout the major regions of the world.

***Cosmopolites sordidus* (Germar):** Belongs to order Coleoptera and family Curculionidae. It is an insect pest of banana, commonly known as banana weevil.

Cosmopterygidae: Family of super family Tinaeoidea of suborder Ditrysia and order Lepidoptera. They are minute narrow winged moths. The antennae have sight basal pecten and the forewings are linear or lanceolate. The larvae feed variously. Some feed in seeds such as *Anatrachyntis simplex* which is destructive to sorghum grains in South India. Some causes blotch-mines in leaves of plants such as *Cosmopteryx mimetis* on *Cyperus rotundus*, *C.bambusae* on bamboo and *C.phaeogastra* on *Lablabniger*.

Cossidae (goat moths, carpenter moths): Family of super family Cossoidea of suborder Ditrysia and order Lepidoptera. This includes medium sized moths, frequently with bipectinate antennae in both sexes; rarely simple but in some males bipectinate for a portion of their length and filiform distally. Proboscis and maxillary palpi wanting; labial pulpi short and vestigial. R_5 and M_2 close together or stalked beyond the apex of the discal cell in the hindwings. Adults are nocturnal. Larvae bore large galleries in the wood of forest, shade and fruit trees.

Cossoidea: Super family of suborder Ditrysia and order Lepidoptera. This is characterized by Cu_2 being present in both pairs of wings and M being furcated within the cell.

Costa: One of the major longitudinal veins, usually forming the front or anterior margin of the wing and usually abbreviated as C. The costal margin is the front edge of the wing.

Costal area: The portion of the wing immediately behind the anterior margin.

Costal Cell: The cell between the costa and the sub-costal vein.

Costal fold: A narrow, thin membrane folded back on upper surface of the costa of the forewing of butterflies, it contains androconia. An expanded costal area folded over or beneath the forewing in the male, often covering special pheromone disseminating scales or hairs.

Costal margin: Anterior edge of the wing.

Cotton aphid: See *Aphis gossypii.*

Cotton grey weevil: *Myllocerus undecimpustulatus maculosus* Desbr., is an insect pest of cotton; adult weevil eats leaf tissue and make holes in leaves.

Cotton jassid: See *Amrasca biguttula biguttula.*

Cotton leaf roller: See *Sylepta derogata.*

Cotton stem borer: *Sphenoptera gossypii* G. an insect pest of cotton; grub bores into the stem.

Cotton stem weevil: *Pempheres offinis* F. an insect pest of cotton, a small white grub that bores into stem near base and causes gall like swellings; attacked plants lodge with strong wind.

Cotton whitefly: See *Bemisia tabaci* (Genn.)

Cottony cushion scale: See *Icerya purchasi.*

Cotyledon: Seed leaf. Leaf like structures in the embryos of seeds. Grasses have one cotyledon in each seed (monocotyledon), while most other plants have two cotyledons (dicotyledon).

Cotyledon leaves: The first leaf or pair of leaves of the embryo of seed plants.

Cotyledonary leaves: See cotyledon leaves.

Cotyledonary stage: Stage of having seed leaves.

Coumachlor ($C_{19}H_{15}ClO_4$): An anticoagulant rodenticide, chemically it is 3-(1-(chlorophenyl)-3-oxobutyl-4-hydroxy- 2H-1 benzopyran-2- one, marketed under trade name Tomorin and Ratilan. It is more toxic to rats than to mice.

Coumaphos: Low toxicity, coumaphos is an animal systemic (given internally) by veterinarians to control fleas on dogs over two months old but not on cats. Marketed under trade name Co-Ral which is also used to control grubs, lice and flies on livestock animals. It is an animal systemic organophosphate, O-(3-chloro-4-methyl-2-oxo-2H-1-benaopyran-7-yl) O, O-diethyl phosphorothioate; moderately toxic to mammals, oral LD_{50} for rats is 90 to 150 mg kg^{-1}.

Coumatetralyl ($C_{19}H_{16}O_3$): It is an anticoagulant rodenticide; chemically, 4-hydroxy-3-(1,2,3,4-tetrahydro-1-naphthatenyl)-2H-1-benzopyran-2-one and marketed under trade name Racumin, Endox and Endrocid. It can be formulated as mastermix powder concentrate having 0.75 per cent coumatetralyl. In bait, 5.0 per cent of this powder gives 0.0375 per cent coumatetralyl treated bait. It produces 100 per cent kill in squirrels, gerbils and rats by feeding them for 5-10 days in laboratory, whereas in field situation a 10 day baiting removed more than 80 per cent of the field population of rodents.

Counter: Trade name of terbufos.

Counter shading: A form of colouration that tends to reduce the three dimensional (3-D) appearance of an animal by suppressing highlights.

Cover: Proportion of the surface area of the target plant on which the pesticide has been deposited.

Cover crop: A crop grown to maintain plant cover on the land to prevent erosion and leaching. If the cover crop is turned under to improve the soil, it becomes a green-manure crop.

Coverage: The amount of spread of a pesticide over a surface.

Cowpea aphid: *Aphis craccivora* Koch, an insect pest infesting leaves, stems and pods of cowpea.

Coxa (pl., coxae): Basal segment of leg by which leg is attached to the body, often fixed and immovable. It is tubular, broad, triangular, oval or spherical in shape. The proximal part is known as subcoxa. The subcoxa is attached to the pleuron by an articular membrane, the coxal corium. In some insects, mecoptera, Neuroptera, Lepidoptera and Diptera, the anterior part of the coxa differentiates into a lobular structure from rest of the coxal region and is known as the meron. In *Machilis* (Thysanoptera), the coxal segment of each meso and metathoracic legs bears typically a small appendicular structure, the stylus.

Coxal cavity: The countersunk housing into which the coxa fits. These housings differ on each of the thoracic segments and are discussed separately. See Procoxal cavity, Mesocoxal cavity, Metacoxal cavity.

Coxal corium: The articular membrane surrounding the base of coxa.

Coxapodite: The basal segment of an arthropod appendage.

Coxopleurite: The sclerite of a generalized thoracic pleuron adjacent to the dorsal margin of the coxa, bearing the coxal articulation, its anterior part becomes the definitive (eutrochantin, trochantinopleura).

Coxosternum: A sclerite representing fusion of the sternum and the limb bases or the coxites, it is usually designated as coxosternum.

Crambidae (grass moths): Family of super family Pyralidoidea of suborder Ditrysia and order Lepidoptera. Small moths with forewings being narrow and elongated; labial palpi porrect. Eggs are flat and laid in two or more rows overlapping each other and are not covered with hairs. Larvae remain in silken webs or galleries and bore into stems, crowns or roots of graminaceous plants. The sugarcane stem borer *Chilo sacchariphagus indicus,* the sugarcane shoot borer *Chilo partellus* and the dark-headed striped borer *Chilo polychrysa,* the pale-headed striped borer *Chilo auricillia* on rice and the rice root caterpillar *Acigona chrysographella* are some pests of importance in India.

Cramps: An involuntary and painful contraction of a voluntary muscle or group of muscles.

Crane fly: One of a large family of two winged, two legged flies (*Tipula* sp.), commonly called daddy long legs, the larvae of which called leather jackets, are abundant in grassland and are pests, particularly of cereal crops.

Cranium: The hardened capsule of the head that bears the mouthparts, eyes.

Crank type duster: Also called rotary duster, used for application of pesticide dust over large areas; operated by handle connected with gears driving fan at high speed which draws dust from dust chamber and discharges it into delivery pipe from which it is driven out through one or two points of nozzles; capacity 3 to 4 kg of dust; covers 2 to 3 acres per day.

Crawler: The active first instar of a scale insect.

Creamed honey: Honey which has been allowed to crystallize, usually under controlled conditions to produce a tiny crystal.

Creeping locomotion: The peristaltic movement of body by contraction in the direction of locomotion in legless larvae and caterpillars.

Creeping perennial: A perennial plant that reproduces by both seed and asexual means (rhizomes, stolons etc.).

Cremaster: A hooked, or spine like process at the posterior end of the pupa, often used for attachment (Lepidoptera).

Crenulate: With small regular indentations.

Creosite: Oily, colourless liquid obtained from the tar produced by destructive distillation of wood, used as an insect repellant, wood preservative and in medicine; mild insecticide and antiseptic.

Crepuscular: Animals those are active in the twilight, predawn and at dusk in the evening.

Crevice tea mite: See *Brevipalpus phoenicis.*

Cricket: Any of various insects of the family Gryllidae (order Orthoptera). Crickets have long antennae and legs adapted for jumping. The males of many species stridulate by rubbing the front wings together. In this way they produce a characteristic shrill chirping sound.

Crimp-wired foundation: Comb foundation into which crimp wire is embedded vertically during foundation manufacture.

Critical areas: Places where pests are most likely to be troublesome.

Critical concentration: Lowest concentration of a pesticide in an environmental compartment at which adverse effects on organisms are likely to be observed.

Critical load: Amount of a pesticide leading to a critical concentration when received by an environmental compartment.

Critical period: The time span during which crops must be kept weed free to maximize yield.

Crochets: The hooked spines at the tip of prolegs of Lepidopterous caterpillars. Caterpillars of sawflies have prolegs without crochets.

***Crocidolomia binotalis* Zell.:** Belongs to order Lepidoptera and family Pyralididae. It is an insect pest of cabbage, commonly known as the crucifer leaf webber.

Crop: (i) Community of plants grown to provide feed, food, fibre or other useful commodities, (ii) The dilated section of the foregut just behind the oesophagus. This enlarged posterior region of the foregut functions as a site for storing food or food reservoir. There is secretory or digestive activity in this region, however, some digestion of the food may take place because of the mixing of enzymes from salivary glands reaching in the crop along with the food.

Crop diversification: Cropping system where a number of different crops are planted in the same area and may be rotated from field to field, year after year.

Crop hygiene: The removal and destruction of infested or diseased plants from a crop so that they do not form a source of infestation for healthy plants (phytosanitation).

Crop loss: A reduction in the quantity and/or quality of the crop yield.

Crop residue: The unused part of the crop that is not harvested. It is usually returned to the land by ploughing (*e.g.*, straw, corn stalks etc.).

Crop tolerance: The degree or the ability of the crop to be treated with a chemical but not injured.

Cross: Crossing between two individuals or fertilization of gametes of two different individuals to obtain offspring is called cross.

Cross contamination: When one pesticide gets into or mixes with another pesticide accidentally. This usually occurs in a pesticide container or in a poorly cleaned sprayer.

Cross infestation: Lateral infestation of storage commonly commence during the post harvest operations from other operations, like previously stacked and infested crop on the threshing floor, previously used and infested transport and stored grain left from last season harvest in the same storage which is termed as cross infestation.

Cross pollination: When pollen grains are transferred from an anther to the stigma of the flower on another plant of same species, cross pollination is said to have taken place. In cross pollination, male and female gametes of two different plants fertilize leading to possibilities of gene recombination and variation.

Cross resistance: A phenomenon that occurs when the development of physiological resistance to a given chemical results in the simultaneous increase in resistance to other chemicals. The main difference between cross and multiple resistance is that in the former, the resistance against selective agent and other compounds is developed by the same mechanism, whereas, in the latter the resistance to different insecticides is developed by different mechanisms. The multiple resistance develops due to simultaneous or consecutive use of several insecticides under field conditions.

Cross vein: A short vein joining any two neighbouring longitudinal veins or a vein connecting adjacent longitudinal veins. The cross vein run in the vertical plane and are confined to the adjacent longitudinal veins only. From their location, these cross veins are known as: Humeral (h): lying between the costa and subcosta;

Costal (c); Radial (r): lying between the first and second branch of radius; Sectorial (s): lying between the sub-branches of second or radius sector, Radio-medial cross vein (rm): lying between the radius and the media vein; Intermedial or media cross vein (m): lying between the branches of the posterior media; Medio-cubital cross vein (m-cu): lying between the media and the cubitus; Intercubital (cu), and Inter-anal (a). See vein also.

Crotoxyphos ($C_{14}H_{19}O_6P$): A synthetic organophosphate insecticide, dimethyl cis-1 methyl-2(1-phenylethoxy carbonyl vinyl phosphate; moderately toxic to mammals; acute oral LD_{50} for rats is 125 mg kg^1; controls ectoparasites of livestock animals. It was introduced by Shell and marketed under trade name Ciodrin.

Crown: The point where stem and root join in a seed plant. This term is also used to describe the foliage and branches of trees.

Crucifer flea beetle: *Phyllotreta cruciferae* Goeze., an insect pest of crucifers; mustard, turnip, radish, cabbage, cauliflower knol-khol, *Rai, Toria*. White larvae mines into roots; bluish greenish adults feed on leaves and make holes, making young leaves dry up and plants die.

Crucifer leaf webber: See *Crocidolomia binotalis.*

Cruciferous: Belonging to the mustard family which includes cabbage, turnip, mustard, radish and some others.

Crufomate: (i) A synthetic insecticide and anthelmintic, an animal systemic; an organophosphate, 4-tert-butyl-2-chlorophenyl methyl) methylphosphoramidate; moderately toxic to mammals, acute oral LD_{50} for rats is 1000 mg kg^{-1}; product of Dow under trade name Ruelene. (ii) Low toxicity, Ruelene is used chiefly as an animal systemic (given internally) pesticide to control ectoparasites (pests on outside of body) such as fleas, ticks, lice, mites etc.

Crustacea: A class of phylum Arthropoda which includes crabs, lobsters etc.

Cryostat microtome: Microtome having low temperature thermostat to cut sections directly from the tissue without emdedding it in wax or jelly.

Cryotochaetidae: Family of suborder Cyclorrhapha, order Diptera. The flies are characterized in having a short spine or tubercle instead of an arista on the enlarged third antennal segment which reaches nearly the lower edges of the head. The flies are attracted to the eyes of man and thus annoy him by entering his eyes. The larvae are parasitic on coccids. *Cryptochaetum iceryae,* a well known parasite of the fluted scale *Icerya purchasi,* was introduced into California from Australia. The parasite is prevalent on the same host in South India.

Crypsis: Combination of form, colour and pattern that facilitates hiding from predators or parasites or both.

Cryptic species: Those which are genetically and sexually distinct but cannot be differentiated on the basis of general observed characteristics.

Cryptic: Colouring and/ or pattern adapted for the purpose of protection from predators or prey by concealment.

Cryptobiosis: Leading a hidden or concealed life. Certain insects can survive in a state of suspended animation or cryptobiosis, after losing virtually all their water. This happens in the eggs of Collembolan, *Sminthurus*. The best example among the insect larvae is the Chironomid, *Polypedilum* which survives in a desiccated state. With its water content reduced to below 8 per cent, in the dried mud of its breeding places and can be kept in this condition for several years.

Cryptobiotic: Leading a hidden or concealed life.

Cryptocephalidae: Family of superfamily Chrysomeloidea, suborder Polyphaga and order Coleoptera. Family of superfamily Chrysomeloidea, suborder Polyphaga and order Coleoptera. This family is closely related to Clytridae but differs from the latter in having a long, usually filiform, antennae. This includes species of the genus *Cryptocephalus.*

Cryptocerata: A division of suborder Heteroptera, order Hemiptera. The insects belonging to this division have short antennae which are concealed in grooves on the ventral side of the head. They are usually aquatic. This includes the families, Corixidae (water boatmen), Notonectidae (black swimmers), Nepidae (water scorpions), Balostomatidae (giant water bug) etc. These are popularly called the short horned bugs or water bugs.

Cryptohasidae: Family of super family Tinaeoidea of suborder Ditrysia and order Lepidoptera. Moths without antennal pecten. Larvae concealed in shelters or covering or tunnel in wood, or scarping leaves for food. Larvae for *Procometis trochala* feed on dry fallen leaves of sugarcane. The coconut black headed caterpillar, *Nephantis* serinopa is a notorious pest of coconut in South India.

Cryptonephridial system: Distal ends of Malpighian tubules encased by rectal tissues. In many beetles, phytophagous carnivorous, and omnivorous species, as well as those feed on dry substances, such as the mealworm (*Tenebrio*), the upper part of the Malpighian tubules closely invest the rectum, being bound to it by a delicate membrane. The precise physiology of this arrangement is not fully known; but it is clear that it serves to add the absorptive powers of the Malpighian tubules to those of the rectal epithelium. The arrangement seems always to be associated with remarkable powers of drying the excrement.

Cryptophagidae (silken fungus beetles): Family of superfamily Cucujoidea, suborder Polyphaga and order Coleoptera. Adults 1 to 5 mm long, elongate-oval, yellowish brown and covered with a silky pubescence. Tarsi are usually 5-5-4 in male. They mostly feed on fungi and a few occur in nests of wasps of bumble bees. *Leucohimatium elongatum* has been obtained in stored *ragi* in Karnataka State.

Crystalline cone (crystalline body or vitreous body): A translucent structure just beneath the corneal lens of an oval or conical body.

Crystallization: See granulation.

Ctenidium (pl., ctenidia): A row of stout bristles like the teeth of a comb, character of fleas.

Cu: Abbreviation for cubitus vein (5th main longitudinal vein).

Cube powder: The finely ground roots of certain leguminous trees and shrubs, *Lonchocarpus* spp., which contain rotenone, an insecticidal substance.

Cubitus: The fifth main longitudinal vein (Cu), with two branches Cu_a and Cu_b; Cu_a forms the posterior margin of the discal cell.

Cucujidae (flat bark beetles): Family of superfamily Cucujoidea, suborder Polyphaga and order Coleoptera. The beetles are reddish, brownish or yellowish, and extremely flat- often with filiform antennae. They are usually found under bark. Most of them are predaceous on coccids and mites. Some species which are parasitic in the larvae of Cerambycidae and Braconidae, exhibit hyperetamorphosis. *Laemophloeus* species is both predaceous and grain-eating. *Laemophloeus minutus* is a common pest in stored products attacking broken grains secondarily.

Cucujiform aedeagus: Aedeagus in which the phallobase forms a sheath or ring around the penis and the parameres are often reduced in size, fused to the phallobase or completely absent. The combined phallobase and parameres, usually referred to as the tegmen, may be drawn out anteriorly to form a narrow strut. This type of aedeagus includes those known as cleroid, sheath or vaginate and cucujoid, ring or annulate; it is characteristic of most Lymexyloidea, Cleroidea and Cucujoidea, as well as members of Chrysomeloidea and Curculionoidea. In Curculionoidea, this type of aedeagus is called the orthocerous type, although this term also implies that the penis is comprised of a dorsal and ventral lobe.

Cucujoid aedeagus: See cucujiform aedeagus.

Cucujoidea: Superfamily of suborder Polyphaga, order Coleoptera.

Cucullus: A dorsal lobe at the apex of the male valve in certain moths.

Cucurbit bugs: *Aspongopus janus* F. and *A. brunneus* N. Red and pale brown bugs that suck sap from cucurbits.

Cucurbits: Any of various mostly climbing or trailing plants of the family Cucurbitaceae, which includes squash, pumpkin, cucumber, gourd, watermelon, and cantaloupe.

Cues: Stimuli that determine when, where and how a person responds.

Culex: A genera of mosquito.

Culicidae (mosquitoes): Family of suborder Nematocera, order Diptera. The mosquitoes which are included in this family exhibit sexual dimorphism- the females with antennae sparsely haired, while those of males carry conspicuous whorls of hairs. The second and fourth veins of wings are forked, while the third is unbranched and long. The anterior cross vein, unlike as in psychodids, is placed away from wing base. This includes the Anophelini or malarial mosquitoes and the Culicini or filarial mosquitoes which are of considerable medical importance.

Culm: The jointed stem of a grass which is usually hollow except at the nodes or joints.

Cultivar: A cultivated variety (genetic strain) of a domesticated crop plant.

Cultural control: A method of crop protection using careful timing and a combination of agronomic practices such as tillage, planting, irrigation, sanitation, mixed cropping and crop rotation which makes the environment less favorable for the proliferation of certain pests or diseases.

Culture: (i) A laboratory insect population usually maintained under controlled conditions. Although, reserved for populations of micro-organisms, it may be used synonymously with colony or subcolony when referring to insect populations. (ii) Functional determinant of a person's wants and behaviour.

Cumulative effect: Overall adverse change which occurs when repeated doses of a pesticide have biological consequences which are additive.

Cumulative pesticides: Pesticides that tend to accumulate or build up in the tissues of animals or in the environment (soil, water etc.).

Cuneus: A more or less triangular region of the forewing of certain Heteropteran bugs, separated from the corium by a groove or suture.

Cuprocide: A copper compound used as an insecticide.

Cupule: A very short and broad, distally concave and usually glabrous antennomere just before the antennal club. This usually forms part of a mechanism in aquatic beetles for obtaining an air bubble.

Cupuliform antennal club: An antennal club in which the distal one or two antennomeres are more or less enclosed within the preceding one.

Curating: It refers to care and superintendence of museum containing animal collection.

Curative: Having the power to heal or cure.

Curative pesticide: A pesticide that can inhibit or eradicate a disease-causing organism after it has become established in the plant or animal.

Curative treatment: Treatment that reduced disease when infection has already taken place.

Curculionidae (weevil): Family of superfamily Curculionoidea, suborder Polyphaga and order Coleoptera. The adult beetles are minute or large, 0.3 to 5cm in length and colored dull brown, grey, black and a few green. The head is more or less prolonged into a snout which varies considerably in size, shape and length and bears the mouth parts at the tip. A labrum is usually wanting. The eyes are well developed. The antennae are geniculate and clavate and arise about the midlength of the snout. In many the body is clothed with scales. When disturbed the adults beetles fall to the ground and remain motionless. In *Brachyrhinus* parthenogenesis is frequent.

Curculionoid aedeagus: A modification of the cucujiform aedeagus in which the ring is very narrow with a narrow anterior strut, the parmeres reduced and fused to the ring, and the ring sometimes partly membranous, so that it resembles a fork. This type occurs in most derived curculionoids, and when combined with an undivided penis, is called the gonatocerous type.

Curculionoidea: Superfamily of suborder Polyphaga, order Coleoptera. A large group of insects which are mostly phytophagous and hence are of considerable agricultural importance. In most species, excepting Scolytidae and Platypodidae, the head is prolonged into a break or snout. The gula is not or nearly always confluent. Excepting Anthribidae the others lack prosternal sutures. The tarsi are four-segment.

Cursorial: Legs adapted for running, *e.g.*, cockroaches and tiger beetles. These legs are elongated, thin and tubular with a flattened femora, highly muscular coxa but with slender and elongated tibia and tarsi. The modifications bring an efficient running.

Cuspidal: Two segments of curved lines meeting and terminating at sharp point.

Custard apple fruit fly: *Pseudococcus vigatus* C., *P. lilacinus* C., *Ferrisia virgata* Ckll. Small, reddish white mealy bug infesting leaves, shoots, buds and fruits of custard apple, sucking sap from them; secretes honey dew which attracts fungus attack.

Cut comb honey: Comb honey cut into various sizes, the edges drained, and the pieces wrapped or packed individually.

Cut off value: Numerical value set by regulatory authorities representing the limit of acceptability for a property or behaviour of a compound for the final step in tiered assessment schemes.

Cutaneous ventilation: Gaseous exchange directly across the cuticle.

Cuterebridae (robust botflies): Family of suborder Cyclorrhapha, order Diptera. The flies are large, robust and rather hairy. The larvae of these flies are mostly parasitic on rodents. Man is attacked by *Dermatobia hominis.* It oviposits on mosquitoes and when the mosquito feeds on man the larvae penetrate the skin and gain entry. The eggs of the botfly may also be carried to man by stable flies and other muscids.

Cuticle: The non-cellular outer layer of the body wall (integument) of an insect. The epidermis secretes the chemical which forms the cuticle. These substances are chitin, protein and lipids which together form the cuticle. The chitin is a nitrogenous polysaccharide. The cuticle is divisible into two segments: (i) epicuticle (non-chitinous) and (ii) procuticle (chitinous).

Cuticular appendages: The setae, spines and spurs. The seta arises from the special epidermal cells called trichogen cells which penetrate through another cell called tormogen cell.

Cuticular membrane: In some regions of integument, the exocuticle is discontinuous due to lack of tanning and thus such parts of cuticle remain membranous. These membranes are flexible and perform intersegmental, intrinsic or extrinsic articulations according to their location.

Cuticular modifications: The macrotrichia (clothing hairs, glandular hairs, sensory hairs and the wing scales), microtrichia, spines, apodemes and apophyses, cuticualr membranes and ecdysial lines constitute the cuticular modifications. See details of these.

Cuticular respiration: See cutaneous ventilation.

Cuticulin layer: It is a refractile amber coloured layer consisting of a lipoprotein, the cuticulin. It is highly resistant to the mineral acids and most of the organic solvents. It serves as a permeability and growth barrier and determines the surface properties of the integument.

Cutting and chewing type mouth parts: Mouth parts are well adpted for cutting and chewing of the food in members of Orhoptera. These consist of labrum (upper lip), a pair of mandibles, a pair of maxilla, labium (lower lip) and hypopharynx, *e.g.*, grass hoppers and cockroaches.

Cutworm: Caterpillars of certain moths (family Noctuidae, Lepidoptera)) that live in the soil. They emerge at night to eat foliage and stems. Cutworms can be serious pests of root crops and of many other crops during the seedling stage.

CWC: Central Warehousing Corporation. The corporation was started as a statutory body in New Delhi on 2nd March, 1957. The objectives of CWC are to acquire and build up godowns and warehouses at suitable places in India. The functions of CWC are to run warehouses for storage of agricultural produce, seeds, fertilizers and notified commodities for individuals, cooperatives and other institutions; to act as an agent of government for purchase, sale, storage and distribution of the commodities and to arrange the transport of commodities

Cyclic: In chemistry, atoms linked together to form a ring structure.

Cyclodienes: Synthetic insecticides belonging to the cyclodiene group of cyclic hydrocarbons, for example chlordane, heptachlor, aldrin, dieldrin and endrin.

Cyclorrhapha: Suborder of Diptera in which the antennae are three segmented with an arista which is usually dorsal in position. The palpi are one segmented. Head of larvae are vestigial. Pupation takes place in a puparium.

Cyclorrhaphous Diptera: The group of flies which emerges from the puparium through a circular opening at one end of the puparium. These flies belong to the more advanced families.

Cydnidae: Family of superfamily Scutelleroidea of division Gymnocerata, suborder Heteroptera and order Hemiptera. These are oval, dark coloured bugs characterized by the provision of expanded and spiny tibiae adapted for burrowing beneath stones or in the soil and some live in the ant nests, *e.g.*, burrower bugs, *Stibaropus tabulatus* infests the roots and rootlets of tobacco plant.

Cygon: Trade name of dimethoate.

Cyhexatin: Moderately toxic, Cyhexatin is relatively useful against mites. It is also used as a fungicide and effective against some molds, mildews and other household fungi. It is marketed under trade name Plictran.

***Cylas formicarius* Fab.:** Belongs to order Coleoptera and family Curculionidae. It is an insect pest of brinjal, commonly known as the sweet potato weevil.

Cylindrachaetidae: A family of super family Tridactyloidea of order Orthoptera and suborder Caelifera. Apterous, subterranean, burrowing forms with long cylindrical bodies and with fossorial legs. Hind femora not enlarged.

Cymatophoridae (Polyplocidae): Family of super family Geometroidea, suborder Ditrysia and order Lepidoptera. They resemble Noctuidae and are mainly seen in the northern hemisphere. Some common species are *Habrosyne derasa* and *Thyatira batis* in the British Isles.

Cynipidae (gall wasps): Family of super family Cynipoidea, suborder Apocrita and order Hymenoptera. This is a very interesting family which includes three subfamilies, *viz.* Eucoilinae and Charipinae comprising parasitic forms and Cynipinae comprising gall makers and gall inquilines. *Euoila eucera* is a parasite of *Oscinella frit, Cothonaspis* sp. are parasitic on puparia of *Dacus* sp. in India. *Dieucoila indica* is parasitic on puparia of the syrphid *Sphaerophoria scutellaris.*

Cynipoidea: Super family of Suborder Apocrita, order Hymenoptera. Cynipoidea are small or often minute insects, mostly black or dark colored and sometimes apterous; in winged forms the venation is greatly reduced. The antennae are filiform and not elbowed. The antennae are filiform and not elbowed. The pronotum extends back to the tegulae and usually trochantellus is not developed. Pygostyles are wanting. The ovipositor arises from anterior to apex of the abdomen. The eggs have a usually elongate pedicel. The larvae are apodous and maggots like. A cocoon is absent. Its members are gall makers, gall inquilines or parasites.

Cynosis: A dark bluish or purplish coloration of the skin and mucous membrane due to deficient oxygenation of the blood in the lungs or to an abnormally great reduction of the blood in its passage through the capillaries.

Cypermethrin: A synthetic pyrethroid, chemically (+) alpha-cyano-3-phenoxybenzyl (+) –cis, trans-3-(2,-2 dichlorovinyl)-2, 2-dimethylcyclopropane carboxylate. It is a viscous yellow liquid miscible in most of the organic solvents. In water, its solubility is 0.2 ppm. It is contact and stomach poison with low mammalian toxicity. It is highly toxic to bees and fishes. It is highly effective against Lepidoptera but also effective against Coleoptera, Hemiptera and Diptera. It has little or no effect on mites and nematodes. It acts against all stages of pests (egg, larva, adult). It is marketed under trade names, Cymbush, Ripcord etc.

Cyst: (i) A resting structure containing a juvenile stage of an organism, such as a cyst of a nematode containing larva. (ii) A sac normal or abnormal, especially one containing a liquid or semisolid. (iii) It is used here for a group of eggs contained within a tough protective envelope, the dormant stage of some species of nematode or eelworm. This cyst is a dead, swollen female.

Cystic stage: The stage in the development of a parasite during which it is enclosed in a sac which separates it from the tissues of the host.

Cystocytes: The cells enclosing the germ cells in a gonodial tube; the follicle cells of the ovary, the cells or the testis.

Cytogenetics: It is the comparative study of chromosomal mechanisms in behaviour in populations and taxa and their effect on inheritance and evolution.

Cytology: Gr., *Kytos*= hollow. This is a branch of biology which deals with the structure, function and life history of cells.

Cyton: See neurocyte.

Cytoplasm: The fluid or gelatinous content of living cells as distinct from the reproductive nucleus.

Cytoplasmic incompatibility: It refers to non-productive crosses between the morphologically alike but reproductively isolated species or sibling species; plasmatic incompatibility.

***Cytoplasmic polyhedrosis virus*:** The cytoplasmic polyhedrosis virus multiplies in cytoplasm of the gut cells and the polyhedra are never seen in the nucleus.

D

Dactylopiidae: A family of superfamily Coccoidea, suborder Homoptera and order Hemiptera.

Dactylopodite: Terminal segment of a generalized appendage; claw like in form, with levator and depressor muscles arising in the propodite. It is represented by a median claw of the pre-tarsus in some apterygote insects.

***Dacus dorsalis*:** An insect pest of mango, commonly known as mango fruit fly.

Damage: The adverse effect on plants or crops due to biotic or abiotic agents, resulting in economic loss (reduction of yield and/or quality).

Damage boundary (DB): The lowest level of injury where the damage can be measured is called the damage boundary.

Damage threshold: Population of insect pests above which the crop losses are visible.

Dammar bee: It is stingless bee. The species like, *Melipona* sp. and *Trigona* sp. are found in several parts of India; the yield of honey is small and difficult to extract.

Damp wood termite: Termite group which makes colonies in living trees or dead wood.

Damping off: Disease of seedlings characterized by the rot of seedlings near soil level. This can happen prior to emergence (pre-emergence) or after emergence (post-emergence).

Damselfly: Any of various insects of the suborder Zygoptera (order Odonata). Damselflies are often brightly coloured, predatory insects. They have a long slender body and elongated membranous wings that fold together when the insect is at rest.

Danaidae (Euploeiinae, Limnadidae): Family of super family Papilionoidea of suborder Ditrysia and order Lepidoptera. Butterflies of this family have scaleless antennae ending in antennal club often but little pronounced. A corrugate knob is present at the terminal end of the very small forelegs. The larvae are strikingly colored with black and yellow, green or red and have two or four pairs of fleshy processes variously distributed on the mesothorax and on one or more of the abdominal segments. The pupa is suspended by its caudal extremity.

***Danais chrysippus* L.:** Belongs to order Lepidoptera and family Nymphalidae. It is an insect pest of ornamental plants, commonly known as the ak butterfly.

Danger: Risk, hazard.

Dark headed stripped borer: See *Chilo polychrysa.*

Darwinism: An evolutionary theory propounded by Darwin.

Dasanit: A synthetic organophosphate insecticide and nematicide, O, O- diethyl O-p[(methylsulfinyl) phenyl] phosphorothioate; highly toxic to mammals; acute oral LD_{50} for rats is 2 to 11 mg kg^{-1}; controls soil insects and nematodes in corn, onions, tobacco, ornamental plants etc.

Dascilloidae: Family of superfamily Discilloidea, suborder Polyphaga and order Coleoptera. These are moderate sized beetles and their larvae generally feed on their roots of grasses. In South India, *Ora picta* occurs in large numbers on paddy but it does not cause any damage. However, the grubs are aquatic and appear to feed on minute particles of food such as algae or protozoa or bits of organic debris in paddy fields.

Dascilloidea: Superfamily of suborder Polyphaga, order Coleoptera. The Dascilloidea are characterized by filiform or slightly serrate antennae, more or less conically projecting forecoxae with their cavities completely open behind and the hindcoxae excavated to receive the femora. Three families, *viz.*, Dascillidae, Helodidae and Eucinetidae are recognized.

***Dasineura citri* Grover:** Belongs to order Diptera and family Cecidomiidae. It is commonly known as citrus blossom midge and is a pest of many varieties of citrus.

***Dasyneura lini* Barnes:** Belongs to order Diptera and family Cecidomyidae. It is an insect pest of sesame, commonly known as the linseed gall midge.

Data warehouse: A collection of current data captured, organized and stored in a company's contact centre.

Date palm scale: See *Aspidiotus destructor.*

Dauerlarva: An infective non-feeding juvenile stage of some insect parasitic nematodes which is the most resistant stage to external conditions and easiest to store in the laboratory.

Dayfly: See mayfly.

DDD ($C_{14}H_{10}C_{14}$): Dichloro diphenyl dichloroethane or TDE tetrachloro diphenyl ethane is a DDT analogue. It is less effective than DDT but has advantage of being less toxic to warm blooded animals and fishes. It is superior to DDT for the control of mosquito larvae.

DDT ($C_{14}H_9Cl_5$): Common name of a widely used synthetic insecticide, Dichloro dimethyl trichloroethane, a contact and stomach poison and belongs to organochlorine insecticide group. It was first synthesized by Zeidler (1874) but insecticidal properties were discovered by Paul Muller (1939). It has diphenyl methane structure. DDT is a white to cream coloured amorphous powder and has variable composition of about 14 compounds of which the isomer p-p′ DDT (65-80 per cent) and o-p′ DDT (15-20 per cent) form the chief constituents. The proper chemical name is 1,1,1-trichloro-2,2-bis-(p-chlorophenyl) ethane. DDT has several analogues such as DDD, methoxychlor and dicofol. DDT is marketed under trade name Guesarol and Neocid under various formulations such as 5 and 10 per cent dust, 25 EC and 50 WP.

DDT, metabolism: There are five principal routes of DDT metabolism in various organisms. (i) Oxidation to DDA**:** In vertebrates, DDT is metabolized to DDA (dichlorodiphenyl acetic acid) by alternate reduction and dehydrochlorination followed by hydration and oxidation. (ii) Oxidation to dicofol**:** It takes place in Drosophila (Tsukamoto, 1960). (iii) Oxidation to DBP**:** In cockroaches and larvae of fruit fly, DDT is metabolized to p,p′-dichlorobenzophenone (DBP). (iv) Dehydrochlorination to DDE (dichlorodiphenyl dichloroethylene)**:** In man, DDE is the principal storage form of ingested DDT. (v) Reductive dechlorination to DDD**:** In plants, DDT is metabolized to DDD (dichlorodiphenyl dichloroethane) which in vertebrates is further metabolized to DDA.

DDVP: See dichlorvos.

Deacetyl azadirachtinol: It is a minor neem chemical which paralyze the swallowing mechanism and so prevent the insect from feeding. This ingredient, isolated from fresh fruits, appears to be as effective as azadirachtin in assays against the tobacco budworm but it has not yet been widely tested in field practice. Two compounds related to salannin, 3-acetylsalannin, recently isolated from neem, also act as antifeedant.

Dead heart: A symptom of insect damage in which the central shoot of plant dies.

Dealate: Wingless as a result of the insect casting or breaking off its own wings, as in newly mated queen ants and termites.

Dealation: The shedding off the wings by the queens (and also males in the termites) during or immediately following the nuptial flight and prior to colony formation.

Debris: Trash or remnants of unwanted plant parts.

Decamethrin: A synthetic pyrethroid, acts on insects by contact and has a direct insecticidal effect on biologically active instars of pests. It is not broken down rapidly by the enzymes in the insects and larvae die after a few hours. It is 100, 80, 50 and 40 times more powerful than the DDT, carbaryl, malathion and parathion, respectively. Decamethrin shows resistance to leach out during heavy rain or sprinkler irrigation. The acute oral toxicity (LD_{50}) to rat is 128.5-138.7 mg kg^{-1}. It is formulated as 2.8 EC and marketed under trade name Decis.

Deccan wingless grasshopper: *Colemania sphenarioides* Bol.; an insect pest of sorghum, maize, pearlmillet, *Ragi*, lablab, blackgram, redgram and groundnut; adults and nymph feed on crop foliage and in severe attack, entire plants are defoliated.

Deciduous: Those plants that are perennial in habit but lose their leaves during the spring season.

Decomposer: An organism that utilizes dead organic matter as food, decomposing it into more simple substances.

Decontaminate: To make safe by removing any pesticide from equipment or other surfaces as directed on a pesticide label or by an agricultural authority.

Decontamination: The removal or breakdown of any pesticide chemical from any surface or equipment.

Decoy hive: A hive placed to attract stray swarms.

Decticous: Possessing articulating mandibles in the pupa used to free the adult from the cocoon.

Decumbent: Bent downward, *e.g.*, decumbent hairs of Coleoptera

Defend: Trade name of dimethoate.

Deffraction mechanism: Burt and Catton (1962) noted in *Locusta*, the formation of a series of superposition images at progressively greater depths in the retina of eye due to diffraction from the lens systems besides a first apposition image. These are the diffraction images which are formed due to the diffraction of the light received from the adjacent facets. The diffraction images are very complex and distorted in comparison to either apposition or superposition images. Their functional significance remains unknown.

Deficiency disease: A disease resulting from the lack of one or more essential nutrients.

Definite or definitive host: An insect in which a parasite attains sexual maturity.

Deflocculating agent: Material added to a spray suspension to delay sedimentation.

Defoliant: Spray which induces premature leaf fall.

Defoliating beetle: See *Anomala dimidiata* (Hope).

Defour's gland: An exocrine gland in the Hymenoptera which secretes the alkaline element of the poison carried by sting.

Degenerate legs: In Papilionoidea and Nymphalidae, the tarsi and pretarsi are lacking in the forelegs. The female coccids have extremely reduced spindle like legs or sometimes they become completely apodous. The parasitic female Strepsiptera are also without legs. The Dipteran larvae are exclusively apodous, while larvae of leaf mining Lepidoptera and Coleoptera are also apodous. The larvae of large number of Hymenoptera are often devoid of legs.

Degenerative and filter feeding mouth parts: (i) Larvae of aquatic Nematocera: True mouth parts are entirely lacking. External mouth parts form brush like structures that brings a current of water and suspended particles into the pharynx. The pharynx develops into a pump and provided with internal screens. The comb like screens are closely arranged with spacing that permit removal of very fine suspended material. (ii) Larvae of Cyclorrhapha: They bear mouth hooks attached to a permanent pharyngeal sclerite. Mouth hooks move parallel to each other in tearing motion. Some flies such as *Musca domestica* and *Stomoxys calcitrans* have only a single mouth hook. True mouth parts are wanting. Hypopharynx, clypeus and associated parts, and sclerites of head collectively form the cephalopharyngeal skeleton. (iii) Filter feeding: Filter feeding occurs in some aquatic larvae which filter suspended particulate matter from water with the help of fans, combs, brushes or sieves (Diptera, Ephemeroptera and Trichoptera) or spun silk nets (some Trichoptera).

Degradability: The ability of chemical to decompose or breakdown into less complex compounds or elements.

Degradate: Chemical product resulting from degradation of a pesticide.

Degradation: Process by which a pesticide is broken down to simpler structures through biological (microbes) and abiotic mechanisms (water, air, sunlight or other agents). Breakdown and decomposition are the terms used synonymously for the same.

Degree days: Insect development depends on temperature. Below a certain temperature threshold there is no development. Also there is a maximum temperature above which development stops. For a certain day, the number of degree days can be calculated using the threshold and the maximum temperature. If the average temperature of that day is one degree higher than the threshold this will count for one degree day (two degrees above the thresholds is two degree days, etc.). Calculating degree days over a period of time can be used to predict when the development of the insect will be completed.

Degree of exposure: The amount or extent to which a person has been in contact with a toxic chemical.

Degree of freedom: (i) Term used to denote the least number of independent variables defining the state of the system (*e.g.*, the temperature and pressure in case of gas) which must be given definite values before this state is completely determined. (ii) The number of independent ways in which a molecule may possess translational, vibrational or rotational energy. (iii) The divisor n—1 for the sum of squares of deviations is termed the dgree of freedom.

Delayed action: With some herbicidal chemicals delayed response is expected and considerable time may elapse before maximum effects can be observed. Usually treated plants stop developing soon after treatment, and then gradually die.

Delayed dormant spray: A spray in orchards applied during the period from swollen bud to late green tip of bud development; often called swollen bud in stone fruits.

Delayed sterility: The irradiated parents produce offspring that survive to adult stage, but a high proportion of gametes of these new adults have chromosomal deficiencies or duplications which results in lethality and sterility in the next generation.

Delayed toxicity: The effect of a toxic substance may become evident some time after exposure which may delay treatment.

Deleterious: Harmful or injurious.

Delnav: Trade name of dioxathion.

Delphacidae (Araeopidae): A family of superfamily Fulgoroidea of suborder Homoptera and order Hemiptera. It is the largest family of superfamily Fulgoroidea, characterized by large flattened spur at the apex of hind tibiae. They are generally small and many possess reduced wings. The *Perkinsiella sinensis;* brown backed plant hopper, *Nilaparvata lugens* and white backed plant hopper, *Sogatella furcifera* infests rice and *Peregrinus maidis* on millets in India.

Deltamethrin: Same as decamethrin.

Delusory parasitosis: A psychological illness in which the affected persons imagine themselves to be infected with living parasites.

Demaree: The method of swarm control that separates the queen from most of the brood within the same hive.

Deme: An assemblage of taxonomically closely related individuals; a local population of species; the community of potentially interbreeding individuals at a given locality.

Dendrogram: A diagram indicating the relationship of items in a classification.

Dendrons or dendrites: Finely ramifying branches given off from a nerve cell.

Dengue: A virus disease of man marked by severe pain in head, eyes, muscles, and joints and transmitted by certain mosquitoes.

Density: The size of a population within a given unit of space.

Density dependent: A proportionate increase in mortality (or decrease in fecundity) as population density increases.

Density dependent factor: In ecology, a factor which changes in intensity with changes in population density (*e.g.*, lethal effects of a factor intensify as population numbers increase).

Density independent: The mortality or the survival varies independently of population density.

Density independent factor: A factor regulating activity or population of living organisms but those influences are independent of the population density of that particular organism.

Density, population: Number of individuals of a species that habitat or live in per unit area (generally per square kilometer) of any geographical region.

Dentate: Toothed or with tooth like projections.

Denticulate: Bearing very small tooth like projections.

Deoxygenation: Depletion of oxygen.

Depleted: Exhausted or reduced.

Depolarization: The reduction of the electrical potential of a neurone's membrane.

Deposit: Quantity or amount of initially laid down insecticidal chemical on a unit area of plant, plant part or other surface at a given application. It may refer to the deposit of the total spray preparation, or it may refer only to the amount of chemical left after the water evaporates.

Deposition velocity: Velocity at which the spray impinges on the target.

Depressed: Flattened dorsoventrally.

Dequeen: To remove a queen from a colony.

Derbidae: A family of superfamily Fulgoroidea of suborder Homoptera and order Hemiptera. It includes tropical species with delicate wings. Terminal segment of rostrum is as long as wide. The insect pest, *Proutista moesta* is prevalent commonly on sugarcane, maize, sorghum, grasses etc. in India.

Dermal glands: Modified epidermal cells or groups of cells on the outer surface of body that secrete wax, scent, silk, irritant poisons etc. It may also be applied to the hypodermal glands that secrete moulting fluid containing enzymes which liquefy the endocuticle prior to moulting.

Dermal toxicity: Toxicity of a chemical substance as a result of contact with the skin.

Dermaptera: Insect order (*Derma*= skin, *ptera*= wings), *e.g.* earwigs. They are characterized by short, hardened wing covers, chewing mouthparts, pincer like abdominal cerci, and an incomplete metamorphosis.

Dermestid beetle: See silkworm pests.

Dermestidae: Family of superfamily Dermestoidea, suborder Polyphaga and order Coleoptera. Small or moderate sized oval or hemispherical beetles often clothed with fine scales or hairs.

Dermestoidea: Superfamily of suborder Polyphaga, order Coleoptera. Dermestoidea are characterized by short antennae, generally clubbed and rarely serrate; one or two ocelli often present and the prothorax is not hood-like. This includes four families, *viz.* Derodontidae, Nosodendridae, Dermestidae and Thorictidae.

***Derris* powder:** A commercial insecticide made up of powdered roots of the plants of *Derris* spp. which contains the poison, rotenone. Used as a fish poison also.

***Derris* sp:** The most important plant sources of rotenone, containing in roots. It is grown principally in Malaya and East Indies. U.S. industry now depends on *Lonchocarpus* species from South America, mostly Peru.

Descendant: Offspring of an earlier form.

Desert locust: *Schistocerca gregaria* Forsk. is the most important among three species of locusts found in India. Visits India periodically in big swarms; dreaded because of its ability to cause heavy damage to crops. Threatens 64 countries or territories extending over 12 million square miles and affects millions of people. These can mass multiply, migrate to long distances, and destroy any vegetation that comes its way. Breeds in North-West desert parts of India through Pakistan, Saudi Arabia and other Middle-East countries, East Africa, Sudan and parts of North and West Africa. Indian Government keeps watch over locust population and its control through network of Scheduled Desert Areas. Not possible to prevent entry into India, hence control operations aim at destroying swarms, hoppers and eggs. Also see locusts.

Desiccant: Chemical that kills vegetation by water loss. These chemicals promotes drying or removal of moisture from plant tissues. See defoliant for difference in these two terms.

Desiccation: Condition of dehydration. When exposed to high temperature, an insect may be killed by temperature itself or by the desiccation consequent upon evaporation. If the insect is sufficiently large, it will be able to lower its temperature below the fatal level (for short periods at any rate) by evaporating water from its body, and thus exercise a limited degree of temperature control. Most insects die when water content is reduced to about 50 per cent of the total body weight.

Desorption: Depletion of one or more components in an interfacial layer.

Detergents: Wetting agents. Substances that increase the spreading of spray liquids and enable wettable powders to disperse readily in water. Detergents can be used as surfactants in some pesticide sprays.

Detoxification: The act or process of removing poisonous or toxic properties of a substance in the body; the detoxification of secondary plant chemicals is either by excretion or metabolism or sequestration in the system.

Detoxify: To make an active ingredient in a pesticide or other poisonous chemical harmless and incapable of being toxic to plants and animals.

Detritus food chain: The passage of nutrients and energy through an unstructured assemblage of decomposer organisms. This type of food chain starts from dead organic matter which is consumed by detritus feeding organism (detritovore). These in turn are eaten by some other predators. Thus such type of food chain does not depend on solar energy but on organic matter.

Detrivores: Animals that feed on dead plant or animal material.

Deutocerebral neuropile: These are two spherical neuropile masses; each situated in each lateral region of the deutocerebrum and are called the antennal lobes. The antennal lobe is divided into two halves, the dorsal sensory and the ventral motor. Both antennal lobes behave as the seat of antennal nerves. Each is a mixed type of nerve containing a group of sensory and that of motor fibres.

Deutocerebrum: The middle pair of lobes of brain that innervates the antennae. It gives out a pair of lateral antennal nerves innervating the antennae (nerves antennalis).

Deutonymph: The second nymphal instar of mites.

Deutoplasm: See yolk.

Deutorotoky (Ampherotokous): These are parthenogenetic parasitoids exhibiting thelyotoky but also produce few males. Such individuals are again uniparental, *e.g.*, *Trichogramma brasiliensis* when reared at temperatures exceeding 25 °C also produce males while beyond 30 °C mainly males.

Deutovum: The quiescent undeveloped larval stage from which after six or seven days, active, six legged larva emerges.

Development: The development of an insect, *i.e.,* from egg to adult can be divided into two parts, early embryonic development and post embryonic development. The early embryonic development in broad sense takes place inside the egg while the post embryonic development is the latter part occurring outside the egg. The early embryonic as well as the post-embryonic development may undergo a short or long period of development arrest at any time called the diapause which may be an obligatory or facultative feature of an insect species.

Development cycle: The period from the birth of the egg to the eclosion of the adult insect.

Developmental threshold temperature: The minimum temperature (cardinal temperature) to initiate development.

Dewlap: The loosely hanging skin under the throat of cow.

Dextrose: One of the two principal sugars found in honey; forms crystals during granulation.

Diabroctins: These are the compounds of insecticidal activity, isolated from bacteria *Bacillus subtilis* and *B. cereus*. These are effective against Colorado potato beetle, *Diabrotica undecimpunctata*.

***Diacrisia obliqua* Wlk.:** Belongs to order Lepidoptera and family Arctiidae. It is an insect pest of groundnut, turmeric, ornamental plants etc., commonly known as the Bihar hairy caterpillar.

Diagnosis: Identification of the nature and cause of a problem.

***Dialeurodes citri* (Ashmead):** Belongs to order Hemiptera and family Aleyrodidae. Commonly known as citrus whitefly and is a pest of citrus, persimmon, *Melia azadirach* etc.

Diamond back moth: *Plutella xylostella* (Linn.) is an insect pest of crucifers; small pale green caterpillar with fine black hair, feeds on leaves and makes holes.

Diapause factor: See diapause.

Diapause hormone (DPH): Secretion of diapause hormone by the neurosecretory cells of suboesophageal ganglion during pupal stage of the silkworm, *Bombyx mori* was investigated by Hasagawa. It plays important roles in the life of insect by stimulating (i) transport of 3-hydroxy-kynurenine from blood into the pupal ovaries, (ii) elevating its accumulation in the developing eggs which results in ommatochrome formation in the diapause eggs, (iii) enhance trehalase activity in the pupal ovaries causing heavy accumulation of glycogen in the diapause eggs, and (iv) blocking the activity of esterase-A in diapause which is carrying the mobilization of yolk necessary for embryonic development in non- diapause eggs.

Diapause: A period of suspended animation of regular occurrence in the lives of many insects, especially in the young stages. Many insects, particularly, in temperate latitudes, periodically enter a state of arrested growth which is not readily eliminated when conditions again become favourable. For this deep seated arrest, the term diapause is commonly used. The immediate cause of diapause is the lack of hormones necessary for growth and development; if the brain from the lepidopterous pupa which is just coming out of diapause after exposure to a period of chilling, is implanted into a pupa that is still in diapause, this pupa will resume its development. In case of silkworm, there is a secretory centre in the suboesophageal ganglion which liberates a diapause factor which in turn influences the developing eggs so that they become hibernating or diapause eggs.

Diaphacinone: Chlorophacinone and diphacinone are anticoagulant rodenticides similar in potency and are significantly more toxic than the anticoagulant compounds developed earlier. Chemically, it is 2-diphenylacetyl-1, 3-indiandione and marketed under rade name Ramik and Ditrac. Consequently, they are formulated at lower concentrations. Chlorophacinone and diphacinone may kill some rodents in a single feeding but multiple feedings are needed to give adequate control of an entire population.

Diaphoretics: Chemicals which increase perspiration, *e.g.*, pilocarpin.

***Diaphorina citri* Kuwayama:** Belongs to order Hemiptera and family psyllidae. Commonly known as citrus psylla is a pest of all the species of citrus, other members of family Rutaceae and some ornamental plants.

Diaphragm: A horizontal membranous partition of the body cavity. The abdominal cavity in some insects is divided into two horizontal membranes, the dorsal and ventral diaphragms; consequently the three sinuses: the pericardial sinus containing the heart, the perineural sinus containing the nerve cord, and visceral sinus containing alimentary canal.

Diaspididae: A family of superfamily Coccoidea, suborder Homoptera and order Hemiptera.

Diastasis: A period of rest between successive heart beats. See mechanism of circulation also.

Diastole: When the alary muscles contract, heart muscles relax so ostial valve at anterior end and auricular valve at posterior end will open and blood enter in the heart. Thus this is relaxation phase of the heart beat cycle. See systole also.

Diazinon ($C_{12}H_{21}N_2O_3PS$): A very good general purpose pesticide; contact and stomach poison with less fumigant action, effective against many insects and arachnid pests, both indoors and outdoors. Diazinon is marketed under trade name Basudin and is slightly more toxic than either carbaryl or malathion. It is moderately toxic, broad spectrum and general usage insecticide that is effective against a wide range of insect pests, including household insects, field insects, fruit insects, vegetable and ornamental plant insects. Diazinon is also effective against houseflies and their larvae (maggots) and against ticks and other arachnids. However, diazinon emulsion should be used soon after it is made up because its effectiveness is lost when left standing for too long. It is both insecticide and acaricide; chemically, O, O-diethyl O-(2-isopropyl-4-methyl-6-pyrimidyl) phosphorothionate; moderately toxic to mammals, acute oral LD_{50} for rats is about 150 mg kg^{-1}; phytotoxic to Stephanotis and African violets and certain other varieties of apples. Insecticide slowly hydrolyzed in water and in presence of acidic material; safer to handle than parathion; available as 2-5 per cent dust, 5-10 per cent granule, 25 per cent emulsifiable concentrate and 10 per cent water dispersible powder; used in control of San Jose scale, wooly aphids, thrips, jassids and bugs.

Dichloropropylene: A soil fumigant and nematicide.

Dichlorvos ($C_4H_7O_4Cl_2P$): It is a synthetic insecticide and space fumigant; an organophosphate, 2, 2-dichlorovinyl dimethyl phosphate (DDVP); moderately toxic to other mammals, acute oral LD_{50} for rats is 50-80 mg kg^{-1}; phytotoxicity none to wide variety of plants at insecticidal concentrations. Though highly toxic to human, this insecticide emits a vapour killing power that is effective against a variety of flying and household insects. Highly toxic and is a good general use insecticide for use in enclosed areas. It is particularly useful in dog and cat flea/tick collars and in vapour release resin strips for the control of flying insects indoors. Dichlorvos can also be used effectively to control pantry, stored food, carpet and fabric pests, as well as bedbugs and cockroaches.

Dichlorvos, metabolism: It undergoes two types of break down, one by the action of esterases which converts DDVP to O,O-methyl phosphoric acid and dichloroacetaldehyde. As a result of oxidation of dichloroacetaldehyde, dichloroacetic acid is formed, whereas, as a result of reduction dichloroethanol is formed. The dichloroethanol is converted into glucuronide which is excreted. In the second type break down, demethylation takes place under the influence of S-alkyl transferase and dimethyl DDVP is obtained. Dimethyl DDVP is converted into O, methyl phosphoric acid which is further demethylated to inorganic phosphoric acid.

***Dichocrocis punctiferalis* Guenee:** Belongs to order Lepidoptera and family Pyralididae. It is an insect pest of castor, tamarind, turmeric, and ginger, commonly known as the castor capsule borer.

Dichoptic: The compound eyes widely separated by the front.

Dichotomous: Double branching as in a taxonomic key.

Dichotomous key: An identification tool to assist a person in identifying an insect (or other organism). It uses paired statements or questions to guide the user to the solution.

Dichthadiform ergatogyne: A member of an aberrant reproductive caste, limited to army ants which is characterized by the possession of a wingless alitrunk, a huge gaster and an expanded postpetiole.

***Dicladispa armigera* (Ol.):** Belongs to order Coleoptera and family Chrysomelidae. It is an insect pest of cereals and millets, commonly known as the rice hispa.

Dicofol ($C_{14}H_9Cl_5O$): A synthetic acaricide, chlorinated hydrocarbon and relative of DDT (DDT analogue); chemically, 1, 1-bis (p-chlorophenyl)-2,2,2-

trichloroethanol; moderately toxic to warm blooded animals, acute oral LD_{50} for rats is around 750 mg kg^{-1}; little evidence of phytotoxicity except on egg plant and avocado, and harmless to bees. It was introduced in 1955 under the trade name Kelthane.

Dicondylic: A joint with two points of articulation.

Dicot: A plant that has two seed leaves or cotyledons; the broad leaf plants.

Dicotyledon: Plant having two cotyledons in each seed, *e.g.*, beans. Dicotyledons include pulses, many flowering plants, the majority of annual weeds and many perennial weeds.

Dicrotophos ($C_8H_{16}O_5NP$): Dicrotophos is a highly toxic organophosphate insecticide having effective systemic and mild contact action that kills sap sucking insects that feed on plant juices. Normally, caterpillars and other chewing pests are less controlled by dicrotophos because they eat the plant tissues rather than sucking the sap or juices. Chemically, it is dimethyl-cis-2 dimethylcarbamoyl-1 methylvinyl phosphate; highly toxic to mammals; acute oral LD_{50} for rats is 22 to 75 mg kg^{-1}; controls certain pests of cotton and ornamental trees. It is brown to yellow liquid miscible with water, acetone and alcohol. It is safer to natural enemies but toxic to honey bees and marketed under trade name Bidrin and Carbicorn as soluble concentrate.

Dictyopharidae: A family of superfamily Fulgoroidea of suborder Homoptera and order Hemiptera. Medium sized insects with anteriorly much prolonged head. In India, *Dictyophara sauropsis*, *D. walkeri* are important.

Dictyoptera: Insect order containing the cockroaches and mantids.

Dieback: Necrotic symptom of disease in which death of shoot tissues begins at the tip and progresses backwards to the main stem.

Dieldrin ($C_{12}H_8Cl_6O$): A highly residual insecticide; a chlorinated hydrocarbon of not less than 85 per cent of 1, 2, 3, 4, 10, 10-hexachloro-6, 7-epoxy-1, 4, 4a, 5, 6, 7, 8, 8a-octahydro-1, 4-endoexo-5, 8-dimethanonaphthalene, somewhat more toxic to mammals than DDT, acute oral LD_{50} for rats is 100 mg kg^{-1}; phytotoxicity none when properly formulated but some crops are sensitive to solvents in certain formulation. It is an insecticide ten times more powerful than DDT, highly poisonous to birds and fish banned from use in sheep dips in 1966 and as a winter wheat dressing in 1975. Its use is now limited under the pesticides safety precautions scheme to specified circumstances where no suitable alternative exists. It is effective against soil insect pests.

Diesel fuel: Although, these common petroleum products generally are not considered pesticides, but can be used effectively to kill or repel a variety of pests in certain cases. For example, diesel fuel has been used for many years and continues to be used by health departments and mosquito abatement districts to control mosquitoes. Sprayed as a thin film on water in ditches, swamps, storm drains, containers, discarded tires etc. Diesel or motor oil causes suffocation of mosquito larvae and other water breeding insect larvae, thus killing them before they mature to flying adults. This control method is both economical and effective.

Thus either diesel fuel or crankcase motor oil can be used around the home to effectively control insects, especially mosquitoes and midges, that breed in water filled containers such as fish ponds, ditches, flower pots, discarded tires and other water holders. Another effective use of diesel, fuel oil, or motor oil is to control or repel the migration of ticks and other crawling anthropods into yards or other confined area. A 3-4 inch wide band or barrier of diesel, fuel or motor oil, applied continuously around the perimeter of the area to be protected, is in most cases, quite effective in repelling ticks and other crawling invaders. Likewise, such a barrier applied around the perimeter of house (or doghouse) may be effective in preventing the entrance of ticks and crawling pests into these structures. Still another little known use of diesel fuel, but not motor oil, is in controlling subterranean termites beneath buildings. It must be emphasized; however, that diesel does not normally kill the termites, at least not in great numbers, but rather repels them. In order to repel effectively, however, the soil beneath the building must be soaked and saturated with diesel fuel. While this method is usually effective and certainly less expensive than the chemical treatment, it cannot be sure simply because the termites are repelled rather than killed.

Difenacoum ($C_{31}H_{24}O_3$): Chemically, it is 3-(2-p-diphenyl-1, 2, 3, 4-tetrahydronapth-1-yl)-4-hydroxycoumarin, marketed under trade name Ratak is an effective rodenticide. 1:19 ratio of 0.1 per cent powder concentration and bait when mixed, give rise to 0.005 per cent difenacoum in the bait.

Differential diagnosis: A statement of important characters which are used for clear cut differentiation of a given taxon from other specifically mentioned equivalent taxa.

Differentiation: Increase in visible distinctive morphology.

Differentiation centre: In all insect eggs, there is a differentiation centre lying in the anterior third of the germ band region which acts as the leading point in all the succeeding stages of embryonic development.

Diffraction: The splitting light into its component spectral colours by grooves or ridges.

Diffraction mechanism: Burtt and Catton (1962) noted in *Locusta,* formation of a series of superposition images at progressively greater depths in retina of eye due to diffraction from the lens system besides a first apposition image. These are the diffraction images which are formed due to the diffraction of light received from the adjacent facets. The diffraction images are very complex and distorted in comparison to either apposition or superposition images. The functional significance remains unknown.

Diffusion tracheae: Cylindrical tracheae having non collapsible walls.

Diflubenzuron: It is an insect growth regulator which is derivative of urea. The Dutch product marketed under trade name Dimilin (25 WP) shows a great promise in the control of forest defoliators, mosquito larvae and other insects. Dimilin is highly suitable for aerial spraying against gypsy moth. It has very low toxicity to fishes (96 hours LC_{50} being 660 ppm for blue gills and 240 ppm for rainbow trout).

Digestion: The digestive system is mainly concerned with changing the food in an assimilable form with the help of a large number of enzymes secreted by epithelial cells and its associated glands. The digestive products are absorbed by special cells and transferred to haemolymph. In some special cases like termites and wood boring beetles, the process of digestion is supplemented by microorganisms of the gut. The pH of gut has great influence on the activity of digestive enzymes and on the absorption of food which varies from 6-8 in most of the insects. In Lepidoptera and Trichoptera, it is 8-10, in predaceous forms weakly acidic, 9.9 in cloth moth and 8.4 in wax moth. Both wool and wax are readily degraded under alkaline conditions. The food is retained for more time in the crop which acts as temporary reservoir in cockroach for mechanical processing by gizzard. There are three types of movement of the food: (i) Peristaltic (forward), (ii) Antiperistaltic (backward), (iii) Gross contraction.

Digestive enzymes: Different types of enzymes isolated from insect gut are (i) carbohydrases: which act upon disaccharides and polysaccharides, (ii) proteases: which act upon proteins and (iii) lipases: which act on fats and oils. The digestive enzymes of insects are, in general, such as might be predicted from the nature of their food. Omnivorous forms, like the cockroach, have a full complement of enzymes capable of digesting all the common food stuffs. Insects which take protein diet, such as the exclusively blood sucking species, have little but proteolytic enzymes. Where an insect feeds only on nectar, as in the adults of Lepidoptera, only an invertase is present; whereas the phytophagous larvae of same forms have a protease, lipase, amylase, maltase and invertase. The cockroaches, *Blaberus* and *Periplaneta* are describrd in current terms as producing alpha-glucosidase (acting on sucrose and maltose), beta-glucosidase (cellobiose), alpha-fructofuranase (raffinose and sucrose), alpha-galactosidase (melibiose), and beta-galactosidase (lactose). Many insects secrete chitinase into the mid gut; this enzyme seems to be identical with the lysozyme which dissolves the cell walls of microorganisms and responsible for the bacteriolytic activity of the gut contents. Most of the enzymes are secreted in the mid gut; the salivary glands usually secrete only amylase, and often no enzymes at all; but enzymes varied in the salivary glands of some plant sucking Hemiptera. Hind gut epithelium does not secrete digestive enzymes.

Digestive system: The digestive system is represented by a tube, the alimentary canal or gut of variable length or diameter which extends from a mouth to an anus. It is supplemented by a pair of labial or salivary glands in many insects, gastric caecae in some insects and the Malpighian tubules which are exclusively the excretory organs in almost in all insects. The alimentary canal is merely a simple tube or completely degenerated as the nutrients are absorbed directly through general body surface exceptionally in parasitic Hymenoptera.

Dihybrid cross: It is a cross between two pairs of contrasting characters (allele).

Dilated: Expanded or widened.

Dilated muscles: Any of the muscles that widen a body part.

Diluent: Any liquid or solid material serving to dilute or carry an active ingredient. Diluents may aid in mechanical application of a formulated pesticide but do not directly influence their toxicity, *e.g.*, water in sprays, oils in sprays, finely-ground inert materials in dust. Most commonly used diluent is water for spray application.

Dilute: To make less concentrated. To mix a pesticide chemical with water, oil and other materials before it can be safely and correctly used or containing large amount of solvent and water (See concentrated).

Dilution rate: The amount of diluent that must be added to a unit of a pesticide to obtain the desired dosage.

DIMBOA: 2, 4-dihydroxy-7-methoxyl-l, 4-benzoxazine-3-one, a compound in corn that confers resistance against first generation European corn borer.

Dimecron: Trade name of phosphamidon.

Dimefox: Organophosphorus compound, N, N, N, N-tetramethyl phosphorodiamic fluoride; a volatile contact insecticide and acaricide; applied on trunk against sap feeding insects like mealy bug of cocoa or on soil as capsules; not considered safe to apply on plant surfaces.

Dimethoate: A systemic and contact organophosphorus insecticide with acaricidal action, chemically, O, O-dimethyl S (N-methylcarbamoyl methyl) phosphorothiolothionate ($C_5H_{12}O_3PS_2N$); used for controlling aphids, spider mites and leaf miners. It is a snow white crystalline solid with melting point 51-52 °C. Technical material is more toxic due to the impurities. It is highly soluble in water (39 g l^{-1}) and most organic solvents but slightly soluble in paraffin hydrocarbons. Moderately toxic, dimethoate is used chiefly as a residual spray to control flies and fly larvae (maggots) and as a soil applied plant systemic to control a broad spectrum of plant feeding insects on fruits, field, vegetable and ornamental crops. Moderately toxic to mammals, acute oral LD_{50} for rats is 245 mg kg^{-1}; slight phytotoxic to some fruit and field crops. It is metabolized in insects and vertebrates similar to malathion. It is marketed under the trade name Rogor.

Dimethoate, metabolism: It is a carboxyamide, therefore cannot be hydrolysed by carboxyesterase as in malathion. Dimethoate is therefore, hydrolysed by separate group of enzymes known as amidases. Dimethoate is hydrolysed by carboxyamidase to dimethoate acid. Other hydrolytic products are phosphoric acid derivatives.

Dimethrin: Low toxicity, synthetic pyrethroid insecticide that is very similar to pyrethrin in the natural form.

Dimetilan: Highly toxic, like Vapona and DDVP, dimetilan kills insects with its toxic vapour that is emitted slowly from plastic or resin strip in which it is usually impregnated. It is especially useful against flies and other flying insects inside buildings.

Dimex: Trade name of dimethoate.

Dimite: A trade name of synthetic chlorinated hydrocarbon acaricide and relative of DDT; chemically 1, 1-bis (*p*-chlorophenyl) ethanol; moderately toxic to mammals and acute oral LD_{50} for rats is 500 mg kg^{-1}.

Dimorphic: Having two different forms of the sexes or two forms of one sex (distinct in colouration and structure), *e.g.*, in some butterflies.

Dimorphism: Occurrence of two morphs or phena (morphological types) in a single population. The difference occurs in size, form, or colour, between individuals of the same species, characterizing two distinct types.

Dinitrophenols: Those compounds are commonly known as dinitrophenols which are derivatives of 4,6 dinitro 2 alkylphenols and their salts and esters. Most of them are acaricides. The examples of these compounds are DNOC, DNOCHP, DINOCAP, binapacryl etc.

DINOCAP ($C_{18}H_{24}N_2O_6$): It is an acaricide (dinitrophenolic compound) and fungicide available as 25 WP, 76 EC and 100 EC and marketed under trade names Karathane and Crotothane. It is chemically, 2-(1-methyl-n-heptyl)-4,6-dinitrophenyl crotonate or 4,6-dinitro-2-caprylphenyl crotonate. Its acute oral LD_{50} for rats is 980-1190 mg kg^{-1}. It was introduced by Rhom and Hass Company in 1946 with trade name Karathane.

Dioecious: Having the male and female organs in different individuals.

Diopsidae (stalk-eyed flies): Family of suborder Cyclorrhapha, order Diptera. Diopsidae flies are extraordinary in having long stalked eyes borne on a snout-like head. The subcosta is either wanting or poorly developed and R_1 ends before the middle of the wing. They are saprophagous or phytophagous. *Diopsis thoracica* and *D.tenuipes* damage rice (*Oryza sativa*) in Cameroon, Rhodesia, Nigeria and Sierra Leone.

Dioptric apparatus: The parts of an ommatidium consisting of cornea and subcorneal crystalline body responsible for the collection of light.

Dioxycarb: An insecticide of carbamate group (phenyl carbamate).

Dip net: A net used to collect or sample the aquatic organisms.

Dip treatment: The application of a liquid pesticide to a plant by momentarily immersing it, wholly or partially, under the surface of the liquid to coat the plant with chemical.

Diphacinone ($C_{23}H_{16}O_3$): Chemically, it is 2-(diphenylacetyl)-1H-indene-1,3 (2H)-dione and marketed under trade name Ramik or Promar is an effective rodenticide. 1:19 ratio of 0.1 per cent powder concentration and bait when mixed give rise to 0.005 per cent diphacinone in the bait.

Diplatyidae: Family of order Dermaptera, suborder Forficulina and superfamily Pygidicranioidea. Paired aedeagus with four gonopores, *e.g.*, *Diplatys*.

Diploid: Cell containing a double set of chromosomes which are arranged in homologous pairs within the nucleus. Most cells in the body are diploid except gametes which have a single set of unpaired chromosomes.

Diplopoda: A class of arthropods which includes millipedes.

Diplura: An insect order (*Dipl*= two; *ura*= tail) to which the campodeids and japygids belong.

Diprionidae: Family of super family Tenthredinoidea, suborder Symphyta (Chalastogastra), order Hymenoptera. These are known as conifer sawflies and are medium sized with serrate antennae in females and pectinate or bipectinate antennae in males.

Dipsocoroidea: Superfamily of division Gymnocerata, suborder Heteroptera and order Hemiptera. These small delicate insects live among moss, dead leaves etc.

Diptera: Insect order (*di*= two, *ptera*= wings), *e.g.*, true flies. This insect order contains flies (gnats, mosquitoes, true flies) is characterized by a complete metamorphosis and by the possession of only the mesothoracic pair of wings. The hindwings are reduced to a pair of short, knobbed structures called the halteres. The adults and larvae have a variety of habits and their mouthparts are adapting for piercing and sucking or rasping and lapping and in some as in the robberflies, the botflies etc., they are vestigial. Some of them are predaceous, others blood sucking and yet others feed on nectar or plant sap and decaying animal and vegetable matter.

Dipteran piercing and sucking mouth parts (Dipteran PSM): These are three types: (i) Mosquito PSM**:** These are found in the female mosquitoes. There are six stylets; two mandibles, two maxillae and one each hypopharynx and labrum epipharynx. The labium forms elongated beak or proboscis. The maxillary palpi are conspicuous structures. The flattened lobe like structures form distal portion of labium and are called the labella. They represent the paraglossa. The fascicle is greatly flexible. Males lack the mandibular stylets and donot feed on the blood. The food channel is formed by the labrum epipharynx and the hypopharynx, but recently it is believed that the mandibles seem to be interposed between these two structures and form food channel. Hudson (1970), however, noticed that the mandibular stylets are positioned outside the maxillary stylets. The stylets (except maxillae) are not protractile and retractile but are embedded at the time of insect bite by a downward or forward thrust of the body. The labrum is shortest stylet forming the food canal. The maxillae are principal piercing devices. (ii) Horse fly PSM**:** They are found in the female horse fly only. Total number of stylets are six similar to that of mosquitoes. The stylets are flattened blade like. The labium is conspicuous. Its median portion is loosely ensheathing the blades and terminating into a pair of large lobes, the labellae. The membranous under surface of each labella is marked by transverse channels called the pseudotracheae. The mandibles and maxillae are lacerating the tissues. The mandibles and maxillae are long slender, tapering, serrated blade like structures. The mandibles cut the tissues like a scissor and the maxillae thrust and retract repeatedly. The blood vessels are ruptured by the teeth armed at the tip of mandibles. The movement is controlled by the antagonistic muscles of the head. (iii) Muscoid PSM**:** The stable fly, *Stomoxys* sp.; horse fly, *Haematobia* sp. and tsetse fly, *Glossina* sp. Bear this type of rigid and non-retractille mouth parts. The piercing organ is proboscis, consisting the haustellum. It is elongated, rigid and swollen at the base to accommodate the contained muscles and tapering towards the extremity. The labium is terminating in the labella. The labellar lobes are small, flat and horny bearing rasping denticles or prestomal teeth. The labrum is almost spherical and firmly locked within the upturned edges of the labium, within which lies the hypopharynx traversed by the salivary channel. The beak like proboscis is formed into the flesh of the victim by a strong thrust of the head and the body of a fly and blood is sucked up directly through the food canal. They mostly feed on haemorrhagic pool released from cutting and rasping the host tissue and blood vessels by prestomal teeth. The saliva is anticoagulant and mandibles are lacking.

Direct flight mechanism: Generation of wing movement by means of muscles at base of the wings.

Direct marketing (Zero level channel): The use of consumer direct channels (CD) to reach and deliver goods and services to customers without using marketing middleman.

Direct pest: An organism which causes immediate and direct damage to a marketable item, such as a fruit, even at low population density. See indirect pest.

Direct wing muscles: Muscles that attach to the wing bases.

***Dirhinus giffardi* Silv.:** A pupal parasitoid of melon fruit fly.

***Dirhinus luzonensis* Rohw.:** A pupal parasitoid of melon fruit fly.

Disaster: Disaster may be defined as an event that may occur suddenly or continuously and is responsible for the loss of life or property, destruction of infrastructure and it adversely affects the environment as well. Various types of disasters which have been recurrent phenomena are earthquakes, psunamis, cyclones, floods, droughts, landslides, nuclear, chemical, biological, fire and environmental hazards, air accidents, railway accidents, nuclear accidents etc.

Disc of pronotum: The area of the pronotum which is visible dorsally and usually delimited laterally by the two lateral carinae. Contrasted with the paired pronotal hypomera, which extend onto the ventral surface.

Discal carina of pronotum (pl., Carinae): One or more sublateral, paramedian or median, longitudinal carinae on the pronotal disc; contrast with the lateral carinae which form the edges of the disc.

Discal: The central portion of a wing from the costa to the inner margin.

Discal Cell: Name given to a prominent and often quite large cell near the middle of the wing. The discal cell of one insect group may not be bounded by the same veins as that of another group.

Discoidal cell/ area/ field: Especially in Odonta, MA(+) and Cu_b (-), unbranched, with the supplement Maspl (-) between them, enclose the discoidal field which ends proximally in the discoidal cell immediately distal to the posterior segment of the Arc. This cell is never divided longitudinally in Zygoptera in which it is quadrilateral. In Anisoptera, this cell is a triangle.

Discrimen of meso or metaventrite: Median longitudinal groove in the metaventrite and occasionally the mesoventrite, representing an area of invagination of the original sternum. It is also called the discriminal line. This term is usually applied to metaventral discrimen only.

Disease: The condition of the living plant or animal body or of one of its parts that impairs normal functioning. A disease is characterized by pathological responses called symptoms.

Disease incidence: The frequency of diseased or damaged individuals or their parts. Usually it is the portion of plant units diseased.

Disease intensity: The total amount of disease present.

Diseased: Unhealthy or abnormal. A plant or animal suffering from an infection by a fungus, bacteria, nematode, virus, other pest or poor growing conditions which disturb normal activity. An opposed to an insect injuring a plant or animal by chewing or sucking.

Disinfect: To kill or inactivate pests or pathogens present upon the surface of plants or plant parts, or in the immediate vicinity (*e.g.*, in soil).

Disinfectant: Chemical which kills or inhibits the growth of microorganisms.

Disk: A type of plough having the rolling disks.

Dislodgeable residue: Portion of a pesticide residue on treated vegetation that is readily removable and may be used as an index for risk to farm workers. Generally, measured by the residue removed from leaf discs by shaking briefly in water.

Dispersal: The movement and spread of individual organisms.

Dispersion: The spatial arrangement of individuals.

Disposal: Refers to discarding the unused spray material, emptied containers and expired pesticides.

Disruption of communication (**confusion technique**)**:** The use of pheromones and pheromone analogues to saturate the environment of a pest and block the use of some important signal (usually reproductive).

Disruptive colouration: A colour pattern that breaks up the outline of the body in such a way that it does not stand out against a variable background.

Dissemination: The transport of inoculum or pest from a diseased to a healthy plant.

Dissipate: To get rid of pesticides by spreading out.

Dissipation time 50 per cent (DT_{50}): Time required for a pesticide to dissipate one-half of the initial quantity or concentration from a system.

Dissipation: In nature, the disappearance of residue takes place in two ways. The first phase is the initial phase in which the disappearance of residue is fast, this phase is called dissipation. In the second phase, decline of pesticide residue is slow from the treated surface which is known as persistence. The dissipation follows the law of first order kinetics and the rate of disappearance is related to the amount of insecticidal deposit. Persistence does not follow the law of first order kinetics because of storage of translocated insecticides and degradation at various rates.

Dissolve: Refers to getting solids into suspension.

Distad: In a direction away from the body.

Distal: Concerning that part of an appendage furthest from the body.

Distal process: The peripheral branch or one of several distal branches of a sensory nerve cell.

Distiproboscis: See sponging type mouth parts.

Distribution: The amount and way in which the pesticide is spread on plant, animal or other surface. It also refers to the geographical area inhabited by a pest species or other species in the world.

Distribution coefficient: See partition coefficient.

Disulfoton or thiodemeton ($C_8H_{19}O_2PS_3$): Highly toxic, colourless oily liquid with characteristic odour, boiling point 62 °C, systemic organophosphatic soil insecticide and acaricide; chemically, it is O, O-diethyl S-2-(ethylthio) ethyl phosphorodithioate; effective against sap sucking pests of patato, vegetables, ornamental plants and flowers. It is also effective against sugarcane root maggot. Highly toxic to mammals, acute oral and dermal LD_{50} for rats are 2-9 mg and 50 mg kg^{-1}, respectively. High dosage is injurious to seed. It is formulated as 5 per cent granule and sold in the trade names of Disyston and Solvirex.

Disulfoton or thiodemeton: A systemic organophosphate insecticide with acaricidal property; chemically, S-2 ethylthioethyl diethyl phosphorothiolothionate ($C_8H_{19}O_2PS_3$); colourless liquid with unpleasant odour, slightly soluble in water but highly soluble in organic solvents. It is metabolized in plants to sulfonium compounds which are very toxic to mammals. It is effective against soil insects and is marketed under the trade names Disyston and Dithiosystox.

Ditrysia: Suborder of order Lepidoptera. Adults do not possess functional mandibles; maxilla without lacinia; larva possesses not more than five pairs of prolegs; pupa obtect; female with a copulatory aperture on sternite; VIII and an egg pore on sternite IX; male with saccus on sternite IX; wings not aculeate; venation of fore and hindwings are not alike.

Ditrysian: Having separate copulatory and oviposition apertures in the female, with the bursa copulatrix connected to the vagina by an internal duct, the ductus seminalis.

Diuretic hormone (DH): The insect neurohormone which enhances water excretion. It is widely distributed neurohormone controlling diuresis in several insects. It is synthesized by the neurosecretory cells of either brain or ventral thoracic ganglia and released into the blood *via* their neurohaemal organs. In *Rhodnius prolixus* and *Glossina,* it is secreted by thoracic ganglionic mass and found to be chemically mimic of 5-hydroxytryptamine (5HT). This hormone could be inactivated by chymotrypsin enzyme.

Diurnal: Active during the day time.

Divergent: Becoming more separated distally.

Diversion baiting: A little used practice is to make plantings especially for animals in an attempt to divert them from the main cash crop. This rarely achieves its objective, as it usually draws more animals into an area which will turn on to the main crop when the supplemental feeding area is depleted.

Diversity of insect fauna: According to survey (State of Art paper, 281, 1980 and Mayr, 1969) about 7.15 lac insect species are recorded throughout the world, while 75 thousand species are reported from India comprising 10 per cent of insect fauna. The Coleoptera is the largest order while Zoraptera the smaller one. In India, Strepsiptera is represented only with two species. About six orders, namely, Coleoptera, Lepidoptera, Hymenoptera, Diptera, Hemiptera and Orthoptera can be considered as the major orders on the basis of number of species they contain.

Diversity: Multiformity or miscellany; unlikeness; various kinds.

Diverticula: Blind sacs that form as evaginations from the alimentary canal, *e.g.*, oesophageal diverticula of mosquitoes.

Diverticulum (pl., diverticula): An off shoot from a vessel or from the alimentary canal usually blind or sac like; the caecal tubes or pouches; any extensions or evaginations of the hypodermis.

Dividing: Separating a colony to form two or more units.

Division board feeder: A wooden or plastic compartment which is hung in a hive like frame and contains sugar syrup to feed bees.

Division of labour: See polyethism.

Dixidae: Family of suborder Nematocera, order Diptera. Closely resembling the culicids in wing venation, but the wings are not scaled. Antenna is elongate and with 16 segments. Larva metapneustic, aquatic. *Dixia maculata* is an Indian species.

Dizziness: It refers to describe the various symptoms in patients, like faintness, giddiness, light headedness or unsteadiness.

D.N.A: An abbreviation for deoxyribonucleic acid a large molecule which stores the data in genes in the form of a character code. D.N.A. is a self replicating molecule. Deoxyribose nucleic acid is genetic material of living cells. Plant viruses are devoid of DNA and have only ribose nucleic acid (RNA) which behaves like hereditary material, DNA. Friedrich Meischer (1869) discovered DNA in the nucleus of pus cells and named nuclein to the substance. Presently, it is known as deoxyribose nucleic acid. Nucleic acids are two types: DNA and RNA.

DNOC ($C_7H_6N_2O_5$): Dinitro-ortho-cresol is a dinitrophenolic compound, a selective herbicide as well as insecticide and acaricide applied as foliar spray. It is also used against locust swarms in semidesert areas. It is a contact and stomach poison having ovicidal properties.

DNOCHP ($C_{12}H_{14}N_2O_5$): 4,6 dinitro-ortho-cyclohexyl phenol is a dinitrophenolic compound. Substitution of the CH_3 group by cyclohexyl (C_6H_{11}) group in DNOC gives DNOCHP which has the advantage of lesser phytotoxicity than DNOC and a powerful acaricidal action. It is marketed under the trade name Dinex and is used against red spider mite on fruit trees.

Dockage: Foreign matter in the harvested grain, such as weed seeds, chaff pieces, dust and pebbles.

Dodecyl methyl ether: A juvenile hormone mimic which has the carbon skeleton similar to farnesol. Krishnakumaran and Schneiderman confirmed the juvenile hormone activity of this mimic. They found that it has prothoracicotropic activity in brainless (decapitated) moth pupae.

Dolomite ($CaCO_3MgCO_3$): It occurs in substantial quantities in soils inherited from parent rocks and formed paedogenetically. It is a white powder used as inert carrier in the insecticide dust formualations.

Domestic quarantine: See quarantine laws.

Dominant characters: The expressed character in the offspring as a result of cross between two pure individuals of allele character.

Donaciidae (long-horned leaf beetles): Family of superfamily Chrysomeloidea, suborder Polyphaga and order Coleoptera. The beetles are elongate, slender, metallic black, greenish or coppery with long antennae. They are characterized by antennae widely separated at base but not separated by the entire front of head, head produced and with prominent eyes, prothorax laterally rounded and narrow prosternum and pygidium not exposed beyond elyra. *Donacia aeraria* in Japan and *D. lenzi* and *D. provosti* in Korea attack rice roots in their larval stage.

DOPA: Dihydroxy phenyl alanine ($C_9H_{11}NO_4$); formed in the liver from tyrosine and converted to dopamine in the brain.

Dopamine receptors: The slow acting insect biogenic amine neurotransmitter receptors which represents the majority of membrane receptors; the receptors traverse the plasma membrane and depend for functions on GTP binding protein and an adenylite cyclase enzyme.

Dormancy: A condition of quiescence or inactivity.

Dormant spray: A spray operation applied during the dormant season.

Dormant: Alive but not growing; buds with an unbroken cover of scales; quiescent; inactive; a resting stage.

Dorsad: Pertaining to the back or upper.

Dorsal (dorsad): An adjective (or adverb) denoting position near or movement toward the upper side of the body or a body part.

Dorsal blastoderm: The extraembronic part of blastoderm.

Dorsal blood vessel: There is only one blood vessel which runs along the middle of the back, the posterior segment of this pulsating vessel, the heart is provided with a series of valved openings or ostia through which blood can enter; the anterior segment, the aorta is a uniform contractile tube. After passing through the brain above the oesophagus, the aorta ends more or less abruptly and from this point the blood simply percolates slowly backwards through the tissues. The pulse rate of dorsal vessel varies due to many causes. It is slower in the less active stages, such as the pupa; in *Sphinx ligustri*, it is 40-50 per minute in the larva to 20 per minute in the pupa.

Dorsal diaphragm: The membranous and muscular sheets of tissues extending from dorsal blood vessel to the laterodorsal parts of body wall, separating the dorsal sinus from the perivisceral sinus.

Dorsal lobe of penis: The upper lobe of penis which is divided horizontally into dorsal and ventral lobes (as in Dascillidae, some Scirtidae and some Psephenidae). See Tectum.

Dorsal longitudinal trunk: See longitudinal trunk.

Dorsal nectary organ: In the larvae of many species of Lycaenidae (blue butterflies), a gland located in the dorsal region of the 7th abdominal segment, it secretes a sweet substance which is attractive to ants.

Dorsal ocellus: The simple eye in adult insects, nymphs and naiads. They are found in the nymphs of hemimetabolous insects and in almost all adult insects. The insects bear three ocelli- one median and two laterals, forming a sort of inverted triangle anterodorsally on the head (frons). They are shifted to the vertex in Diptera and Hymenoptera. The median ocellus is formed due to fusion of paired ocelli or in some cases the dorsal ocelli are completely degenerated.

Dorsal organ: In an insect embryo, the amnioserosal sac arising from the amnion or a mass of cells on the dorsal side produced by the invaginated serosa.

Dorsal shield scutum: Sclerotized plate covering most of the dorsal surface of males and the anterior part of females, nymphs and larvae of hard backed ticks.

Dorsal shield: The scutum or sclerotized plate covering all or most of the dorsal surface in males and the anterior portion in females, nymphs, and larvae of hard-backed ticks.

Dorsal sinus: The space of definitive body cavity above the dorsal diaphragm in which heart lies, also known as pericardial sinus.

Dorsal trachea: The dorsal segmental trachea originating at a spiracle.

Dorsal tracheal trunk: A longitudinal dorsal trunk utilizing the series of dorsal tracheae.

Dorsal vessel: It is a mesodermal tube occurring throughout the postembryonic development and functions as a principal pulsatory organ. It retains the mid dorsal position above the dorsal diaphragm except in cephalic region where it runs along the alimentary canal. It is typically differentiated into two parts (ii) the anterior part or the aorta, and (ii) the posterior much dilated part or the heart in the insects except in Thysanura.

Dorsocentral bristles: The two rows of bristles running along the thorax of a fly on the outer side of the acrostichal bristles.

Dorsolateral: Towards the sides of the dorsal (upper) surface.

Dorsomesal: At the top and in the mid line.

Dorsopleural line: The line of separation between the dorsum and the pleural region, most often marked by a fold or groove.

Dorsoscutellars: A pair of bristles on the dorsal part of the scutellum, one on each side of the median line as in Diptera.

Dorsoventral: Running from the dorsal (upper) to the ventral (lower) surface.

Dorsum: The dorsal or back region of the insect above the pleural region.

Dorylophile: Belonging to the tribe Dorylini, these are obligatory guests of one of the army ants.

Dosage: When the quantity of pesticide applied is unknown. Often spraying the chemical, we do not know how much amount has been taken up through we have created concentration in the environment.

Dosage mortality curve: The curve resulting from plotting percentage mortality of test insects over a period of time against dosage of insecticide. To draw the dosage-mortality curve usually at least four or five doses are used at logarithmic intervals, *e.g.*, dosage 1, 2, 4, 8, 16).

Dose: The quantity of pesticide applied per individual (plant or animal), or per unit area, or per unit volume, or per unit weight. The quantity of insecticide is known in terms of animal system, *e.g.*, microgram/g or g/ kg body weight.

Dose effect relationship: Graded relationship between the dose of a pesticide to which the organism is exposed and the magnitude of a defined biological effect either in an individual organism or in a population.

Dose ratio: Ratio between successively increasing doses.

Dose response relationship: Correlation between dose of a pesticide and the defined biological effect on an exposed population.

Double citation: Citation of the name of both the original author of a species group name and the person who transferred the taxon to which it applies to another genus, the latter author's name is placed after the parentheses that enclose the name of the original author, in new combinations.

Double screen: A wooden frame, 1/2 to 3/4 inch thick with two layers of wire screen to separate two colonies within the same hive, one above the other. An entrance is cut on the upper side and placed to the rear of the hive for the upper colony.

Double tegmen: Tegmen which has three anterior struts, two on one side of the penis and one on the other (usually two dorsally and one ventrally).

Doubling time: The time needed for a population to double in size (number of individuals).

Downwind: Refers to the direction of wind towards which it is blowing.

Dragonfly: Any of various insects of the suborder Anisoptera (order Odonata). Dragonflies are large insects with a long slender body and two pairs of narrow, net-veined wings. Usually the wings are held outstretched while resting. They feed as predators on small flying insects. Sometimes, they are referred to as 'mosquito hawk' or 'snake doctor'.

Drained honey: The honey obtained from uncapped and broodless honey combs.

Drawn combs: Combs with cells built out by honey bees from a sheet of foundation.

Drench treatment: Application of a liquid chemical to an area until complete saturation, *e.g.*, saturation of soil with an insecticide or oral treatment of an animal.

Drepanidae [Drepanulidae] (hook tips): Family of super family Geometroidea, suborder Ditrysia and order Lepidoptera. These are found in the Indo-Malaysian region of the Orient and include the genera *Drepana, Cilix* etc.

Drift: (i) Method of applying pesticide aerosol sprays for the control of flying insects. (ii) Movement by the wind of pesticide droplets or dust beyond the intended or target area of application.

Drift control agent: Formulant that control the distribution of spray droplet sizes and prevents production of excessive fines.

Drift retardant: The drift retardant if mixed to a spray mix, may reduce spray drift by reducing the number of small drift susceptible droplets (adjuvant oils). The chemicals act by increasing the viscosity of the spray mix.

Drifting of bees: This refers to failure of bees to return their hive in an apiary which has many colonies. Young bees tend to drift more than older bees and bees from small colonies tend to drift into larger colonies.

Driver ants: African legionary ants, *Anomma* sp. belonging to tribe Dorylini.

Drone: The male honey bee unable to feed himself and is dependent on the workers for food. The sole function of this caste is to fertilize the virgin female, when she leaves the colony on her nuptial flight. The drones follow in swarm and the individual which mate with the queen dies immediately after copulation. Soon after the swarming season is over, the surplus drones are driven out of the hive by the workers.

Drone comb: Comb measuring about four cells per linear inch that is used for drone rearing and honey storage.

Drone layer: An infertile or unmated laying queen.

Drop spectrum: Distribution by number or volume of drops of spray into different droplet sizes.

Droplet density: Number of droplets required per unit area (cm^2) for effective coverage of target area. This should be 10-20 droplets in high volume spray (normal spray), 20-30 droplets in herbicide spray, 30-50 droplets in low volume spray and 50-70 droplets in ultra low volume spray.

***Drosicha mangiferae* Green:** Belongs to order Hemiptera and family Coccidae. It is commonly known as mango mealy bug, an important sucking insect pest of mango and 62 other host plants.

Drosophilidae (pomace flies or fruit flies): Family of suborder Cyclorrhapha, order Diptera. Dorsophilidae are small yellowish flies with red eyes. The larvae of most species feed on decaying fruit and fungi and a few mine into leaves or fruits. Some are predaceous on aleyrodids and coccids. The larvae of the fly *Acletoxenus indicus* are predaceous on the *aleyrodids Trialeurodes ricini, Siphoninus phillyreae, Aleurocanthus spiniferus* etc.

Drumming: Pounding on the sides of a hive to make the bees ascend onto another hive placed over it.

Drupelet: A small individual fleshy fruit as in blackberries and raspberries.

Dry weight basis: Pesticide residue reported in dry matter of the test sample. Residues in soils and feeds and maximum residue limits (MRLs) for feedstuffs are expressed on dry weight basis.

Dryinidae: Family of super family Bethyoidea, suborder Apocrita and order Hymenoptera. Dryinidae are parasites of cicadellids, membracids and fulgorids. They possess ten-segmented antennae and in the female the foretarsi are pincers like. In a few cases polyembryony occurs. In several species the females are wingless.

Dryopoidea: Superfamily of suborder Polyphaga, order Coleoptera. The Dryopoidea are mostly subaquatic and possess serrate antennae and a prosternal spine which fits in a mesosternal cavity. Some exhibit plastron respiration and show special adaptations.

Duct: A tube carrying the secretions from a gland to the point of discharge.

Ductless glands: These are endocrine glands not joined by tubular organs that secrete hormones in the body which are transmitted throughout the body via the blood stream.

Ductus ejaculatorius: The median outlet tube of male reproductive system, ectodermal in origin.

Dulosis: The act of slave making in ants; a species which makes a slave of another is often referred to as Dulotic. The social parasite, *Formica sanguinea* has remarkable habit of slave making or dulosis. The host ant is *F. fusca*. The queen of *F. sanguinea* steals herself offsprings and pupae (workers) from the *F. fusca* in a corner of the nest. Here, *F. sanguinea* produces her brood and the *F. fusca* workers attend the parasite's offspring. The *F. fusca* colony is gradually out numbered by the parasite workers as they are continuously produced. When the number of *F. fusca* workers is insufficient or low, then *F. saguinea* workers press-gang others into their service by compelling in raiding another nest of *F. fusca* and seizing larvae and pupae, which they carry back to their own nest and rear. The resulting *F. fusca* offsprings work for the host. Repeated slave making raids are undertaken during the existence of a *F. sanguinea* colony.

Dun (Sub-imago): Winged but sexually immature mayfly which has to undergo moulting to be sexually mature.

Dung: Animal excrement.

Dunnage: Wooden planks or loose wood of any kind or wheat straw used to insulate the stored grain against the seepage of moisture from the floor as well as the out side.

Durable resistance: Long lasting resistance of a plant against a pest that is rarely overcome by the pest.

Durofume: It is a mixture of EDB and methyl bromide in the ratio of 1:1 w/w, first prepared by CFTRI, Mysore. The mixture is effective against all stages of insects including resistant strains of *Tribolium* and *Trogoderma*. The dosage of Durofume is 1-2 lbs per 100 ft.^{3} with exposure period of 48 hrs. If the temperature is above 26°C, the exposure period may be reduced to 36 hrs. If the temperature is below 18°C, the mixture may be altered to 1:3. For oilseeds, the mixture in the ratio of 1:2 is used with dosage of 1.5-2 lbs per 100 m^3 and exposure period of 36-48 hrs. and for milled cereals, the mixture in the ratio of 1:3 is used @ 2-3 lbs per 1000 ft.^{3} with exposure period of 48-72 hrs. Residues of this mixture reported on different commodities were far below than that of EDB or MB used alone.

Dursban: Trade name of chlorpyriphos.

Dusky cotton bug: *Oxycarenus laetus* Kirby is an insect pest of cotton; the nymphs and adults feed on the open bolls and spoils the lint.

Dusky cotton bug: See *Oxycarenus laetus.*

Dust formulation: Formulation prepared by mixing of insecticides to the finely ground dry particles of earth, talcum or others. Such formulation, generally, contains low amount of active ingredient, generally 2-10 per cent and applied without dilution.

Dust: An insecticide (or pesticide) which is formulated to be used as a dry powder. Same as dust formulation.

Dustable powder: Free flowing powder pesticide formulation suitable for dusting.

Duster: Equipment for applying pesticides that are formulated as a dust.

Duster, bellow type: A duster having small dust container or chamber and bellow for producing air stream.

Duster, hand shake: A small hand operated duster with perforations sufficient to allow the dust out of the container when jerked by hand or lever. It is suitable for basal application of dusts on the plants of low height.

Duster, mist blower: A type of mist blower converted to duster with the help of Y shaped nozzle for directing the dust to both sides of the operator. A wide spreader boom type dusting attachment is used for dusting on low covers of long area. It can also be used for granular application.

Duster, plunger: It consists of an air pump, a dust chamber and a discharge assembly. The air blast by hand operated air pump creates dust cloud which passes through the delivery vent onto the target.

Duster, power: A commercial duster with boom of spreaders mounted on a trolley or tractor and operated by power engine. The dust is blown out of the spreader with the air pressure created either from an air compressor or a high speed fan blower.

Duster, push: A perforated drum on two wheels that can be pushed between the rows of crops so that the dust inside can falls on the cover when the drum rotates while being pulled on the wheels.

Duster, rotary: A dusting implement having a hopper (for holding dust) connected to a crank and a fan enclosed in a metal or plastic chamber. The high speed fan blows the air over the hopper and carries the dust cloud through the spreader nozzle onto target. The spray liquid joins the dust when it is passing out of the blower. Mist blowers with dusting attachment can also be used for wet dusting.

Duster, wet: It is a combination of a duster and a sprayer for use in the semi arid and arid ecosystems where water paucity is a problem.

Duster-cum-sprayer: A machine in which the features of both sprayer and duster are combined is called duster-cum-sprayer or spray duster. In this machine, a hopper for the dust and a pump with suction and delivery hoses are provided.

Dusting: The application of insecticides, fungicides etc. to crops in a dry, powdered state; less effective and requires about double amount of chemical per unit area than spraying.

Dwindling: The rapid dying off of old bees in the spring, sometimes called spring dwindling or disappearing disease.

Dyar's law: It was observed long ago by Dyar, working with the larvae of Lepidoptera that the linear dimensions of the hard parts increased at each moult by a constant ratio, usually of the order of 1.4. More recently, it has been shown that various hemimetabolous insects (*Sphodromantis*, *Carausius*) approximately double in weight from one moult to the next, while their linear dimensions increase 1.26 (*i.e.*, 3"2) times. This growth factor of 2 was first demonstrated by Przibram and Megusar; it is perhaps best regarded simply as a special case of Dyar's law; but it is sometimes referred to as Przibram's factor. The physiological basis proposed by Przibram to explain his factor was at each moult (or at each latent division) every cell in the insect body divides into two. But what histological evidence exists is definitely opposed to this improbable hypothesis.

Dyfonate: A synthetic organophosphate insecticide and acaricide, *O*-ethyl *S*-phenyl ethyl phosphorodithioate; highly toxic to mammals, acute oral LD_{50} for rat is 8-16 mg kg^{-1}; incorporated in the soil for the control of corn rootworms, wireworms, sugarbeet root maggots and other soil pests.

Dylox: Trade name of Trichlorfon.

Dynamics: In population ecology, the study of the reasons for changes in population size. Pest dynamics is the study of changes in pest population size.

Dynastidae (rhinoceros beetles, unicorn beetles or elephant beetles): Family of superfamily scarabaeoidea, suborder Polyphaga and order Coleoptera. These beetles are very large to medium sized insecta, smooth, shiny, brilliantly-colored and sometimes sculptured. Horns and tubercles are present on head and pronotum and they exhibit great sexual dimorphism. The characteristic feature is the mandibles which are expanded and visible from above. In India the beetles *Oryctes rhinoceros, Xylotrupes gideon* and *Phyllognathus dionysius* are known as pests of palms (including coconut), and paddy respectively. *Heteronchyus porpoygus* has been obtained from roots of paddy in Karnataka State.

Dyscrasia: A diseased state resulting from the presence of toxic material in the blood.

***Dysdercus koenigii* (Fab.):** Belongs to order Hemiptera and family Pyrrhocoridae. It is an insect pest of cotton and ornamental plants, commonly known as the red cotton bug.

Dysentery: Abnormal condition of adult bees characterized by severe diarrhoea and usually caused by starvation, low quality food, moist surroundings or nosema infection.

Dytiscidae (true water beetles or diving beetles): A family of superfamily Caraboidea, suborder Adephaga, order Coleoptera. Aquatic beetles mostly found inhabiting running and standing water, a few live in thermal spring and other inhabit brackish water.

E

***Earias insulana* (Boisduval):** Belongs to order Lepidoptera and family Noctuidae (Agrotidae). It is an insect pest of cotton, commonly known as the spotted bollworm.

***Earias vitella* (Fab.):** Belongs to order Lepidoptera and family Noctuidae (Agrotidae). It is an insect pest of cotton, commonly known as the spotted bollworm.

Early embryonic development: In broad sense, it refers to development in the egg. The early embryonic development of insects, like in other animals, starts with the cleavage followed by blastoderm formation, vitellophages, germ band formation, germinal layer formation, formation of amniotic cavity and extra embryonic membranes, blastokinesis, segmentation of appendages, organogenesis and eclosion.

Earwig: Any of various insects of the order Dermaptera. Earwigs have an elongate body with a pair of pincer like appendages protruding from the rear of abdomen.

E-business: The use of electronic means and platform to conduct a company's business.

EC_{50}: It is the concentration of a chemical resulting sterility in 50 per cent of test animal.

ECD: Electron capture detector, used in gas chromatography.

Ecdysial glands: These were first described in 1772 by Lyonet but their endocrine nature was investigated quite late by Fukuda (1940). These play important role in moulting and metamorphosis. These are transparent paired cellular strings or loose masses of variable shape and size and located in the ventrolateral areas of prothorax of Hemiptera and Holometabola and are commonly called the prothoracic or thoracic glands. In rest of the Hemimetabola and Apterygota, these are situated in the ventrocaudal region of head and termed as the ventral glands. The hormone secreted by these glands is called ecdysone or moulting hormone.

Ecdysial line: When the insect has digested its old cuticle and laid down new one, it has fresh problems to solve. In the first place, it had to rupture the old skin and escape from it. Now any cuticle that has to be shed, that is, the cuticle in every stage except the adult, has an ecdysial line or line of weakness- some times in the head, most often in the middle of thorax. Along this line, the exocuticle is wanting and the endocuticle extends upto the very delicate epicuticle. Consequently, when endocuticle is completely dissolved (and the process of moulting is so timed that this happens just when the new cuticle is ready), the old skin needs the slightest force for it to split along the line of weakness.

Ecdysis: The terms moulting and ecdysis in insects are referred originally to the casting or shedding of the old skin of larval arthropods from one stage of development to another. The final moult leads to the formation of puparium or chrysalis. But both terms come to be used in two senses (i) for the entire process of growth which leads up to the next stage, or stadium; (ii) for the final act of shedding the cuticle. Some authors limit the use of term moulting to the former sense and ecdysis to the latter, but there is no uniformity in this. Two useful terms have been introduced by Hinton: apolysis for the separation of epidermis from the old cuticle and the pharate state for the period between apolysis and ecdysis when the new cuticle is in process of formation below the old.

Ecdysoid: Any synthetic or natural chemical of plant or animal origin exhibiting biological activity similar to the true molting hormone, ecdysone.

Ecdysone: It is secreted by prothoracic gland and responsible for moulting in the insects. Ecdysone has been isolated and crystallized in pure form from larvae of *Calliphora* and pupae of silkworm; it is steroïd closely related to cholesterol.

Ecdysone as insecticides: When beta-ecdysone is injected or topically applied, the insects die soon due to toxic effect or may undergo abnormal moulting due to hyperecdysonism as found in various insects, like silkworm, *Tenebrio* etc. It inhibits DNA synthesis in epidermal cells, internal tissues and organs except the blood cells and connective tissues. The exogenous ecdysones, therefore, can be used as the insecticides against all stages from larva to adult. These can be applied topically in various solvent and penetrate the cuticle and can also be used as fumigant. They are quite stable and act as antiecdysone substances and block moulting. Many steroids have been isolated from different plants showing alpha-ecdysone and beta-ecdysone activity and called phytoecdysones.

Ecdysterone: A more active form than ecdysone has one additional hydroxyl group.

***Echidnophaga gallinacea* Westwood:** Belongs to order Siphonaptera and family Pulicidae. It is an insect pest of farm animals, commonly known as the poultry stickfast flea.

***Echinocnemus oryzae* Mshll.:** Belongs to order Coleoptera and family Curculionidae. It is an insect pest of cereals and millets, commonly known as the paddy root weevil.

Echinophthiriidae: A family of order Anoplura. Body stout; invested with thick setae, occasionally modified with scales and with a pair of spiracles on metathorax. These insects are found on marine mammls.

Ecitophile: An obligatory guest of one of the army ants belonging to the tribe Ecitonini, especially of the genus *Eciton*.

Eclosion: The process of hatching from the egg or Emergence of the adult or imago from the pupa.

Eclosion hormone: The eclosion hormone is secreted by the cerebral medial or posterior neurosecretory cells and transported *via* corpora cardiaca to the haemolymph. The eclosion hormone is found in the silk moth, *Hyalophora cecropia*, initiates eclosion behaviour by triggering the release of a specific sequence of behaviour patterns pre-programmed in the central nervous system which induce the function of extricating the moth from its pupal case. It also causes plasticization of the cuticle of wings at the time of eclosion, initiates programmed cell death of the abdominal intersegmental muscles; switch off the pupal behavioural patterns and to switch on the adult behaviour.

Ecoclimate: Climate within the plant (crop) community.

Ecological efficiency: Per cent of energy in biomass produced by a trophic level that is incorporated into biomass by the next higher trophic level.

Ecological homologues: (i) Two allopatric species of organisms with identical or very similar ecological niche requirements, especially with regards to some environmental provision present in only a limited supply. (ii) Species that have the same fundamental ecological requirements and, therefore, cannot occupy the same area indefinitely without one species displacing the other.

Ecological niche: The functional habitat of a species that provides with the resources such as food, shelter, mates etc.

Ecological pyramid: A rational study of various trophic levels in ecosystem is termed as ecological pyramid. This study is based on various factors, *viz.*, number of individuals, biomass or energy in a particular trophic level. Generally, producers form the base of pyramid due to large constituent. After that it decreases in size as go to the successive trophic level and become tapered at the top. Ecological pyramids are of three types: (i) pyramid of number (ii) pyramid of biomass, and (iii) pyramid of energy.

Ecological pyramid of biomass: Total amount of living material in any organism is known as biomass. It is measured in terms of dry weight of organisms per unit area. In terrestrial ecosystem, there is gradual decrease in biomass at different trophic levels. But in aquatic ecosystem, inverted biomass pyramids are also seen.

Ecological pyramid of energy: This pyramid is found always in upright pattern because approximately 10 per cent energy is lost at each successive trophic level. Producers have higher energy than herbivores and carnivores. Energy is represented in terms of calories.

Ecological pyramid of number: This pyramid is based on the number of individuals per unit area at each trophic level. In this pyramid, generally number of individuals decreases successively and it show relationship between producers, herbivores and carnivores and it is known as upright pyramid. But, number of pyramid may be inverted and spindle shaped. Inverted pyramid shows increased number of individuals at successive levels and spindle shaped pyramid shows both sides tapered.

Ecology: Gr., *oikos*= house or place to live in. The science dealing with living organisms and their relation to the environment. In IPM, the study that deals with the effect of environmental factors, such as soil, climate, natural enemies etc. on the occurrence, severity and distribution of plant pests (See also agroecology).

Eco$_2$fume: A fumigant developed by M/s Cytee Industries, USA. It is a non-inflammable, premixed cyliderized mixture of phosphine and CO_2 (in the ratio of 1:49 w/w) which provide highly effective fumigation in sealed and unsealed storage facilities. It is used to fumigate grains, dried fruits, cocoa, coffee, beans, nuts, tobacco, cut flowers and other items in structures and buildings.

E-commerce: E-commerce or electronic commerce means trading through internet. Purchasing and selling things through internet service is known as e-commerce. The e-commerce comprises the branches, like e-banking, e-business, e-shopping, e-trading and e-cheque.

Economic control: Control of pests in such a way that expenditure on control measures is less than the increased value of yield.

Economic damage: The amount of crop injury which will justify the cost of artificial control measures.

Economic entomology: It refers to exploring the biology, bionomics, population dynamics and management of various insect species which are directly or indirectly concerned with the economy and welfare of human society. In this context some insect pests can be referred as harmful or destructive and other ones as the harmless or beneficial insects.

Economic injury level (EIL): The lowest population density that will cause economic damage. This is density of a pest population at which the cost of pest control equals the amount of crop loss caused by the pest.

Economic pest: A pest causing a crop loss of about 5-10 per cent according to definition.

Economic threshold level (ETL): ETL is the pre-alarming situation obtained where damage to the crop or livestock can no longer be tolerated. The density of a pest population at which control measures are justified to prevent an increasing pest population from reaching the economic injury level. The pest population level at which control measures should be started to prevent it from reaching the economic injury level. The economic threshold may be expressed in pest numbers or amount of crop damage. (**Note:** It is not easy to define economic threshold, as growth of pest populations and resulting damage does not only depend on the size of pest population but also on other factors, such as crop stage, crop condition, presence of natural enemies, weather etc. All these factors should be considered together when deciding if control measures are needed).

Ecosystem: Assembly of populations of different species (often interdependent on and interacting with each other) interacting with their surroundings within a specified physical location and a functional entity. The term ecosystem was first introduced by Sir Arthur Tansely (1935) with the meaning of 'interrelationship of living organism with the environment for satisfying their basic needs.' The ecosystem may be categorized as (i) natural (terrestrial and aquatic) and (ii) artificial or man made ecosystem (*e.g.*, agroecosystem).

Ecosystem biodiversity: When variation takes place in the structure and function of ecosystem, it is known as ecosystem biodiversity. It explains niches, trophic levels, food chain, food web, energy flow etc.

Ecosystem, components: The components of ecosystem are classified into biotic components (plants and animals) and abiotic components (physical environment, *viz.*, climatic factors like temperature, rainfall, light, wind, humidity etc. and edaphic factors like soil pH, topography, minerals etc.).

Ecotoxicologically relevant concentration (ERC): Concentration of a pesticide (active ingredient, formulation and relevant metabolites), *i.e.*, likely to affect determinable ecological characteristics of an exposed system. It is related to the toxicity characteristics, generally the no observable effect concentration to the most sensitive species or groups of species.

Ecotoxicology: The study of toxic effects of chemical substances in living organisms, especially on populations and communities within defined ecosystems. Eco-toxicology includes transfer pathways of these chemicals and their interaction with the environment.

Ecotype: A specific strain of a species adapted to a particular set of environment.

Ectad: Outward relative to midline of the body.

Ecto: Outside; external.

Ectoderm: The outermost of the three primitive germ layers that give rise to the body wall, foregut, hind gut, nervous system, respiratory system and many glands.

Ectohormone: A substance secreted by an animal to the outside of its body causing a specific reaction, such as determination of physiological development in a receiving individual of the same species.

Ectognathous (Ectognathy): Having mouthparts that are exposed and not somewhat pulled or deep into the head.

Ectohormones: See pheromones.

Ectoparasite: Parasite that lives on the external surface of its host, *e.g.*, insects which live on the skin of cattle such as ticks, lice and fleas.

Ectoparasitoid: A parasitoid that develops outside its host. It obtains nutrition by penetration of the host's body wall.

Ectosymbiont: A symbiont that associates with the host colonies during at least part of its life cycle in some relationship other than internal parasitism.

Ectotherm: Organism that relies on heat from the external environment for their survival.

Ectothermic: Deriving heat from without the body, cold blooded.

ED_{50}: In such cases where death rate is not required but the effects are measured on the efficiency of vigour and reproduction of test animals, then the expression ED_{50} is used. For example, chemosterilants reduce the vigour and make the test animal sterile. The dose of such a chemical which brings sterility in 50 per cent population in test animals will be the value of ED_{50}.

EDB: See ethylene dibromide.

EDB^+: It is a mixture of ethylene dibromide and carbon tetrachloride in the ratio of 1:8 w/w. It has a greater penetration power and hence can be used for bagged storage under gas proof covers. It is available in ampoules of size 11 ml, 22 ml and 30 ml suitable for small quantity of food grains stored in the rural storage structures. The dosage is 11 ml q^{-1} grain with exposure period of 7 days.

EDC: See ethylene dichloride.

EDCT: Ethylene dichloride and carbon tetrachloride mixture in 3:1 ratio, fumigant; available as liquid; volatile, vapourizes at normal temperature.

Edge: The exact edge of a structure. Contrasted with the word margin which implies a beaded edge or bordered edge.

Eelworms: Nematodes.

Effective environment: The elements in an ecosystem that have a direct reproduction, survival and movement of a species population.

Effector organs: Two types of well developed effector organs in large number of insects are sound producing organs and light producing organs. (See sound producing organs and light producing organs).

Efferent neurones: See motor neurones.

Efficacy: Capacity for serving to produce effects.

Efficiency of a pest control measure: The more or less fixed reduction of a pest population regardless of the number of pests involved.

Effluent: Liquid flowing out, the term is generally used for unwanted liquid wastes from factories, industries and city drains.

Egg: The insect eggs are internally packed with yolk (centrolecithal) and measure from 0.06 mm (ants) to 8.0 mm (Acrididae) in diameter and occur in variety of forms with wide modifications, such as pedicel (parasitic Hymenoptera), terminal bladders (*Encurtus*), horns (*Drosophila*), Operculum (*Capsus*), long caudal filaments (*Ictinogomphus*) and elongate.

Egg batch: A group of eggs that have been laid singly or in a group by the same female at about the same time.

Egg chamber: One of the compartments or follies of an ovarial egg tube formed of the follicle cells containing an oocyte (Eikammer).

Egg guide: A median process of the subgenital plate behind the genital opening particularly developed in Acrididae.

Egg membrane: Besides the chorion laid down by the follicular cells, the egg is enclosed by inner membranes which are deposited by the oocyte itself. These may amount to more than a thin layer of wax which serves to water proof the egg, and a more or less delicate vitelline membrane beneath; or the blastoderm of the developing embryo may lay down a tough chitinous layer, similar to the endocuticle of the free living insects, which often forms the most substantial covering of egg.

Egg parasite: An insect that lives in the eggs of its host (other insect or animals).

Egg pod: A capsule which encloses the egg mass of grasshoppers and which is formed through the cementing of soil particles together by secretions of the ovipositing female.

Egg sheet: A piece of paper, cardboard, polythene, or other artificial substrate on which eggs are deposited.

Egg shell: The hard outer most layer of egg of some organisms.

EIA: Environmental impact assessment (EIA) can be defined as the study of probable changes in socio-economic and biophysical characteristics of the environment that may arise from the proposed plan.

Ejaculatory duct: A duct formed by joining of the terminal portions of vasa deferentia of each testis and involved in the propulsion of semen.

Elasmidae: Family of super family Chalcidoidea, suborder Apocrita and order Hymenoptera. Elasmidae are small brown or black elongate insects with enlarged, compressed hindlegs. They are primary parasites of lepidopterous insects or hyperparasites of braconids and ichneumonids parasitic on lepidopterous larvae.

***Elasmus johnstoni* Ferriere:** Larval parasitoid (Elasmidae, Hymenoptera) of pink bollworm of cotton.

Elateridae (click beetles, wireworms): Family of superfamily Elateroidea, suborder Polyphaga and order Coleoptera. Elateridae are minute or medium-sized, grey, black or brown, elongate, slightly flattened beetles with 11-segmented, usually pectinate, antennae inserted near eyes. The prothorax is loosely attached to mesothorax and its sternum is lobed in front. Its process is movable and is received in the mesosternal cavity and returned suddenly; if the beetle happens to be on its back it gets hurled into the air. The tarsi are five-segmented and end in pectinate, denate or simple claws. The abdominal segments are free. A few beetles belonging to *Pyrophorus* which are mainly neotropical are phosphorescent and emit bright light in their egg, larval and adult stages.

Elateriform larva: A larva resembling a wireworm with a slender body, heavily sclerotized with short thoracic legs and only a few body bristles.

Elateroidea: Superfamily of suborder Polyphaga, order Coleoptera. The forecoxae are small with concealed tronchantins. The hindangles of pronotum are acutely projecting and a transverse metasternal suture is absent. The hincoxae are contiguous. This superfamily comprises of the families Cebrionidae, Elateridae, Trixagidae, Cerophytidae, Perothopidae and Eucnemidae of which Elateridae is of some importance.

Elbowed Antenna: Antenna, particularly of ants in which there is a distinct angle between two segments, usually between the first and second segment in which the first segment is usually much longer than the others.

Electric embedder: A device allowing rapid embedding of wires in foundation with electrically produced heat.

Electroantennogram: A graph obtained from a electrophysiological device electroantennograph that analyses the sensory physiology of antennae.

Electrodialysis: In this method, the cations and anions of the salts present in brackish water are pulled out by passing direct current through it.

Electrolyte: Any compound which in solution conducts a current of electricity and is decomposed by it.

Electron microscope: In this, a parallel beam of electrons is used. The object which must be in the form of a very thin film of the material, allows the electric beam to pass through it. The much magnified image is finally received on a fluorescent screen and recorded by using a camera.

Electrophoresis: Cataphoresis; separation of macromolecules or cells according to their overall charges in an electric field.

Elenchidae: Family of order Strepsiptera. The members are parasites of Delphacidae. The male has four-segmented antennae and two-segmented tarsi. In the female, the brood passage opening with broad and semicircular.

Elephantiasis: A parasitic disease of lymphatic vessels characterized by great enlargement of part of the body, generally caused by filaria.

Elevate: To raise an appendage, *e.g.*, leg or wing.

ELISA: Enzyme linked immunosorbent assay; sensitive technique used to detect specific protein by antigenic means.

Elite: Referring to a colony member displaying greater than average initiative and activity.

Elytral epipleuron: See epipleuron.

Elytral suture: The line formed when two elytra in folded or closed position meet along the midline.

Elytron (pl., elytra): The fore wing in Coleoptera, which is more or less uniformly sclerotized and in resting position is longitudinally oriented, usually meeting the opposite elytron along the midline. The elytron is also characteristic feature of earwigs (See also hemelytron).

E-mail: It is the post facility available on computer network. When many computers are linked with each other through a network, then messages can be sent and received by one computer to other. This facility is known as electronic mail or E-mail. E-mail has shortened the distances in the world.

Emamectin benzoate: It is a semisynthetic derivative of avermectin which is highly effective against the lepidopterous pests. It is marketed under the trade name Proclaim.

Emarginate: With a distinct notch or indentation in the margin.

Embiidae: Family of order Embioptera. Tenth tergite in males asymmetrical; wings with complete wing venation with R_{4-5} forked; echinulations are distinct on inner surface of basal segment of left circus. Two second tarsal bladders in all the instars, *e.g.*, *Embia (Parembia) valida*, *E. minor* occurs in nests of termites.

Embioptera: Order (*embio*= lively, *ptera*= wings) Insect order, made up of the embids or webspinners. They are characterized by slender bodies, wings absent or membranous, chewing mouthparts and incomplete metamorphosis.

Embolium: A narrow region along the front margin of forewing in certain Heteropteran bugs, separated from rest of the corium by a groove or suture.

Embryo: Gr., *En*= in+ *bryo*= swell. The young animal at early stage of development, before leaving the body of parent or before emergence from the egg.

Embryogenesis: Development of embryo.

Embryology: The study of growth and development of new individual within the egg.

Embryonated: Term used in reference to a parasite egg containing a small, immature parasite.

Embryonic induction: Different germ layers determine organ formation by means of various developmental interacting processes called the embryonic inductions.

Emergence: (i) The adult insect leaving the last nymphal skin, or pupal case, (ii) Germination of a seed and the appearance of the shoot.

Emery's rule: The rule that species of social parasites are very similar to their host species and, therefore, presumably closely related to them phylogenetically.

Emetic: Substances used to make humans or animals vomit. To induce vomiting 15 g common salt with warm water can be given in case of poisons taken through ingestion.

Emigration: The movement of individuals out of a population.

***Emmalocera depressella*:** Belongs to order Lepidoptera and family Pyralidae. It is an insect pest of sugarcane, commonly known as the sugarcane root borer.

Empididae: Family of suborder Brachycera, order Diptera. Predaceous flies, without metallic sheen, feeding on small insects including the Diptera and with a well developed downwardly projecting horny proboscis, used for piercing. They look very much like asilids, but lack the facial cluster of hairs. Legs often show differences in the two sexes, the males often with thickened femora, tibia or tarsi. Larvae live in the soil, among decaying vegetation. *Empis* is a typical genus.

Empodium (pl., empodia): A pad like or bristle like structure at the apex of last tarsal segment between the claws as in Diptera. It may be bristle like.

Emulsifiable concentrate (EC): Liquid homogeneous formulation of a pesticide with emulsifier in an organic solvent which forms dispersion when added to water as a diluent.

Emulsifier: A substance that promotes the suspension of one liquid in another. Emulsifiers are often added to pesticide formulations to mix oil based pesticide formulations in water.

Emulsion: (i) A stable dispersion of oil droplets in aqueous solution. (ii) Invert suspension of aqueous solution in oil.

Encapsulation or microencapsulation: The encapsulation of a pesticide in a non-volatile envelope of gelatin, usually of minute size, for delayed release.

***Encarsia merceti* Silv.:** Belongs to order Hymenoptera and family Eulophidae. It is a parasitoid of citrus blackfly, *Aleurocanthus woglumi* Ashby.

Encephaliticides: A group of diseases of the central nervous system caused by mosquito-borne viruses.

Encephalitis: *Culex tritaeniorhynchus* transmits the disease encephalitis in Eastern Siberia to West Asia including India by transmitting a virus, *Flavivirus* (Togoviridae). The disease is characterized by inflammation of the brain. Virus caused encephalitides can affect man and horses.

Encyrtidae: Family of super family Chalcidoidea, suborder Apocrita and order Hymenoptera. Encyrtidae are small insects measuring 1 to 2 mm long and are widely distributed.

End chamber: The germarium of a gonadial tube.

End point: Measurable ecological or toxicological characteristics or parameter of the test system (usually an organism) that is chosen as the most relevant assessment criterion (for example, an acute test or tumour incidence in a chronic study).

Endangered species: An animal or plant species threatened with extinction.

Endemic species: A species that occurs only in a restricted area, such as an island or a mountain range.

Endemic: (i) Insect species prevalent or regularly found in a particular area or well defined geographical region. (ii) Restricted to or native to a particular locality; a population in equilibrium with its environment (not an epidemic). (iii) Referring to a diseases that is regularly present in a given region or area but does not necessarily cause significant losses, such a disease may become epidemic.

Endochorion: The inner layer of chorion in insects.

Endocon: Portion of a conjugated metabolite which is derived from natural products of the metabolizing organisms such as sugars and organic acids.

Endocrine: Secreting internally, applied to organs whose function is to secrete into blood or lymph a substance which has an important role in metabolism.

Endocrine glands: Ductless glands whose products (hormones) are released directly into the blood or haemolymph. Circulating hormones are involved in the operation of all systems of the body. This system of hormonal secretions is internal which is known as endocrine system.

Endocrine system: See endocrine glands.

Endocuticle (endocuticula): The inner most and thickest layer of cuticle (part of procuticle), chitinous, soft and flexible, non-stabilised and not involved in tanning.

Endoderm: The innermost of the three primitive germ layers of the developing embryo.

Endodont: Mandibles with one or two teeth curved inwards, their tips touching when closed.

Endogenous: Arising from inside.

Endoparasite: A parasite lives inside its host. Female parasites lay eggs on open vacuities or on body through mouth or after puncturing the skin. Ectoparasites cause intestinal or muscular myiasis, *e.g.*, common horse bot fly, throat bot fly, warble maggots, screw worms etc. Most of the ichneumons are endoparasites during their larval stages.

Endoparasitoid: An insect parasite which develops within the body of its arthropod host and ultimately kills it.

Endopeptidase: An enzyme that catalyzes the splitting of a peptide chain.

Endoperitrophic space: The space enclosed within the peritrophic membrane in the gut.

Endophallus: The internal, extrusible portion of the penis, sometimes bearing spines, hooks or other sclerotized structures.

Endophilous: Requiring the constant presence, provisions and artificially stable microenvironment for survival.

Endopodite: The mesal branch of a biramous appendages; the main shaft of the limb beyond the basipodite.

Endopterygota: Subclass of the insects including orders with a complete metamorphosis. In these insects, the external and internal changes during the life history are the greatest. The eggs hatch into larvae which feed actively during the different instars. The larvae may or may not have legs. The development of wings is not visible during the larval stages. After several moults, a pupa is formed. A pupa is an inactive stage, it does not feed and it does not move. Sometimes, the pupa is protected by a cocoon of silk, or it is found in an earthen cell in the soil. During the pupal stage, big changes take place internally. After the pupal stage, a highly active winged adult appears. Often, the larvae and the adults live in different types of habitat and use different types of food.

Endopterygotes: Refer to insects in which the wings develop internally during the larval stage, *i.e.*, insects that undergo complete metamorphosis; immature instars dissimilar to adults.

Endoskeleton: (1) It is the internal framework of rigid processes of the body wall. Such a framework in the head is called tentorium. (2) Chitinous processes extending inward into the body cavity from the body wall for attachment of muscles.

Endosperm: A food storage tissue in seeds.

Endospore: A highly resistant dormant stage in the bacterial cell and which after liberating is capable of infecting a new host.

Endosulfan ($C_9H_6Cl_{16}O_3S$): A synthetic insecticide and acaricide; chlorinated hydrocarbon, 6, 7, 8, 9, 10, 10-hexachloro-1, 5, 5a, 6, 9, 9a-hexahydro-6, 9-methano-2, 4, 3-benzodioxathiepin 3-oxide; moderately toxic to mammals, Finkenbrink (1956) described the insecticidal properties; acute oral LD_{50} for rats is 110 mg kg^{-1}; no phytotoxicity experienced on citrus and deciduous trees and on many raw crops. Wide spectrum effectiveness against a variety of insects and mites found on vegetables, fruits, flowers, and in green houses. Moderately toxic, endosulfan is a broad spectrum insecticide; less harmful to the natural enemies and marketed under trade name Thiodan, Endocel etc.

Endosulfan, metabolism: Endosulfan undergoes oxidation to form a primary insecticidal metabolite, endosulfan sulphate (in plants and mice) which is as toxic as endosulfan. In addition to endosulfan sulphate, four other metabolites which are practically non-toxic are found. The formation of endosulfan sulphate is followed by hydrolysis and endosulfan diol is formed. Further, by stepwise oxidation and cyclisation, endosulfan hydroxyether, endosulfan ether and lactone are formed. All these four metabolites are interconvertible.

Endotherm: Organism that metabolically generates heat required for thermoregulation.

Endothermic reaction: A chemical reaction which takes place by absorption of heat is called an endothermic reaction.

Endothermy: Ability of an organism to regulate the body temperature.

Endothorax: A large number of apodemes are produced inside the thorax by the exoskeletal elements, tergites, sternites and the pleura. These endoskeletal structures constitute a sort of compact thoracic endoskeleton, the endothorax. Owing to the origin, the endothorax is composed of three types of endoskeletal processes, the endotergites, endosternites and endopleurite. The endotergites are also known as phragmata. They are lacking from the prothorax, while often occur in the meso and metathorax of pterygote insects.

Endotoxin: Any of a group of toxic substances found in certain disease producing bacteria and liberated by the disintegration of the bacterial cell. A toxic substance formed by certain bacteria and retained within their vegetative cells. In *Bacillus thuringiensis*, the endotoxin occurs as a part of the crystal shaped parasporal body. Also see exotoxin.

Endrin ($C_{12}H_8Cl_6O$): 1, 2, 3, 4, 10, 10-hexachloro-exo-6, 7-epoxy-1, 4, 4a, 5, 6, 7, 8, 8a-octahydro-1, 4, 5, 8-endo-dimethano-napthalene; insecticide, acts as contact or stomach poison; available as emulsifiable concentrate and dust formulation; more toxic than dieldrin to warm-blooded animals. Highly toxic, endrin is a close chemical relative of aldrin and dieldrin (endo-endo isomer of dieldrin) and like its chemical cousins; endrin usage too has been restricted. Besides its general insecticidal effectiveness, endrin is also useful for controlling pigeons and other perch-resting birds inside buildings. A highly residual insecticide, particularly effective against lepidopterous larvae; a chlorinated hydrocarbon, the endo-endo isomer of dielrin; highly toxic to mammals, acute oral LD_{50} for rats is 10-12 mg kg^{-1}; phytptoxicity none when properly formulated but some crops are sensitive of solvents in certain formulations. Endrin is readily isomerised in the light to form non-toxic ketone. Like endrin, isodrin is also an endo-endo isomer of aldrin which is not used in agriculture but used in production of endrin which is epoxide of isodrin.

Energy flow: Energy is defined as the ability to do work. Flow of energy between trophic level is an essential requirement for survival of ecosystem. In ecosystem, producers or green plants consume solar energy and converts into organic food or chemical energy, *viz.*, carbohydrate through photosynthesis with the help of carbon dioxide and water in presence of chlorophyll. Then this energy is transferred to next trophic level with gradual loss of energy through respiration, decomposition etc. Energy flow is always unidirectional. It can be further explained with the help of two laws of thermodynamics. The first law deals with the conservation of energy. It says that energy neither can be produced nor destroyed but it can be transformed from one form to another, *e.g.*, burning of fuel converts the potential energy into kinetic energy. The second law states that whenever one form of energy is converted into another form, some amount of useful energy is lost always in the form which is not useful for general purpose, *e.g.*, when electricity is converted into light; some amount of energy is lost in the form of dispersed heat to the surroundings. In ecosystem, only 10 per cent energy is transferred to next trophic level, 90 per cent energy is lost in useless manner.

Energy: Energy is the ability to do work. It is present in a number of forms (such as mechanical, thermal, chemical etc.) which differs basically from one another, but all together constitute the physical reality of our universe. Energy is necessary for anything we do, *i.e.*, for all physical processes energy is needed.

Enhanced degradation: Increased rate of degradation of pesticide in soil or other environmental matrix by a population of microorganisms that has adapted to metabolize it through previous exposure to it or a similar chemical. Enhanced degradation or accelerated degradation are synonymous.

Enhancement: The term denotes the use of measures that increase the longevity and reproduction of natural enemies in the field.

Enkephalin: See vertebrate hormones in insects.

Ensifera: A suborder of Orthoptera. Antennae longer, many segmented, tympanal organs when present are on the foretibiae.

Entad: Inward relative to the midline of body.

Enteric: Relating to the digestive canal.

Enteric caeca (Gastric caeca): See hepatic caeca.

Enterohepatic circulation: Cyclical process in which a pesticide residue is absorbed and transported to the liver, metabolized (often including conjugation), transported to the intestine by the bile, reabsorbed (often after deconjugation) and transpored to the liver for further metabolism.

Entognathous (Entognathy): Having mouth parts somewhat pulled or withdrawn into the head.

Entoleter: A centrifugal force machine to kill insects infesting grain or their products. This machine revolves at a speed of 3000-4000 rpm which destroys the insects by centrifugation of their haemolymph. This device is used by roller flour mills and some rice mills but its operation is not economical in rural storage system.

Entomobryidae: A family of order Collembola. Third abdominal tergite not reduced and abdominal segments 4-6 free; fourth abdominal segment longer than third; a group of setae on trochanter called trochanteral organ present. For example, *Entomobrya* sp., *Lepidocyrtus* sp.

Entomobryoidea: A super family of order Collembola and suborder Arthropleona. Pronotum reduced or absent and lack setae.

Entomogenous: Growing in or on an insect, for example, certain fungi.

Entomologist: A person who studies the biology of insects.

Entomology: Gr., *Entomon*= insects, *logus*=discourse or study. The branch of zoology which deals with insects or it is the science of study of insects.

Entomophagous: An organism which feeds upon insects.

Entomophilous: Insect loving.

Entomophily: Transfer of pollens by insects, a type of pollination. Insects, like bees, flies, moths, butterflies, wasps, beetles etc. help in the pollination which is called entomophily. About 80 per cent of insect pollination is done by bees. Usually, the flowers of insect pollinated plants are brightly coloured, possess sweet fragrance and produce nectar.

Entomophobia: An irrational, pronounced fear of insects.

Envelope: The carton or wax made outermost coated layer of the nest of social insects, particularly, those of wasps.

Environment: The sum total of all of the physical, biological and chemical surroundings in which an insect lives. The environment consists of four segments: (i) Atmosphere: It is the blanket of gases, suspended solids and liquids which sustains life on the earth and separating it from the outer space. The atmosphere consists mainly four layers, *viz.*, troposphere (lowest layer of atmosphere in which living organisms operate, 12 kms thick), stratosphere (about 50 kms above the surface of earth, it contains the ozone layer called ozosphere which is formed by oxygen through photochemical reaction), mesosphere (extends nearly 80 kms above the earth, about 35 km thick, temperature is about -95 °C and low atmospheric pressure exists), thermosphere (extends upto 500 kms above the earth surface, has higher temperature than mesosphere) and exosphere (extends upto 32,190 kms from the earth). (ii) Hydrosphere: includes all types of water resources. (iii) Lithosphere: is the outer mantle of earth including earth's crust and soil. (iv) Biosphere: includes living organisms and their interaction with the environment.

Environmental health effects due to pesticides: (i) Bioconcentration- movement of a chemical against a concentration gradient (ii) Biomagnification- movement of a chemical through a food chain (iii) Bioaccumulation- bioconcentration plus biomagnification. This is a rare combination but few exogenous chemicals bioconcentrate. It is not only the ability to bioconcentrate by the chemicals but the chemical is also typically fat soluble and persistent. Examples, DDT, PCBs, dioxin

Environmental impact assessment: Assessment of the potential releases of a pesticide to the environment and their potential effects upon the environment and its components including man.

Environmental pollution: Undesirable levels of pesticides in air, soil, water, vegetation, animals and man.

Environmental resistance: The integration of physical and biological factors which limit the growth of an insect population.

Environmental risk: Probability that an adverse effect on human or the environment will be observed for a given exposure to a pesticide based on the frequency of the occurrence and sensitivity of the system.

Environmental toxicology: A branch of toxicology which deals with the metabolism, transport, translocation and physicochemical transformation of poisons in all forms of biological system..

Enzyme: A protein molecule, highly specific (biological catalyst secreted by a cell) which initiates and catalyzes or accelerates a biochemical reaction. See also coenzymes.

Eosentomidae: A family of order Protura. Members of this family have tracheae and spricals. The abdominal appendages are two-segmented and the eighth abdominal segment does not possess dorsolateral combs. For example, *Eosentomon* sp.

EPA: A number assigned by EPA (European Pesticide Association) to a product when it is registered. It must appear on all labels for that product as EPA registration number followed by the company number and a product number.

Ephemeroidea: A super family of Ephemerida. M_{1+2} and Cu_1 of forewing diverging strongly at base; four or less movable tarsi on hind legs and a fifth segment if present is fused with the tibia. This includes the families Palingeniidae, Polymitarcidae, Ephemeridae and Potamanthidae.

Ephemeroptera (Ephemerida): An insect order (*ephemero*= for a day, *ptera*= wings) consisting of the mayflies. They are characterized by membranous wings, nonfunctional mouthparts, two or three abdominal cerci, and an incomplete metamorphosis.

Ephydridae (shore flies): Family of suborder Cyclorrhapha, order Diptera. Ephydridae are small, black or darkly colored, long legged flies found in marshy and damp places. The larvae of *Psilopa petrolei* is known to live in pools of crude petroleum in California. The larvae are mostly aquatic and are found in decaying vegetable matter, saltish or brackish pool etc. A few are either phytophagous or predaceous on eggs of spiders.

Epicranial suture: (Coronial plus frontal suture). It is a composite suture. Inverted Y-shaped suture whose stem begins from back of head across the vertex. The dorsal part of the suture is called coronial suture (the base of Y) and the anterior branches are called frontal suture. This suture is completely reduced (apterygote insects) or partially reduced in some insects. Snodgrass (1935) and Du Porte (1957) stated that the epicranial suture represents merely the line of weakness and can be said in true sense as the ecdysial suture. It does not bear the skeletal ridge on the inner surface and, therefore, does not provide either a site for muscle attachment or mechanical support to the head capsule.

Epicranium: The epicranial suture separates the epicranium from the frons. It represents the entire upper region or top of the head. It extends from the anterior frontal to the posterior occipital suture. Above the frons, it is divided by the coronal suture into two identical lateral plates known as the parietals. The parietals are characterized by bearing the antennae, lateral ocelli and the compound eyes. The posterior undivided region of the epicranium is known as the vertex.

Epicuticle (epicuticula): Unpigmented outer layer of the insect cuticle which consists of three layers depending on the insect species: (i) Cement layer (tectocuticle) (ii) Lipid epicuticle (wax layer), and (iii) proten epicuticle. This layer is non-chitinous and thinner as compared to procuticle.

Epideictic pheromone: An intraspecific chemical messenger that elicits dispersal from potentially crowded food sources, also called the spacing pheromone.

Epidemic: A widespread and severe, temporary increase in the incidence of an infectious disease, particularly within a season.

Epidemiology: The science of disease in populations. The study of the development and spread of disease and of the factors affecting these processes.

Epidermis (hypodermis): The cellular layer of the integument that underlies the cuticular parts. Some of the cells are enlarged to form glands which open outside through a duct. These glands produce cuticular material, moulting fluid and poisons for defensive purposes. Some of the cells are variously modified to form sense organs. This cellular layer of the integument secretes or deposits a comparatively thick cuticle on its outer surface.

Epigaeic: Living or foraging primarily above ground (the opposite of hypogaeic).

Epigastric furrow: A transverse ventral suture near the anterior end of the abdomen, along which lie the openings of the book lungs and the reproductive organs (spiders).

Epigynum: The external female genitalia of spiders.

***Epilachna duodecastigma* (Mulsant):** Belongs to order Coleoptera and family Coccinellidae. It is an insect pest of brinjal, commonly known as hadda beetle.

Epileptic convulsions: Convulsions owing to nervous disorder or mental disturbance.

Epimere: A dorsal process of the phallobase.

Epimeron: The posterior part of the side wall of any of the three thoracic segments.

Epimorphic development: Immatures possess same numbers of body segments as adults.

Epinasty: Increased growth on upper surface of a plant organ or part (especially of leaves) causing it to bend downwards.

Epineural sinus: The channel of the embryonic haemocoel beneath the yolk or alimentary canal.

Epinotum: The first abdominal segment when it is fused with the last thoracic one, relating to the higher thin waisted Hymenoptera. Also called propodeum.

Epiphallus: A sclerite in some Orthoptera in the floor of genital chamber proximal to the base of phallus (pseudosternite).

Epipharyngeal apparatus: It acts as the sound producing mechanism in the death's head moth, *Acherontia* sp. The proboscis acts as an amplifier, the pharynx as an air pump and the epipharynx as a vibrating instrument. The pharyngeal muscles operate the air pump (pharynx).

Epipharyngeal wall: The membranous inner surfaces of the clypeus and 1abrum.-

Epipharynx: A component of many insect mouth parts which is attached to the posterior surface of labrum or clypeus. In chewing insects it is usually a small lobe but in the fleas it is greatly enlarged and used for sucking blood.

Epiphysis (pl., epiphyses): A movable pad or lobe like process on the inner surface of front tibia as in Lepidoptera.

Epiphyte: Non-parasitic plant that is attached to another plant for mechanical support only. For example, orchids, lichens and mosses that are growing on trees.

Epiplemidae: Family of super family Geometroidea, suborder Ditrysia and order Lepidoptera. A small group of inconspicuous species found in all continents. The moths at rest resemble spiders.

Epipleuron (pl., Epipleura): A lateral, infolded portion of the elytra which is separated from the elytral disc by a distinct fold or carina and which usually fits against the lateral portions of the abdomen.

Epiproct: An appendage arising from the mid line of the last abdominal segment, just above the anus. In the bristletails and some mayflies, it is very long and forms the central tail.

Epipyropidae: Family of super family Tinaeoidea of suborder Ditrysia and order Lepidoptera. Larvae of epipyropids are parasitic on jassids and fulgorids. *Epipyrops melanoieuca* is parasitic on *Pyrilla perpusilla* on sugarcane. *Epipyrops eurybrachys tomentosa* and *E.fulginosa* on mango jassids, *Idioscopus* sp. are other parasitic epipyropids in South India.

***Epipyrops melanoleuca* Tams:** It belongs to family Epipyropidae, order Lepidoptera and is a natural enemy of mango hopper, *Amritodus atkinsoni.*

Epistemological: Scientific study of knowledge, its acquisition and communication.

Episternum (pl, episterna): Anterior portion of the thoracic pleuron. The anterior part of the side wall of any of three thoracic segments.

Epistoma: See clypeus.

Epistomal suture (or sulcus): The sulcus between the frons and the clypeus. It bears laterally the anterior tentorial pits and internally bears strong inflection. The inflection strengthens the lower part of head against stretching of mandibles.

Epithelium: The layer of cells that covers a surface or lines a cavity.

Epithet: A descriptive name or title.

EPN: A synthetic acaricide with insecticidal property also; an organophosphate, *O*-ethyl o-*p*-nitrophenyl phenyl phosphorothionate; highly toxic to mammals, acute oral LD_{50} for rats is 35-45 mg kg^{-1}; white crystalline substance with melting point 36 °C. It is insoluble in water. Phytotoxic and may injure Mclntosh and related varieties of apples. It has less residual effect than the parathion.

Equine bot flies: There are three species of bot flies, *viz., Gasterophilus intenstinalis* (common bot fly), *G. nasalis* (throat bot fly) and *G. haemorrhoidalis* (nose or lip bot fly. These belongs to family Oesridae (Diptera), adults are 15-18 mm long and brown in colour; cosmopolitan in distribution; infests equines, rarely dogs, pigs and sometimes man. Intestinal myiasis and large ulcers are produced in various parts of gut.

Eradication: The complete elimination of weeds, insects, disease causing organisms or other pests from an area.

***Eretmocerus serius* Silo:** Belongs to order Hymenoptera and family Eulophidae. It is a parasitoid of citrus blackfly, *Aleurocanthus woglumi* Ashby.

Ergatogyne: Any form morphologically intermediate between the worker and the queen.

Ergatoid male: See ergatomorphic male.

Ergatoid reproductive: A supplementary reproductive termite without a trace of wing buds, usually larval in external form and with a distinctively rounded head (same as third form reproductive, tertiary reproductive and apterous neoteinic).

Ergatomorphic male: An individual (social insect) with normal male genitalia and a worker like body.

Ergonomics: The quantitative study of work, performance and efficiency.

Ergotism: Poisoning by a toxic substance contained in the sclerotia of the fungus, which grows on many grasses and on some grains, particularly, rye.

Erinium: A growth of hair in dense patches on plant leaves resulting from the attack of certain Eriophyidae (Acarina).

Erinose: Any plant disease in which an abnormal growth of hair occurs in patches on the leaves such as caused by the attack of certain gall mites (genus, *Eriophyes*).

Eriocraniidae: Family of super family Eriocranioidea, suborder Montrysia and order Lepidoptera. The adults have a short proboscis formed of galeae, the laciniae being absent. The apodous larvae of *Eriocrania* sp. is a leaf miner on birch, hazel, oak and chestnut.

Eriocranioidea: Super family of suborder Montrysia and order Lepidoptera. Characterised by venation of both pairs of wings alike. Female has a single genital aperture.

***Erionota thrax* L. (Lepidopera; Hesperidae):** It is a minor insect pest of banana.

***Eriophyes* sp.:** See *Aceria* sp.

Eriophyid mite: See *Aceria jasmini.*

Eri-silk: Creamy white silk obtained from the cocoons of the eri silkworm.

Erode: To wear away.

Erosion: The wearing away of the land surface by water and wind.

***Eruca*:** Latin word for caterpillar.

Eruciform larva: Caterpillar or a larva with a cylindrical body, well developed head, with thoracic legs and some abdominal prolegs (polypodous), *e.g.*, larva of Lepidoptera, Mecoptera and Symphyta.

***Erycia nymphalidaephaga* Bar.:** Larval parasitoid (Hymenoptera) associated with citrus caterpillar, *Papilio demoleus* Linn.

***Erythroneura* sp.:** Belongs to order Hemiptera and family Cicadellidae. It is commonly known as grapevine leaf hopper and is a pest of grapes.

Essentialistic species concept: The concept of Plato and Aristotle and latter of Linnaeus in which the observed diversity of the universe reflects the existence of a limited number of universals or types and the variations are not taken care of.

Ester: A compound formed by the union of an organic acid and an organic base (alcohol), *e.g.*, 2,4-D and isooctyl alcohol to form the isooctyl ester of 2,4-D.

Estimated daily intake (EDI): Prediction of the daily intake of a pesticide residue, based on most realistic estimation in food items and the best available food consumption data for a specific population. Residue levels are estimated taking into account the known uses of the pesticide, the proportion of the commodity treated and the quantity of contaminated commodities. The EDI is expressed in milligrams of residues per person.

Estimated Environmental concentration (EEC): Predicted concentration of a pesticide within an environmental compartment based on estimates of quantities released, discharge pattern and inherent disposition of the pesticide (fate and distribution) as well as the nature of the specific receiving ecosystem.

Estimated maximum daily intake: Prediction of maximum daily intake of a pesticide residue, based on the assumptions of average daily food consumption per person and maximum residue in the edible portion of a commodity, corrected for the reduction or increase in residues resulting from preparation, cooking or commercial processing. The EMDI is expressed in milligrams of residues per person.

Ethane: A colourless gas present in the gases flowing from gas wells and in refinery off-gases.One of the major raw materials for producing ethylene (Ethane, propane and butane comprise LPG).

Ethane dinitrile: Ethane dinitrile is a colourless gas with an almond like odour. It has got potential fumigant property for the control of termites, wood borers and stored grain insects. It is highly phytotoxic and hence useful for weed control and sterilization of grains. In the laboratory tests at CFTRI, it was noted that *Khapra* larvae and the developmental stages of rice weevil were highly tolerant and require 32.8 g per m^3 for 48 hours to achieve 100 per cent mortality. The fumigant is likely to be registered under trade name Sterigas.

Ethers: Ethers are organic compounds in which two hydrocarbon radicals are joined through an atom of oxygen.

Ethion ($C_2H_{22}O_4P_2S_4$): A synthetic insecticide and acaricide; an organic phosphate, *O, O, O′, O′*- tetraethyl *S, S′*-methylene bis phosphorodithioate; moderately toxic to mammals, acute oral LD_{50} for rats is 96 mg kg^{-1}; phytotoxicity: safe on all crops except defoliation may result when applied to Wealthy apples. Highly toxic, colourless liquid insoluble in water but soluble in most organic solvents, hydrolyses both in alkaline and acidic conditions. Ethion is particularly useful for controlling mites on fruits, vegetable plants, flowering plants, ornamental plants and also for controlling scales and onion maggots. It is generally formulated as 50 per cent EC, 25-50 per cent WP and 2 per cent solution in spray oil.

Ethocline: A series of different behaviors observed among related species and interpreted to represent stages in a single evolutionary trend.

Ethology: The branch of biology which deals with comparative study of animal behaviour.

Ethoprop (Mocap): Highly toxic, Mocap is especially useful against corn rootworms and wireworms.

Ethyl formate: It is liquid fumigant having low boiling point (54.1 °C) with a pleasant odour. It is naturally present in certain raw and processed food at low level (0.05-1.0 mg kg^{-1}). It has rapid action against insect pests and known to break down into non-poisonous natural products, ethanol and formic acid. It is generally recognized as the safe fumigant. It has been used for individual package fumigation of dry fruits in Australia, South Africa and other countries. In Australia, Vapormate, a formulation containing 16.7 per cent ethyl formate in liquid CO_2 has been developed for insect control in food processing facilities (dosage 420 g/m^3, 24 hrs exposure). CO_2 serves the dual function of reduction in flammability risk of ethyl formate and synergistic effect. The effective dosage ranged from 300-400 g per m^3 with a 72 hrs exposure period for treating different cereals, pulses, spices and dry fruits.

Ethyl guthion: It is the trade name of azinphos ethyl.

Ethylene dibromide: EDB is a moderately toxic nematicide and grain fumigant insecticide. It controls nematodes and their eggs, as well as soil insects when injected into the soil of seed beds, flower beds, potting soil and green house soil. When EDB is mixed with ethylene dichloride and carbon tetrachloride, it is an effective grain fumigant against stored grain insect pests. It is used as a soil and commodity fumigant, insecticidal and nematicidal; chemically 1, 2-dibromoethane ($CH_2Br.CH_2Br$); moderately toxic to mammals; phytotoxic, product of several companies and sold under trade names, Minifume, Orthofume, Bromotox, Fumogen, Bromofume, Escobrome-D and Dowfume W-85. It is a colourless, heavy organic liquid with mildly sweet chloroform like odour with boiling point 131-132 °C. The specific gravity is 6.49 (gas) and 2.17 (liquid) at 20 °C and highly inflammable, the molecular weight of which is 187.88. It is one of the toxic chemical to the insect pests damaging the stored grain pests. The use of EDB, filled in ampoules has been found suitable for fumigation of stored grains stored in household structures, holding small quantities which are made reasonably air tight. Its dosage is 3 ml q^{-1} grain with exposure period of 3 days in small receptacles. It can be used in all food grains except milled products, oilseeds and moist grains.

Ethylene dichloride (EDC): This is a fumigant recommended first by Cotton and Roark in 1927. It is 1, 2-dichloroethane ($CH_2Cl.CH_2Cl$) which is a colourless liquid, chloroform like odour and gas being highly inflammable and soluble in organic solvents. Its boiling point is 83.5 °C, molecular weight 98.97, specific gravity of gas is 3.42 and that of liquid 1.257 at 20°C and vapour pressure 78 mm Hg at 25 °C temperature. It is a slow acting and death of insect takes place from one to three days after exposure and toxic to plants and human beings. It is available under the trade names of Borocide and Destruxol.

Ethylene dichloride and carbon tetrachloride (EDCT): It is a general mixture of three parts of ethylene dichloride and one part of carbon tetrachloride by volume. This mixture was evolved by US Department of Agriculture. It is colourless liquid with chloroform like odour and both gas and liquid are non-inflammable. All cereals, millets and pulses are safely fumigated at the dosage of 30-40 kg per

100 m^3 space with exposure period of 48-72 hrs, depending upon intensity of infestation, moisture percentage of the commodity, temperature and humidity. The chemical present on the grains gets reduced by turning of the grains and residues completely lost during cooking of the bread.

Ethylene oxide (Carboxide): Moderately toxic, highly inflammable and explosive. Ethylene oxide can be bought mixed with 1 part ethylene oxide to 9 parts carbon dioxide which reduces the inflammability danger. Ethylene oxide is useful for fumigating grain and stored food products. It has bactericidal properties also, thus making it useful for disinfecting individual items as well as entire building. This chemical is used as fumigant, odourless, heavier than air with good penetrating properties. Used at low temperature, toxicity enhanced by mixing with 9 parts by volume of carbon dioxide, used for fumigating dried fruits and oilseeds or commodity containing vegetable oils. It is also used for making chemical derivatives such as glycol esters, ethanolamines and synthetic detergents.

Ethylene: An olefin, a basic chemical most widely used petro-chemical building block in the world. A colourless, sweet smelling gas made principally from ethane and propane in the United States and from naptha in the other countries.

***Etiella zinckenella* Treitschke:** Belongs to order Lepidoptera and family Phycitidae. It is an insect pest of pulse crops, commonly known as the lentil pod borer.

Etofenprox: The non-ester synthetic pyrethroid compound, a nerve poison which specifically inhibits the transport of Na^+ ions across the nerve membranes. It is marketed under trade name Trebon (10 EC).

***Eucalyptus camaldulensis* Hook:** A plant, the oil of which is effective in controlling the insect pests of stored grains.

Eucaryotic: Having a membrane bound nucleus.

Eucharitidae: Family of super family Chalcidoidea, suborder Apocrita and order Hymenoptera. The insects are black or metallic blue or green with somewhat hump-backed thorax having spines on the scutellum and petiolate abdomen. These are tropical insects and are ectoparasites of pupae of ants.

Eucone eye (Eucone ommatidium): Eye in which each ommatidium bears a crystalline cone just beneath the corneal lens. See ommatidium, types.

Eucosmidae (Olethreutidae): Family of super family Tortricoidea of suborder Ditrysia and order Lepidoptera. Adults are generally small and brownish or grey with patterns on the wings. Fringes of long hairs arise on the basal part of cubitus in the hindwing which is a characteristic feature of this family. Larvae web or roll up the leaves and bore into the fruits. The coding moth *Enarmonia (Carpocapsa) pomonella* is one of the well known pests of apple and other fruits in the United States.

Eulophidae: Family of super family Chalcidoidea, suborder Apocrita and order Hymenoptera. Eulophidae are small insects measuring about 1 to 3mm long and are parasitic in the eggs, larvae and pupae of a wide variety of insects, especially of Homoptera, Coleoptera, Lepidoptera and Diptera. They possess four or five segmented tarsi and the axillae extend in front of anterior margin of scutellum and usually beyond the tegulae. The antennae are pectinate in males of many species.

Eumastacidae: A family of order Orthoptera, suborder Caelifera and super family Acridoidea. Tropical group with pronotum usually flattened dorsally.

Eumecoptera: Suborder of order Mecoptera. Wings narrow and slender; male genitalia swollen and body generally cylindrical.

Eumolpidae: Family of superfamily Chrysomeloidea, suborder Polyphaga and order Coleoptera. The beetles are usually oblong, convex, either metallic or yellowish with spots. The beetles are characterized by prothorax narrower than elytra and legs not compressed. The genera included under the family are: *Colaspis, Scelodonta, Eryxia, Adoxus, Chrysochus, Colasposoma, Bromiodes, Eurypelta, Nodostoma, Pachynephorus* etc. in India *Scelodonta strigicollis* is a known pest of grapevine. *Nodostoma pubicolle* and *Colasposoma metallicum* are other common species.

Eupelmidae: Family of super family Chalcidoidea, suborder Apocrita and order Hymenoptera. These insects are characterized by flat mesonotum and presence of parasidal furrows. Many are capable of jumping. Apterous or short winged forms occur in some. These are parasitise a wide variety of insects and spiders.

Euphagous: Feeding on specific host plant.

Euplantulae: Pad like structure on the ventral surface of the tarsal sub segments (tarsal pulvilli).

Eupleuron: See anapleurite.

***Euproctis lunata* Wlk.:** Belongs to order Lepidoptera and family Lymantriidae. It is an insect pest of castor and ornamental plants, commonly known as the castor hairy caterpillar.

Eupsocida: A suborder of Psocoptera. Tarsi two or three segmented, antennae mostly with 13 segments and with single segmented labial palpi.

***Eupterote cardamomi* Renga:** Belongs to order Lepidoptera and family Eupterotidae. It is an insect pest of spices, commonly known as the cardamom hairy caterpillars.

European bee: *Apis mellifera*, a species of honeybee found in Europe, Italian variety considered best; similar to the Indian honey bee in habits; average honey yield from its colony is 100 to 400 lbs.

European corn borer: See *Ostrinia mubilalis.*

European foulbrood: An infectious brood disease of honey bees caused by *Streptococcus pluton.*

European mealworm: *Tenebrio molitor* L., a black beetle, the grub of which feed on flour, meal, bran, broken grains etc. Usually found in dark, damp and dingy places and godowns.

Eurybrachidae: A family of superfamily Fulgoroidea of suborder Homoptera and order Hemiptera. The Eurybrachys tomentosa is a common hopper found in the sandal and *Calotropis gigantea* in South India.

***Eurytoma* sp.:** Larval parasite of citrus leaf miner, *Phyllocnistis citrella* Stainton.

Eusocial: Applied to the condition or to the group possessing it in which individuals display all of the following three traits: cooperation in caring for the young, reproductive division of labour, with more or less sterile individuals working on behalf of individuals engaged in reproduction, and overlap of at least two generations of life stages capable of contributing to colony labour. This is the formal equivalent of the expressions truly social or higher social which are commonly used with less exact meaning.

Eusociality: A type of social behaviour involving overlap of generations, cooperative brood care and reproductive division of labour in which many colony members are sterile, *i.e.*, truly social.

Eusternum: The intrasegmental ventral plate of a thoracic segment, exclusive of the spinasternum but usually including the sternopleurites.

Eustheniidae: A family of order Plecoptera. Primitive forms characterized by wings having entire margin without any notch and anal lobe with mesh work of fine veins or archedictyon, *Eusthenia* sp., *Stenoperla* sp.

Eutrichosomatidae and Tanostigmatidae: Family of super family Chalcidoidea, suborder Apocrita and order Hymenoptera. These are parasites of coleopterous insects,especially weevils.

Eutrophication: The process of enrichment of a body of water in dissolved nutrients either naturally or artificially as by fertilization.

***Euzophera perticella* Rag.:** Belongs to order Lepidoptera and family Phycitidae. It is an insect pest of brinjal, commonly known as the brinjal stem borer.

Evagination: An out pocketing or sac like structure on the outside; eversible, capable of being turned outward.

Evaniidae (ensign wasps): Family of super family Evanioidea, suborder Apocrita and order Hymenoptera. The Evaniidae have the antennae inserted well above the clypeus, 1 or no m-cu cross-vein on forewing, an anal lobe on hindwing, short propleuron, long and abrupt petiole, compressed and short gaster and short unexerted ovipositor. On forewing 2r-m cross-vein is absent. These are black insects and carry the gaster almost like a flag and hence the common name 'Ensign wasps'. They are parasitic in the oothecae of cockroaches. Some common species found in India are: *E.antennalis, E.albitarsis, Evania appendigaster* and *E.curvicarinata.*

Evanioidea: Super family of Suborder Apocrita, order Hymenoptera. These insects are characterized by a wider costal cell and forewings and the gaster being attached by a slender petiole to the top of propodeum above the base of the hindcoxae.

Evaporation: The process of a solid or liquid turning into gas.

Evergreen: Plants that retain their functional leaves throughout the year.

Eversible: Capable of being averted or turned outward.

Evolution: L., *e*= out + *voluo*= roll. It refers to origin and differentiation of animal life.

Evolutionary divergence: The degree of divergence at the intra and interspecific levels of two or more populations which presumably have evolved from a common ancestor.

Evolutionary: The process of biological and organic change in organisms by which descendants come to differ from their ancestors.

Exarate pupa: A pupa in which the appendages are free and not glued to the insect body.

Excavate: Hollow in the coxa of many beetles. The coxa is hollow to allow the housing of the femora when the legs are folded.

Exclusion: Control of disease or pest by preventing its introduction (*e.g.*, quarantine) into the disease or pest free areas.

Excretion: The elimination of the waste products of metabolism from the body. The eliminatory substances are solid, liquid or gaseous in form. The principal gas expelled is carbon dioxide and liquid is water. The solid excreta are given out as crystalline masses or aqueous solutions which includes nitrogenous and non-nitrogenous salts.

Excretion, mechanism: The terrestrial insects are the urotelic animals like reptiles and birds and their principal excretory product is uric acid while the aquatic insects are often uricotelic excreting mainly ammonia in the haemolymph, the uric acid combines with sodium and potassium carbonates or bicarbonates to form sodium and potassium urate, respectively. Sodium and potassium urate salts are highly soluble in water and are easily absorbed by the Malpighian tubules as they can cross the cell membrane in the solution. In insects, like *Rhodnius*, the tubules are differentiated into two parts, proximal having the brush border and distal with honey comb border. The aqueous solution of urates is always absorbed by the distal part. The urates are gradually accumulated in the lumen of tubules at distal part and slowly reach the proximal part. During their transport from distal to proximal part, the uric acid is precipitated. In alkaline pH, enzyme carbonic anhydrase splits urates to regenerate the bicarbonate and uric acid. The carbonic anhydrase is secreted by epithelial cells of Malpighian tubules and subsequently water and carbonates or bicarbonates are reabsorbed into the blood so that the uric acid is left into the lumen of tubules. The water and carbonates combined with uric acid, form urates which can be excreted in the similar way in a next excretory cycle. The semisolid uric acid particles move into the lumen of hindgut. Inside the hindgut, again the precipitation of uric acid takes place through reabsorption of water and few other metabolites by ileac or rectal wall into the haemolymph.

Excretory system: The mechanical devices developed in the organism for excretion of waste metabolic products is known as excretory system. The excretory organs are: (i) Malpighian tubules, (ii) tracheal system, (iii) integument, (iv) wall of alimentary canal and (v) accessory organs like nephrocytes, fat bodies, oenocytes, cephalic glands etc.

Exelastis atomosa **Walshingham:** Belongs to order Lepidoptera and family Pterophoridae. It is an insect pest of pulse crops, commonly known as the plume moth.

Exerted: Protruding or projecting from the centre of body.

Exite: An outer lobe of any limb segment.

Exocon: Portion of a conjugated metabolite which is derived from the parent pesticide.

Exocone eye: Eye in which each ommatidium bears a cone like, internal projection of the cornea which thus forms a lens.

Exocrine gland: A gland that discharges its products to the outside.

Exocuticle (or exocuticula): Stabilized layer of cuticle between the endocuticular and epicuticular layers; resistant to moulting fluid and often pigmented. It is hard, sclerotized and dark in colour due to tanning.

Exodont: Mandibles with 3 to 4 teeth or lobes, usually directed outward, their tips not touching when closed. See endodont also.

Exogenous rhythms: Activity patterns governed by variations in the external environment.

Exogenous: Arising from outward.

Exopeptidase: An enzyme that catalyzes the splitting of terminal peptides from a peptide chain.

Exopodite: The layer of the branch of a biramous appendage.

Exopterygota: Subclass of the insects including the orders with a simple or incomplete metamorphosis. In moulting from egg, *via* the nymphal stages to an adult, there is a gradual change in the external appearance. The late nymphal stages already show the development of wing pads. But only in the last moult functional wings are developed. The nymphs usually have the same feeding habits as the adults.

Exopterygotes: Refers to insects in which the wings become externally apparent in the instars (immatures resemble adults), *i.e.*, undergo simple metamorphosis or hemimetabolous development.

Exoskeleton: A skeleton or supporting structure on the outside of body, present in all members of the Arthropoda. Collectively, the external plates of the body wall constitute exoskeleton.

Exothermic reaction: A chemical reaction in which energy is liberated usually in the form of heat is known as exothermic reaction, *e.g.*, $N_2+3H_2 \rightarrow 2NH_2$ (93.33 kJ).

Exotic species: An organism that evolved in one part of the world and that now occurs either accidentally or intentionally in a new region. Opposite of native species.

Exotic: Introduced from another country.

Exotoxin: A soluble toxic substance produced by certain bacteria and found in their surrounding growth medium; also called endo-toxin.

Exploratory trail: An odour trail laid more or less continuously by the advance workers of a foraging group. This kind of communication is used regularly by army ants (opposite of recruitment trail).

Exposed surface: Surface that can be successfully attacked by a pest. Surface that need to be protected by a pesticide chemical or other agent.

Exposure assessment: Process of measuring or estimating concentration, duration and frequency of exposures to pesticide present in environment or estimating hypothetical exposures that might arise from release of pesticide into the environment.

Exposure: Concentration or amount of a pesticide that reaches the target population, organism, tissue or cell, usually expressed in terms of concentration, duration or frequency. Also the process by which a substance becomes available for absorption by the target population, organism, tissue or cell by one route.

Exposure period: The length of time something has been under attack by a pest. Also, the length of time a pest is in contact with a pesticide chemical.

Extant: Still existing.

Extension: All methods, techniques and activities aimed at the dissemination and proliferation of agricultural information in rural communities which make such information more easily accessible to desired target groups such as small farmers, pest control operators and agricultural students.

Extermination: Complete extinction of a species over a large continuous area such as an island or a continent.

External feeder: The group of stored grain insect pests which feed the germ and endosperm from outside the seed. They may attack the whole seed and damage the germinal portion or feed on the seeds which has already been damaged or infested by other insect pests or broken mechanically. These insects and their different stages are generally visible among the seeds; for example, *Khapra* beetle, Indian meal moth, red rust flour beetle, saw toothed grain beetle etc.

External genitalia: The outside reproductive structures concerned with mating and deposition of eggs.

External respiration: The process of transferring the respiratory gases through the body wall; taking place in insects through thin areas of the ectoderm, either at the body surfaces or in the walls of invaginations (gills) or evaginations (tracheae).

Extinct: Having no living descendants; no longer in existence.

Extinct insect orders: The palaeontological studies revealed some extinct insects, mostly from their well preserved wings in the fossils of coal, lignite, peats and amber impregnated with resin of Pennsylvanian or Permian to the Lower Carboniferous age. The earliest Collembolan, Rhyniella precursor has been recorded from the middle Devonian fossil peat bag of Rhynie in Scotland while Palaeodictyoptera, Protorthoptera, Protodonata, Megasecoptera and

Protoblattoidea are obtained from the fossils of carboniferous period. There are about eleven insect orders classified based on wing venation and widly considered as the early ancestors of several existing insect vectors: Subclass Apterygota: (i) Monura; Subclass Pterygota: (ii) Palaeodictyoptera, (iii) Megasecoptera, (iv) Diaphanopterodea, (v) Protodonata, (vi) Protorthoptera, (vii) Miomoptera, (viii) Caloneurodea, (ix) Protecoleoptera, (x) Protoperlaria, (xi) Glosselytrodea.

Extractability: Degree to which a pesticide residue may be removed from matrix (*e.g.*, soil, plant etc.) through use of appropriate extraction techniques.

Extracted honey: Honey removed from the comb by centrifugal force.

Extraction: Extraction is a process by which the toxicant is transferred from the treated bulky biological material into a solvent. Since, the residue of insecticide is present in minute quantities in or on the substrate, the selection of a suitable solvents or combination of solvents for extraction of samples is very important which depends on the physical or chemical properties of the toxicant which is to be extracted and the type of substrate where it persists.

Extraintestinal digestion: Many predaceous beetles (larvae of *Dytiscus*, *Carabus* and *Cincindela*, and the adults of these) which lack salivary glands, as well as the ectoparasitic larvae of certain Hymenoptera, eject their intestinal secretions from their mouth, allow much of the digestion of prey to take place outside the body and then reabsorbs the fluid products. This process is termed as extraintestinal digestion. The flesh eating maggots achieve the same results by discharging proteolytic enzymes in their excrements. The assassin bug (*Platymeris*) discharges toxic and strongly lipolytic and proteolytic saliva which quickly paralyses the prey and liquefies its tissues.

Extraneous residue limit: Maximum concentration of a pesticide residue arising from environmental sources (including former agricultural uses), other than from those of a pesticide applied directly or indirectly on the commodity, that is recommended to be permitted in or on a feed or food commodity.

Extrinsic superiority: Refers to the relative competitive ability of the beneficial organisms in the environment, especially regarding efficiency in host location and attack; also called intrinsic superiority.

Exudate: Matter diffused from a cell.

Exuvia: The cast-off outer skin of insect or other arthropods.

Exuvial glands: Glands of the epidermis supposed to secrete the exuvial or molting liquid.

Eye-cap: Hood formed by the base of antenna and partly covering the eye in certain small moths.

Eye, compound: The organ of sight in insects, the term is generally applied to compound eyes only.

Eye, simple: See ocellus.

F

F_1: The first filial generation produced by cross between two individuals.

F_2: The second generation of given individuals.

Facets (sing., facet): (i) The individual units that compose a compound eye. (ii) Surface of an ommatidium.

Facial carinae: A pair of carinae (ridges) running between the lateral ocelli and the clypeus.

Factitious (Surrogate) host: An insect that is not the preferred host but is used as a host for the production of parasitoids. Such hosts are usually used because they are easier to rear in large numbers than the preferred host, *e.g.*, *Corcyra cephalonica* St. is a factitious host of *Trichogramma* sp.

Faculative parthenogenesis: Parthenogenesis in which a species reproduces with or without males. The offsprings are females if the mother is unmated. See bligatory parthenogenesis.

Facultative mutualism: Capable of forming but not obliged to form a mutualistic relationship.

Facultative parasite: A parasite that can complete its life cycle without invading a host but can also take advantage of host if one is available.

Faeces: Excrement or the material passed from the alimentary tract through the anus. The excrement of insects varies from character from dry pellets to a copious clear liquid according to the amount of fluid in the diet. In plant sucking coccids and aphids, it contains much unabsorbed organic matter, particularly carbohydrates and amino acids; the dried residue collects on the surface of plants to form honey dew like products.

Falcate: Hooked or curved like a sickle.

Fallow: Ploughed but left unseeded during a growing season.

False legs: The prolegs.

Family: A taxonomic division consisting of one or more genera agreeing in one or a set of characters and so closely related that they are apparently descended from one stem. It is subdivision of an order, suborder or superfamily, containing a group of related genera, tribes or subfamilies. Family names of animals end in idae, *e.g.*, Reduviidae.

Family group: Consisting of family, subfamily, tribe and subtribe.

Farmyard manure: Cattle droppings (faeces) mixed with straw or similar material used as bedding in sheds, barns or night yards. Farmyard manure is often heaped and left to rot well before using it as an organic fertilizer in crop fields. It helps to improve water retaining properties of soils.

Farnesol: A sesquiterpenoid alcohol, $C_{15}H_2OH$ with weaker juvenile hormone activity in insects. The materials of this kind are widely distributed in small amounts in animals and plants. In 1961, Schmialek isolated it from the faeces of yellow mealworm. Schmialek subsequently tested a variety of terpene derivatives and found that farnesyl methyl ether and farnesyl dimethylamine were more active than farnesol.

Fascicle: A small bundle; the bundle of piercing stylets of insects with piercing sucking mouthparts.

Fasiculate antennæ: Every joint possessing a distinct long hair.

Fastigium: With Orthoptera, anterior part of vertex of the head projecting forward beyond the eyes. (a Latin word means a slope, roof.)

Fat body: A meshwork of loosely organized mesodermal cells in the body which involved in intermediate metabolism, storage and storage excretion. The fat body is composed of three types of cells: (i) Adipocytes or trophocytes are the predominant metabolic storage cells resembling the vertebrate brown fat, (ii) urocytes contain uric acid for storage excretion, and (iii) mycetocytes contain various types of symbionts, mostly the bacteria. Sometimes, the oenocytes of obscure function are found in conjuction with the fat body.

Fatigue: Exhaustion from stimuli following each other too rapidly to permit complete recovery of nerve tone or irritability.

Fauna: All of the animals found in a given area at a given time.

Faunistic: Pertaining to fauna of a region; or animal life characteristics of a particular region or environment.

FCI: Food Corporation of India. The FCI came into being as an autonomous body under the administrative control of the ministry and started functioning from 1st January, 1965 with the policy of distribution of food grains throughout the country at reasonable price, managing satisfactory level of operation and buffer stocks of food grains to ensure national food security and safeguards farmers' interests by effective price support operations.

Fecundity: The number of eggs or young ones produced by an individual or species.

Feeding deterrent: A chemical which prevents feeding of the pests on treated material.

Feeding organs: The insects are adapted for various types of feeding habits mostly because of extreme evolution of their mouth parts. Following types of feeding organs or mouth parts are found in various insects: (i) Chewing-biting mouth parts, *e.g.*, Orthopteroid insects; (ii) Piercing-sucking type mouth parts, *e.g.*, Hemipterous type (bugs), Anopleuran type (biting lice), mosquito type (female *Anopheles*), horse fly type (horse fly), muscoid type (stable fly), Siphonapterous type (fleas); (iii) Sponging type (house fly); (iv) Rasping and sucking type, *e.g.*, thrips; (v) Siphoning type, *e.g.*, butterflies and moths; (vi) Chewing and lapping type mouth parts (honey bees) and (vii) degenerative and filter feeding type, *e.g.*, mosquito larvae.

Feigning death: Playing dead, *i.e.*, going into a rigid motionless position when disturbed, also called as thanatosis.

Feldspar ($KAlSi_3O_8$): It is anhydrous silicate of aluminium and potassium. Feldspar forms the important constituent of the acid rocks like granite. The powder is pinkish white in colour. It is used as an inert carrier in the insecticide dust formulations. Feldspar is the most common mineral in the Earth's crust and are the major component in nearly all igneous rocks found on the Earth, on the Moon, and in some meteorites. They also are common in metamorphic and some sedimentary rocks. Their complex chemical and structural properties make them useful for interpreting the origins of rocks. Natural feldspars can be divided into alkali and plagioclase feldspars.

Felt chamber: Atrium of a spiracle lined with fine hairs.

Female reproductive system: The female reproductive system consists of a pair of ovaries (except some collembolan which possess a single ovary with one ovariole), a pair of lateral oviducts, a median (common) oviduct, a spermatheca and accessory sex glands. In Ephemeroptera, the lateral oviducts do not open into the median oviduct as found in most of the insects but both open directly through paired gonopores on the seventh sternum. In other insects, the median oviduct opens through a gonopore on the eighth sternum. In Lepidoptera (other than Microlepidoptera), two genital openings (on the eighth sternum forming a bursa copulatrix and on the ninth sternum) are found amongst which the latter gonopore represents a true genital orifice.

Femur (pl., femora): The third long segment of the leg found between the trochanter and tibia. It varies in shape and size among insects. It constitutes the largest part of leg and in Orthoptera, it is highly developed for leaping or jumping.

Fenestrae: Openings in the dorsal diaphragm through which hemolymph can circulate.

Fenitrothion ($C_9H_{12}O_5NPS$): It is an organophosphate insecticide with contact and stomach poison; chemically, dimethyl 3-methyl-4-nitrophenyl phosphorothionate; pure form is clear liquid with unpleasant odour, the technical material is 95 per cent pure yellowish brown oily liquid with faint smell. It is insoluble in water (30 mg l^{-1} at 20 °C) but soluble in alcohol and ether. Its insecticidal properties were discovered in Japan. It is less toxic to mammals but highly toxic to bees. Its chemical properties were similar to methyl parathion. It is marketed under trade name Sumithion and Folithion.

Fenpropathrin: A synthetic pyrethroid marketed under trade name Meothrin (10 EC).

Fenthion: An organophosphate insecticide, avicide and acaricide. Trade name is quelitox.

Fenvalerate: A synthetic pyrethroid discovered by Sumitomo Chemical Co., Japan in 1972. It is chemically, alpha-cyano-m-phenoxybenzyl alpha-isopropyl-p-chlorophenyl acetate ($C_{25}H_{22}ClNO_3$). It is yellowish oily liquid, practically insoluble in water but soluble in organic solvents. It is a contact and stomach poison, effective against lepidopterous larvae, marketed under trade names Sumicidin and Fenval.

Fermentation: (i) The process of growing a selected organism, usually a bacterium, mould or yeast on substrate so as to bring about a desired change or to generate products of the cell metabolism. For example, ethanol and carbon dioxide are produced from yeast fermentation. (ii) A chemical breakdown of honey caused by sugar-tolerant yeast and associated with honey having high moisture content.

Fermentation chamber: Certain lamellicorn larvae (*Cetonia, Oryctes, Osmoderma*) that feed on pine needles and ingest with their food those microorganisms which ordinarily ferment cellulose in nature. These flourish in the dilated hind intestine (the so called fermentation chamber) of these larvae, and form an important, perhaps an essential, aid to digestion, being themselves digested latter by the proteolytic enzymes of their host. Other insects that feed on wood, harbour yeast like organisms, often within the cells of the gut; and it was natural to attribute to these also, the function of digesting cellulose. But in fact some beetles that have such symbiosis are unable to digest cellulose and others which are without them can do so.

***Ferrisia virgata* (Cockerell):** Belongs to order Lipidoptera Hemiptera and family Pseudococcidae. It is an insect pest of potato, jack fruit, coffee etc. commonly known as the white-tailed mealy bug.

Fertile: Capable of producing offspring.

Fertile queen: A queen inseminated instrumentally or mated with a drone which can lay fertilized eggs.

Fertility: The reproductive potential of an individual or population as measured by the ability or rate at which fertilized eggs (zygotes) are produced.

Fertilization: (i) The application of nutrients (fertilizer) to soil in order to promote growth and development of crop plants, (ii) The penetration of the spermatozoa through the micropyle of a union with its cell nucleus.

Fertilizer: Substance added to soil to make it more fertile. Natural or artificial substance containing the chemical elements that improve growth and productiveness of plants. Fertilizers enhance the natural fertility of the soil or replace the chemical elements taken from the soil by previous crops. The use of manure and composts as fertilizers is probably almost as old as agriculture. Modern chemical fertilizers include one or more of the three elements most important in plant nutrition: nitrogen, phosphorus, and potassium. Of secondary importance are the elements sulfur, magnesium, and calcium.

Fetotoxin: A substance that can poison the foetus (child developing in the womb).

Fibre tracts: Groups of axons with a ganglion that run parallel to one another fidelity. The accuracy with which populations estimate follows actual numbers in a filament. A slender thread like structure.

FID: Flame ionization detector for gas chromatography or HPLC.

Field bees: Worker bees at least three weeks old that work in the field to collect nectar, pollen, water, and propolis.

Field drainage: Removal of excess water from soil and transport to surface water in order to improve soil productivity.

Fig midge: See *Anjeerodiplosis peshawarensis.*

Fig or almond moth: See *Cadra* (=*Ephestia*) *cautella* (Walker).

Figitidae: Family of super family Cynipoidea, suborder Apocrita and order Hymenoptera. These insects are parasites of Diptera and Neuroptera.

Filament: A thread like structure, especially one at the end of an antenna.

Filariasis: Infection with parasitic nematodes, *Wuchereria bancrofti* (Bancroftian filariasis) and *Brugia malayi* (Brugian filariasis) often transmitted by mosquito, *Culex pipiens quinquefasciatus.*

File transfer protocol (FTP): With the help of FTP, a file may be transferred from one computer to another computer. When we transfer any file from other computer to our own computer, then it is called file down loading and when a file is sent from our computer to any other computer, then it is called file uploading.

Filiform: Hair like or thread like antennae in which all the segments are nearly uniform in size and there is no prominent constriction between the two segments. This type of antenna is found in grass hopper, tiger beetle, ground beetle and Blattidae.

Filler: Inert component of pesticide dust or granule formulation.

Filter chamber: Due to the fusion of anterior hindgut with the anterior or posterior most part of midgut or end of foregut the so called filter chamber is formed in many Homoptera (Cercopidae, Sternorrhyncha etc.) This modification facilitates quick removal of excess water from the lumen of foregut or anterior region of midgut to the hindgut so that it avoids dilution of concentration of enzymes involved in digestion in the middle region of midgut and also in turn avoids dilution of the haemolymph. The anterior midgut thus modifies into a bladder. The water passage directly from the anterior midgut to the hindgut along the osmotic gradient and posterior midgut receives concentrated fluid suitable for enzyme activity. The structure in different families of Homoptera undergoes complex modifications.

Filter feeding: Feeding by straining small particles from water occurs in some aquatic larva, *viz.*, culicid larva (Culicidae, Diptera), some Ephemeroptera and Trichoptera. The suspended matter from water are strained with the help of fans, combs, brushes or sieves (Diptera, Ephemeroptera and Trichoptera) or spun silk nets (some Trichoptera). See degenerative mouth parts also.

Fin theory: Bradley (1942) framed the fin theory of wing origin. According to him, the preflight wings were too small to support the insect in air and later they have been evolved into the true wings by acquiring essential characters such as increase in size, development of venation, flight musculature and basal articulation. The ancient terrestrial Pterygotes have become amphibiotic adapted for excursions in the water, they were, however, capable of leaving the water for mating and dispersal. In his opinion, the preflight wings or prowings could have been highly useful as the propulsion organs or the fins in the water. In the process of using prowings as fins, the hinge and flight musculature might have started to develop. As the aquatic nymphs of Ephemeroptera and Odonata have closed type of respiratory system, the spiracles are mechanically closed and the fin like structures have been developed into the wings. It is supported by Grant (1945) while opposed by Ross (1955).

Fipronil: An insecticide belonging to a new group phenyl pyrazoles. It is broad spectrum, used in foliar and soil application and marketed under the trade name Regent 5 SC.

First filial: See F_1.

Fitness: The contribution of an individual to the gene pool through its offspring or success of its kind.

Fixation: Chemical treatment to check biochemical changes in cells at a particular stage for histological studies.

Fixed action pattern: A segment of behaviour performed in a stereotyped, species specific manner.

Flabellate: Antennae with fan-like processes or projections or special apparatus known as flabella, *e.g.*, stylopids.

Flacherie: See silkworm diseases.

Flagellum (pl., flagella): The long, apical antennal segments beyond the scape and pedicel. It is also known as clavoke.

Flange: A term sometimes used for a broad, projecting or convex margin; here used as a synonym of collar (when referring to the prothorax).

Flareback (Pest resurgence): The rapid increase of a target insect pest population subjected to an insecticidal treatment which also destroys its associated and possibly regulative natural enemies.

Flash heater: A device for heating honey very rapidly to prevent it from being damaged by sustained periods of high temperature.

Flatidae: A family of superfamily Fulgoroidea of suborder Homoptera and order Hemiptera. Widely distributed but mostly tropical, large moth like beautiful insects often delicately pigmented wedge shaped with wings held vertically at sides of the body at rest. Costal area of the forewings with numerous cross veins. The pale green *Flata ocellata* is a pest on *Jasminum sambac*.

Flavines: Greenish yellow pigments found in insects.

Flea: Any of various insects of the order Siphonaptera. Fleas are small, wingless, blood sucking insects that have legs adapted for jumping. They are parasitic on warm blooded animals.

Flea beetle: Any of various beetles of the family Alticidae (order Coleoptera). These are small leaf feeding beetles that have hind legs adapted for jumping.

Flight muscles: These muscles help the insects in the flight and can be grouped into four categories: (i) indirect muscles: include dorsal longitudinal muscles and dorsoventral muscles, (ii) direct muscles: include basalar- subalar muscles and the flexor muscles, (iii) accessory indirect muscles: include the pleurosternal muscles and tergopleural muscles, and (iv) accessory dorsoventral muscles.

Floatation method: The process used to separate microorganisms from soil by causing them to float.

Flora: All of the plants found in a given area at a given time.

Flowability: Property of flowing possessed by dusts.

Fluorescent: Tracer fluorescent material added to a spray to aid the assessment of spray deposits on plants.

Fluted scale: See *Icerya purchasi.*

Fly: Any of numerous two-winged insects of the order Diptera, especially any of the family Muscidae. Examples of true flies are the house fly and the tsetse fly. The true flies (order Diptera) should not be confused with other insects that are called 'flies', such as caddisfly, butterfly, and dragonfly that belong to other insect orders.

Fog treatment: The application of a pesticide as a fine mist for the control of pests.

Foliar application: Application of a pesticide to the leaves or foliage or plants.

Foliar spray fertilizers: Nutrients sprayed in solution on to leaves.

Follicle: Structure composed of oocyte, nurse cells and follicular epithelium (follicle cells, the inner epithelial cells of ovariole).

Follicular epithelium: Layer of cells that invests a developing oocyte that secretes the chorion.

Follower board: A thin board used in place of a frame usually when there are fewer than the normal number of frames in a hive.

Food chain: The transfer of food energy from the producers through various organisms as they depend for food on one another (herbivores to carnivores to decomposers) with repeated eating and being eaten is known as food chain. Thus it refers to sequence of species within a community. Each member serves as food for the next higher species in the chain, *e.g.*, plants-herbivores-carnivores; grass-cow-man, cabbage- caterpillar- bird. In any food chain, producers constitute the first trophic level. The energy stored in the green plants is used by the herbivores (primary consumers) which constitute the second trophic level. Carnivores (secondary consumers) eat the herbivores and constitute the third trophic level. In longer food chain, the highest consumer will get less energy because 90 per cent energy is lost during each transfer in the form of heat etc. The types of food chains are: (i) predator or grazing food chain, (ii) parasitic food chain, and (iii) detritus food chain.

Food chamber: A hive body filled with honey for winter stores.

Food factor: It is average fraction of the total diet made by the food or class of food under question. For example, if total food consumed by a man per day is 2 kg and the treated food (containing 1 ppm pesticide) be 500 g, the food factor will be 500/2000= 0.25.

Food pyramids: Various biotic components of ecosystem have nutritional relationship between producers and consumers which can be shown graphically by means of ecological pyramids. The first or producer level constitutes the base of the pyramid and successive levels constitute the apex. Ecological pyramids are of three types: (i) Pyramid of numbers, (ii) pyramid of biomass, and (iii) pyramid of energy.

Food web: Food chains in any ecosystem are not simple and do not operate in isolated sequences. The food chains in an ecosystem are interconnected with each other and forming a sort of complex network which is known as food web. Energy flow in a food chain is although unidirectional but through various path in a food web. With more complex food web in an ecosystem, the system will be more stable because consumers can obtain their food by various organisms. If there is a scarcity of a particular kind of organisms, then other organisms can be used as a food in complex type of food web.

Foot rot: Rotting that involves the lower part of the stem-root axis, but not the distal parts of the roots.

Forageal legs: See corbiculate legs.

Foramen magnum: Posterior opening of the head capsule through which internal (alimentary canal, dorsal nerve cord etc.) pass to the thorax.

Foregut: The anterior portion of the alimentary tract from the mouth to the midgut. It is also known as stomodaeum, ectodermal in origin and differentiated into pharynx, oesophagous, crop and gizzard or proventriculus. These structures differ in morphology as well as in function.

Forensic entomology: The application of the science of entomology to answer questions of interest to the legal system, for example in relation to a human death, crime or to a civil action.

Forficula: A forked process; an osmeterium; fork shaped structure on the abdomen of the springtail that engages the tenaculum and is used to propel the insect through the air *i.e.*, the springing apparatus. A pair of backwardly directed appendages which overlies in a forked position at the base of the supra anal plate in Orthoptera.

Forficulidae: Family of order Dermaptera, suborder Forficulina and superfamily Forficuloidea. Aedeagus unpaired, tergum of tenth segment well developed, eleventh tergum and telson reduced or absent, second tarsal segment bilobed and dilated, *e.g., Forficula auricularia, Allodahlia*.

Forficulina: A suborder of Dermaptera. Mandibles are dentate at apex and base. Forceps well sclerotised, wings well developed, free living forms with well developed eyes.

Forficuloidea: Superfamily of order Dermaptera and suborder Forficulina.

Form: A subdivision within a species or variety.

Forma specialis: Special form. Biotype (or group of biotypes) of a species of pathogen that differs from others in the ability to infect selected genera or species of susceptible plants. Forma specialis can be abbreviated as f. sp.

Formic acid: The secretion of ants (family, Formicidae; order, Hymenoptera) serves as a venom, as a repellent against the enemies, and as a pheromone for communication within the society.

Formicoidea (Ants): Super family of Suborder Apocrita, order Hymenoptera. This superfamily includes a single family Formicidae, the members of which are very common and widespread insects and called ants. These are social insects and elaborate investigations have been carried out on their behaviour.

Formothion ($C_6H_{12}O_4PS_2N$): It is a contact and systemic insecticide with acaricidal action belonging to organophosphate group. Chemically, it is S-N formyl- N-methylcarbamoyl- methyl dimethyl phosphorothiolothionate; clear liquid, soluble in water and organic solvents. It is hydrolysed to dimethyl thiophosphoric acid and phosphoric acid. It is marketed under trade name Anthio.

Formulant: Any added material in a pesticide formulation other than the biologically active ingredient. This may include carrier or other substances which may enhance the biological activity or physico-chemical properties of the formulation.

Formulation: Way in which basic pesticide is prepared and sold for use. A formulation contains the active ingredient(s) and other substances such as carriers and stickers. Examples of formulations include: emulsifiable concentrates, wettable powders, suspension concentrates, dusts, baits, fumigants, aerosols and granules.

Fortuitous biological control: The accidental movement of exotic natural enemy to new area where pest population suppression eventually results, *e.g., Aphytis lepidosaphes* Comp. on purple scale of citrus, *Lepidopsaphes beckii* (Newm.), indigenous to the areas of China has spread to California and other countries along with the import of citrus plantings.

Fossil: Any hardened remains or traces of plant or animal life of some previous geological period, preserved in rock formation in the earth's crust.

Fossil fuel: Millions of years back when plants and animals present at that moment were suppressed in the earth due to the excessive pressure and temperature in the earth; these were converted into fossil fuel like coal, petroleum and gas. This whole process is the result of lacs of years.

Fossorial: Legs bearing sclerotized claws for digging, *e.g.*, mole cricket, Gryllotalpidae, Cylindrachetidae (Orthoptera) and nymphs of *Cicada*. The foretibiae are broad, highly sclerotized but stumpy with terminal digital processes or teeth while the tarsi and pretarsi are completely reduced. The foretibial digital processes are used for digging as the insects are adapted to the subterranean habit. In the scarabaeid beetles, the tibiae are toothed and are used in excavation.

Fossulae spongiosae: In Hemiptera, expanded, spongy, setose pads on the foretibia are present which assist in holding prey.

Fovea: A raised blister like area of wing membrane.

Foveola (pl., foveolae): One of the paired depressions on each side of the vertex in grasshoppers.

Fowl tick: See *Argas persicus.*

FPD: Flame photometric detector for gas chromatography.

Frame: Four pieces of wood designed to hold honey comb, consisting of a top bar, a bottom bar and two end bars.

Frass: Plant fragments made by plant-feeding insects, usually mixed with excrement.

Freezing tolerance: Survival despite freezing of body fluids from the antennae.

Frenulum: A bristle or group of bristles on the hind wings of certain moths and butterflies that hold the forewings and hind wings together during flight (a type of wing coupling apparatus).

Fresh weight basis: Pesticide residues are reported on the laboratory sample as soon as it is received, with no allowance of the moisture content. Maximum residue limits (MRLs) and pesticide residues in food commodities are expressed in this way.

Freundlich isotherm: Empirical relationship describing the adsorption of a solute with a liquid or gaseous phase with a solid in which the quantity of material adsorbed per unit mass of adsorbent is expressed as a function of equilibrium concentration of the sorbate.

Fringed wings: Wing margins fringed with long setae; wing lamina reduced in size, *e.g.*, wings of thrips (Thysanoptera).

Frons: Anterior sclerite or frontal region of the head which often lies between the arms of epicranial suture. In orthopteroid insects, it possesses the median ocellus. In some cases, it may bear the lateral ocelli also. In true flies (Diptera), it occupies almost the entire front surface of the head apart from the eyes.

Frontal bristles: The two vertical rows of bristles running down the face of a fly from the ocelli to the antennae.

Frontal ganglion: The median ganglion of the stomodaeal or stomatogastric nervous system that lies above the oesophagous.

Frontal lunule: A small crescent shaped sclerite located just above the base of antennae and below the frontal suture.

Frontal region: Usually referring to the area between the anterior portions of the eyes and in front of the eyes, whether there is a frontoclypeal suture or not.

Frontal sutures: Two sutures that diverge ventrally across the anterior portion of the head.

Frontal-orbital bristles: Short row of bristles on a fly's head between the frontal bristles and the eye.

Frontoclypeal region: It is an important modification in head capsule in some insects. The epistomal suture is wanting in cockroaches and in other insects in which the clypeal region can be separated from the frontal region by the position of anterior tentorial pits. Sometimes the frontal sutures are also indistinct. In Hemiptera, Hymenoptera and Lepidopteran larvae, the frons is greatly reduced or suppressed. In these insects, on the basis of position of dilators of cibarium or that of buccal cavity, the clypeal region can be recognized. Similarly, the frontal region can be identified by the position of a median ocellus and origin of labral muscles.

Frontoclypeal suture: A transverse groove separating the frons and clypeus and forming internally an epistomal ridge. Also called the epistoma.

Frontogenal suture: See subocular suture.

Fronto-orbital bristles: The short row of bristles on each side of a fly's head between the eye and the frontal bristles.

Fructose: The predominant simple sugar found in honey; also known as levulose.

Fruit fly: Any of various small flies of the family Drosophilidae (order Diptera). The larvae feed on ripening or fermenting fruits and vegetables or any of various flies of the family Tephritidae (order Diptera). The larvae hatch in plant tissue. See *Dacus dorsalis.*

Fruit sucking moth: See *Ophideres conjuncta.*

Fulgoridae: A family of superfamily Fulgoroidea of suborder Homoptera and order Hemiptera. Tropical insects with large size, brilliantly coloured, ocell are two, rarely three or may be wanting. Three segmented antennae terminate in filaments and inserted beneath the eyes. Easily identifiable by the reticulate anal area of the hind wings. In some peanut like proboscis or prolongation in front of the head is very much characteristic which was once believed to be luminous and hence the name lanternflies. *Pyrops chennelli* is a common species in India.

Fulgoroidea: A superfamily of suborder Homoptera and order Hemiptera. Antennae arise beneath the eyes on the sides of the head. Numerous sensillae present on the pedicel and a large sense organ is present on the basal segment of the flagellum. Recognised by the presence of tegulae; mid coxae separated and elongated; hind coxae immovable. The families covered in this superfamily are: Tettigometridae, Cixiidae, Delphacidae (Araeopidae), Derbidae, Meenoplidae,

Kinnaridae, Dictyopharidae, Fulgoridae, Flatidae, Acanaloniidae, Issidae, Ricaniidae, Eurybrachidae, Lophopidae etc.

Fuller's earth: See attapulgite.

Full-sib: Having both parents in common.

Fumarin ($C_{17}H_{14}O_5$): Chemically, it is 3-(1-(2-furanyl)-3-oxo-butyl)-4-hydroxy-2H-1-benzopyran-2pme, marketed under trade names Fumarin, Ratafin and Coumafuryl. It is used against the rats in the ratio of one part of the poison and 19 parts bait material.

Fume board: A rectangular frame, the size of a super, covered with an absorbent material such as burlap, on which is placed a chemical repellent to drive the bees out of supers for honey removal.

Fumidil-B: The trade name for Fumagillin, an antibiotic used in the prevention and suppression of nosema disease.

Fumigants, classification: A large number of chemicals used as fumigants could be broadly grouped under three categories: (i) Liquid fumigants: Chemicals having higher boiling point and exist in liquid form at normal room temperature, *e.g.*, EDCT mixture, ethylene dibromide and carbon disulphide etc. (ii) Solid fumigants: Chemicals available at tablets or powered form, react with atmospheric water vapour to release some toxic gas, *e.g.*, aluminium phosphide tablets, calcium cyanide etc. (iii) Gaseous fumigants: Chemicals having low boiling point and highly volatile at normal room temperature. Such chemicals are stored in metallic cylinders strong enough to withstand pressure exerted by the gas, *e.g.*, methyl bromide etc.

Fumigants, mode of action: The poisonous gas enters into the larvae, pupae and adult insects through the spiracles and into the eggs through chorion. The toxic action of the fumigants to insects and warm blooded animals results from its ability to combine with iron atom present in the iron containing respiratory enzyme, such as cytochrome oxidase, succinic dehydrogenase resulting in the mortality of the insects and animals.

Fumigants, properties: The fumigants have simple formulae compared to many other pesticides and possess the following properties: (i) These exist in the gaseous state at ambient or other temperature and pressure required. They diffuse readily in air and through granular commodities and materials such as soil. (ii) All these acts as respiratory poisons and potentially lethal to all stages (eggs, larvae, pupae and adults) of pest organism provided that adequate concentration of gas is achieved for a long enough period. (iii) These are generally non-persistent and present in treated materials for a relatively short period of time. However treatment at high dosage or repeated treatments can cause the accumulation of significant residues. It happens when the fumigant reacts chemically with the constituents of the treated products.

Fumigants: Liquid or solid chemicals that form gas and kill organisms by entering through the spiracles and tracheae. Fumigants are often used in the treatment of areas difficult to penetrate with sprays or other pesticidal formulations. It is as a volatile pesticide that enters an insect *via* the tracheal system or a pesticide exhibiting toxicity in the vapour phase, *e.g.*, ethylene dibromide, EDB^+, durofume, hydrogen cyanide or hydrocyanic acid, carbon bisulphide, carbon tetrachloride, ethylene dichloride, EDCT mixture, Eco_2fume, sulfuryl fluoride, grain-o-cide, methyl bromide, chloropicrin etc.

Fumigation: The application of a fumigant for disinfestation of an area or it is the process of exposing the pest or pest affected stocks to the fume of a chemical at a lethal dose in an air tight space.

Functional spiracle: A spiracle having a distinct opening and tracheal connection. Non-functional spiracles may be represented by a minute opening or scar and a blocked tracheal remnant. Primitively, there are 8 functional abdominal spiracles, but 8 is lost in many taxa and 7–8, 6–8 or 4–6 may be lost in some others.

Fundatrices: Wingless or apterous, viviparous parthenogenetic female aphids which emerge in the spring from the hibernated eggs. The antennae, legs and sense organs are not well developed in comparison to succeeding apterous generations.

Fundatrigeniae: Different life stages seen in the life cycle of aphids which are apterous, viviparous, parthenogenetic females, they are progeny of the fundatrices and live on the primary host.

Fungicidal: Fungi killing.

Fungicide resistance: A decrease in sensitivity to a fungicide. Fungicide resistance is the result of selection or mutation following exposure to the fungicide.

Fungicide: Substance that kills or inhibits the development of fungus spores or mycelium.

Fungivorous: Feeding on fungi.

Fungus (pl., fungi): Non chlorophyll bearing plants of a lower order than mosses and liverworts. Fungi often show mycelial, spreading growth. Some fungi can be disease causing organisms of plants. Others live on dead plant material and play a role as decomposers. Examples of fungi are rusts, smuts, mildews, moulds and yeasts.

Furcula: Forked spring of a springtail.

Furrow application: Placement of pesticides with seed in the furrow at the time of sowing.

Fuscous: Smoky grey-brown in colour, normally applied to wings.

G

Galea: The outer lateral lobe of a maxilla, provided with a muscle arising in the stipes.

Galecron: A synthetic insecticide and acaricide; chemically, N-(4-chloro-o-tolyl)-N, N-dimethyl formamidine; moderately toxic to mammals; acute oral LD_{50} for rats is 162-170 mg kg^{-1}; kills mites and their eggs infesting apple trees but restricted to non bearing trees or trees in dormancy.

***Galerucella birmanica* (Jacoby):** Belongs to order Coleoptera and family Chrysomelidae. It is an insect pest of brinjal, commonly known as the singhara beetle.

Galerucidae (cucumber beetles): Family of superfamily Chrysomeloidea, suborder Polyphaga and order Coleoptera. These are soft-bodied insects, 2.5 to 11 mm in length; yellowish with dark spots or stripes and characterized by closely approximated antennae, deeply bilobed third tarsal segment, globose forecoxae, tibiae not grooved and hindfemora not swollen. The larvae feed on roots and underground stems of cucurbitaceous plants. Some feed on leaves of millets by scraping the leaf tissue. The adults of the cucumber beetle *Acalymma vittata* is known to be a vector of cucurbit wilt in the USA. This family includes genera like *Aulacophora, Monolepta, Lema, Diabrotica, Hallirhotius, Galerucella, Galerucida* etc.

Gall: An abnormal growth of a plant caused by the presence in its tissues of a young insect or some other organism. Aphids, gall wasps, and gall midges are among the major gall causing insects.

***Galleria mellonella* Linn.:** Belongs to order Lepidoptera and family Galleriidae. It is an insect pest of bee wax, commonly known as the greater wax moth.

Gallery: A gallery or tunnel is relatively a small, usually circular passage made in the bark or in the wood by an insect for feeding or for oviposition.

Gamete: A sexual reproductive cell (an egg or sperm) that will fuse with one another before development. Gamete is a haploid cell formed from germ cells during gametogenesis.

Gametocyte: A sex cell stage of *Plasmodium vivax*, the malarial parasite.

Gamma irradiation: The use of rays from a radioactive source to control insect pests.

Gamma taxonomy: A taxonomic level which deals with various biological aspects of taxa, ranging from the study of intraspecific populations to the studies of specifications and evolutionary rates and trends.

Ganglion: A nerve mass that serves as a center of nervous influence.

Gardona: A synthetic organophosphate insecticide; chemically, 2-chloro-1-(2, 4, 5-trichlorophenyl) vinyl dimethyl phosphate; slightly toxic to mammals; acute oral LD_{50} for rats is 4000 to 5000 mg kg^{-1}; controls corn earworm and armyworm on corn seed crop; residual wall spray is done to control flies and flies in poultry cages.

Gas mask: A device which filters out chemicals in the spray, dust or gas from air breathed by the wearer. A full face gas mask must be worn to protect from gases; it should be equipped with adequate canisters if absorbent materials (or with oxygen supply). Simple respirators protect from spray and dust without covering the eyes but not from poisonous gases.

Gaster: The hymenopteran abdomen apart from the first segment (the propodeum) which is fused to the thorax. The front part of the gaster often forms a narrow waist.

Gasterophilidae (botflies): Family of suborder Cyclorrhapha, order Diptera. These are medium sized, stout flies with bare arista and often spotted wings. The larvae live in the alimentary canal of horse, mule, ass, rabbit, rhinoceros, elephant, rarely dog and man. The horse botflies are well known. *Gasterophilus intestinalis* is the bot fly of horse, ass and mule.

Gasteruptiidae: Family of super family Evanioidea, suborder Apocrita and order Hymenoptera. These are characterized by having the propleuron forming a long neck, swollen hind tibiae, a long gaster and often a long exerted ovipositor. An anal lobe is wanting in the hindwing. They are mainly parasitic on solitary bees and also on sphecoids and other wasps. Some Indian species are *Gasteruption orientale* and *G.mandibulare.*

Gastric caecum (pl., gastric caecae): The sac like diverticula at the anterior end of mid gut (2-6 in different insects). This is also known as hepatic caeca or enteric caeca. The number of gastric caecae are six in cockroach and wanting in Lepidoptera.

Gastric lavage: To remove the stomach content incase of insecticidal poisoning through ingestion. To induce vomiting 15 g sodium chloride with warm water can be given.

Gastric mill: A grinding structure within the foregut of some orthopteroid insects.

Gastrin: See vertebrate hormones in insects.

Gastrocoele: The gastrulation cavity (Archenteron).

Gastro-enteritis: The inflammation of the membrane of both stomach and intestine.

Gastro-intestinal irritation: An Irritation relating to both stomach and intestine.

Gastrulation: The formation of endoderm, either by invagination of one wall of the blastula or by internal proliferation of cells from the blastoderm.

Gattine: Disease of silkworm, characterized by worms appearing translucent due to complete absence of food in gut; retarded growth. It is caused by high temperature in rearing room.

Gause's law: An ecological principle stating that different species occupying identical ecological niches (*i.e.*, ecological homologues) cannot co-exist indefinitely in the same habitat.

GC-EC: Gas chromatography with electron capture detector.

GC-MS: Gas chromatography with mass spectrometry.

GC-MSD: Gas chromatography with mass selective detection (usually low resolution mass spectrometry using selected ions.

Gelechiidae: Family of super family Tinaeoidea of suborder Ditrysia and order Lepidoptera. This family consists of a large number of small to minute moths which are mostly cryptically colored. The forewings are trapezoidal and narrower than hindwings. Feedind habits of larvae vary considerably and often they are destructive to crops. The groundnut leaf miner *Stomopteryx subsecivella* often causes serious damage to groundnut crop. The larvae of well known potato tuber moth *Phthorimaea operculella* mine the leaves and burrow into the tubers of potato in the field and in store

Gena (pl., genae) and subgenal areas: The cheek or lower part of the head beneath the compound eyes and laterad of gular sutures. On either side, the genae extend from the compound eye to the mandibular trochantin. The hinder part is developed into the socket to accommodate the mandibular condyle. Posterior to the occipital suture, each lateral gena is commonly called as the post-gena. Each post-gena provides the condylar articulation for the maxillae. The subgenal area above and below the mandibles can be referred to as the pleurostoma and the hypostoma, respectively. These areas in some insects are separated by the hypostomal suture. Similarly, the pleurostomal suture separates the pleurostoma from the base of mandibles.

Genal comb: (i) A row of strong spines borne on the anteroventral border of gena of insects belonging to order Siphonaptera. (ii) Row of spines on the lower gena of certain fleas.

Genal process: An anterior projection of the gena, sometimes visible from above beyond the mandibles.

Gene: A particular sequence or cistron or nucleotides along a molecule of DNA (or on occasion RNA; certain viruses) which represents a functional unit of inheritance.

Gene cloning: Gene cloning is the step concerned with transformation of a suitable host with a chimaeric vector and cloning of transformant cell. It is also called DNA cloning.

Gene flow: The spread of genes from one breeding population to others owing to the dispersal of gametes or zygotes; it may give rise to changes to the allele frequency and thus a factor of evolution.

Gene for gene resistance: See vertical resistance.

Gene library: The collection of DNA molecules constituting the entire genome of an organism is called genome library and collection of the copies of DNA is called cDNA library. Gene library is also known as gene bank, DNA library or gene store.

Gene mutation: Sudden change in the structure of a gene is called gene mutation or point mutation which leads to the appearance of new characters.

Gene pool: The total genetic information encoded in the sum total of the genes in a breeding population existing at a given time.

Genera: Plural of genus.

General equilibrium position (GEP): GEP is the mean value of pest density around which the pest population tends to fluctuate as changes occur in the biotic and abiotic components of the environment without accompanied by a permanent modification in the composition of environment. It is long term average density of populations.

General use pesticide: A pesticide that can be purchased and used by the general public (See also, restricted-use pesticide).

Generation: The group of individuals of a given species that have been reproduced at approximately the same time; the group of individuals of the same genealogical rank. In other words, it is the period from a given stage in the life cycle (usually the egg) to the same stage.

Generation time: The length of time necessary to complete a life cycle, *e.g.*, from the time an individual first lays its first egg.

Genetic biodiversity: It refers to presence of variety of genes within the same species. Each gene is responsible for individual characteristics and the number of genes differ from one species to another.

Genetic control: A method of pest control which makes use of selected strains of the target species that possess genetic abnormalities. When released into the target population they mate with wild individuals and produce sterile offspring. See also sterile insect technique.

Genetic drift: Change in allele frequency as a result of random occurrences, variations in fecundity and mortality in a population.

Genetic technique: See autocidal control.

Genetically modified organism (GMO): An organism whose genes have been deliberately manipulated.

Genetics: Gr., *genesis*=origin. It is a science of heredity and variation.

Geniculate: Elbowed. A type of antenna in which first segment is large and following segments are small forming an angle like bent knee, *e.g.*, honey bees, ants, wasps, Chalcid flies etc.

Genital appendages: The genital segments 8-10 in male and 8th and 9th in female bear a group of appendages commonly called the genital organs and constitute well defined male and female external genitalia, respectively. These are involved in mechanism of copulation and perform sperm transfer from the male while copulation in the female. The main genital organ in the male is, therefore, the intromittent organ or terminalia or penis while in the female, it is ovipositor. The other genital organs in true sense are the accessory structures. According to Snodgrass (1935), the accessory genital organs can be differentiated into two groups. The organs of one group constitute the median intromittent apparatus and called the phallic organs and those of the other group, which are peripheral accessory structures called the periphallic organs. The phallic organs are involved in coitus while the periphallic organs perform the function of holding or grasping the female during copulation. The accessory genital organs of the male include the paired periphallic organs such as claspers, grasping hooks (harpagones) of the ninth segment and even the cerci of the eleventh segment in some species. The secondary copulatory organs situated on the second abdominal segment of Odonata represent a group of phallic organs and a penis. In Diplura, the penis is rudimentary. In female, however, besides the primary genital organ or egg laying tube, the ovipositor which develops from the appendages of eighth and ninth segments, no remarkable accessory external genital organs are found.

Genital appendages, post: The post-genital appendage is the cerci located on segment XI. They develop in between the epiproct and paraprocts. They are found in Aperygota, all hemimetabola except Hemiptera and are also reported in Mecoptera and Hymenoptera (Symphyta) from Holometabola.

Genital appendages, pre: The styliform appendages, collembolan appendages, larval leg like or gill like appendages and prolegs of caterpillars form the pregenital appendages. Apterygota commonly have the styliform pregenital appendages. Thysanura have the styli on II to IX abdominal segments, particularly in Machilidae. The abdominal appendages are in the reduced condition. The pre-genital abdominal appendages in collembola include collophore, tenaculum and forficula. The abdominal segment I to VII of the larva of *Sialis* (Neuropteran) bears paired lateral long appendages (legs) which are six segmented structures, tapering, attached basally to the lateral pleural lobe or coxapodite which is innervated by the muscles for movement of the appendages. The caterpillars of Lepidoptera bear five pairs of prolegs, four pairs on the abdominal segment III to VI and fifth pair on the X segment. The prolegs are considered as the true segmental appendages as found to be developed from embryonic limb rudiments and perform locomotory function.

Genital capsule: A capsule more or less enclosing the aedeagus. This capsule normally consists of the combined ninth and tenth abdominal segments but occasionally it may be formed by the eighth segment, in which case the pregenital segments are reduced to a fork like spiculum gastrale. See pre-genital segments.

Genital chamber: See vagina and bursa copulatrix.

Genital claspers: Organs of the male genitalia which serve to hold the female during copulation.

Genital ridge: Embryonic tissue of mesodermal origin that together with the germ cells are the precursors of gonads.

Genital segment: Specifically the ninth segment of the abdomen in the male insects, though other segments are frequently associated with ninth in the genital complex (Gonosomite).

Genitalia, female: (i) Ovipositor: The female genitalia mainly consist of simple egg laying tube called ovipositor. It is basally communicated with the female gonopore which is located on either the 8th or 9th segment. It is of gonapophysial origin. The paired gonapophyses which develop on the sterna of the 8th and 9th segments form this tubular structure. It is mainly found in most of the Thysanura and Pterygota, while absent in Collembola, Diplura, among other Apterygota and Ephemeroptera, Odonata, Plecoptera, Mallophaga, Anoplura, tubuliferous Thysanoptera, Coleopteran, Neuropteran, Mecoptera, Trichoptera, Lepidoptera and Diptera among Pterygota. (ii) Substitutional ovipositor: The tubuliferous Thysanoptera, Mecoptera, Lepidoptera, Coleopteran and Diptera bear the substitution ovipositor in absence of the true ovipositor. This substitution ovipositor represents simply a telescopic tube which is formed due to the modification of the posterior abdominal segments into a tubular extension which acts as the egg laying tube, like a true ovipositor. See ovipositor structure also.

Genitalia, male: The male genitalia exhibit wide variation among different orders. (i) The penis is formed due to fusion of paired appendages of the 9th segment. It is single median tubular structure or a phallus in most of the insects while paired in Ephemeroptera and Dermaptera. In few higher insects, few accessory phallic structures after establishing a close association with penis constitute a complex median intromittent organ. The median phallic organs are median genital outgrowths of the 9th segment around the gonopore and develop either into the lobes, the phallomeres or into a median tubular penis or phallus. The phallic organ is formed from three phallomeres in Dictyoptera. The typical phallus is a conical or tubular structure but variably modified in various groups of insects. There is only a single tubular phallus in Thysanura and Plecoptera. Each phallus or penis is divided into a proximal part, the phallobase and terminal part, the aedeagus, while in undifferentiated condition entire phallus is called aedeagus.

Genitalia: All the genital structures are collectively known as genitalia. These are sexual organs and associated structures, but usually applied to external genitalia.

Genite: Moderately toxic, genite is chemically similar to Aramite, Ovex and Tedion. It is a good general purpose miticide that is also effective against some insects. It is a synthetic acaricide, a sulfonate, 2, 4-dichlorophenyl ester of benzenesulfonic acid; acute oral LD_{50} for rats is approximately 1650 mg kg^{-1}; phytotoxicity may cause leaf yellowing and damages to apples when applied after pre-blossom stages.

Genome: The genetic material of an organism, *i.e.*, one haploid set of chromosomes with their constituent genes.

Genotype: The total genetic character of an organism, *i.e.*, all its D.N.A. or genes. Genetic make up or constitution of characters in an individual or taxon, *e.g.*, TT (pure tall), Tt (hybrid, impure tall) and tt (pure dwarf) etc.

Genovertical plate: An area on the head above the antenna, next to the compound eye as in Diptera.

Genus (pl., genera): A group of closely related species; subdivision of a family. Genus names are written with a capital and should be printed either in italics or underlined. The name of the genus is incorporated into the scientific names of all the member species: *Pieris napi* and *Pieris rapae*, for example, both belong to the genus *Pieris*.

Genus group: Consisting of genus and subgenus.

Geocline: A gradual or continuous change in a character over a considerable area as a result of its adjustment to changing geographical conditions.

Geographical isolation: Prevention of gene exchange between a population and others by geographical barriers.

Geology: The study of internal structure of earth and hypothesis of its present form. The various substances, like minerals, fossils, sediments and rocks found in the earth are also studied in this branch of science.

Geometridae (true loopers, carpets, waves, pugs etc): Family of super family Geometroidea, suborder Ditrysia and order Lepidoptera. This is a very large family, the members of which have a slender body with relatively large wings. The moths are not strong fliers and often keep the wings horizontal at rest. Generally the moths have proboscis and frenulum and in a few cases either the one or the other may be absent. Females of some genera possess degenerate wings. *Alsophila, Apocheima, Eranius, Operoptera* and its allies are apterous. The larvae are generally elongate and slender and possess prolegs on the sixth and the tenth abdominal segments.

Geometroidea: Super family of suborder Ditrysia and order Lepidoptera. Adults with vestigial or atrophied maxillary palpi and tympanal organs in abdomen. This superfamily includes the family: Axiidae, Drepanidae, Cymatophoridae, Geometridae, Uraniidae, Epiplemidae and Sematuridae.

Geothermal energy: Energy present in depth of the earth is called geothermal energy. The temperature of earth increases with the increasing depth. Temperature in the earth at the distance of 10 km is about 120 °C and it increases to 300 °C at the depth of 320 km.

Geotrupidae (dor beetles): Family of superfamily scarabaeoidea, suborder Polyphaga and order Coleoptera. These beetles are characterized by distinct scutellum, approximated midlegs, two spurs on tibia on hindlegs and the three terminal segments of 11-segmented antennae ending in a lamellate club. They stridulate by rubbing together a file on the hindcoxae and the sharp edge of the coxal cavity. The common Indian genera are *Geotrupes, Athyreus, Ceratophyses* and *Bolbocerus.*

Germ band formation: With the formation of blastoderm, the egg attains the blastula stage of development. The blasoderm initially appears as a egg, it undergoes cell division repeatedly so that it is differentiated into a dorsal thin and ventral thick ports. The ventral thick part f the blastoderm is called the germ band. The germ band alone, latter on develops into the embryo. The dorsal thin part of the blastoderm remains as the extra embryonic membrane and is often termed as the serosa. The germ band is also called the germinal disc or the primitive band.

Germ cells: Cells that differentiate into gametes, *i.e.*, eggs and sperm.

Germarium: The apical tissue of a testicular follicle or ovariole from which the germ originate. In female reproductive system, it is second part of ovariole which contains primordial germ cells, prefollicular cells in some insects, and nutritive cells. The primary oocytes are compactly arranged in the germarium and active site of the oogenesis.

Germinal layer formation: The embryonic stage which involves formation of the germinal layer is commonly referred as the gastrulation. In insects, however, blastulation and gastrulation stages cannot be separated from one another. According to some workers, the blastulation stage is suppressed by gastrulation during the embryonic development of insects and the blastula recognized by some workers is in the real sense of the gastrula.

Gerrididae: Family of superfamily Gerroidea, suborder Heteroptera and order Hemiptera. Predaceous bugs which live on the surface of water, running or staking over the water film. Head is elongate and slender with a four segmented rostrum. The forelegs are short used in capturing the prey, whereas, the long middle and hindlegs are used in locomotion. The tarsi are two segmented and clothed with fine hairs which do not get wet and thus enable the insect to move about on the water surface. Scent gland is present on the metathorax. Winged and wingless forms are noticed in many species. The examples are *Halobates* and *Gerris tristan.*

Gerroidea: Superfamily of division Gymnocerata, suborder Heteroptera and order Hemiptera.

Gestalt theory: Most of the information on colour vision of insects has been obtained by training experiment: stimuli which the insect can learn to associate with the presence of food, or with the location of its nest, are judged to be perceptible. such experiments, of course, need the greatest care in their interpretation; for the insect may be guided by some other stimulus that has been over looked by the experimenter; indeed, in the course of a single experiment, it may cease to be orientated by one stimulus and come to depend upon another; and there are many who hold that the whole perceptual experience of the organism is integrated, as it were, into a pattern, and that it responds to general changes in this pattern, and not to the isolated stimuli of which it is composed.

Ghujhia weevil: See *Tanymecus indicus.*

Giant axon: These large nerve fibres conduct impulses at a rapid rate; they carry stimuli from sense organs in the abdomen and enable the insect to give a very quick stereotyped response with about the same speed as the eye blink in man. Insects which are capable of very quick escape reactions, such as the cockroach or larvae of dragon flies, possess giant axons.

Giant chromosomes: Chromosomes those are much larger than ordinary chromosomes.

Gill (Bronchial respiration): Breathing or respiratory organs possessed by many aquatic creatures, including numerous young insects. Insect gills are usually very fine outgrowths from the body and they contain numerous air-tubes, or tracheae. Oxygen passes into the tubes from the water by diffusion.

Gill cover theory: The Gill cover theory of origin of insects has been proposed long back in 1906 by Woodworth. It bridges the spiracular flap theory and the fin theory. According to this theory, the origin of wings is assumed to be conditioned by a change of habitat from terrestrial to aquatic to aerial. Woodworth (1906) considered the abdominal gills as the gill covers found in some mayfly nymphs which actually function as the opercula for the filamentous gill clusters. These covers are stiffed by a real venation and the veins are even provided with rows of spines as in the flying wings and in the final moult, these covers develop into the wings.

Ginglimus: See clypeus.

Girdle: A silken thread supporting the mid section of a pupa.

Gizzard: See proventriculus.

Glabrous: Smooth, without hairs. See subglabrous.

Glandular hair: These are macrotrichia (cuticular modifications). These occur around the opening of various epidermal glands and can be seen in a few Lepidopteran larvae.

GLC: Gas liquid chromatography.

Glial sheaths: Throughout the nervous system the neurones, both cell bodies and axons are invested and insulated from one another by cellular sheaths formed by the glial cells (Schwann cells). Only at the synapses do the nerves fibres come into intimate contact with one another. In the central ganglia, modified glial cells named perineurium cells surround the entire nervous system and secrete over it a tough connective tissue sheath, the neural lamella.

Gliding: Mechanism to reduce heat production in insects by alternate flying and gliding.

Global warming: The slow increase in the earth's surface air temperature caused by man made CO_2 in the atmosphere, trapping excessive amounts of solar radiation which would otherwise be reflected back into space (also called greenhouse effect).

Globose, globular: Spherical or nearly so.

Glomeruli: Spheroidal collections of neuropile in each antennal lobe of the brain.

Glossa (pl. glossae): One of a pair of lobes at the tip of labium or lower lip: usually very small, but long in honey bees and bumble bees, in which the two glossae are used to suck up nectar.

Glyphipterygidae: Family of super family Tinaeoidea of suborder Ditrysia and order Lepidoptera. Small moths, the larvae of which feed on leaves. The larvae of *Phycodes radiata* and *P.minor* web and feed on leaves of *Ficus* sp. South India.

Gnathal appendages: The gnathal appendages consist of paired mandibles, maxillae and labium. The labium is formed due to the fusion of paired structures, the second pair of maxillae. All these structures undergo variable modifications and are adapted for different modes of feeding.

Gnathocephalon: Three postoral appendage bearing segments fused to the procephalon to provide mouth parts.

Gnathosoma: The anterior part of the body of mites and ticks which bears the mouth and mouthparts.

***Gnorimoschema (=Phthorimaea) operculella* (Zeller):** Belongs to order Lepidoptera and family Gelechiidae. It is an insect pest of potato, commonly known as the potato tuber moth.

Golden nematode: See *Heterodera rostochiensis.*

Gomphidae: A family of order Odonata. Includes clubtails. Example, *Gomphus* sp.

Gonadal endocrine cells: The apical cells or apical complex in fire fly, *Lampyris noctiluca* and follicle wall cells in the testis of cockroaches and other insects secrete the male sex hormone, the steroid, 20-hydroxyecdysone. Similarly, in oocyte follicle cells in the mature ovary produce the female sex hormone in a large number of insects (Lagueux *et al.*, 1981; Hoffman *et al.*, 1981). Similar to the male sex hormone, the ovarian hormone is also steroidal substance, 20-hydroxyecdysone and it is deposited in the eggs along with the yolk. For details of gonadal hormones, see ovarian steroid hormone and testis steroid hormone.

Gonopore: The external opening of the reproductive organs.

Good agricultural practice (GAP): Nationally authorized safe uses of pesticides under actual conditions necessary for effective and reliable pest control. It encompasses a range of levels of pesticide applications up to the highest authorized use, applied in a manner that leaves a residue which is smallest practicable amount. Authorized safe uses include nationally registered or recommended uses that take into account the public and occupational health and environmental safety considerations. Actual conditions include any stage in the production, storage, transport, distribution and processing of food commodities and animal feed.

Good laboratory practice (GLP): The formalized process and conditions under which laboratory studies on pesticides are planned, performed, monitored, recorded, reported and audited. Studies performed under GLP are based on the national regulations of a country and are designed to assure the reliability and integrity of the studies and associated data. The USA-GLP definition also covers field experiments.

GPC: Gel permeation chromatography.

Gracillariidae: Family of super family Tinaeoidea of suborder Ditrysia and order Lepidoptera. These insects are recognized by their habit of resting with the fore part of the body upraised. Larvae are apodous and mine into leaves. Adults are very small, delicate with narrow long fringed wings. The larvae of *Acrocercops syngramma* on mango, *A.coerulea* on *Lablab niger, Gracillaria soyella* on redgram *(Cajanus cajan)* and *G.theivora* on tea cause blotch mines.

Grade: The unit of biological improvement from an evolutionary point of view comprising a group of individuals similar in their level or organization.

Gradual metamorphosis (Simple metamorphosis): The newly hatched individuals resemble the adults in general body form but lacks wings and external genital appendages. Young individuals are called nymphs and undergo series of feeding instars instantaneously, *e.g.*, grasshopper, cockroach and aphids.

Grafting: (i) Removing a worker larva from its cell and placing it in an artificial queen cup in order to have it reared into a queen, (ii) A method of plant propagation by transplantation of a bud or a scion of one plant on another plant. The joining of cut surfaces of two plants so as to form a living union.

Grafting tool: A needle or probe used for transferring larvae in grafting of queen cells.

Grainages: These are the places where silk seed is produced at large scale. The healthy disease free cocoons of pure race are selected after testing the pupae for pebrine disease. Deformed flimsy, stained and dead cocoons are removed and quality cocoons are selected. Floss from cocoons is removed so that moths can emerge from them. They are placed in grainage in a single layer in wooden trays and can be preserved at 23-25 °C ad 70-80 per cent RH. Moths generally emerge out during early morning. The virgin females are transferred into plastic bowl like trays and then the male moths of required race are put in the same trays and are covered with black cloth. Males are removed after copulation. The females lay eggs after 12 hours at a rate of 400-600 per moth. The egg cards along with eggs are washed with 2 per cent formalin solution for one hour and latter by water and dried in room. The eggs are transferred to incubation chamber for supplying to the sericulture centres.

Grain-o-cide: It is a mixture of CS_2 and CCl_4 in the ratio of 4:1. The efficacy of mixture is directly proportional to its CS_2 content. The CCl_4 is mixed to reduce the fire hazards of CS_2. It is well suited grain fumigant under Indian conditions and also used as a soil fumigant. It is liposoluble and hence a nerve poison. It is used at a dosage of 30-40 kg per 1000 q grains or 2.2 l per m^3 with exposure period of 7 days.

Gram cutworm: See *Agrotis flammatra.*

Gram dhora: See *Callosobruchus chinensis.*

Gram pod borer: See *Helicoverpa armigera.*

Gramineae: A family of plants. Grasses (including cereals).

Granular cells: See oenocytoids.

Granular formulations: See granule.

Granulation: The formation of sugar (dextrose) crystals in honey.

Granule applicator: Machine designed to apply measured quantities of granules.

Granule: Coarse particle of inert material (pumice, Fuller's earth, rice husks) impregnated or mixed with a pesticide. Used mainly for soil application but sometimes for foliar application (pumice formulation). Solid formulation comprising particles of defined size, *i.e.*, more than 80 μm diameter for application generally to soil and plant whorls without further dilution.

Grapevine beetle: See *Synoxylon anale.*

Grapevine leaf hopper: See *Erythroneura* sp.

Grapevine leaf roller: See *Sylepta lunalis.*

Grapevine sphinx: *Hippotion celerio.*

Grapevine thrips: See *Rhipiphorothrips cruentatus.*

Grapevine: See *Arboridia viniferata.*

Grasping mandibular mouth parts: The mandibles are provided with a pair of large tubular mandibular glands and small abductors and large adductors. The maxillae, labium and hypopharynx are united to form a large composite structure projected behind the mouth like a thick upper lip called the maxillolabial hypopharyngeal complex. It bears spinneret at its tip. Basally, the organ is supported on hypostomal lobes of post-genae. The wall of under lip complex is membranous. Two lateral lobes represent maxillae distinctly. The median lobe is formed of labium and hypopharynx. The hypopharynx bears distally spinneret, a hollow spine having orifice of silk gland duct at its extremity. Example, lepidopteran larvae.

Grasping sucking mouth parts: The mandibles are long, curved fangs hinged to lateral angles of head by dorsal and ventral articulations and thus work in a horizontal plane. They are found in some Neuropterans and Coleopterans such as Lampyrid or Dytiscid larvae.

Grass: Plant belonging to the family Gramineae.

Grasserie: See silkworm diseases.

Grasshopper: Any of numerous insects of the families Acrididae and Tettigoniidae (order Orthoptera). Grasshoppers usually have long, powerful hind legs adapted for jumping.

Gravid: Pregnant.

Grazing food chain: The food chain is starting from the green plants and going through the herbivores to the top carnivores.

Grease band: Adhesive material (*e.g.*, resin in castor oil or sticker) applied as a band around a tree to trap ascending wingless female mealy bugs or ants.

Grease band: Adhesive material applied as a band round the tree trunk to trap insects.

Greasy cutworm: See *Agrotis ypsilon.*

Greater wax moth: See *Galleria mellonella.*

Green bee eater: See *Merops orientalis.*

Green borer: See *Raphimetopus ablutellus.*

Green bug: (i) The common name for certain species of aphids that feed on crop plants. (ii) It is a phytophagous bug (*Nezara viridula*) feeding on many vegetable and crop plants.

Green house effect (Global warming): Rainforest canopies absorb carbon dioxide from the atmosphere. When the rainforests are burned and cleared, the carbon dioxide is released. Also when the trees are cut down, they can no longer absorb carbon dioxide. This means more carbon dioxide will be present in the atmosphere. Carbon dioxide allows the heat through the atmosphere. However, it does not allow the reflected energy to escape from the atmosphere. This is called green house effect and causes global warming. See also green house gases.

Green house gases: The atmospheric gases which are permeable to short wave solar radiation like U.V. or visible but are strongly absorber of long wave radiations emitted from the surface of earth and donot allow their escape from the atmosphere which are known as green house gases. These include CO_2, CH_4, nitrous oxide (N_2O), chlorofluorocarbons (CFCs), ozone (O_3) and water vapour (H_2O). These gases in the lower level of atmosphere behave like glass covering of a green house. Green house is basically a glass chamber in which glass operates in a fashion similar to CO_2, O_2 and H_2O vapour allowing the sun rays to pass into green house but blocking the reverse radiation. Certain green house gases do not allow the passage of these radiation out side the atmosphere. But not letting the solar rays to escape into outer space, green house gases add to the heat that is already present on the earth's surface which results in an increased temperature, commonly known as green house effect.

Green lacewing: See *Chrysoperla carnea.*

Green leafhopper: See *Nephotettix virescens and N. nigropictus*

Green manure: A crop grown and ploughed under to improve the soil.

Green nettle slug caterpillar: See *Thosea aperiens.*

Green semi-looper: See *Anomis flava.*

Gregarious parasitism: It is the phenomenon of parasitization of an individual host by more than one larva of single parasitic species but all survive is gregarious parasitism, *e.g.*, *Cotesia* (*Apanteles*) *flavipes* Cameron is a gregarious parasitoid of sorghum stem borer, *Chilo partelluls* (Swinhoe).

Gregarious: Living in groups.

Ground water: Water present in the saturated subsurface zone of the soil profile where all open spaces or pores in the sediment and rock are filled with water.

Groundnut aphid: See *Aphis craccivora* Koch.

Groundnut leaf miner: See *Stomopteryx nertaria.*

Groundnut stem borer: See *Sphenoptera perotetti* G.

Groundnut white grub: See *Holotrichia consanguinea* Blanchard.

Group rearing: The rearing of more than one individual (usually of the same species) in a single rearing container of individual rearing.

Growth ratio: See Dyar's law.

Growth regulators: Organic substances those are effective in minute amounts for controlling or modifying growth processes in plants or insects. For example, plant hormones.

Growth stage: A morphologic phase of plant or crop development that can be easily recognised in the field. For example, seedling stage, tillering stage, flowering stage etc.

Grub: A thick-bodied larva of a beetle with thoracic legs and a well developed head. The body is curved, C-shaped.

Grub, white: A white scarabaeiform larva; thick-bodied, with a well developed head and thoracic legs, without abdominal prolegs, usually sluggish in behaviour. The white grubs, *Holotrichia consanguinea* and *H. serrata* are phytophagous pests.

Grubbing: Digging out roots.

Gryllacrididae: A family of order Orthoptera, suborder Ensifera and super family Tettigonioidea. Wingless forms of grasshoppers.

Gryllidae: A family of order Orthoptera, suborder Ensifera and super family Grylloidea, *e.g.*, crickets.

Grylloblattodea: An order. Head prognathous with a complete tentoriun. Compound eyes are reduced or absent, ocelli absent, mouth parts are biting type ectognathous. The insects are secondarily absent. Arolium and pulvilli are wanting. Abdominal segments are ten, eleventh vestigial. Example, *Grylloblatta campodeiformis* sp.

Grylloblattodea: Insect order containing the rock crawlers.

***Gryllodes sigillatus* Linn.:** Belongs to order Orthoptera and family Gryllidae. It is an insect pest of household, commonly known as the house cricket.

Grylloidea: A super family of order Orthoptera and suborder Ensifera. Tarsi three segmented, *e.g.*, crickets and mole crickets.

Gryllotalpidae: A family of order Orthoptera, suborder Ensifera and super family Grylloidea, *e.g.*, mole cricket.

Gryopidae: A family of suborder Amblycera of order Mallophaga. Ectoparasites on rodents. Legs generally modified for clasping. Example, *Gyropus ovalis* on guinea pigs.

Guideline level: Maximum concentration of a pesticide residue in or on a feed or food commodity resulting from use of good agricultural practice but an acceptable daily intake has not been estimated.

Guideline value: Maximum level of pesticide residue in an environmental medium that ensures aesthetically pleasing air, water or food and does not constitute a significant risk to the users.

Gula: It is an important modification in head capsule. It is a ventral sclerotic plate of the head capsule separating the foramen magnum from the base of labium. It is well developed in insects having prognathous head in order to place the mouth parts in forward direction. The gula is separated from the laterally expanded post genae by the gular sutures. The gular sutures are infact modified post-occipital sutures. The gula varies in length and width in different species of Coleoptera and Neuroptera.

Gum: Sticky substance formed by cells in reaction to wounding or infection (usually in trees).

Gummosis: Production of gum by plant tissues.

Gurdaspur borer: See *Acigona steniellus.*

Gymnocerata: A division of suborder Heteroptera, order Hemiptera. The insects belonging to this division have long antennae which are not concealed in grooves on the ventral side of the head.

Gynandromorph: An individual creature with a mixture of male and female characteristics. One half of the body may be male and the other half female. This is particularly noticeable among the blue butterflies and others in which the sexes are differently coloured.

Gypsum ($CaSO_4.2H_2O$): It is material, grey in colour and is characterized by high bulk density. It is used as an inert carrier in the insecticide dust formulations.

Gyrinidae (whirligig beetles): A family of superfamily Caraboidea, suborder Adephaga, order Coleoptera. Small to large, ovoid or elliptical, more or less flattened, steely black or bronze-lustred aquatic beetles which are mostly gregarious. They keep part of the body submerged and swim on the surface of the water round and round rapidly and often also dive under, carrying a bubble of air attached to hind end. The short, thick, 11-segmented antennae are inserted beneath the front. The eyes are divided into dorsal and ventral portions, the former adapted for aerial vision and the latter for vision under the water surface. The forelegs are long and prehensile, in the male the dilated tarsi are with suckers. The mid and hind legs are adapted for swimming. The hind legs with rigid coxae are greatly flattened.

H

Habitat: The sum total of environmental conditions of a specific place that is occupied by an organism, population or community.

Habitat loss: Fragmentation or destruction of the environment with the characteristics necessary to support a given species, genus etc.

Habituation: A form of associative learning in which a given stimulus no longer elicits a response.

Habitus: Body build, general appearance.

Hadda beetles: See *Epilachna duodecastigma.*

Haematophagous: Blood feeding.

Haematopinidae: A family of order Anoplura. Sometimes with indistinct tergal and sternal plates and found on ungulates, dogs and rodents, *e.g.*, buffalo louse, *Haematopinus tuberculatus* and dog louse, *Linognathous setosus.*

***Haematopinus eurysternus* Mitzsch.:** Belongs to order Siphunculata and family Pediculidae. It is an insect pest of farm animals, commonly known as the sucking cattle louse.

Haemocoel: Body cavity of the insect in which the viscera are located and the haemolymph circulates. It is divided into three sinuses: (i) dorsal pericardial sinus, (ii) middle perivisceral sinus, and (iii) ventral perineural sinus. The dorsal pericardial and middle perivisceral sinuses are differentiated by dorsal diaphragm and middle perivisceral and ventral perineural sinuses are differentiated by ventral diaphragm.

Haemocyte differentiation: It is found that the primary types of haemocytes are the prohaemocytes which develop first from the embryonic mesoderm and thereafter they produce various types of haemocytes. During post-embryonic development, the prohaemocytes aggregate at some place and form distinct or some naked encapsulated haemopoietic organ. In Lepidopteran larvae, they are located behind the prothoracic spiracles, wing discs, fat bodies etc. and in third instar larvae they produce almost all types of haemocytes. The haemopoietic organs undergo complete disintegration at pupal stage, so they could not be seen in adults. The phagocytic organs of Orthoptera are equivalent to the haemopoietic organs of Holometabola. The role played by the haemocytes includes phagocytosis, encapsulation, storage and transport, coagulation, wound healing, connective tissue formation, immunity and detoxication.

Haemocytes: The insect haemolymph contains a variety of mesodermal, amoeboid, nucleated cells called haemocytes or blood cells. They are homologous with the vertebrate leucocytes and circulate freely in the haemocoel. The different types of blood cells are proleucocytes (prohaemocytes), plasmatocytes or phagocytes (amoebocytes), granular leucocytes, adipohaemocytes (adipocytes), coagulocytes (cystocytes), spherule cells, oenocytoids (crystal cells), podocytes (stellate cells), and vermiform haemocytes (nematocytes).

Haemocytopoietic organs: Organs that produce blood cells. See haemopoietic organs.

Haemoglobin: The red pigment (respiratory pigment) in the blood of vertebrate, certain invertebrates and very rarely in insects like the small red blood worms, aquatic larvae of chironomid gnats.

Haemogram: A count of the total number of haemocytes per cubic mm (THC) and an estimate of the differential haemocyte count (DHC) in an insect at a given time.

Haemolymph: The extra-cellular fluid containing blood cells or haemocytes and plasma, believed to represent blood, found in the haemocoelic body cavity of certain invertebrates. In insect body, it makes 15-75 per cent of the volume. The haemolymph is slightly acidic (pH 6-7) and includes organic ions, amino acids, other substances and pigments. It is, in general, a colourless or greenish yellow fluid. The green colour is due to chlorophyll and red colour in chironomid larvae is due to haemoglobin.

Haemopoietic organs: Many phagocytic organs are concerned with producing and liberating new haemocytes within the circulating blood which are known as haemopoietic organs.

Haemostasis: Inhibition of haemolymph loss; coagulation.

Hair plate: Cluster of tiny trichoid sensilla that have a proprioceptive function.

Hair: A cuticular extension, also called as macrotrichium or seta.

Hair pencils: Eversible structures covered with fine hairs providing extensive sun which facilitates dispersal of pheromones.

Half life ($t_{1/2}$): Time taken by the concentration of a pesticide in a compartment to decline by one half of its original value. Usually an estimate based on observed dissipation over several half lives that can be described by first order kinetics with rate constant k, $t_{0.5} = 0.693/k$.

Half-sib: Having one parent in common.

Halicitidae (mining bees): Family of super family Apoidea, suborder Apocrita and order Hymenoptera. Halictids are widely distributed, small to moderate-sized, often metallic bees which are mostly solitary and some are social with well developed workers. They visit flowers. The antennal socket is connected with the frontoclypeal area by a single suture. The submentum and mentum are wanting. The first sector of vein M on forewing is nearly always strongly curved. Usually prepectus is present and the post-scutellum is horizontal.

Halictophagidae: Family of order Strepsiptera. The insects belonging to this family are parasites of Homoptera Auchenorrhyncha and Tridactylidae. The adult males possess seven-segmented antennae and three segmented tarsi. In the female, the brood passage opening is narrow, linear or oblong.

Haliplidae: A family of superfamily Caraboidea, suborder Adephaga, order Coleoptera. Aquatic beetles with ten segmented, bare antennae and legs adapted for swimming. The hindcoxae are produced backwards in a plate partly covering the abdomen. The larvae carry long or short fleshy processes on the body and feed on algae in running and standing water. Some common species are *Haliplus angustifrons* and *H. pulchellus.*

Hallucination of arthropod infestation: A condtition in which a person imagines he or she is being molested by small, difficult to locate forms which localize on the body despite extra-ordinary preventive measures.

Haltere (pl., halteres): Hind wings modified to club shaped structures or small knob like balancing organs which are important and characteristic of Diptera (true flies).

Hamulate coupling: A wing coupling mechanism used in Hymenoptera. A row of small hooks (hamuli) on the anterior margin of the hind wing, engage with a sclerotised anterior margin of the fore wing.

Hamuli *(sing.,* hamulus*)*: Row of tiny hooks on the costal margin of the hind wings and a thin cuticular fold along the hind margin of forewings form the hamulate type wing coupling apparatus among Hymenoptera.

Hand duster: A manually operated device used for applying pesticides in the form of dust. In a hand duster, a plunger expels a blast of dust laden air. The dust chamber may be at the end of plunger tube itself or an enlargement at the end or it may be located below the plunger tube.

Hand shake duster: See duster, hand shake.

Haplodiploidy (facultative parthenogenesis): Reproduction *via* a combination sexual and asexual modes resulting in both haploid offspring (from unfertilized eggs) and diploid offspring (from fertilized eggs).

Haploid: Having only a single (maternal) complement of chromosomes.

Hard pesticide: A pesticide which persists for a long time in the environment (*e.g.*, an organochlorine insecticide, like DDT).

Hard water: Water with minerals such as calcium, iron and magnesium dissolved in it. Some pesticides added to hard water will curdle or settle out.

Harmful organism or pests: Includes vertebrate and invertebrate pests, pathogens and weeds.

Harmonic growth: A growth pattern in which all parts of the insect body and the body as a whole increase by the same ratio at each moult, also called the isogonic growth.

Harpagones: Movable per phallic processes of the ninth segment individually with muscles, probably derivatives of the gonostyli, usually having a clasping (termed herpes in Lepidoptera).

Harrow: A farm implement used to level the ground and crush clods, to stir the soil and to prevent and destroy weeds. Five principal kinds are the disk, spike tooth, spring tooth, cross harrow and soil surgeon.

Harvesting ants: Ant species that store seeds in their nests. Many taxonomic groups have developed this habit independently in evolution.

Hatch: To break out of the egg.

Hatching: The breaking of the egg shell by an insect in the process of eclosion.

Haustellate: Those insects having the haustellum (proboscis) or sucker, a portion of the mouth.

Haustellum: See sponging type mouth parts.

Hazard: (i) The probability that a substance will cause harm under conditions of use or exposure or handling. (ii) The probability that injury will result from use of a substance in a proposed quantity and manner, the sum of the toxicity plus the exposure to a pesticide. (iii) It refers to exposure to risk associated with the handling of pesticides. It is a function of two, primary toxicity and exposure.

Hazard assessment: Determination of likely effects of a hazard such as mechanism of toxicity, dose effect relationship and worst case exposure levels. This is prelude to the risk assessment.

Hazard minimization: The ways to reduce pesticide hazards include: (i) To adopt regulatory measures to ensure safety during manufacturing, formulating, sale, transport, storage and use of pesticides. (ii) Educating people through training, demonstration and publicity on the safe and judicious use of pesticides. (iii) Minimizing the reliance on pesticides by promoting integrated pest management approach for pest control and (iv) Developing and promoting safer pesticides or safe alternatives to pesticides (like attractants, repellents and chemosterilants etc. for use only in confined space).

Hazardous distance for most sensitive effect (HDSE): Statistically determined safety margin corresponding to a distance from treated areas at which protection of the terrestrial environment can be adequately achieved as measured by the most sensitive non target species.

HCN: See hydrogen cyanide.

Head: The anterior body region of insects which bears the mouthparts, eyes, and antennae. The insect head is hard, highly sclerotized compact capsule, most anterior of the three main body divisions of insects that bears the feeding appendages around the mouth, and antennae and internally a major centre of neuroendocrine coordination.

Head capsule measurement: The width of the head capsule (usually in mm) measured laterally across the widest part of the gena or subgena. Such measurements are used to distinguish between larval instars.

Head capsule, modifications: The anterior frontoclypeal and the posterior, lateral as well as ventral regions of the head capsule are greatly modified in some insects. See details of frontoclypeal region, hypostomal bridge and gula.

Head sclerites and areas: The cuticular areas can be seen on the external surface of the head. They become well evident after dechitinization when treated with aqueous KOH at high temperature. These areas of the head capsule are also known as head sclerites. The sclerites represent the intersutural areas. The head of generalized insect is typically exhibiting the labrum, clypeus, frons, epicranium, occiput, post-occiput, genae and subgenal areas, and ocular sclerites.

Head segmentation: The head consists of two parts: (i) an anterior procephalon and (ii) a posterior gnathocephalon. The procephalon represents preoral region of the head and bears two paired structures, the eyes and the antennae. The procephalon is thought to be derived from the ancestral head, the archencephalon or blastocephalon, after an incorporation of atleast one posterior trunk segment of the ancestor. The brain is considered as a fused ganglion of the procephalon.

Head sutures: The head capsule of adult insect can be differentiated into several regions by the sutures. The sutures are nothing but secondarily developed fine grooves. Although, the true nature and the morphological significance of the sutures is still incompletely understood, they are inflections or merely external impressions of the internal stiff cuticular ridges which provide mechanical support to the cranial wall. Some workers (Ferris, 1943; Matsuda, 1956) have taken into account the sutures as the intersegmental lines but now it is well established that these structures are the product of sclerotization and except the post-occipital suture, others have no metameric significance. Various sutures of head are: clypeolabral suture, clypeofrontal or epistomal suture, epicranial suture, occipital suture, post-occipital suture, frontogenal suture or subocular suture, subgenal suture, ocular suture and antennal suture. See details of these.

Heading stage: Growth stage of grain crops when the seed head of a plant begins to emerge from the sheath.

Health advisory level: Estimate of upper concentration limit for a pesticide in drinking water that can be consumed for life time without adverse effects. HALs generally do not have formal legal significance but have been used particularly in USA for preliminary risk assessment.

Healthy insect: An insect that is able to develop normally, and is fertile and disease free.

Heart: The chambered, pulsatile portion of the dorsal blood vessel. It extends from the hind region of the thorax to the terminal end of abdomen. It is divided into a series of dilations or chambers which are separated from each other by distinct constrictions. The number of chambers varies among different groups of insects from 12 in cockroaches to a single in Mallophaga, Anopleura and Hemiptera. The single chambered heart seems to be a composite structure as it bears two or three pairs of ostia in lice and three pairs of ostia in *Nezara* (Hemiptera), while multichambered heart, each chamber is typically provided with only a single pair of ostia.

Heart chamber: One of the segmental swellings of heart.

Heat: Heat is the energy transferred from a hot body to a cold body without involving any mechanical work. Heat is measured in unit of joule (J) or kilojoule (kJ)

Heat generation: Besides generating heat by muscular activity, many insects make use of the sun rays to warm their bodies to the optimum temperature, orienting them differently according to the temperature attained. Butterflies spread their wings in the sun when their temperature is too low. It has often been suggested that the tendency of insects in the cold climates to be dark in colour is related to absorption of radiant heat from the sun.

Heat treatment: Most of the stored grain pests die at 55-60 °C within a period of 10-20 minutes. Therefore, flour mills and other types of food processing plants have arrangements of super heating by infra-red heaters for control of insects.

Heat unit: See calorie.

Hebridae: Family of superfamily Aradoidea of division Gymnocerata, suborder Heteroptera and order Hemiptera. The bugs are very small, stout and semiaquatic. The velvety hairs found on the body prevent it from becoming wet. It is capable to run and walk on the surface of water, *e. g.*, velvet water bugs.

Hectare: In the metric system a land measure equal to 10,000 square meters. One hectare is equivalent to 2.471 acres.

Hecto: Prefix for denoting one hundred times, *i.e.*, one hectometer is one hundred meters.

***Helicoverpa armigera* (Hub.):** Belongs to order Lepidoptera and family Noctuidae. It is an insect pest of pulse crops, potato, cotton, sun flower, tomato etc., commonly known as the gram pod borer.

Heliodinidae: Family of super family Tinaeoidea of suborder Ditrysia and order Lepidoptera. Family of super family Tinaeoidea of suborder Ditrysia and order Lepidoptera. Small moths possessing narrow lanceolate fringed hindwings. The hindlegs are elevated above the wings at rest. Represented by a few members only. Some are known to mine leaves. Larvae of *Stathmopoda theoris* occur on sorghum ears in Coimbatore. Some species of *Stathmopoda* are predaceous on coccids.

Heliothripinae: A subfamily of Thripidae, family of superfamily Thripoidea of suborder Terebrantia order Thysanoptera. Terminal antennal segments are long and thin, dorsum of body deeply reticulate with polygonal areas, *e.g.*, *Rhipiphorothrips cruentatus*, *Caliothrips indicus* etc.

Heliotroplasm: Curvature to the stimulus of sun light. See also phototropism.

Heliozelidae: Family of super family Incurvarioidea of suborder Montrysia and order Lepidoptera. Larvae are apodous and mine into leaves of plants.

***Hellula undalis* Fab.:** Belongs to order Lepidoptera and family Pyralididae. It is an insect pest of cabbage, commonly known as the cabbage borer.

Helmet scale: See *Saissetia coffee.*

***Helopeltis theivora* Waterhouse:** Belongs to order Hemiptera and family Miridae. It is an insect pest of tea, commonly known as the tea mosquito bug.

Hematophagous: Feeding or subsisting on blood.

Hemelytron (pl., hemelytra): The forewing of true bugs (Heteroptera) the basal portion is thickened and the apical portion membranous.

Hemerobiidae: Family of superfamily Hemerobioidea, suborder Planipennia, order Neuroptera. Cotal area with numerous branched veinlets. R_1 and R_3 fused. Subcosta free. Eggs devoid of pedicels and larvae fusiform and smooth. Small delicate insects with moniliform antennae and without ocelli and ovipositor, *e.g.*, brown lacewings.

Hemerobioidea: Superfamily of suborder Planipennia, order Neuroptera. Mandibles simple not stout, body without dolichasters.

Hemi-elytron (pl., hemi-elytra): The forewing of a heteropteran bug, differing from the beetle elytron in having the distal portion membranous. Same as hemelytron.

Hemimeridae: Family of order Dermaptera and suborder Hemimerina. The thoracic segments are laterally expaned broader than abdomen, the prothorax being very broad. Viviparous, developing a placenta. Forceps or cerci unjointed, feebly sclerotised, apterous, flattened, ectoparasitic mostly on rats, eyes absent, *e.g.*, *Hemimerus* sp.

Hemimerina: A suborder of Dermaptera.

Hemimetabola or Hemimetabolous development (Incomplete metamorphosis): The insects showing incomplete metamorphsis comprising a simple process of gradual or partial development (Hemimetabolous metamorphosis comprising egg- nymph and adult) during which the immature nymphs become imago by development of external wing buds into perfect wings and thus showing the exopterygote condition. Immature stages are called nymphs or naiads. Naiads (Odonata, Ephemeroptera and Psocoptera) spend their life in water and respiration takes place through the tracheal gills or general body surface. Examples of simple metamorphosis are Odonata, Ephemeroptera, Psocoptera, Dictyoptera, Orthoptera, Dermaptera, Isopteran, Anoplura, Mallophaga, Psocoptera and Hemiptera. It can be divided into palaeometabola, heterometabola and neometabola. See details.

Hemimetabolous: Insects having an incomplete metamorphosis, like that in the Ephemeroptera, Odonata, Psocoptera, Dictyoptera, Orthoptera, Dermaptera, Isoptera, Anoplura, Mallophaga and Hemiptera.

Hemiphlebioidea: A super family of order Odonata. Arculus placed farther away from base; usually only two antenodals present. Post nodals not in a line with cross veins behind.

Hemipneustic: One or more spiracles are non-functional or reduced in comparison with most generalized insects. It may be amphipneustic (prothoracic and posterior abdominal spiracles are open, *e.g.*, Dipteran larvae); propneustic (one pair of spiracles open at prothorax, *e.g.*, pupa of Diptera and metapneustic (one pair of abdominal spiracles open, larval instars of Diptera).

Hemiptera: (*Hemi*= half, *pteron*= a wing). Insect order. Members are commonly referred to as bugs, aphids, coccids, mealy bugs etc. They are widely distributed in the world but abundant in tropical regions. Mouth parts are piercing and sucking types. The forewings or hemelytra of heteropterous insects are characterized by the proximal area being well sclerotized and the smaller distal portion being membranous. The hind wings are always membranous. Abdomen is eleven segmented.

Hemipteran piercing and sucking mouth pats: The beak is usually long and slender segmented and formed by terminal labium. The labrum is short and covers the basal part of beak. The beak encloses four stylets only, two mandibular and remaining two maxillary. Both the mandibles are symmetrically developed and their apical parts are modified into long slender stylets. They can be retracted or protracted by the retractor and protractor muscles, respectively. The lateral plates of maxillae are mostly incorporated into the head capsule. The palpi are lacking and the maxillary stylets are long slender needles like the mandibular ones. The salivary syringe is a powerful pump for ejecting the saliva. The salivary duct opens on the tip of hypopharynx. Both food canal and salivary canal of a beak lie between the closely apposed inner surfaces of the maxillary stylets. The food channel or functional mouth opens into the cavity of sucking pump or cibarium.

Hemipteroid group: See taxonomic grouping of orders.

Hemorrhagic: Fever resulting from the infection of blood cells by a parasite.

Hepatic caecum: See gastric or enteric caecum.

Hepialidae (swift moths): Family of super family Hepialoidea of suborder Montrysia and order Lepidoptera. They are generally medium sized moths with short antennae and vestigial mouthparts. Some are relatively gigantic. A jugate type of wing coupling apparatus is present. Tibial spurs are wanting. The moths are variously colored, some are bright. About 300 species are known and they are well represented in Australia. They are distributed in Australia and South Africa. The larvae are internal wood feeders and bore into stems of plants. *Sahyadrassus malabaricus* is found in the Nilgiris boring into tea bushes. The other families included under superfamily Hepialoidea are Prototheridae, Palaeosetidae and Anomosetidae.

Hepialoidea: Super family of suborder Montrysia and order Lepidoptera. Members of this super family have similar venation on both pairs of wings and females possess two genital apertures on segment IX.

Heptachlor ($C_{10}H_5Cl_7$): Moderately toxic, heptachlor is chemically related to DDT and the other chlorinated hydrocarbon pesticides. It is highly effective for subterranean termite control beneath buildings. Heptachlor is more volatile and thus, emits greater vapour toxicity than the other members of this chemical group. Chemically, it is 1, 4, 6, 7, 8-heptachloro 3a, 4, 5, 5a-tetrahydro-4 7-endomethanemethanoindene. It possesses higher toxicity than BHC as contact and stomach insecticide, available as dust, water dispersible powder and emulsifiable concentrate. Used in control of grasshoppers, locusts, thrips beetles, caterpillars, white grubs; also against termites and cutworms.

Heptageneoidea: A super family of Ephemerida. M_{1+2} and Cu_1 of forewing are parallel at their bases; tarsi five segmented and free. This includes the families Baetiscidae, Siphlonuridae, Ametropodidae and Heptageniidae.

Herbicide: A substance that kills herbs particularly weeds, also called the weedicide.

Herbivore: Plant eating; an organism that feeds on plant material (see also carnivore and omnivore).

Herbivorous: Feeding on plants (see also, phytophagous and carnivorous).

Herbs: Plant with no persistent woody stem above grounds.

Hereditary: Inherited; Inheritance from progenitors; organic resemblance based on descents; organic relation between successive generations.

Hereditary sterility: See delayed sterility.

Hermaphrodite: (i) Bisexual; in flowering plants, both stamens and carpels are present in the same flower; (ii) In animals, an individual with both male and female gametes. There are a few insects which are functional hermophrodites. The best known example is the Californean race of the coccid, *Icerya purchasi* (fluted scale or cottony cushion scale). The gonads produce both sperms and oocytes. Most of the eggs are fertilized by sperm from the same individual, but these hermaphrodite females may copulate with the rare males.

Hermaphroditic: Containing the sex organs of both sexes in one individual.

Hermaphroditism: The phenomenon of production of both male and female gametes by an individual.

Hermetic or air tight storage: Food grains kept in air tight sealed structures remain insect free, reason being that during storage the grain respire and liberate carbon dioxide. Excessive CO_2 accumulation and depletion of O_2 do not allow the insects to survive. Under ground storage of paddy in Eastern India is a type of hermetic storage. Seeds cannot be stored by this method due to loss of viability.

Hesmosis: Same as colony fission.

Hesperiidae (skippers): Family of super family Hesperiidae, suborder Ditrysia and order Lepidoptera.The member of this family are known as skippers in view of their erratic darting flight. These are widely distributed insects but most of them are tropical. The antennae are widely separated at their bases and each antenna is apically prolonged beyond the club into a hook or a small recurved point. The larvae are moderately stout and taper at both ends.

Hesperioidea: Super family of suborder Ditrysia and order Lepidoptera. The adults are without maxillary palpi but without more or less rough haired labial palpi. Antennae have apically dilated club which often ends in a hook which is characteristics of the members of this superfamily. Frenulum is wanting in hindwing excepting in male of *Euschemon*. This superfamily includes a single family, *viz*. Hesperiidae.

Hetero: Other, different.

Heterochromatic: Containing heterochromatin (chromatin rich in nucleic acid) and involved in nuclear and cytoplasmic nucleic metabolism.

Heterocyclic: Containing more than one type of atom joined in a ring.

***Heterodera avenae*:** Belongs to order Tylenchida and family Heteroderidae. It is a pest of wheat, commonly known as the molya nematode.

***Heterodera rostochiensis*:** Belongs to order Tylenchida and family Heteroderidae. It is a pest of potato, commonly known as the golden nematode of potato.

Heterodynamic life cycle: A life cycle in which there is a period of dormancy.

Heterogametic: Possessing the XY chromosome configuration.

Heterogamy: Alternation of bisexual reproduction with parthenogenetic reproduction.

Heterogenous: Composed of dissimilar or non-uniform constituents.

Heterogony: Same as allometry.

Heteromerous: The three pairs of tarsi differing in the number of segments (Coleoptera, *e.g.*, with a tarsal formula of 5-5-4).

Heterometabola: Aleyrodids and male coccids of Hemiptera though basically belong to Hemimetabola, show a tendency towards the Holometabolous type of metamorphosis. Earlier instars appear as larvae, while latter as nymphs. Nymphs differ from adults coneogenetically, *i.e.*, in secondary nymphal characters. The wings develop gradually. It can be divided into: (i) Archimetabola: In which nymphs are aquatic with tracheal gills, *e.g.*, Plecoptera and Odonata, and (ii) Paurometabola: in which nymphs are terrestrial insects with wing pads, nymphs more or less like adults but differing secondarily. The adult characters appear gradually but in the last moult, *e.g.*, Orthoptera, Dictyoptera, Embioptera, Psocoptera, Hemiptera.

Heteromorphous/ Heteromorphic regeneration: Describes occurrence of a regenerated appendage different in form from the appendage it has replaced.

Heteroneurous venation: Lepidoptera, in which the number of veins in the hind wing is less than the forewing. It is considered to be the more advanced condition. See homoneurous.

Heteroptera: *Hetero*= different, *ptera*= wings. A suborder of Hemiptera, *e.g.*, true bugs. Heteroptera is the suborder of Hemiptera. The insects belonging to Heteroptera are characterized by wings overlapping the abdomen, rostral base not contacting precoxae, gular portion sclerotized and tarsi usually three segmented.

Heterosis: Hybrid vigor; the increased ability of hybrid offspring to overcome environmental resistance because of increased size, fecundity, survival rate etc.

***Heterotermes indicola* (Wasmann):** A termite species wide spread from Afghanistan to North India and causes serious damage all over the range to wood work in buildings.

***Heterotermes* spp.:** The common Heterotermes species of termites are *H. gertrudae* Roonwal which occurs in the sub-Himalayas; *H. malabaricus* Synder in Peninsular India; *H. ceylonicus* in Ceylon; *H. philippinensis* damages buildings in the Philippines.

Heterothripidae: A family of superfamily Thripoidea of suborder Terebrantia order Thysanoptera. Antennae without sense cones or with short triangular ones. Antennae nine segmented. Foretarsus usually with a claw like appendage at base of second segment. *Adiheterothrips jambudvipae* and *Holarthrothrips* sp.

Heteroxenous: Describes a species which requires the use of more than one host species to successfully complete its annual life cycle. See monoxenous.

Heterozygote: Having two different alleles of a given gene.

Heterozygous: Containing two forms (alleles) of a gene, one from each parent.

HETP: A synthetic insecticide and acaricide; an organophosphate, a mixture of ethyl polyphosphates containing 12-20 per cent of tetraethyl pyrophosphate; extremely toxic to mammals, acute oral LD_{50} for rats is 7 mg kg^{-1}; Hexaethyltetraphosphate is effective in control of aphids, thrips, scale insects, mealy bugs and mites. The dust formulation hydrolyses and loses its efficacy in the presence of water. It is available as 1 per cent dusting powder and 20-40 per cent emulsifiable concentrate; extremely toxic to mammals.

Hexachlorophene: A chemical compound which is used as an insecticide.

Hexalure: Synthetic pink bollworm moth attractant; cis-7-hexadecen 1-ol acetate; synthetic female pheromone attractive to male moths of pink bollworm; used in traps to survey extent of infestations; product of USDA.

Hexapod: An animal possessing six legs, more specifically the parent group that contains insects and their close kin.

Hexapoda: Insects or a class in the phylum Arthropoda.

Hibernate: Overwinter; spend the winter in a dormant condition.

Hibernation: A type of dormancy that occurs during the winter.

Hierarchy: The system of ranks in an animal classification indicating the categorical levels of various taxa.

***Hieroglyphus banian* Fab.:** Belongs to order Orhtoptera and family Acrididae. It is an insect pest of cereals and millets, commonly known as the rice grasshopper.

High pressure high volume sprayer: Hydraulic sprayer used for spraying tall trees and dense crops; similar to multipurpose farm sprayer, except for pump which is multi-cylinder, plunger type. It has larger tank size; can cover 10 to 20 acres a day.

High pressure sprayer: Same as hydraulic sprayer.

High volume spraying: Spraying when the spray material deposited over surface of leaves as in large drops, when they spread over surface as film, accumulate at edge, finally run off to ground. Plants are sprayed thoroughly to cover them to the point of run-off drip, foliage is thus completely wetted. It requires 500-1000 litres ha^{-1} of the spray fluid in the case of field the amount of the pesticide deposit is determined by its concentration in the spray fluid; full covers spraying.

Higher category: A taxonomic category from subgenus and above.

Highly volatile: A liquid that quickly forms a gas or vapours (evaporates) at room temperature.

Hind wings: Posterior wings that developed on the third thoracic segment.

Hindgut: The posterior region of insect alimentary canal between the midgut and anus, also called the proctodaeum. It is ectodermal in origin and lined with intimal layer. The anterior part of hind gut is called pylorus possessing pyloric valve at the junction of hind gut and mid gut. Externally pylorus is marked by the presence of Malpighian tubules. Pylorus opens in a chamber called ileum. The posterior most part of hind gut is known as rectum which generally thin walled chamber is containing various muscular pads which play important role in reabsorption of water, essential ions and amino acids from urine. During excretory activities of Malpighian tubules these essential substances are removed from haemolymph.

***Hippobosca maculata* Leach:** Belongs to order Diptera and family Hippoboscidae. It is an insect pest of farm animals, commonly known as the Horny or leathery fly.

Hippoboscidae (louse flies): Family of suborder Cyclorrhapha, order Diptera. These are minute to small, dark brownish, flat, leathery flies with sunken head in the anterior part of the thorax, a retractile proboscis with sheath formed of palpi, short and stout legs, wings either present or absent. They are commonly found on cattle and dogs. These insects are ectoparasites of birds and mammals.

***Hippotion celerio* (Linnaeus):** Belongs to order Lepidoptera and family Sphingidae. It is commonly known as grapevine sphinx and is a minor pest of grapevine.

***Hirsutella thompsonii*:** A species of entomopathogenic fungi causing disease in many insect species. It is marketed under trade name, Mycar for the management of insect pests.

Hispidae (wedge-shaped or leaf-mining leaf beetles): Family of superfamily Chrysomeloidea, suborder Polyphaga and order Coleoptera. Small elongate beetles, 4 to 7 mm in length, with the head free and not retracted beneath prothorax, antennae are usually short and not inserted on frontal prominences, mouth is inferior, tibial spurs are absent, body is narrow anteriorly and broad and truncate posteriorly, and elytra with ridges and spines. The larvae are mostly leaf-miners and also feed by scraping leaf tissue. The genera included in the family are *Dicladispa, Hispa, Asmangulia, Rhadinosa, Trichispa, Dorcathispa, Polygonia, Leptispa, Platypria, Onchocephala, Dactylipsa, Oediopalpa* and *Callispa.*

In India, the rice hispa *Dicladispa armigera* is a well-known pest. The other common insects noticed are: *Leptispa pygmoea* on paddy, *Onchocephala tuberculata* on sweet potato, *Asmangulia cuspidata* on sugarcane and *Callispa* sp. on coconut.

Histeridae (steel beetles): Family of superfamily Histeroidea, suborder Polyphaga and order Coleoptera. Minute or medium-sized, broadly oval, compact, hard, shining, brown or black, deeply striated beetles; some with red markings and a few are metallic. The antennae are geniculate and clubbed strongly and rest in groove beneath the pronotum. The legs are stout, short and fossorial. The elytra are truncated behind leaving the last two abdominal segments exposed.

Histeroid aedeagus: Aedeagus in which the phallobase and parameres form a tube surrounding the penis. This type is found only in Sphertididae, Synteliidae and Histeridae.

Histeroidea: Superfamily of suborder Polyphaga, order Coleoptera. Histeroidea are characterized by geniculate antennae, the three to four terminal segments forming a compact club, truncated elytra and the last one or two abdominal segments exposed. This superfamily includes three families. Synteliidae, Sphaeritidae and Histeridae of which the last one is a large family of some importance.

Histidine conjugation: In such conjugations fumigants like methyl bromide result in alkylation of one or both the nitrogen in histidine nucleus.

Histochemistry: The chemistry of living tissues.

Histogenesis: See histolysis.

Histology: Gr., *Histos*= tissue. It is the study of microstructure of tissues.

Histolysis: A breakdown of body tissues. In the extreme case of metamorphosis of Diptera, many larval organs (Malpighian tubules, certain muscle groups) are remodeled without much change to form those of the adult. Other organs undergo extensive break down (histolysis) and new organs are built up from the embryonic cells of the imaginal discs (histogenesis). There has been much controversy about the physiology of this process. It is now clear that when metamorphosis begins, certain of the larval cells die and disintegrate.

Histosipathology: The study of abnormal microscopic changes in the tissue structure of diseased organism.

Histosiphon: Same as stylostome. The tube formed by the host as a result of the feeding of a chigger secreting salivary fluids, the chigger partially digests skin tissues, which induces the host to form a proteinaceous tube walling off the injury.

Hive: A man made home for bees.

Hive body: A wooden box which encloses the frames.

Hive odour: See nest odour.

Hive stand: A structure that supports the hive.

Hive tool: A metal device used to open bee hives, pry frames apart and scrape.

Hodotermitidae: A family of order Isoptera. Wood inhabiting with pronotum narrower than head, tarsi imperfectly five segmented and without anl lobe on hindwings. Example, *Hodotermopsis, Archotermopsis, Zootermopsis.*

Hold back: To retard or prevent the development of an immature stage using low temperature.

Hollow cone: Spray jet with a core of air breaking to give drops in an annular pattern.

Hollow cone nozzle: See nozzle types.

Hollyhock tinged bug: See *Urentis eunymus.*

Holoblastic cleavage: Method of blastoderm formation characteristic of Collembola and certain parasitic wasps and associated with eggs that contain little yolk; the entire zygote divides (compare with meroblastic cleavage).

Holometabola (Complete, indirect or complex metamorphosis): The group of insects undergoes complete metamorphosis is known as indirect and holometabolous metamorphosis. The life cycle includes four developmental stages egg, larva, pupa and adult. The larva differs greatly in form from the adults into which it develops through a series of moults. This type of metamorphosis is characterized by a preimaginal resting and non feeding pupal stage. The wings develop as buds which are internal showing the so called endopterygote condition. External rudiments of wings appear only in the pupal stage. Example, Mecoptera, Neuroptera, Trichoptera, Coleoptera, Lepidoptera, Diptera, Strepsiptera, Hymenoptera.

Holometabolous: Insects having a complete metamorphosis as in the Diptera, Hymenoptera, Coleoptera, Lepidoptera etc.

Holophyletic: A monophyletic group including all the progeny of the original ancestor.

Holopneustic: The ten pairs (two on thorax and eight in first abdominal segments) of spiracles are open in this type of tracheal respiration, *e.g.*, order Diptera and Hymenoptera.

Holoptic: *Holo* means whole. With the eyes touching or almost touching on the top of the head; used mainly when describing flies (Diptera).

***Holotrichia consanguinea* Blanchard:** Belongs to order Coleoptera and family Melolonthidae. It is an insect pest of chillies, ground nut, ornamental plants etc., commonly known as the groundnut white grub.

***Holotrichia insularis* Brenske:** Belongs to order Coleoptera and family Melolonthidae. It is an insect pest of tamarind, commonly known as the white grub.

Holotype: The single specimen selected by the author of a species as its type or the only specimen known at the time of description. The type specimen of a species is the actual insect from which the original description of that species was produced. If several specimens were used for this purpose, one of them should have been designated as the type. Because the type can be of only one sex, it is usual to designate a certain individual of the opposite sex as the allotype. The original type specimen is then called the holotype.These type specimens are very important in taxonomy and classification.

Homeostasis: Biological systems remaining in a state of equilibrium while maintaining balance between favourable and unfavourable factors.

Homo: Alike or similar.

Homodynamic life cycle: A life cycle in which there is a not a period of dormancy and there is continuous development.

Homodynamic: Homologous or homology in metameres.

Homogametic sex: Possessing two X chromosomes.

Homogenous: Uniform in structure or compsition.

Homolog: A series of compounds which express similarities of structure but regular differences of formulae.

Homologous genes: Genes from different species that encode proteins with a similar sequence and functions.

Homologous: (i) Characters in two or more species those are fundamentally similar and attributable to common ancestry. (ii) Organs or parts which exhibit similarity in structure, in position with reference to other parts, and in mode of development, but not necessarily similarity of function, are said to be homologous.

Homology: A characteristic feature in two or more taxa which can be traced back to the same feature in their common ancestor.

Homometabola: See neometabola.

***Homona coffearia* (Nietner):** See tea tortrix.

Homoneurous venation: Wing venation in which the fore and hind wings are similar.

Homonym: A scientific name which has been given to two different species. When such an instance comes to light one of the species must be given another name.

Homoplasy: Correspondence between organs or structures in different organisms acquired as a result of evolutionary convergence or of parallel evolution.

Homoptera: *Homo*=uniform, *ptera*= wings. A suborder of Hemiptera, *e.g.*, cicadas, aphids, leafhoppers, scale insects, mealy bugs and relatives. They are characterized by uniformly leathery or uniformly membranous forewings, sucking mouthparts and an incomplete metamorphosis. Wings sloping over the abdomen, rostral base contacting forecoxae, gular portion absent or if present membrane like and tarsi one to three segmented etc.

Homozygote: Having two similar alleles of a given gene.

Homozygous: A condition of having inherited a given genetical factor from both parents, therefore, of producing gametes of only one kind as regards that factor.

Honey: A thick, sweet fluid prepared by worker bees in a bee hive from nectar collected from flowers and stored in the honeycomb (see comb) for feeding to the larvae. Different honeys with characteristic flavours are produced from nectar derived from different plant species. The product solely produced by the honey bee; contains 65 per cent invert sugar, 17 per cent water, small amount of mineral matter, vitamins, proteins, and enzymes.; specific gravity 1.413 at a temperature of 20 °C.

Honey bee: See *Apis dorsata, A. mellifera, A. indica, A, adamsoni, A. florae.*

Honey bee, predators: The death head moth, *Acherontia styx* enter the bee hive in the night and sucks honey from comb cells. The wax moth, *Galleria mellonella* and *Achroia grisella* lay eggs in the comb cells and emerging larvae feed on stored pollen and tunnel the combs and invade them with webs and develop into the pupae. Wasps, *Vespa orientalis* are predators of bee larvae whenever the workers go out from the nest for foraging. The green bee eaters, ants, lizards, toads, frogs, crows and bears are other natural enemies of honey and honey bees.

Honey comb: The waxy structure in which honey is stored by bee.

Honey dew: A sweet liquid excreted by aphids, leaf hoppers, and some scale insects that is collected by bees, especially in the absence of a good source of nectar. It is a liquid discharged by anus of many Hemipteran insects that is high in sugar content.

Honey extractor: A machine which removes honey from the cells of comb by centrifugal force.

Honey flow: A time when nectar is plentiful and bees produce and store surplus honey.

Honey gate: A faucet used for drawing honey from drums, cans, or extractors.

Honey house: Building used for extracting honey and storing equipment.

Honey pot: A container made by sting less bees or bumblebees from soft cerumen and used to store honey.

Honey pump: A pump used to transfer honey from a sump or extractor to a holding tank or strainer.

Honey stomach: An organ in the abdomen of the honey bee used for carrying nectar, honey, or water.

Honey sump: A clarifying tank between the extractor and honey pump for removing the coarser particles of comb introduced during extraction.

Hooked trichomes: Minute, sharp, hooked hairs found on certain plants that provide defense against insect attack.

Horizon: A layer of soil, roughly parallel to the surface of the soil, with distinct characteristics which were produced by soil forming processes.

Horizontal infestation: Some stored grain insect pests, like Angoumois grain moth and lesser grain borer infest grains right from the standing mature crop and their infestation migrates to storage through transportation, crop drying, threshing and grain drying during post-harvest operations which is termed as horizontal infestation.

Horizontal resistance: Resistance that does not involve a gene-for-gene relationship. Opposite of vertical resistance.

Hormone: A chemical substance formed in some organ of the body, secreted directly into the blood, and carried to another organ or tissue where it produces a specific effect.

Hormones: Internally secreted chemical messenger substances produced by endocrine glands which influence or activate other organs or physiological process of same organism. These are neurohormones, juvenile hormones, ecdysones, gonadal hormones and midgut hormones. On the basis of chemical nature, they can be classified into three classes (i) All neurohormones and midgut hormones as the peptides or polypeptides, (ii) Juvenile hormones as the isoprenoids (lipids) and (iii) ecdysones and gonadal hormones as the 27-carbon steroids.

Hormones in pest control: A new approach which captures worldwide attention is the use of analogues and antagonists of insect growth regulators such as juvenile hormones and ecdysones to control pest populations. See juvenile hormones and ecdysones as insecticides.

Horn worm: Larvae of the family Sphingidae (Hawk moths) with dorsal spine or horn on the last abdominal segment.

Horny or leathery fly: See *Hippobosca maculata.*

Horny: Thickened or hardened.

Horse flies: See *Tabanus* spp.

Horse fly: The common horse flies are *Tabanus, Chrysops* and *Haematopota* consisting about 2000 species and pests of veterinary importance. In India, *T. rubidus, T. viva, C. dispar* and *H. javana* are common.

Horse fly transmitted diseases: The horse fly transmit *Trypanosoma evansi* protozoan parasite causing the well known fatal disease, Surra. Other diseases of horses due to these flies include filariasis, tularaemia, caliber swellings, anthrax, and equine infectious anaemia. Even in non-pathogenic condition, the animals become weak, loose their weight and are often restless.

Horse ked or cattle ked: Three species of *Hippobosca, i.e., H. equina, H. rufipes* and *H. maculata* commonly attack dog, horse and cattle (Hippoboscidae, Diptera). They transmit *Trypanosoma theileri* to cattle and bacteria to horses causing anthrax.

Host: (i) The plant on which an insect feeds or an organism that furnishes food, shelter or other benefits to another organism of a different species. For example: Cabbage is a host for the diamondback moth. The rice meal moth, *Corcyra cephalonica* Staint. is a host of many natural enemies. (ii) The organism in or on which a parasite lives.

Host plant resistance: A method of pest control in which resistant crop plants are used. With the advancement of science, a lot of crop varieties have been developed that show varying degree of resistance against different insect pests. The resistance is due to one or more biochemical, physiological, genetical and histological characters of the plant. Presently there are many varieties of cereals and pulses that show varying degree of resistance to insects.

Host range: The various kinds of host plants that may be attacked by a pest.

House crickets: See *Gryllodes sigillatus.*

House crow: See *Corvus splendens.*

House fly transmitted diseases: The flies are known to transmit more than 100 species of pathogenic organisms including the causative organisms of amoebic dysentery, typhoid fever, cholera, shigellosis (diarrhoea), salmonellosis (typhoid), anthrax, yaws, trachoma, poliomyelitis, infectious hepatitis, carrying and spreading the eggs of Helminthes like pinworm, whipworm, hookworm and tapeworm.

House fly: *Musca domestica* and *M. nebulo* (Muscidae, Diptera) are pests of medical importance. The flies are abundant and very active during summer and rainy seasons. Other species of *Musca* found in Asiatic countries are *M. vicina, M. autumnalis* and *M. sorbens.*

House sparrow: See *Passer domesticus.*

HPLC: High performance liquid chromatography.

HPTLC: High performance thin layer chromatography.

HRGC: High resolution gas chromatography (GLC with narrow bore capillary columns).

Hulocrine secretion: Cell products released by complete breakdown of a cell.

Human flea: *Pulex irritans* (Pulicidae, Siphonaptera) is a pest of medical importance and cosmopolitan in distribution. It transmits the bubonic plague from man to man. The orient rat flea, *Xenopsylla cheopis* (Dolichopsyllidae) is a vector spreading plague (caused by bacterium, *Yersinia pestis*) from rat to man.

Human health effects due to pesticides: (i) Acute: High dose, short term exposure, rapid onset (headache, nausea, vomiting, respiratory failure, death); (ii) Chronic: Low dose, long term exposure, outcome not seen for many years (cancer, reproductive damage, immune dysfunction).

Human louse: *Pediculus humanus* L. (Pediculidae, Anoplura) is an ectoparasite of man and pest of medical importance transmitting many diseases. It is world wide in distribution, abundant in overcrowded slum areas and prisons. The lice include (i) *P. humanus capitis* (head louse), (ii) *P. humanus corporis* (body louse), and *Phthiris pubis* (crab louse).

Human louse transmitted diseases: The louse, *Pediculus humanus* causes the disease, epidemic typhus in Africa, Europe, Asia and Central and South America by transmitting the pathogen, *Rickettsia prowazekii*. It also transmits the pathogen, *Rochalimaea quintana* (*Rickettsia*) causing trench fever in the same countries. The epidemic relapsing fever in Ethiopia and Europe caused by *Borrelia recurrentis* (Spirocheta) is also transmitted by *P. humanus humanus*.

Humectant: Material added to a spray to delay evaporation of the water carrier.

Humeral angle: The front basal part of the wing, close to its attachment to the body.

Humeral lobe: It is a wing coupling apparatus. The jugal lobe at the base of forewing and a humeral lobe at the base of costal margin of hind wing form a simple or primitive type of apparatus in some Mecoptera. Both lobes contain setae. The humeral lobe specially bears the frenular bristles. From this primitive type of wing coupling mechanism, complex types (jugate and frenate types) have been evolved in some Neuroptera, Trichoptera and Lepidoptera with the development of long jugal lobe on the forewing or that of frenulum and retinaculum like bristles.

Humeral vein: A small cross-vein running from the costa to the sub-costa in humeral (basal) region of the wing.

Humidity: The presence of water vapour in the air is known as its humidity. The amount of water vapour present in one cubic metre of air at a given temperature is called the absolute humidity. It is expressed in terms of kg/m^3. Air cannot hold more than a definite amount of water vapour present at a given temperature. When maximum amount of water is present in the air, it is called saturated otherwise unsaturated. The ratio of amount of water vapour present in a given volume of air (m) to the amount of water vapour required to saturate the same volume at the same temperature (M) is called the relative humidity. The relative humidity (RH) is expressed in per cent, therefore, $RH = (m/M) \times 100$.

Humoral immunity: Humoral encapsulation is an immunity system in certain Diptera (*Chironomus* larvae) against foreign bodies. It is a fast and highly efficient immunity reaction. The humoral capsules are generally made up of melanin like substance. The two types of humoral immune factors generally found in the haemolymph are; (i) phenol oxidase and (ii) lectins.

Humped slug caterpillar: See *Spatulicraspeda castaneiceps.*

Humus The well-decomposed, more or less stable part of the organic matter of the soil.

Hyaline: Transparent, colourless.

***Hyalomma anatolicum* Koch:** Belongs to order Acarina and family Ixodidae. It is a pest of farm animals, commonly known as the cattle tick.

Hybrid: (i) The offspring of two plants or animals of different races, varieties or species. (ii) A cross between two individuals of different traits gives a hybrid trait (offspring).

Hybrid sterility: A potential source of sterile insects for mass release programmes resulting from the successful hybridization of certain pairs of insect's species which produce fully competitive but partially or completely sterile offspring not reproductively isolated from the parental species.

Hybridization: The pairing of complementary DNA or RNA strands. Hybridization is the procedure by which two genetically different parents are crossed and new offspring is obtained.

Hydrated: The addition of water (Chemically combined).

Hydraulic: Pertaining to water or other liquids.

Hydraulic agitation: It refers to the agitation of the spray fluid caused from the return flow of some spray fluid from the pump.

Hydraulic agitator: A device which keeps the tank mix from setting out by means of water flow under pressure.

Hydraulic sprayer: (i) Power sprayer of hydraulic type, in which spray liquid under pressure created by pump is forced out through nozzle in small droplets; consists of engine, pump, tank with agitator, framework with or without wheels pressure regulator with relief valve, gauge control valves, suction and delivery hoses, strainers, lance with nozzle, over-flow pipe and booms; mounted on 2 or 4 wheels. (ii) A machine which applies pesticide by using water at high pressure and volume to deliver the pesticide to the target, same as high pressure sprayer.

Hydro. Water; fluid; hydrogen.

Hydro pneumatic sprayer: Sprayer in which the liquid is carried in pressure tank and spraying pressure developed by engine powered air compressor, either skid type or mounted on wheels; not commonly used for spraying crops.

Hydrocarbon: A chemical whose molecule contain only carbon and hydrogen atoms. The simplest hydrocarbon is methane (natural gas), each of whose molecules contains one atom of carbon and four atoms of hydrogen. Crude oil is a mixture largely of hydrocarbons.

Hydrocarbon oils: Petroleum (mineral) oil and coaltar oils are the oils which are used in insect control. In these oils, the presence of unsaturated compounds makes them phytotoxic. These unsaturated compounds when treated with strong sulphuric acid leave unsulphonatable residue (U.R.), the percentage of which indicates the purity of the oil. Based on the U.R. rating, two types of oils are recognized, summer oils with U.R. rating 90-96 per cent and dormant oil with U.R. rating 50-90 per cent.

Hydrofuge hairs: The effect of surface forces becomes more evident when hairs and bristles are present on the body. When a fine pile of these hairs is set vertically on the body surface, they hold the water at a distance, so that when insect goes below the water surface, can carry with it a film of air.

Hydrogen cyanide (Hydrocyanic acid): It was introduced by American Cynamid Co. in 1783 and first employed as fumigant by D.W. Coquillet in 1886 for control of the cottony cushion scale, *Icerya purchasi*. It is a volatile colourless liquid with a bitter almond odour and a specific gravity of 0.699 at 20 °C and boiling point 26 °C. The gas has specific gravity of 0.9483 and is highly inflammable and explosive in mixtures above 5.6 per cent or above 4 lbs. per 1000 ft^3. It is one of the toxic insect fumigant and may be employed for fumigating dry food stuffs, grains and seeds. In the market, two forms of CaCN are available. Cyano gas contains 40-50 per cent CaCN and the gas is liberated slowly. The Calcium cyanide dust contains 88 per cent CaCN. It is available under trade names Calcid, Cyanogas, Cymog and Cyclon. The dose for general fumigation of sodium cyanide is 10-16 g, sulphuric acid 20-30 ml and water 40-60 ml per m^3. HCN does not react with the constituents of food stuffs having normal moisture content. Extremely toxic gas (killing both animals and plants), this is in fact, the gas that is used most often to execute condemned prisoners in penitentiary gas chambers. Even a few inhaled breaths of this gas are sufficient to render a person or animal unconscious. Death follows quickly from respiratory muscle paralysis. Its toxic action to insects and warm blooded animals is due to its ability to combine with the iron atoms present in the enzymes such as cytochrome oxidase. HCN is used only by certified pest control operators who employ it most often to fumigate ground burrows for such pests as rats, moles and gophers.

Hydrogen ion concentration: A measure of the acidity and alkalinity. The hydrogen ion concentration is expressed in terms of the pH of the solution. For example, a pH of 7.0 is neutral, below 7.0 is acidic and above 7.0 is alkaline.

Hydrogen phosphide: See aluminium phosphide.

Hydrological cycle: This cycle begins with evaporation which requires energy. The energy can be obtained by direct solar energy or heat energy from soil, ocean and lakes. By evaporation, moisture from various water bodies converts into clouds. One more route for cloud formation is transpiration. Through it, water from plants gets converted into clouds. The combined process of evaporation and transpiration is known as evapotranspiration. The evaporated water is condensed in the form of clouds and come back towards earth by rainfall or precipitation. This water is stored in water bodies and soil. Large amount of water in soil is tightly stored which is not easy to remove but small amount of this is extracted by plants by their roots and again releases in the atmosphere by transpiration. The whole cycle is represented by water balance equation, *i.e.*, P=N+E+S (Here, P= precipitation or rainfall, N= percolation or entrapped in soil by gradual filtration, E= evaporation, and S= storage change in plants).

Hydrolysis: Reaction in which a chemical bond formed with the oxygen atom of a molecule of water.

Hydrometridae: Family of superfamily Aradoidea of division Gymnocerata, suborder Heteroptera and order Hemiptera. These are small, slender, grayish bugs resembling tiny walking sticks. The long head bears the eyes which conspicuously bulge out laterally. These bugs move slowly on the surface of stagnant water, *e.g.*, *Hydrometra* sp.

Hydromineral homeostasis: See osmoregulation.

Hydrophil: A substance or system which attracts or is attracted to water is hydrophilic in nature.

Hydrophilic: Having water attracting characteristics, wettable.

Hydrophiloidea (water scavengers): Family of superfamily Hydrophiloidea, suborder Polyphaga and order Coleoptera. These are tropical insects inhabiting principally decomposing vegetable matter. They are aquatic or terrestrial. *Helophorus* inhabits glacial ponds on the Himalaya at 5400m elevation. The beetles are minute or large, black or dull colored oval somewhat convex or flattened with short six to nine-segmented antennae on the head sunk into prothorax. Maxillary palpi are often longer than antennae. The foretarsi in the male are adapted for clinging to the female. The mid and hindlegs are fringed with setae for swimming. In many species the metasternum is prolonged into a sharp spine posteriorly.

Hydrophiloidea: Superfamily of suborder Polyphaga, order Coleoptera. Hydrophiloidea are charcterised by elongate maxillary palpi which are tactile in function and antennae concerned with respiration in aquatic species.

Hydrophily: Transfer of pollens by means of water, a type of pollination.

Hydrophobe: A substance or system which repels or is repelled by water is hydrophobic in nature.

Hydrophobic: Having water repellent characteristics; non wettable.

Hydroponics The growing of plants in aqueous chemical solutions

Hydroprene: A juvenile hormone analogue.

Hydrostatic balance: Maintenance of body position in the water by aquatic insects.

Hydrostatic organ: In pelagic larva of *Corethra,* there are four developed air sacs which function solely as hydrostatic organs. This larva can adjust its specific gravity to that of the water. It is by varying the capacity of the air sacs, apparently by some active change which causes expansion or contraction of their cuticular walls.

Hydrostatic skeleton: Turgid structural support provided by fluid pressure which maintains contractions on a fixed volume of liquid, especially in larvae.

Hygrophilus: Moisture loving.

Hygroreception: Perception of moisture in the air.

Hygroreceptor: A sense organ involved in the perception of moisture.

***Hylemya antiqua* (Meigen):** Belongs to order Diptera and family Anthomyiidae. It is an insect pest of onion, commonly known as the onion maggot.

Hymenoptera: *Hymen*=membrane, *ptera*= wings. Insect order containing bees, wasps, ants, and sawflies. These are characterized by membranous wings, chewing or chewing-lapping mouthparts and complete metamorphosis. This is a large order comprising of a great many insects which are beneficial to man. Most of them are parasites or predators on other insects and the part played by these insects in biological control of notorious and introduced pests of crops needs no emphasis. The bees are well known for their role in pollination of plants and honeybees yield honey and wax, in addition.

Hypandrium: See subgenital plate.

Hyperestrogenic syndrome: A condition in which the estrogenic hormone is present in undue high proportion.

Hypermetabola: The group of insects having distinct types of larvae. See hypermetamorphosis.

Hypermetamorphosis: A kind of complete metamorphosis in which the different larval instars represent two or more different types of larva. During various larval stages, the morphology can change, for example from that of a campodeiform larva (elongate and flattened body with well developed legs and antannae) to a scarabaeiform larva (grub like) or to a vermiform larva (maggot like). In blister beetle, the larvae are caterpillar like and maggot like. The pupa develops into a subimago stage prior to imago stage, *e.g.*, Carabidae (Coleoptera), Strepsiptera, Bombilidae (Diptera).

Hyperneustic tracheal system: More than standard numbers (ten pairs) of spiracles are open, *e.g.*, Diplura (*Japyx* sp.) with four pairs of thoracic and seven pairs of abdominal spiracles (total eleven pairs).

Hyperosmotic: Having a salt concentration greater than that of the surrounding medium.

Hyperparasite: A parasite whose host is another parasite (Any category of parasitism beyond primary).

Hyperparasitism: It is the phenomenon of parasitization of a host individual when a parasite attacks and develop on primary parasite, *e.g.*, *Aphanogmus fijiensis* is a hyperparasitoid on cocoons of *Apanteles flavipes*, a primary parasitoid of *Chilo partellus* (Swinhoe).

Hyperparasitoid: An parasite whose host is also a parasite.

Hyperparasitoidism: A parasitoid serving as a host for another species of parasitoid.

Hyperpneustic tracheal system: It represents more than ten pairs of spiracles. Diplura, like *Japyx* bear four pairs of spiracles on thorax and seven pairs on abdomen.

Hyperpolarization: Increase in the electrical potential difference across a cell membrane.

Hypo: Under or lower.

Hypertrehalocaemic hormone: It is secreted by the intrinsic cells of corpora cardiaca. It is either blocked octapeptide or blocked decapeptide. It is isolated and characterized in several insects and found as even 29 amino acids long, glucagons-like hormone. It causes fuel mobilization and results into the hypertrehalosaemia and reduction into the glycogen content in the fat body. It occurs 1 ng per gland in *Manduca sexta*. It stimulates cAMP synthesis.

Hypertrophy: Abnormal growth and hyperplasia; under development.

Hypha (pl., hyphae): One of the simplest branched filaments of the mycelium of a fungus that is composed of one or more cylindrical cells and that increases in length by growth at its tip. New hyphae arise as lateral branches.

Hypocerebral ganglion: Ganglion located just posterior to the brain and dorsal to the foregut.

Hypocotyl: The stem of the embryo or young seedling below the cotyledons.

Hypodermis: See epidermis.

Hypogastruridae: A family of order Collembola. Lack pseudocelli, more or less prognathous head; furca, if present rarely reach ventral tube. For example, *Hypogastrura* sp.

Hypoglycemic: An abnormally small concentration of glucose in the circulation blood, *i.e.*, less than the minimum of the normal range.

Hypognathous: The long axis of the head is vertical and mouth parts are directed downwards, *e.g.*, grass hopper.

Hypomere: A ventral process of the phallobase.

Hypomeron (pl., hypomera): That portion of the pronotum which is visible from the ventral side; when there is a lateral pronotal carina, this is the portion below that carina (pronotal disc being above it). The post-coxal projection in Polyphaga is part of the hypomeron.

Hypopharynx: A component of the insect mouth parts arising behind the mouth and just in front of the labium or lower lip. Usually short and tongue like in species with biting jaws, but often drawn out to form a tube for the salivary duct in those species with sucking mouth parts.

Hypopleural bristles. A curved row of bristles on the side of the thorax of certain true flies just below and in front of the haltere and just above the base of the hind leg.

Hypopleuron (meropleuron): It is a plate on the lower side of the thorax in Diptera. It may be bare or with weak hairs or strongly bristled.

Hypopneustic tracheal system: It represents a partially degenerated or a typical tracheal system in which one or more pairs of spiracles disappear completely and insect, therefore, bears often less than ten pairs of spiracles, *e.g.*, Mallophaga, Siphunculata (one pair of thoracic and six pairs of abdominal), coccids, Collembola etc.

Hypopus: A nymphal stage in the development of certain mites in which the organism is small and has developed suckers or claspers for grasping and thereby affecting dispersal.

Hypostoma: The part of a margin of the cranium posterior to the mandible, usually narrow but as a hypostomal bridge in the ventral wall of head.

Hypostomal bridge: It is an important modification in head capsule which acts as the closure of occipital foramen by fusion of the posterior subgenae.

Hypostomal suture: See sub-genal suture.

Hypostome: In ticks, the median ventral dart-like mouthpart that is immovably attached to basal part of the capitulum.

***Hypothenemus hampei* Ferr.:** Belongs to order Coleoptera and family Scolytidae. It is an insect pest of coffee, commonly known as the coffee-berry borer.

Hypotrehalocaemic hormone: Duvae *et al.* (1977) and El-Salhy *et al.* (1980) noticed that Dipteran hypotrehalocaemic hormone and the vertebrate insulin like peptide are one and the same. It is further confirmed in Lepidopteran, Hymenopteran, Orthopteran and other Dipteran species. Insulin peptides have been identified by immunological, immune-histochemical and biological assays in 18 species of insects. Insulin serves as a hormone of fuel storage.

Hypsidae: Family of super family Noctuoidea, suborder Ditrysia and order Lepidoptera. The moths have a well-developed proboscis. The larvae have long, dense hairs and pupate in a very light silken cocoon. The insects are found in Africa, the Orient and Australia. The common Indian species are: *Hypsa complana, H. alciphron, H. ficus, Argina argus, A. syringa* and *A. cribraria.*

Hysterosoma: In mites, the posterior part of the body when there is a demarcation of the body between the second and third pair of legs.

Hythergraph: A polygonal diagram resulting from plotting temperature means against rainfall.

I

Ibaliidae: Family of super family Cynipoidea, suborder Apocrita and order Hymenoptera. These are relatively large insects which are parasites of Siricidae.

***Icerya purchasi* Maskell:** Belongs to order Hemiptera and family Coccidae. Commonly known as cottony cushion scale or fluted scale is a pest of citrus, apple almond, walnut, peach, apricot, fig, grapevine, guava, pomegranate, wattle etc.

Ichneumonidae: Family of super family Ichneumonoidea, suborder Apocrita and order Hymenoptera. The ichneumons are large, slender, black, yellow or reddish yellow insects with longer antennae having 16 or more segments. These are parasitic on many groups of insects and spiders and a few are hyperparasites. The adults are active on warm sunny days and frequent the flowers especially of Umbelliferae.

Ichneumonoidea: Super family of Suborder Apocrita, order Hymenoptera. This is a very large group and members of this superfamily are all parasitic on other insects and occasionally on other Arthropoda. Economically these insects are important as they exercise check over the multiplication of the injurious species and also in view of the possibilities of exploiting them for biological control of crop pests.

Identification: Process of unambiguously determining the chemical identity of a pesticide or metabolite in experimental or analytical situations. It also refers for determining the identity of a species.

***Idioscopus clypealis* (Lethierry):** Belongs to order Hemiptera and family Cicadellidae. It is same as *Idiocerus clypealis* commonly known as mango hopper.

Igneous rock: Rock masses generally accepted as being formed by the solidification of magma (molten rock) injected into the earth crust or extruded on its surface.

Ignition temperature: The temperature at which the substances starts burning.

IGR: Insect growth regulator.

IGSI: Indian Grain Storage Institute.

Illegal residue: Residue that is in excess of a pre-established government enforced safe level.

Image formation: There are two principal mechanisms of image formation: the apposition and the superposition. See apposition and superposition mechanism.

Imaginal disc: In the larva of some insects, notably in Diptera and Lepidoptera, many of the organs of adult are already recognizable as clusters of embryonic cells, the so called imaginal discs of Weismann. But these are not a necessary feature of metamorphosis.

Imaginal instar: Pertaining to adult stage or imago.

Imago (pl., imagoes): The adult or reproductive stage of an insect.

Imbricated: Overlapping.

Imidacloprid: Imidacloprid, a novel insecticide derived synthetically from a nitromethylene insecticidal chemical, nithiazine, is closely related to the nicotinoids in mode of action and structure activity relationship. It belongs to chloronicotinyl group of neonicotinoids and effective against sucking insect pests. It is chemically, 1-(6-chloro-3-pyridylmethyl)-N-nitroimidazolidin-2-ylideneamine, marketed under trade name Confidor 17.8 SL and Gaucho 70 WS. It inhibits nicotinic acetylcholine by binding with nicotinic acetylcholine receptor (nAChR).

Imidan: A synthetic insecticide and acaricide; an organophosphate, *O, O*-dimethyl *S*-phthalimidomethyl phosphorodithioate, moderately toxic to mammals; acute oral LD_{50} for rats 147 to 299 mg/ kg; controls certain insects and mites infesting deciduous fruit trees. It is a product of Stauffer Chemical Company, New York.

Immature stages: A developmental stages preceding the adult stage. For example, the eggs, nymphs, caterpillars and pupae are immature stages.

Immigration: The movement of individuals into a population.

Immiscible: Incapable of being mixed to form a homogenous phase, usually applied to liquids, *e.g.*, oil and water are immiscible (also see incompatible).

Immobilisation: (i) Process leading to restricted mobility of a pesticide in plant or soil due to strong binding. (ii) Incorporation of terminal pesticide degradates into complex organic forms in microbial or plant tissue.

Immune: Exempted from infection.

Immunized: To render invulnerable to the toxicant, usually by injecting the toxin in small quantities at short intervals without appearance of severe symptoms.

Immunoassay: Ligand binding assay based on antibodies capable of specific binding to the pesticide analyte. Most commonly used in a competitive binding format where analyte molecules compete with a specific antigen complex labeled for detection using a radioisotope (radioimmunoassay or RIA) or enzyme (enzyme linked immunoassay or ELISA).

Immunogen, insect: Substance that develops immune response in insects (similar to antigen for other animals).

Immunology: The branch of science dealing with immunity to the different diseases and insect pests.

Immunotoxin: A substance that damages the immune system.

Impermeable: Not capable of being penetrated. Semi-permeable means permeable to some substances but not to others.

Impervious: Hard to penetrate or soak through the condition of soil when it will not soak up water.

Impurity: By-product of the manufacture or storage of a pesticide. Impurity requires definition, evaluation and regulation (if toxicologically significant).

Incipient species: A population which has not attained all the attributes of reproductive isolation from parent population of a species.

Incisors: Cutting or fore teeth; teeth in the forepart of the mouth.

Inclusion body: The proteinaceous or crystal like structure produced in insect cells infected with certain viral pathogens; it occurs in various shapes and size and usually encloses a number of replicated viruses.

Inclusion viruses: See occluded viruses.

Incompatibility: The inability of a pesticide to mix with other pesticides without producing undesirable effects.

Incompatible: Not compatible. Not capable of being mixed or used together in the case of pesticides, the effectiveness of one or more is reduced or they cause injury to plants or animals.

Incomplete metamorphosis or Simple metamorphosis. Metamorphosis in which the wings (when present) develop externally during the immature stage and there is no prolonged resting stage (*i.e.*, pupa) preceding the last moult; stages included are the egg, nymphal, and adult. It is also called gradual or partial metamorphosis, and paurometabolous development. An incomplete metamorphosis is characteristic for insect orders belonging to the exopterygota.

Incrassate antenna: An antenna in which the antennomeres gradually increase in width towards the apex or increase and subsequently decrease in width. When the increase is marked, the antenna looks club like and is often called clavate.

Increment sample: An individual portion (unit) of material collected by a single operation of a sampling device from bulk materials or large units.

Incurred residue: Residue in a commodity resulting from specific use of a pesticide, consumption by an animal or environmental contamination in the field as opposed to residues from laboratory fortification of samples.

Incurvariidae: Family of super family Incurvarioidea of suborder Montrysia and order Lepidoptera. In females the antennae are shorter than in males. Males possess larger eyes which are approximated dorsally. Larvae are leaf miners, a few live in cases and feed on vegetable refuse. However, there exits an interesting biological relationship between the yucca plants. The female moths collect the pollen with their modified maxillae and deposit the pollen on the stigma of the yucca flowers and lay the eggs in the ovaries of the flowers.

Incurvarioidea: Super family of suborder Montrysia and order Lepidoptera. Characterized by reduced wing venation in hindwing and female possessing sclerotised ovipositor and a long cloacal duct.

***Indarbela quadrinotata* (Wlk.):** Belongs to order Lepidoptera and family Arbelidae. It is commonly known as bark caterpillar and is a pest of citrus, mango, guava, jamun, mulberry, loquat, pomegranate, ber, drumstick, litchi, amla, rose and a number of forest and ornamental trees.

***Indarbela tetraonis* M.:** Belongs to order Lepidoptera and family Arbelidae. It is an insect pest of fruit trees, commonly known as the bark eating caterpillar.

Index: A list of records used to arrange based on some attributes.

Indian honeybee: *Apis indica* F., the common honeybee reared throughout India; several strains known, lives in cavities of trees, hollows of rocks, hollowed-out logs, boxes, kerosene empties, earthen pitchers. Makes a series of parallel combs, has gentle temperament and is easy to handle, yields 8 to 10 lbs of honey per colony, given to heavy swarming, absconding and robbing.

Indian meal moth: *Plodia interpunctella* Hub. is a moth with fringed wings that lays eggs on stored food materials. The caterpillar prefers to feed on flour and other milled products and grain. The caterpillar stage lasts for 30 to 35 days; whole life cycle for 5 weeks.

Indicator organisms: Organisms which tend to accumulate pesticides from ambient medium like air, soil and water, *e.g.*, earthworms, snails and oysters. The selection or choice of an indicator organism is based on (a) the degree of concentration of pesticide in the ambient medium, (b) tolerance pesticide, and (c) tendency of organism to filter the medium or degree of the accumulation of chemical. For example in case of water samples, oysters and other selfish that filter a lot of water to obtain food; in case of soil, earthworms and certain beetles that fly larger distance are useful indicator organisms for soil samples.

Indigenous: Native to a particular region or country.

Indigo caterpillar: See *Spodoptera exigua.*

Indigo psyllid: *Arytaina punctipennis* Cr., a small light brown insect with spotted wings that sucks the sap from the indigo plant; leaves curl up; knot formed at top of stem.

Indirect pest: An organism which causes delayed or insidious degradation of marketable produce through generalized attacks on the host which may reduce the growth, vigour or yield; see direct pest.

Indoxacarb: An insecticide belonging to new insecticide group, indoxacarb. It inhibits the flow of sodium ions into nerve cells in insects that cause paralysis and death. It is contact and stomach poison, effective against borers and marketed under trade name Avaunt 14.5 SC.

Induced resistance: Resistances of a plant to attack of pest or disease brought about by condition under which it is growing and varying in degree according to these conditions; generally brought about by farm practices and nutritional means.

Inductive: Leading to inferences by method of logical induction.

Industrial toxicology: The branch of toxicology that deals with the safety of industrial workers from the toxic effects of poisons.

Inert ingredient: A substance having no biological action or formulant which does not add materially to effectiveness for the purpose of which the preparation is intended, *e.g.*, solvent, emulsifier, diluent, carrier.

Infect: To enter and establish a pathogenic relationship with a plant (host); to enter and persist in a carrier.

Infection: The introduction or entry of a parasite or pathogenic micro-organism into a susceptible host, resulting in the presence of that organism within the body of the host, whether or not this causes pathologic effects.

Infectious disease: A disease caused by a living organism.

Inferior appendages: The lower one (Anisoptera) or two of the terminal abdominal appendages used in grasping the female at the time of copulation (male Odonata and Zygoptera).

Infest: To occupy and cause injury to either a plant or a stored product.

Infestation: Presence of animal pests (insects, rodents etc.) on the plant crops.

Inflorescence: The flowering portion or stalk of group of flowers of a plant.

Information technology: In this combined technology, processing and transmission of information is studied. Computing, telecommunication and microelectronics are also included in this.

Infra-red rays: Invisible heat radiation or radiant heat or infra red radiation which possess high power of penetration and scatter the ordinary visible light.

Ingest: To eat or swallow.

Ingluvial: Pertaining to the crop, the distensible middle portion of the foregut in many species of insects in which liquid food is stored.

Ingluvial ganglion: A paired ganglion of the stomodaeal nervous system in some insects situated on side of the crop (gastric ganglion).

Inhalation toxicity: Poisoning through the respiratory system.

Inhibition: (i) It is a special mode of reproduction, in which the maturation of gonads is prevented. Under certain natural conditions, the reproductive system is adversely affected so as to inhibit or interfere with the production of ova or sperms. This leads to the castration of individuals. The physiological cause of castration is nutritional deficiency, *e.g.*, in larval worker of honey bee. (ii) Artificial induction of castration by active gamma rays or by use of chemicals (chemosterilants) has great potentiality in insect control. Large populations of insects are artificially bred and subjected to render them sterile without adversely affecting their mating or other behaviour. The males are liberated in the field which compete with the natural population of males and inseminate females with non-viable sperms resulting in heavy reduction in population of successive generations.

Inhibitor, growth: A chemical, usually of the regular type, that prevents or suppresses growth or other physiological processes in plants, 2-amino-1, 2, 4-triazole (ATA) is an example of a growth inhibitor.

Injector: A device for ejecting a pesticide below the soil surface, or into the transport system of a tree.

Injurious: Harmful, can cause damage.

Injury: Damage of a plant which impairs growth, functioning or appearance but not necessarily resulting in loss of yield or quality.

In life phase: Phase of a study following treatment in which the test system is alive/ growing.

Inner cover: A light weight cover used under a standard cover on a beehive.

Inoculation: The act of introducing an insect, often at the egg or larval stage over a food source in a rearing container.

Inoculative release: The establishment of a population of exotic entomophagous insects for several generations per season using laboratory reared insects.

Inoculum: Incoming fungus spores or other agents of disease which initiate the development of a parasite within the host.

Inorganic compound: A compound which does not contain carbon atoms except CO_2, which is considered an inorganic compound.

Inorganic-organic cycles: See biogeochemical cycles.

Inquiline: Animal that shares a home with an unrelated animal with no apparent adverse effect on the animal.

Inquilinism: The relation in which a socially parasitic species spends the entire life cycle in the nest of its host species. Workers are either lacking or if present, are usually scarce and degenerate in behavior. This condition is sometime referred to loosely as permanent parasitism. Inquilinism is a special form of commensalism. An inquiline is a commensal that lives in a very close spatial relationship with its host for its shelter or rest. It not only feed upon its host but does not frequently destroy it. The gall wasp, *Synergus umbraculus* and *S. rainhardi* are inquilines of the gall maker, *Andricus kollari*. The oak marble gall is the larval home of a gall wasp, *A. kollari*. The synergus wasps are themselves incapable of inducing gall formation and are fully dependent upon *A. kollari* for the provision of their larval food. They are smaller than *A. kollari* and one gall can support about 20 *S. umbraculus* and 10 *S. reinhardi* oak gall wasps.

Insect (pl., Insects): (i) An animal of class Insecta or Hexapoda (Phylum, Arthropoda) with bodies divisible into a head, thorax and abdomen; the head bearing a pair of feelers or antennae; the thorax three pairs of legs and wings. (ii) An air-breathing animal of phylum Arthropoda with a distinct head, thorax and abdomen, has one pair of antennae, three pairs of legs, usually two pairs of wings. Some infests plants and animals; some pollinate plants, some produce edible or otherwise useful products. (iii) Any of the numerous small invertebrate animals generally having segmented bodies and for the most part belonging to the class insects, comprising six legged, usually winged forms. (iv) Insects are tracheate arthropods belonging to the class Hexapoda. They have a segmented body with three distinct regions, the head, the thorax, and the abdomen. The head bears one pair of antennae, compound eyes, and the mouthparts. The thorax bears three pairs of legs (one on each of the three thoracic segments). The thorax bears often one or two pairs of wings (on the second and third thoracic segments). A gonopore is present at the posterior end of the abdomen. Exceptionally, some insects are apterous (Collembolans, Proturans, Thysanurans and Diplurans) and Proturans do not possess antenne.

Insect control: The biological or chemical inhibition or killing of insect pests.

Insect growth regulator: Chemical substance which disrupts the action of insect hormones that control processes such as moulting, development from pupa to adult etc.

Insect parasites: Insects which lay eggs on the body of other insects, when the emerging larvae feed internally on them and killing them.

Insect pathology: The study of insect diseases, their causes and treatments.

Insect pest management: An ecologically based strategy of maintaining insect pest population below the economic injury level by the use of any or all control techniques that are economically, ecologically and socially acceptable.

Insect predators: Insect which catch other insects and feed on them.

Insect rearing laboratory: Any room or place used to rear insects. In an insectary, there is some degree of environmental control which is favourable for insect growth and development.

Insect rearing management (IRM): It is the efficient utilization of resources for the production of insects of standard quality to meet the research goals.

Insect rearing management system (IRMS): The integration of insect rearing management techniques and production processes and their application to rear a particular insect species.

Insect sociology (Insect sociobiology): The study of social behavior and population characteristics related to social behavior in insects.

Insect toxicology: A branch of toxicology that deals with such poisons which are used in killing insects without appreciable effect on mammals.

Insect traps: Traps are the simple devices by which the insect pests can be captured, killed and their population can be reduced. The insect traps may be of various size, shape, simple or complex and constructed in such a way that insects can easily enter the receptacle but unable to escape from it. The attractant agents are mostly the lantern, lamp, aromatic chemicals etc. These are mechanical traps, suction type trap, grid type trap, glass trap, pheromone trap etc.

Insect vector: An insect which carries virus, bacterium or spores or mycelium of a pathogenic fungus and inoculates susceptible plants or animals.

Insecta: A class of phylum Arthropoda distinguished by adults having three body regions head, thorax and abdomen and by having the thorax three-segmented with each segment bearing a pair of legs.

Insectary: A place where insects are housed or propagated or bred in large numbers; also called insectarium.

Insecticidal: Pertains to an insecticide or its lethal properties.

Insecticide: (i) An insecticide is a toxic substance which kills and controls insects. Most of those in farm use today are synthetic organic compounds and include chlorinated hydrocarbons, organophosphorous compounds, carbamatesc pyrethroids. They are available in granular, liquid and powder forms and may be applied in various ways. (ii) Chemical prepared with the purpose of controlling insect pests; active ingredients can be gas, liquid or solid, formulated in liquefied gases, smokes, dusts, water-dispersible powders, emulsions and solution. (iii) A substance or mixture of substances intended to prevent or destroy any insect which may be present in any environment. Use of insecticides is the first line of defense in controlling the insect outbreaks. These control the large populations of insect pests effectively and immediately.

Insecticide resistance: (i) Resistance to insecticides or pesticdes is the ability of a strain of insects to tolerate doses of toxicants which would prove lethal to a normal population of insects of the same species. Resistance results when occasional resistant individuals arise in a population and survive the pesticidal treatment. The survivors then reproduce and confer the resistance to their offspring in their succeeding generations. Insecticide resistance is a dynamic phenomenon dependent on biochemical, physiological, genetic and ecological factors. (ii) The degree to which a species of insect tolerates a toxic substance. The degree of resistance is based on LD_{50} values of a population that has been previously exposed to insecticide, compared to those of an unexposed population.

Insecticide solution: Insecticide whose active ingredient is dissolved in water or other cheap solvent; contains higher percentage of ingredient and, therefore, more expensive than emulsion.

Insecticide solvents: Substances in which insecticide in the form of crystal or powder is dissolved so that it can be used as spray; water is commonly used solvent; acetone, ethyl alcohol, benzene, xylene, linseed oil, groundnut oil, cotton seed oil, turpentine, kerosene and petroleum are other solvents; methyl naphthalenes are superior solvents.

Insecticides act: To minimize pesticide hazards and regulate the use of pesticides in public health and agriculture, Government of India in 1968 passed a comprehensive legislation called Insecticides Act, 1968. However, the act, along with the rules framed with the Act started functioning (implemented) from 1st

August, 1971. Various regulatory provisions made in the act include compulsory registration of pesticides at the central level; issue of license for manufacture or formulation and sale at the state level; inter departmental or ministerial or organizational coordination in the various fields of matters arising out of the implementation of this act; establishment of enforcement machinery like Insecticide Analysts, Insecticide Inspectors etc. under various sections of the act two statutory bodies namely, the Central Insecticide Board and the Registration Committee for pesticides have been constituted.

Insectivorous: Feeding on insects.

Insemination: The transfer of sperm to the female which takes place at mating, an elaborate process. The aedeagus or penis of male after entering the vagina may actually penetrate up the duct of the receptaculum so that spermatozoa mixed with the secretion of accessory glands are ejaculated directly into the spermatheca. But more often they are discharged into the vagina or bursa copulatrix which arises from it.

Instar: Stages of growth, each separated by a moult or the form of an insect between successive moults; the first instar being the stage between hatching and the first moult. Thus, a newly hatched insect which has not yet moulted is said to be a first instar nymph or larva. The adult (imago) is the final instar.

Instrumental insemination: The introduction of drone spermatozoa into the genital organ of a virgin queen by means of special instruments.

Insular: Living or inhabiting an island.

Insulin: A protein hormone which is produced by the islets of Langerhans of the pancreas.

Integrated control: The combination of different methods of pest control.

Integrated Pest Management (IPM): The FAO of United Nations Organization established an FAO panel of experts on Integrated Pest Control (IPC) which defined it as 'a system that in context of the associated environment and the population dynamics of pest species, utilizes all suitable techniques and methods in as compatible manner as possible and maintains the pest populations at levels below those causing economic injury'. However, the publication of Council of Environmental Quality document entitled as 'Integrated pest management' in 1972, there has been wide spread use of the term IPM. Here for all purposes, IPM has precisely the same meaning as IPC. Often, the term IPM includes all elements contributing to an effective, safe, sustainable and economically sound crop protection system. The abbreviation IPM stands for this term.

Integrated pest management model: The simple definition of integrated pest management model is that 'a model is an imitation and representation of the real world' (Ruesink, 1975). It is a computer model which truly imitates and represents all occurrences in particular locality at any given time from planting till harvesting the crop. It helps in understanding the ecosystem-crop plant development changes due to fertilizer use, irrigation, pest development and abundance and thus it helps in selecting an appropriate combination of crop plant variety, fertilizers, pest control methods etc., to achieve optimal yield of the crop. If the model develops through a collective task of interdisciplinary research teams including entomologists, plant pathologists, agronomists, meteorologists, economists etc., then it can provide amost realistic blue print on the basis of which IPM can be established. The models can independently be prepared as (i) ecosystem or conceptual model, (ii) population models, and (iii) economic and decision models or (iv) even a single simulation model representing summation of all the three. The simulation model includes various components or subsystems, such as plant sub-system, local plant environment, plant growth and insect pests, parasites and predator sub-systems and their interactions.

Integrated pest suppression: An approach to compatible utilization of all available forms of pest suppression, including mechanical, biological, chemical and natural control in systematic fashion with the primary goal of safe, effective and economical pest population reduction. It may be directed at a single important pest species by combining a variety of measures against the species or at a complex of pests, integrating the individual protective measures applied against each so as not to interfere one with the other.

Integument: It is that part of ectoderm which externally surrounds the insect body and its appendages. The general structure of integument consists of cuticle, epidermis (hypodermis) and basement membrane.

Integument excretion: Moulting is considered a type of excretion of nitrogenous waste products and others deposited in the exuviae. Chitin constitute 30-40 per cent of insect cuticle.

Integument respiration (Cuticle respiration): In Proturans and Chironomid larvae, the respiration takes place by circulation of air through integument.

Interantennal suture: A suture extending between the bases of the two antennae (Siphonaptera).

Interbreeding: Phenomenon of actual or potential gene exchange in individuals by hybridization.

Intercalary segment: The premandibular segment of the insect head, in the embryo insect.

Intercalary vein. An additional longitudinal vein arising at the wing margin and running inwards but not directly connected to any of the major veins.

Intercoxal process: See Abdominal process or prosternal process.

Intercropping: The growing of two or more crops simultaneously in the same field.

Interfacetal setae of eye: Setae arising between adjacent eye face..

Interference: In colours, iridescent colours produced by variable reflection of light by narrowly separated surfaces (as scales of Lepidoptera), in intra and interspecific interactions -between predators and parasitoids, a reduction in the profitability of otherwise high resource density.

Intermediate host: The host which harbours the immature stages or the asexual stages of a parasite; a separate organism to that which harbours the sexual stage.

Intermittently: Starting and stopping from time to time.

Internal feeder: The internal feeders include borers and leaf miners. Larvae of some species of Lepidoptera, Coleoptera and Diptera are borers. Many caterpillars (Lepidoptera), grubs (Coleoptera) and maggots (Diptera) are adapted as borers. They bore into the root, stem, shoot, bud, flower, boll and seed. The larvae of few species of Coleoptera, Diptera and Lepidoptera mine the leaves of some crop plants. Some stored grain insect pests, like rice weevil, pulse beetle, granary weevil, Angoumois grain moth etc. lay their eggs inside or on the seeds and only emerge in the adult stage. They contribute significant loss of germ portion due to which the seeds lose viability which is not detectable outside.

Internal respiration: The process of oxidation accompanying metabolism in the cells of body tissues.

International unit: An arbitrarily set basis for comparing the efficacy of insect-pathogenic *Bacillus thuringiensis* preparations. An international unit is one thousand of the amounts of insecticidal activity contained in one milligram of a preparation of the primary standard E-16 strain of *B.thuringiensis* as measured by bioassay against certain lepidopterous larvae (*i.e.*, comparative $LD\text{-}_{50}$). A *B. thuringiensis* formulation with a potency of 1,000 IU/mg is therefore, equal to the international standard. A secondary reference standard *B. thuringiensis* strain (HD-1-S-1971) is in use in the USA and has been assigned a potency of 18,000 IU/mg against the cabbage looper. *Trichophusia ni* (Hubner).

Interneurones: Neurones found within the ganglia. They make association between the afferent and efferent neurones and are called the interneurones, internuncial or interpolated or association neurones. They bear comparatively small cell bodies mostly occupied by the nuclei and are called the globuli. The giant neurones running the total length of nerve cord, represent themselves as the association neurones. Similarly, the transverse commissures are formed by these neurones.

Internodes: The length of stem between two successive nodes.

Internuncial neurones: See interneurones.

Interplanting: The interplanting of one crop within another, for example with the purpose of trapping or repelling pest insects.

Intersegmental membrane: The flexible conjunctiva between two secondary segments; usually the non-sclerotized posterior part of a primary segment but variable.

Intersegmental muscles: The principal longitudinal muscles running antecostae in insects with secondary segmentation.

Interspecific: Between two or more species.

Intersterility: Cross sterility confined to specific genetically determined group of individuals.

Intersternites: Primary intersegmental sclerites of the ventor, becoming the spinasterna of the thorax.

Interstitial: Situated between two segments.

Intestinal microorganism: The intestinal microorganisms occur in insects (like, cockroaches, locusts and ants) feeding on wood, dry cereals, feathers, hairs, wool, blood and plant sap. Most common microorganisms are the bacteria found in blattaria, Isoptera, Homoptera, Anoplura, Mallophaga, Coleoptera, Hymenoptera and Diptera. Among protozoans, flagellates are widely occurring in hindgut of wood eating cockroaches and termites. The yeasts are found in some Homoptera and Coleoptera and actinomycetes in epithelial crypts of anterior midgut of *Rhodnius*. Other cells of the body housing microorganisms are called mycetocytes and aggregation of mycetocytes form the organs called mycetomes. The mycetomes in different insects are scattered in fat bodies, midgut epithelium, Malpighian tubules etc. The intestinal flagellates carry out the digestion of wood in cockroaches and termites.

Intestine: The anterior portion of the hind gut.

Intima: Chitinous lining of the foregut and hindgut and tracheae continuous with the cuticle$^-$ of integument.

In toto: In it's entirely (totality).

Intoxication: Poisoning or the adverse effect due to the intake or exposure to any substance.

Intra: Within.

Intra-alar bristles: A row of two or three bristles situated on the mesonotum above the wing base, between the dorsocentral and the supra-alar bristles (Diptera).

Intradermal: Within the skin.

Intraspecific: Within a species.

Intraspecific competition: Situation in which individuals within a population compete for limited resource.

Intrinsic articulation: Articulation in which the points making contact are sclerotic prolongation in the articular membrane.

Intrinsic muscles: Both origins and insertions entirely within a given appendage.

Introgression: The incorporation of genes of one species into gene pool of another species by hybridization and back crossing.

Intromittent organ: The phallus; the penis.

Inundative release: The production and release of large numbers of beneficial insects against a particular pest population.

Invagination: An infolding or inpocketing so that an outer becomes an inner surface.

Invasion: The penetration by a microorganism of the integument and other organs of the body of a host organism.

Inversely density dependent: Refers to mortality factors or processes in the environment which destroys a decreasing percentage of the subject population as its numerical density increase.

Inversion: (i) Used in meteorology to describe the state of lower atmosphere when the temperature is lowest near the ground. Frequent at night, the lower atmosphere is non turbulent in inversion in contrast to its turbulent state under the opposite (lapse) condition when the ground is warmed by solar radiation. (ii) When temperature increases with the elevation from the ground, an inversion condition exists.

Invert emulsion: One in which oil is the continuous phase and water is suspended in it.

Invertase: (i) An enzyme produced by the honey bee which helps to catalize transformation of sucrose to dextrose and levulose. (ii) An enzyme that catalyzes the breakdown of sucrose into fructose and dextrose.

Invertebrate: Animals, not having a backbone or spinal column (molluscs, arthropods, worms, etc.)

Inviscate: (i) To trap insects in a sticky substances or on a sticky surface. (ii) To make a substances more sticky by mixing it with viscous matter.

***In vitro*:** A biological process occurring experimentally in isolation from an organism. It is a situation where micro-organisms are kept or propagated in isolation of their natural hosts or other living organisms. *In vitro* means in glass (referring to test tubes); contrast with *in vivo*.

Involucrum: A sheath of soft cerumen surrounding the brood chamber in a nest of stingless bees (Meliponini).

IOBC: International Organization for Biological Control of Noxious Animals and plants (an affiliate of the International Union of Biological Sciences). A global organization of governmental units and individuals interested in biological pest suppression, headquartered in Zurich, Switzerland. Major objectives include dissemination of information, coordination and promotion of research and application of biological pest suppression; publisher of the journal, Entomophaga.

Ionic surfactant: One that ionizes or dissociates in water.

IPC: Isopropyl-phenyl-carbamate; weedicide, insoluble in water and oil and applied as dust; useful in the eradication of some annual grasses; usually applied as pre-emergence weedicide. Does not affect some of the broad leaved plants like cotton, crucifers and legumes, unlike 2, 4-D; some other cultivated plants such as solanaceous and cucurbitous plants are very susceptible and some cereals like barley and wheat susceptible; paddy, maize and sorghum and other millets are affected to lesser extent.

IPM: Integrated Pest Management.

IPS: It refers to the 'International Pollination System'. The IPS is specialized in what we call the 'managed bees', *i.e.*, bees that exist and pollinate in a controlled fashion in man made environments. The most commonly used pollinators include the alfalfa leaf cutting bee, the alkali bee, the bumble bee, the blue orchard bee and the honey bee. Contemporary research is focused on a number of additional bees, which might be proved to be variable managed pollinators in the coming years.

Iridescent colours: Glittering changing colours; also see structural colours.

Iridescent virus: A non-occluded insect virus characterized particularly by its fascinating optical properties; these stem form Bragg reflection, and give purified pellets of virus particles an opalescent appearance.

Iris cell: See ommatidium, structure.

Irritability: The ability to respond to stimuli.

Ischiopodite: The third segment of a generalized limb or second segment of the telopodite (second trochanter, prefemur).

Ischnocera: A suborder of Mallophaga. Antennae three to five segmented, filiform, mandibles vertical, without maxillary palpi Meso and metathorax generally fused.

Iso: Equal; uniform.

Isogonic growth: See harmonic growth.

Isolan: It is a carbamate insecticide, chemically it is 1-isopropyl-3-methylpyrazolyl-(5)N, N-dimethyl carbamate ($C_{10}H_{17}O_2N_3$). It is a colorless liquid soluble in water and other organic solvents. It is systemic insecticide effective against aphids.

Isolating mechanism: Prevention from mating between breeding group owing to spatial, topographical, physiological, genetic behaviour or other barriers.

Isomer: (i) Any of the two or more chemical compounds having like constituent atoms but differing in physical properties owing to difference in the arrangement of the atoms. (ii) In other words, two or more chemical compounds having the same structure but different properties.

Isometry: The condition in which the size of two body parts remains constant relative to each other as total body size is increased (isometry is a special case of allometry).

Isophene: A contour line delimiting area corresponding to a given frequency of variant form.

Isoptera: Order (*iso*= equal; *ptera*= wings). Order of insects to which termites belong. The termites or white ants which are social insects living in organized communities composed of males, females and sterile workers and soldiers. The head is prognathous, oval to rounded; moniliform antennae (bead like antennae) bearing 9-30 segments; eyes are well developed in winged forms and reduced in others; mandibulate mouth parts; cervical sclerites are characteristics of neck; legs are similar and short; wings alike, long and membranous are characteristic of sexually mature males and females and there are 10 segments of abdomen, the tergites 10 and 11 being fused while the sternum of 11th segment is represented by a pair of paraprocts bearing cerci as in the blattids. These are characterized by an incomplete metamorphosis.

Isotomidae: A family of order Collembola. Lack trochanteral organ and scales, mucro short, post-antennal organ present. For example, *Isotoma* sp.

Isotope: Isotope means 'the same place'. Those elements which are kept at one place in the periodic table (Sody, 1913). Those atoms of the same elments whose mass numbers are different but having same atomic numbers are known as isotopes of that element.

Issidae: A family of superfamily Fulgoroidea of suborder Homoptera and order Hemiptera. Widely distributed family. Some possess short wings and beetle like snout; forelegs are enlarged with foliaceous femora. Costa border with many cross veins. Wings held less steeply at rest, *e.g.*, sandal tree issid, *Sarina nigroclypeata*.

Istoblas: A small group of cells not affected by the histolysis of old tissue; the morphological unit or cells characteristic of a particular tissue; the imaginal bud in the histolysis of the larval structures during pupation; imaginal disc.

Istogenesis: The formation of new body tissues.

Itching: A peculiar irritating sensation in the skin that causes the desire to scratch.

Ivermectin: It is a semisynthetic derivative of abamectin used to control parasite of cattle.

J

Jack fruit bud weevil: *Ochyromera artocarpi* M. is an insect pest of jack fruit, the grub stage of which bores the tender buds and fruits.

Jack fruit leaf webber: Jack fruit leaf webber, *Perina nuda* is an insect pest of jack fruit.

Jack fruit shoot borer: *Margaronia caesalis* W. is an insect pest of jack fruit, the caterpillar stage of which bores the shoots, buds and fruits.

Japygidae: A family of order Aptera (Diplura). Styli present on first abdominal segment; cerci unsegmented, sclerotised and forceps like. For example, *Japyx* sp., *Heterojapyx* sp., *Indojapyx* sp.

Jasmine budworm: See *Nausinoe geometralis*.

Jasmine thrips: See *Thrips orientalis*.

Jet: Liquid emitted from a nozzle orifice.

Johnston's organ: A sense organ or a chordotonal organ located in the second antennal segment (pedicel) of most insects; this organ functions in sound perception in Diptera.

Joint: The articulation between two segments, body parts, or appendages.

Jointing stage: Growth stage of grain crops when the internodes of the stems are elongating.

Jugal area: An area of the wing, less constantly present, lying behind the anal area and at the base of wing posteriorly.

Jugal veins (J): These veins occur in the jugal lobe of the wing. Their number may differ variably but commonly are two, the first and second jugal veins. These are short unbranched and concave in their appearance.

Jugum. A narrow lobe projecting from the base of forewing in certain moths and Trichoptera and overlapping the hind wing, thereby coupling the two wings together.

Jute apion: *Apion corchori* Marshall is a pest of jute. The grub stage feeds inside the bark, damages fibre and a mucilaginous substance oozes out forming knots in the fibre.

Jute mealy bug: See *Phenacoccus hirsutus.*

Jute semilooper: *Anomis sabulifera* Guen. is a pest of jute. The greenish caterpillars feed on leaves, apical buds and top shoots and check the plant growth.

Jute stem girdler: See *Nupserha bicolor postbrunnea.*

Jute stem weevil: See *Apion corchori.*

Juvabione: Juvenile hormone mimic or analogue found in the paper products prepared from American balsam fir tree. It was accidentally discovered by Slama and Williams (1966) and is active against European red bug, *Pyrrhocoris apterus* L. They tried to rear the European bug, *P. apterus* in the laboratory but the culture failed repeatedly. The first instar larvae instead of moulting to adult stage produced an extra larval instar or changed to adultoids. It was found due to the paper used to line the cages wherein the bugs were kept and was called paper factor. This paper was prepared from the balsam fir, *Abies balsamea,* a pulp wood. The paper was analysed and the chemical was isolated by Bower (1968). It was found that the chemical present in the paper was similar to juvenile hormone and was named as juvabione. It showed insecticidal properties and has considerable specificity also. The discovery of the juvenile hormone mimic in balsam fir tree initiated search for other mimics of juvenile hormone.

Juvenile hormone as insecticides: JH treated immature insects fail to ecdyse, die soon after ecdysis or fail to reproduce and develop intermediates between larva and pupa, pupa and adult or larva and adult. They are not formed due excess amount as found in ecdysone but are developed due to application of JH at wrong times in their development programmes. Young insects are not killed by JH. The JH can kill insects, if they act on the last instar larva during metamorphosis by inducing production of intermediates. On adults, they function as chemosterilants. The JH may, therefore, provide an opportunity of using them as the safe insecticides. Many substances have been isolated having JH activity. A number of terpenoids, aromatic hybrid compounds have been prepared. Some of them are alkylating agents. They can be subjected to oral, intraperitoneal and subcutaneous administration in acute toxicological trials. JH have been used against various pests and found much effective in suppressing the population of agricultural and livestock pests.

Juvenile hormone: The corpus allatum, small endocrine gland lying just behind the brain secretes the so called juvenile hormione or neotenin which ensures that the larval organism is again produced when the insect moults. It acts at cellular level. In the fifth larval stage, the corpus allatum no longer secretes the juvenile hormone so that when moulting is brought about by the growth and moulting hormone, the latent imaginal organism is able to develop and metamorphosis takes place. If corpus allatum is removed in the young insect, precocious metamorphosis occurs, diminutive silk moths of few milligrams are produced this way; or conversely, if corpus allatum from a young insect is implanted into a late stage, giant larvae is produced. In the adult insect, the corpora allata after stopping their secretion during the moults to the pupal stage, once again begin to secrete and produce juvenile hormone. This time the development of ovaries depends on this hormone and as such it is also sometimes called the gonadotropic hormone. Although, active insect extract of male silk moth (*Hyalophora cecropia*) was prepared by Williams (1956) but its chemical structure was known a decade latter (Roller, 1967) and named Cecropia C18-JH or Methyl 12, 14 dihomojuvenate, also called neotenin. Subsequently, more cercopia hormones were isolated whose structures were identical with cercopia C18-JH.

Juvenile stage: Every post embryonic stage in the development of an insect that precedes the adult stage, *e.g.*, egg, larval and pupal stage of insects.

Juvenile: Immature stage.

Juvenoids (Juvenile hormone analogue or mimic): Juvenile hormone mimics or analogues (JHA's) have synthetic or natural occurrence and exhibit similar activities with that of true juvenile hormone. Compounds tested for juvenile activity are farnesol, farnesyl methyl, farnesal etc.

K

Kairomone: A chemical or mixture of chemicals emitted by an organism (*e.g.*, a plant) that induces a response in an individual of another species (*e.g.*, an insect) that is beneficial to the receiving organism. For example, a plant scent which makes the plant more easily identifiable to an insect pest.

***Kala azar*:** A fatal disease transmitted by blood sucking sand fly. The genus *Phlebotomus* (dipteran family Phlebotomidae) transmits the pappataci fever virus, and in parts of South America, Africa, and Asia it carries the protozoan parasites that cause kala azar, Oriental sore, espundia, and bartonellosis. Aquatic larvae of fly live in the intertidal zone of coastal beaches, in mud, or in wet organic debris.

Kalotermitidae: A family of order Isoptera. Primitive, wood dwelling termites without workers but with pseudergates or pseudoworkers. Pronotum flat, broader than head; wings without anal lobe. Example, *Kalotermes, Neotermes, Glyptotermes.*

Kaolinite: $Si_4A_{14}O_{10}(OH)_8$. It is a mineral powder, pale pink cream in colour. The crystals are hexagonal in shape. It is neither marketedly abrasive nor hygroscopic and is low in ion exchange. It is used as an inert material in the insecticide dust formulations.

Karathane: See DINOCAP.

Karyomorphology: A branch of genetics dealing with structure and form of chromosomes.

Karyotype: Group of individuals having the same number of homogeneous chromosomes and similar linear arrangements.

KD_{50}: It represents the median knock down dose sufficient to kill 50 per cent population of test animal.

Keel: A narrow ridge, also called carina.

Kelthane: Trade name of dicofol. Low toxicity, while primarily an acaricide, Kelthane is also effective against many insects. Its chief use, however, is in controlling a wide variety of mites that infest building, trees, plants, shrubs, flower and lawns.

Kepone: See chlordecone.

Keratin: A protein forming the principal matter of hair, horns and nails.

Kerria lacca: Scientific name of lac insect.

Key: Tabulation of diagnostic characters in dichotomous couplets facilitating rapid identification of insects.

Key factor: The hypothetical causative agent most responsible for population density change. In practice, a variable factor in the environment recognized as most closely linked to changes in population density and measurable by continuous monitoring for use in predicting future population trends (Morris 1959, 1963).

Key pest: This is the most severe and damaging pest whose GEP lies well above DB and EIL. Human intervention in the form of control measures may bring the population temporarily below the EIL but it increases back rapidly and repeated interventions (sprays) may be required to minimize damage, *e.g.* cotton bollworms, diamondback moth, gram pod borer, sugarcane borers and some vectors.

Keystone species: A species whose removal causes marked changes to the community.

***Khapra* beetle:** See *Trogoderma granarium* Everts.

Kin selection: Selection operating between closely related individuals to produce cooperation.

Kineses (sing., kinesis): In the first place the insect may reach its destination without being truly oriented; the direction of movement is not clearly related to the source of stimulus. Reactions of this kind are known as kinesis or undirected reactions. See also klinokinesis, orthokinesis and taxes.

King: Fully sexually developed male among termites and other social insects.

Kinnaridae: A family of superfamily Fulgoroidea of suborder Homoptera and order Hemiptera. A small family recognized by the claval veins being formulate and apical segments of labrum much longer than wide. In Trinidad, the *Bytrois nemoralis* infests cacao.

Kinoprene: Low toxicity, kinoprene is useful for controlling white flies and aphids on flowers and other plants in green houses and for this type of pest control problem outside green houses as well.

Klinokinesis: A more complex form of undirected response is termed klinokinesis. Here, the insect moves in a straight line in a favourable environment but as soon as it enters a mildly unfavourable environment, it begins to make turns, the frequency of turning increases with the strength of the stimulus. If the stimulus is very strong the insect turns aside instantly, giving an avoiding reaction. After a time it becomes adapted to the adverse effect, it then goes straight and continued to do so as long as the stimulus remains the same or diminishes.

Klinotaxis: Sometimes the insect will compare the intensity of stimulus on two sides, either by moving the whole body or by swinging the antennae first to one side and then the other. It then moves in the preferred direction. A striking example of this can be seen in the maggots of flies which retreat from the light. As they move they swing the anterior end of the body where the light sensitive organs are most shaded. This process of making the successive comparisons is termed klinotaxis.

Knap sack duster: A duster carried on the back. It is operated by bellows on the top of a cylindrical dust container, the bellows being actuated by a hand lever at the side of operator.

Knap sack sprayer: A sprayer in which pump is operated constantly to get continuous discharge of liquid; operated by lever handle provided at bottom or on top of tank; builds up more pressure than compressed air sprayer; can spray up to 2 acres per day.

Knock down activity: Ability to manifest paralytic effect on the insects within a short time (4 hours or less after application).

Knock down effect: The rapid falling down of the insects treated with an insecticide.

Knock down insecticide: A poison that knocks down flying insects rapidly but may not kill them.

KT_{50}: It represents the median knock down time required to kill 50 per cent of test animals for a given dose or concentration of toxic chemical.

L

Label: All written printed or graphic matter on or attached to the economic poison or the immediate container or wrapper to the retail package of the economic poison.

Label, radioactive: When a substance is made incorporating a radioactive isotope of one of its constituent's atoms, in order to help trace the movement and chemical reactions of the substances, it is said to carry a (radioactive) label.

Labeling: All information in the form of written, printed or graphic matter upon the economic poison or any of its accompanying containers or wrappers to which reference is made on the label or in supplemental literature accompanying the economic poison.

Labellum (pl., labella): Expanded tip of labium, seen in many Dipteran flies, used to soak up fluids. See also sponging type mouth parts.

Labial disc: A raised area in the cuticle of the most anterior annule of some species.

Labial gland: In Collembola and some Thysanura (*Japyx*), the cephalic region bears large number of pretubular glands which open collectively at base of the labium. Each gland is differentiated into an anterior oval region, the saccule; the middle long coiled tubular region, the labyrinth and a distal short convoluted lobular region, the dorsal loop. The saccule is composed of flattened columnar epithelium, while the other two regions consist of the cells resembling that of Malpighian tubules. The saccule opens into labyrinth while labyrinth and dorsal loop open into the common duct which opens at the base of labium. Their excretory function is experimentally proved. See salivary gland.

Labial palpus: One of the pair of sensory appendages (feeler like and two to five segmented) of the insect labium.

Labial suture: The suture of the labium between the prementum and the postmentum, always distal to the mentum when latter is present.

Labiduridae: Family of order Dermaptera, suborder Forficulina and superfamily Forficuloidea. Aedeagus paired, forefemora not compressed and keeled, telson smaller than pygidium, *e.g.*, ground nut earwig, *Euborellia annulipes*.

Labiidae: Family of order Dermaptera, suborder Forficulina and superfamily Labioidea. Second tarsal segment unlobed and not dilated, unpaired aedeagus, *e.g.*, *Labia* and *Prolabia* sp.

Labioidea: Superfamily of order Dermaptera and suborder Forficulina.

Labiostipites: The prementum or that part of the labium formed by the stipites of the component labial (second maxillary) appendages.

Labium: Lower lip. Unpaired set of mouthparts of insects located ventrally or opposite to the labrum (upper lip) enclosing the mouth from lower side in cutting and chewing type of mouth parts. The labium has following parts: (i) Submentum: Flat leaf like basal part, (ii) Mentum: Central portion which bears three segmented labial palpus on each side. The lobes of mentum at which the labial palpus attaches is known as palpiger, (iii) Prementum: It is the apical portion of labium and has terminal lobes born at distal margin. The median pair is the glossae and lateral pair is paraglossae. The labium has a pair of segmented palpi. In the bug, the labium is modified into 3-4 segmented slender rigid organ known as rostrum or beak. Its anterior surface is deeply grooved to enclose the mandibular and maxillary stylets. Its function is to protect the delicate stylets and to search the spot for feeding.

Labral segment theory: Wiesmann (1926) reported two preantennal segments with two paired rudimentary sacs in the embryo of the stick insect, *Carausius morosus*. The first of these is the labral segment with the labrum as a fused appendage while the second one is preantennal segment without appendages in the adult stages. Roonwal (1937) and Jhingran (1947) also found two preantennal coelomic sacs in the embryos of the locusts, *Locusta* and *Schistocerca*.

Labrum: It is freely attached sclerite to the lower margin of the clypeus by the clypeolabral suture, generally hanged over the mouth or the preoral cavity and also termed as an upper lip, particularly in the mandibulate insects. In the fluid feeding insects, it undergoes various modifications. Internally, it forms anterior part of the epipharyngeal wall of preoral cavity. It is internally lined with the gustatory receptors and can be moved upwards, downwards and be pulled and pushed to some extent due to the elastic nature of suture and muscle innervation. The labrum is innervated by following three types of muscles: (i) Compressor labral: These may occur in a pair on lateral sides or a single median one. These are attached by one end on the anterior and by the other end on the posterior wall of labrum. The compressors bring about the retraction, prolongation and contraction of the labrum, (ii) Anterior labral: These are paired but unbranched labral muscles. These originate from the inner surface of the frons. The muscles pull the labrum backward, (iii) Posterior labral: These are paired and branched labral muscles and originate from the anterolateral regions of the frons. The muscles pull the labrum backward.

Labrum epipharynx: A mouthpart composed of the labrum and epipharynx and usually elongate.

Lac: Shellac is the byproduct secreted by a scale commonly called lac insect, *Kerria lacca*.

Lac culture: The practice of producing the lac from lac insect is known as lac culture. There are two strains of lac insects grown in India. The *Kusumi* grown on *Kusum* (*Schleichera trijuga*) and *Rangeeni* on *Palas* (*Butea monosperma)* and *Ber* (*Zizyphus mauritiana)* and other plants. The other plants include *Babul, Acacia arabica; Khair, Acacia catechu; Peepal, Ficus religiosa; Cajanius indicus* and *Shorea talurea*. Both strains have two generations per year and, therefore, two crops per year can be taken. The period of lac crops of *Kusumi* and *Rangeeni,* moreover, differ each other and named after Hindi months during which the crops are harvested. (a) *Kusumi* (i) *Jethwi* (Jan.-Feb. to June-July), (ii) *Aghani* (June-July to Jan.- Feb.); (b) *Rangeeni* (i) *Katki* (June-July to Oct.-Nov.) (ii) *Baisakhi* (Oct.-Nov. to June-July). The steps in lac culture involve inoculation, lac secretion and lac processing.

Lac insect, predators: There are two important predators of lac insect, *viz., Eublemma amabilis* larvae and moth, *Holocera pulverea* larvae. The life cycles of both predators coincide with each other. The eggs are laid near the lac encrustations.

Lac insect: See *Kerria (= Laccifer) lacca*.

***Laccifer communis* Mahdihassan:** Belongs to order Hemiptera and family Lacciferidae. It is a minor pest of grapevine.

Lacciferidae: A family of superfamily Coccoidea, suborder Homoptera and order Hemiptera.

Lacinia (pl., laciniae): Distal, mesal lobe of the maxilla that bears teeth on its term lamella (pl., lamellae). It is a leaf like plate.

Lacquer: Pesticide incorporated into a lacquer or varnish to achieve slow release over a length of time.

Lacrimation: The secretion of tears especially in excess.

Lady bird beetles: See *Coccinella septempunctata* Linn., *C. rependa, Cheilomenes sexmaculata, Brumus suturalis* (Fab.). Bright coloured beetles in the Coccinellidae family (order Coleoptera). Most species of ladybirds are predatory insects (both the adults and the larvae). They feed on aphids and other small insects and play an important role in biological control of certain pests.

Laemobothridae: A family of suborder Amblycera of order Mallophaga. Infesting aquatic birds. Head not enlarged at temples and not evenly expanded behind. Example, *Laemobothrion*.

Lag phase: Period which may precede commencement of rapid degradation of a pesticide by a microbial population. It is the period needed either for induction of microbial enzymes or for growth of the microbial population to adequate size.

Lamella: Thin plate like structure.

Lamellate antennae: Plate like or leaf like structures or segments are found in this type of antenna, *e.g.,* Scarabaeid beetle.

Lamellate antennal club: Antennal club composed of broad, flattened, plate-like segments, which are capable of lying flat against one another.

Lamiinae: A subfamily of Cerambycidae, Coleoptera. Black or mottled grey beetles, over 2.5 cm long, antennae with first segment bearing a scar like area near its tip and measuring sometimes twice as long as body in male and about as long as body in female. Usually these infest freshly cut logs and living trees are rarely attacked. This subfamily includes the common genera such as *Anoplophora, Batocera, Coelosterna, Monochamus, Sthenias, Linda, Oberea, Apomecyna, Coptops, Olenecamptus* etc.

Lampyridae (glow worms; fireflies): Family of superfamily Cantharoidea, suborder Polyphaga and order Coleoptera. These are small or medium-sized somewhat flat, elongate, yellow, black, brown or red colored mostly luminous beetles. They are nocturnal. The antennae are 11-segmented and may be pectinate or flabellate in male. Eyes are large in male and much reduced in female. The head is more or less concealed by pronotum. The males are winged and the females are often apterous and vermiform. The abdomen with photogenic organs in segments 6 and 7 in male and in segment 7 only in female.

Lance: An accessory in sprayers; a metal tube of uniform cross section on one end of which the spray nozzle is mounted while the other end is connected to the liquid container with a hose pipe or pressure tubing.

Lanceolate cell: A cell in the anal area of the wing (Hymenoptera).

Lanceolate: Spear shaped, tapering at each end.

Land races: Varieties of plant species selected by farmers on a local basis over many years, which are strongly adapted for local conditions

Languriidae (lizard beetles): Family of superfamily Cucujoidea, suborder Polyphaga and order Coleoptera. These are small, narrow and elongate (5 to 10mm long) beetles with reddish pronotum and deep, blue, green or black elytra. The coxal cavities of forelegs open behind. Tarsi never 5-5-4. The claws are simple. The adults feed on leaves and stems of plants. The larvae bore into stems and cause damage. The grubs of the beetles *Anadastus parvulus* bore into stems of *tenai* (*Setaria italica*) and cause dead hearts or drying up of ears.

Lannate: Trade name of methomyl.

Larva (pl., larvae): An immature stage of the insects (between egg and pupa) having complete metamorphosis. Larvae are of following types: protopod, oligopodous (scarabaeiform and campodeiform), polypodous (eruciform) and apodous (maggot). See also, nymph.

Larvicide: Chemical substance that destroys larvae of a pest.

Larviform: Larva like.

Larviparity: A form of viviparity in which the female gives birth to larvae rather than depositing the eggs.

Larviparous: Giving birth to larvae instead of eggs as in certain Diptera.

Lasiocampidae (eggars, lappet moth): Family of super family Bombycoidea of suborder Ditrysia and order Lepidoptera. The moths are stout, moderate to large sized insects densely covered with scales. They are abundant in the tropics. Ocelli are absent and the proboscis is atrophied. On the hindwings the humeral lobe is prominent. In the females the anal end possesses tuft of hairs. Larvae are stout with densely covered secondary hairs which obscure the primitive setae and often laterally possess downwardly projecting hair flanges and dorsally on the anterior segments hairy tufts.

Latency: Dormancy or the period within a host during which no symptoms of infection are evident even though a pathogen is present.

Latent: Concealed; not clearly visible; dormant.

Latent heat: The heat utilized in the change of phase (from solid to liquid). The latent heat of fusion is the amount of heat necessary to convert one kg of the solid to one kg of liquid without a change in temperature. The latent heat of vaporization is the amount of heat required to convert one kg of substance from its liquid state to vapour state at constant temperature (boiling temperature).

Latent infestation: After storing grains, the insect pests infest grains through their latent or hidden stages from the wall cracking and surface of the storage containers or bags called the latent infestation.

Latent learning: Learning without apparent reward or punishment.

Laterad: Towards the side away from midline of the body.

Lateral: Towards the side, away from the midline of body.

Lateral arms: Paired lateral projections of the metendosternite, often bearing laminae and anterior tendons.

Lateral lobes of aedeagus: See accessory lobes, paramere.

Lateral longitudinal trunk: See longitudinal trunk.

Lateral movement: Chemical movement in a plant or in the soil to the side or horizontal movement in the roots or soil layers.

Lateral nerve cords: The lateral strands of nerve tissue produced from the ventral neuroblasts.

Lateral ocellus: The simple eye in holometabolous larvae, also called stemma (pl., stemmata).

Lateral oviduct: In insects, one of the paired lateral ducts of the female genital system connected with the ovary.

Lateral plates: The embryonic lateral areas of the germ band after differentiation of the middle plate.

Lateral pronotal carina (pl., carinae): A sharp lateral edge on the prothorax separating the pronotal disc above and the hypomeron below.

Lateral salivary duct: Duct that carries secretion products from a salivary gland.

Lateral tracheal trunk: The usual longitudinal tracheal trunk on each side of the body closely connected with the lateral spiracles.

Lateral vessel: In some insects, the heart bears the lateral vessels arising from the excurrent ostia. The heart of cockroaches contains six pairs of lateral vessels, two thoracic and four abdominal. The terminal end of the blood vessel is blind in a large number of insects but it is branched into three slender branches entering the cerci and a median caudal filament in the ephemerid larvae.

Laterosternite: The lateral part of a definitive thoracic sternum apparently derived from ventral arc (sternopleurite) of the subcoxa.

Laterotergite: A lateral sclerotization of the dorsum distinct from a principal median tergite (paratergite).

Lateroventral: To the side (away from the midline of body and below.

***Latoia lepida*:** Belongs to order Lepidoptera and family Cochlidiidae. It is an insect pest of castor, commonly known as the castor slug.

Laws of thermodynamics: Energy flow and its distribution in the ecosystem are according to the laws of thermodynamics which are as follows: (i) First law: Amount of energy is constant in the world and neither energy can be synthesized nor can be destroyed. Energy can be converted from one state to another state. The green plants convert light energy into chemical energy. (ii) Second law: During energy transfer, some part of energy is lost. This energy is diffused in the environment mostly in the form of heat and remaining is stored in the form of tissues.

Laying worker: A worker which lays infertile eggs, producing only drones, usually in colonies that are hopelessly queenless.

LC_{50}: A means of expressing the toxicity of a compound present in air as a dust, mist, gas or vapour as ppm. The $LC\text{-}_{50}$ is the statistical estimate of the dosage or concentration necessary to kill 50 per cent of a very large population of the test species, through toxicity on inhalation under stated condition or by law, the concentration which is expected to cause mortality in 50 per cent of the test animals so treated.

LD_{50}: Lethal dose of toxicant required to kill 50 per cent of a large group of individuals of one species.

Leachate: Aqueous phase percolating through a soil profile or a soil column.

Leaching: (i) The movement of a pesticide or other chemical downward through the soil as a result of water movement. (ii) Movement of substances downward or out of the soil with water as the result of water movement.

Lead arsenate: It was first used as an insecticide in USA in 1892 for controlling the gypsy moth. The commercial lead arsenate is a mixture of acid orthoarsenate ($PbHAsO_4$) and basic orthoarsenate ($Pb_4(PbOH)AsO_4)$) in which acid form predominates. Lead arsenate is relatively weak stomach poison as compared to the other arsenicals with a little contact action. It is insoluble in water, therefore, possesses high degree of safety to foliage. It only affects the insects feeding on the treated material and the useful insects like parasites and predators remain unaffected. It is used in the concentration of 0.25 per cent against lemon butterfly, epilachna beetle and red pumpkin beetle and 0.60 per cent against castor semilooper. An equal quantity of hydrated lime is added to this insecticide to prevent the phytotoxicity.

Leaf curl mite: See *Aceria* (*Eriophyes*).

Leaf insect: Insect belonging to order Phasmida. Stick insects in the same order resemble sticks. Leaf insects look like leaves.

Leaf miner: An insect which lives in and feeds upon the cells between upper and lower epidermis of a leaf; these being larvae of Agromyzidae (Diptera), Gracillariidae (Lepidoptera), Hispidae (Coleoptera) etc.

Leaf roller: Any of several moths of the family Tortricidae (order Lepidoptera). The caterpillars make nests of rolled leaves and silk.

Leaf scorch: Leaf necrosis, usually marginal due to phytotoxicity or nutrient deficiency.

Leaf sheath: The lower part of leaf that encloses the stem.

Leaping: Jumping by many insects is affected by a spring and catch mechanism, the necessary elastic energy being stored in some skeletal structure and the spring being released by a separate muscle. Thus in the flea, the energy is stored in a pad of reslin; in locusts, in the elastic components of muscle system.

Learning behaviour: Behaviour acquired as a result of experience.

Learning: Changes in an individual's behaviour arising from experience.

Leather jackets: See crane fly.

Lectins: A group of secondary plant compounds present in seeds of certain legumes.

Lectotype: One of the syntypes which subsequent to the publication of original description, is selected and designated through publication to serve as a type.

Leg: Legs are jointed, uniramous appendages of the thorax. They develop as the hollow, tubular outgrowths from the lateral or latero-ventral wall of thorax of embryo. They are provided with the well organized set of muscles, nerves and the membranous joints so that the individual segment can move independently. The insect leg is composed of six distinct segments, coax, trochanter, femur, tibia, tarsus and pretarsus.

Leg articulation and articular sclerites: See trochantin and antecoxal piece.

Leg musculature: The extrinsic and intrinsic muscles innervate the legs. The extrinsic muscles include the tergal promoters and remoters, the sternal promoters, remoters and adductor and the pleural abductor and adductor muscles. They are inserted in the coxae and are responsible for the coxal articulation on the pleuron or pleuron and trochantin both. The intrinsic muscles are levator and depressor types. They are inserted in all segments of the legs except coxae.

Legal residue: Residue that is within safe levels according to the regulations.

Legionary ant: See army ant.

Legislative or legal control: The restriction on the movement of commodities being suspected of infested (by new pest or disease) from one country to another or from one region of a country to another region is controlled by legislations. In India, the legislative measures are always taken at the international ports of entry and all goods imported are thoroughly checked under Destructive Insect Pest (DIP) act of 1914. The Directorate of Plant protection Quarantine and Storage (Government of India) has established well equipped quarantine and fumigation stations at Mumbai, New Delhi, Kolkata, Chennai, Cochin and Vishakhapatnam etc. in 1949. Legislation is also made provision for proper fumigation of weeds prior to the transportation and declaration of any insect pest as a pest so that interstate cooperation for eradication of some wide spread pests, such as desert locust, red hairy caterpillar, surface grasshopper, cotton bollworms, cotton stem weevil, sugarcane borer, mango hopper, coconut caterpillar etc. Various state governments have passed Agricultural pests, Diseases and Noxious Weeds Act (APDNWA) to enable them to take measures for the control of pests.

Legs, types or modifications: See ambulatorial (walking type), Cursorial (running), saltatorial (leaping or jumping), natatorial (swimming), clasporial (clinging type), raptorial (preying type), fossorial (digging type), corbiculate (pollen collecting type) and degenerate type.

Legume: Plant belonging to the family Leguminosae. Legumes contain root nodules and are capable of symbiotic nitrogen fixation with bacteria of the genus *Rhizobium*, *e.g.*, beans, peas etc.

Leguminous: Having the nature of or bearing a legume.

Leishmaniasis (pl., Leishmaniases): A disease caused by a member of the protozoan genus *Leishmaniases*. The sand flies, *Phlebotomus* sp. and *Lutzomyis* sp. cause Leishmaniases in forested and arid tropics of new world and subtropical region of the old world by transmitting the pathogen, *Leismaniases donovani*, *L. intantum* (visceral), *L. tropica*, *L. braziliensis*, *L. mexicana* (cutaneous) protozoans. A visceral leishmniasis caused by *L. donovani* is also known as *Kala azar* and occurs in Mediterranean countries, equatorial Africa, Brazil, India, China and Argentina to Mexico due to enlargement of the spleen and liver. The dermal leishmaniasis is also known as oriental sore and uta, espundia and chiclero ulcer in the new world.

Lek: An assembly area to carry on display and courtship behaviour.

Lemon butterfly: See *Papilio demoleus* Linn.

***Lenodera vittata* Wlk.:** Belongs to order Lepidoptera and family Lasiocampidae. It is an insect pest of Spices, commonly known as the cardamom hairy caterpillar.

Lens: A lenticular outer part of the eye generally formed in the cornea.

Lentil pod borer: See *Etiella zinckenella*.

Lepidopsocidae: A family of order Psocoptera and suborder Trogiomorpha. Claws with preapical tooth, body and wings with distinct scales. Head transverse, maxillary palp with sense organ. Example, *Lepidopsocus*.

Lepidoptera: (Gr., *Lepidos*= scales, *ptera*= wings). Insect order containing butterflies and moths. They are characterized by scale-covered wings, coiled siphoning mouthparts and a complete metamorphosis. The coiled tongue is also called proboscis.

Lepidopteran piercing and sucking mouth parts: In contrast to common habit of nectar feeding, a few species of Noctuidae are adapted for piercing and sucking the fruits [*Othreis materna*, *Calpe* (=*Calyptra*) *thalictri*] and blood of buffalo, deer and tapirs (*Calpe eustrigata*). The proboscis is formed from two concave galeae and they protract and retract alternately. The tip of proboscis is pushed against the skin and it drills the skin and penetrates inside the tissue by the rotatory movement at about long axis of 180°. The tip of proboscis is provided with horny curved short hooks, long barbs and fine spines. Penetration is affected by the erect and backwardly pointing barbs resisting retraction but allowing protraction of galeae. This alternating motion of galeae pulls the proboscis into the deep tissue.

***Lepisma saccharina* Linn.:** Belongs to order Thysanura and family Lepismidae. It is an insect pest of household, commonly known as the silverfish.

Lepismatidae: A family of order Thysanura. Minute compound eyes and ocelli wanting. Eversible vesicles usually present on abdominal segments. Styli wanting in thoracic coxae but present on abdominal segments 2-9 or 8 and 9. For example, *Lepisma* sp., *Acrotelsa* sp., *Nicoletia* sp., *Ctenolepisma* sp.

Leprosy: A chronic, transmissible disease due to a specific bacterium, *Mycobacterium leprae*.

***Leptocorisa acuta* (Thunberg):** Belongs to order Hemiptera and family Coreidae. It is an insect pest of cereals and millets, commonly known as the rice bug leafhopper.

Leptoperlidae: A family of order Plecoptera. Possesses six or less anal veins in hindwings; no transverse vein connecting R to Cu near middle of wings and distal portion has cross veins, *e.g.*, *Leptoperla* sp. and *Gripopteryx* sp.

Lesion: A wound or injury, a more or less circumscribed pathologic change in the tissues. One of the individual points or patches of a disease of the skin. Any break in the epidermis of a plant or a localized, diseased or disordered tissue.

Lesmosome: One form of subcellular attachment between the plasma membranes of two adjacent cells.

Lesser grain borer: *Rhyzopertha dominica* (Fab.), a destructive pest of all cereal grains in store; also breeds in flour and dry fruits; larva bores into grain and feed inside.

Lesser wax moth: *Achroia grisella* F., an insect pest in bee colonies; caterpillar attacks stored combs, feeding on outer surface.

Lestidae: A family of order Odonata. Brightly coloures damselflies with wings held upwards and backwards when resting, *e.g.*, *Lestes* sp.

Lestobiosis: The relation in which colonies of a small species nest in the walls of nest of the larger species and enter the chambers the larger species to prey on brood or to rob the food stores.

Lethal: Fatal or deadly.

Lethal concentration 50 per cent: See LC_{50}.

Lethal dose 50 cent: See LD_{50}.

Lethal gene: A gene when expressed is fatal to its carrier.

***Leucinodes orbonalis* Guenee:** Belongs to order Lepidoptera and family Pyraustidae. It is an insect pest of brinjal, commonly known as the brinjal fruit borer.

Leucocytes: See haemocytes.

***Leucopholis coneophora* Burm.:** Belongs to order Coleoptera and family Melonlonthidae. It is an insect pest of coconut, commonly known as the coconut white grub.

Leukocytosis: An abnormally large number of white blood cells.

Levulose: See fructose.

Libellulidae: A family of order Odonata. Skimmers without metallic sheen, *e.g.*, *Pantala*, *Libellula* sp.

Libelluloidea: A super family of order Odonata. Antenodals complete, triangular varying in shape and position.

Lice: Plural of louse.

Licensing: The issuing of a certificate of permission from a constituted authority to carry on a business, profession or service.

Lichen: A regular association of an alga with a fungus.

Life cycle: The sequence events in the development of an insect that occur from birth (hatching of the egg) to reproduction (mating and egg laying).

Life history: Habits and changes undergone by organism from the egg stage to its death as an adult.

Life sciences: Life sciences deal with the study of living objects. These include botany, zoology and medical science.

Life system: A population and the environment which influences it.

Life table: A device for expressing in an orderly fashion, observations on the changing density of an insect population in time and space and the processes which direct those changes especially in relation to the age specific distribution of mortality and its causes. In life table, the separation of a pest population into its different age components is done (*e.g.*, eggs, larvae, pupae, adults).

Light compass response: See menotaxis.

Light producing organ: Light producing organ develops from ventral fat body in Collembola (lighting spring tails), Dipteran larvae (glow worms), some bugs and Coleopteran fireflies, *Lampyris*, *Luciola* (Lampyridae); *Pyrophorus* (Elateridae); *Diplocladon*, *Phryxothrix* (Drillidae); *Phenogodes* (Phenogodidae) etc. Light is produced due to oxidation of luciferin by an enzyme luciferase.

Ligula (pl., ligulae): Name given to the lobes situated at the tip of the labium, usually divided into glossae and paraglossae.

Lily moth: See *Plytela gloriosae*

Limb basis: The primary basal segments of appendages (coxopodite) supporting the telopodite; sometimes subdivided into a proximal subcoxa (pleuropodite or pleuron) and a distal coxa.

Lime sulphur: An insecticide and fungicide made by boiling sulphur and lime together in water which reacts to from both soluble and insoluble salts of calcium polysulphide; irritating to eyes, nose and skin; phytotoxic to peach trees and on sulphur-shy varieties.

Limit of detection (LOD): Lowest concentration of a pesticide residue in a defined matrix where positive identification can be achieved using a specified method.

Limit of reporting: Practical limit of residue quantization at or above the LOQ. The limit of quantization for a defined matrix and method may vary between laboratories or within the one laboratory from time to time because of different equipments, techniques and reagents.

Lindane, metabolism: There are two schools of thoughts regarding metabolism of lindane in mammals. According to one group, lindane is first metabolized to gamma PCCH (pentachlorocychlohexane) which is further metabolized to 1,2,4-TB (1,2,4- trichlorobenzene) and then to phenols, sulphates and glucuronic acid conjugates which are excreted to urine. The second group argues that in lindane metabolism, various chlorophenols are formed directly through hydroxylation by an efficient system and gamma PCCH is not a product. In plants, lindane is metabolized to pentachlorocychlohexane.

Lindane: (i) A 99 per cent pure gamma isomer preparation of benzene hexachloride (BHC); acts as stomach and contact insecticide and also as fumigant. It is named as lindane after the name of Van der Linden a German chemist in 1912 that isolated gamma isomer. It is marketed under trade name Lintox and available as 0.65 and 1.3 per cent dust, 20 S, 10 per cent granules and 20 EC. Lindane is more toxic nerve poison than DDT. It induces tremors, ataxia, convulsions, falling and prostration. House flies treated with lindane show fanning movement of wings, this symptom is not caused by DDT. It attacks nervous system and requires a complete reflex arc to be effective. Lindane is effective against a wide range of insects and arachnids. (ii) Moderately toxic, lindane quickly paralyzes most insect species. Unlike most other insecticides, lindane is stable to heat and thus can be used in and around heat radiators such as heaters, radiators, stoves and boilers without losing its effectiveness. However, lindane is more volatile than DDT and thus provides less residual effectiveness. Lindane is somewhat more toxic to mammals than DDT, acute oral LD_{50} for rats is 125 mg kg^{-1}; non phytotoxic at insecticidal concentration and can damage plants if used in excess.

Lineage: Direct descent from an ancestor; ancestry; family; stock.

Linear: Line like; long and very narrow.

Linear feet: Running feet, example, width and the length of a field is measured in linear feet.

Linear relationship: A relationship existing between two variable quantities which are directly proportional to each other so that a straight line can be obtained by plotting graph between the two variables.

Linnaean hierarchy: When in the categorical ranks of taxa each category of the lowest includes one or more subordinate categories.

Linseed aphid: *Myzus persicae* Sulz. Aphid pest of linseed; congregates in small colonies, the feeding of which resulted in the stunted plant growth.

Linseed cutworm: *Agrotis* sp. is an insect pest of linseed; caterpillar cuts plants near roots under the soil.

Linseed gall fly: *Dasyneura lini* Barnes. is a pest of linseed. Maggot feeds on parts of bud which then does not develop seed capsules.

Linseed gall midge: See *Dasyneura lini.*

Lint: The fibre surrounding the seed of unginned cotton.

Liopteridae: Family of super family Cynipoidea, suborder Apocrita and order Hymenoptera. These are large insects and their habits are unknown.

***Lipaphis erysimi* (Kalt.):** Belongs to order Hemiptera and family Aphididae. It is an insect pest of mustard, commonly known as the mustard aphid.

Lipase: Any enzyme that catalyzes the splitting of fats into glycerol and fatty acids.

Lipid: A class if diverse chemical compounds that includes fats, waxes, phospholipids, carotenoids and sterols that are insoluble in water but soluble in alcohols and ethers.

Lipid epicuticle (wax layer): Second layer of epicuticle underlying the cement layer or tectocuticle.

Lipoid: A fat or fat like substance.

Lipophilic: Fat loving, used to describe a molecule or group of molecules tending to promote oil rather than water solubility, contrast to hydrophilic.

Lipophilicity: Affinity for fat as described by partitioning behaviour between water and an immiscible organic solvent, favouring the latter and which correlates with bioconcentration.

Lipophobic: Used as opposite of lipophilic and therefore also of hydrophobic.

Lipophorin: The haemolymph protein found in lepidopterans and orthopterans with affinity for insecticide of intermediate polarity such as dieldrin.

Liposcelidae: A family of order Psocoptera and suborder Troctomorpha. Body flattened with broad and flattened hind femora. Wings reduced or absent. Meso and metanota fused. Example, *Liposcelis.*

***Liposcelis transvaalensis* Enderlein:** Belongs to order Psocoptera and family Psocidae. It is an insect pest of household, commonly known as the book louse.

Liquefied petroleum gas: Liquefied petroleum gas or LPG is a mixture of ethane, propane and n-butane. But its main component is n-butane. In petroleum industry, this product is obtained as a by-product. It can be liquefied by the application of pressure. It is used in the form of fuel and it burns in blue flame in the air. One gram gas produces 50 kJ of energy.

Liquid: Fluid state of matter in which the particles of substances are so loosely bound that they cannot acquire a definite shape by themselves but may possess definite volume at room temperature.

Listed corn: Corn planted in a furrow or trench or below the general level of the ground.

Lister: A plough with double moldboard that heaps soil on both sides of the furrow; it may also be combined with a drill that plants seed in the same operation.

Litchi bug: See *Chrysocoris stollii.*

Litchi fruit stone borer: *Argyroploce illepida* But. is a serious insect pest of litchi; caterpillar eats stalk of seed and borers into seed or pulp. Attacked fruit is unfit for consumption. Also attacks *Bael* (*Aegle marmelos*), wood-apple, tamarind, *Babul, Amaltas* and *Dhaincha*.

Litter: Straw that is spread especially for animals to lie upon. It is also refers to a brood of an animal.

Little bee: *Apis florea* F., is a species of honeybee found in plains of India, frequently migrating; makes single small comb suspended from branches bushes, hedges and trees.

Live burrow: Burrow that contains living rat.

Lixiviation: Separation of soluble components of a mixture by washing them out with water.

Lobe: Prominent rounded process.

Lobopodial: Pertaining to lobopod.

Local infection: An infection affecting only a limited part of the plant.

Locomotion: The locomotion or mechanism of walking is thoroughly studied in cockroaches. All six legs are grouped into two functional sets. One functional set is formed from the forelegs and hindlegs of one side and the middle leg of other side, resulting in a tripod. Similarly remaining three legs form the other tripod. When the legs of one set are protracted, those of other set are retracted and vice-versa. As the ratio of protraction and retraction is one, three legs of a set remain on the ground at a time to provide support to the insect body. The insect thus performs the locomotion in a zig-zag fashion. In this process, the fore, middle and hind legs of a triangle serve as the organs of the traction, balance and propulsion, respectively. All movements of the legs are provided by the muscles and coordination is maintained by the thoracic ganglia.

Locus: The location of gene on genetic map.

Locust control: The locust control is done at international level by various organizations, such as (i) the Locust Warning organization set up in 1939 in India, (ii) the International Locust Research Centre, (iii) anti-locust operations under FAO (Food and Agricultural Organization of the United Nations) and several other agencies. These agencies alert the people about new outbreaks, swarms and suggest the package schedule of promising cultural, chemical and microbial control methods.

Locust *vs.* grasshopper: Locusts are Acridids that have both a solitary phase and a gregarious phase; in the latter, dense migratory swarms are formed. Certain species of a given genus may be locusts while others are not. Grasshoppers do not form swarms.

Locust: Any of bigger grasshoppers of the family Acrididae (order Orthoptera) that form swarm. Locusts are often migrating in huge swarms that consume vegetation and crops, *Schistocerca gregaria, Locusta migratoria* and *Patanga succincta.*

***Locusta migratoria* (Linn.):** Belongs to order Orthoptera and family Acrididae. It is an insect pest of locusts and grasshoppers, commonly known as the migratory locust.

Lodge: To throw or beat down as growing grain.

Logarithm: The ten base logarithm of a number is the power to which ten must be raised to give that number.

Longevity: Life span of an individual; duration of life.

Longitudinal trunk: In highly developed insects, there is well developed respiratory system. There are three trunks: (i) Lateral longitudinal trunk, (ii) Dorsal longitudinal trunk, and (iii) ventral longitudinal trunk. The transverse, dorsal and ventral commissures connect the system of each side. The dorsal longitudinal trunk gives segmented branches which pass to the heart and dorsal musculature. The visceral branches supply to the digestive canal and reproductive organs. The head and mouth parts are supplied by the branches from anterior most spiracle and the dorsal longitudinal trunk.

Longitudinal: Lengthwise of the body or of an appendage (leg, wing, antenna etc.).

Longitudinal vein: A vein passing through wing longitudinally. See vein.

Looper: (i) A caterpillar of the family Geometrid, with only one pair of abdominal prologs (in addition to the terminal claspers), and which moves by looping its body. (ii) Larva possessing two pairs of abdominal legs. They are located on the sixth and tenth abdominal segments. The larva crawls by bringing the tail to the thorax and forming a loop of the intervening segments. Such larvae are found in the family Geometridae of the order Lepidoptera. Other common names for loopers are measuring worms or Inchworms.

Lophopidae: A family of superfamily Fulgoroidea of suborder Homoptera and order Hemiptera. Characterized by hind trochanters directed backward and a moderately long basal segment in the hind tarsi. Includes *Pyrilla perpusilla, Elasmoscelis platypoda* and *Lophops carinatus.*

Lorsban: See chlorpyriphos.

Lorum: The cheek; a sclerite on the side of the head.

Lot: Quantity of material which is assumed to be a single population for sampling purposes.

Louse: Any of numerous insects of the orders Mallophaga or Siphunculata.

Low concentration solution: A solution which contains a low concentration or a small amount of active ingredient in highly refined oil. The solutions are usually purchased as a stock sprays meant for use in aerosol generators.

Low pressure boom sprayer: A machine which can deliver low to moderate volumes of pesticide at pressures of 30-60 psi. The sprayers are most often used for field and forage crops, pastures etc.

Low pressure low volume sprayer: Hydraulic sprayer; for spraying at pressures less than 100 lbs. psi, used for spraying crops with smaller quantity of spray liquid at low cost. Use limited to sediment free type of spray material. It can be mounted on tractor or wheel mounted trailer covers 6 to 8 hectares a day.

Low volatile: A liquid or solid that does not evaporate quickly at normal temperatures.

Low volatile ester: An ester with a high molecular weight and a low vapour pressure.

Low volume spraying: Spraying, when spray material deposited in small drops do not form pocket. It is obtained by spraying liquid up to 12-120 l per ha; saves labour, active ingredient and time.

LT-$_{50:}$ The time required for a toxic substance to kill 50 per cent of a test population at certain dose or concentration.

Lubricating oils: Fractions of boiling range greater than 300- °C are employed as lubricating oils. The refinement of these oils is done by sulphuric acid treatment or by solvent treatment. Highly refined petroleum oil, therefore, contains high content of saturated oils. Since the colour of the oil disappears during the process of refinement, the terms white, half-white and red are used to designate the degree of refinement of oil. The white oils are, thus, highly refined petroleum distillates. These being safe to leaves, are used for the control of red spider mites and scale insects upon the tender foliage.

Lucanidae (stag beetles, stag-horn beetles): Family of superfamily scarabaeoidea, suborder Polyphaga and order Coleoptera. These are usually large, black or brown beetles measuring about 3.5 to 10 cm long with large mandibles in males which are projected forwards and branched like the antlers of a stag. The three terminal segments of the geniculate antennae are enlarged, flattened and cannot be held tightly together. The labrum is indistinct and the mentum conceals the maxillae and the ligula.

Lucerne caterpillar: *Laphygma exigua* Hubner is an insect pest of lucerne, also attacks linseed, safflower, indigo, lentil, cabbage, onion, maize, coriander, jute and cotton. Completely defoliates plants when attack is severe.

Luminescence mechanissm: Light is produced by the oxidation of luciferin in the presence of enzyme luciferase. Luciferin is first activated by ATP in the presence of the ezyme, luciferase to produce adenylluciferin. The adenylluciferin is oxidized by organic peroxide, again in the presence of luciferase to form so called excited adenyloxyluciferin which decays spontaneously to low energy adenyloxyluciferin with the production of light. The energy for this process is obtained directly from the oxidation process, not from the ATP and it is realized in one large step. The radiation is very efficient, about 98 per cent of the energy involved being released as light.

Luminous organ: A light producing organ usually located just beneath the surface of one or more apical or preapical ventrites, but occasionally elsewhere.

Lycaenidae (blues, coppers, hair streaks): Family of super family Papilionoidea of suborder Ditrysia and order Lepidoptera. Small to moderate sized butterflies with upper surface of wings being metallic blue or coppery, dark brown or orange and undersurface more somber with delicate streaking or dark centered eye spots. Tail like delicate prolongations are seen on hindwings. Each eye is surrounded by a rim of white scales and the antenna ringed with white. Legs are normal excepting the forelegs of males which may possess more or less shortened tarsi and may be wanting in one or both claws. The coloration in both sexes often may show great differences.

Lyctidae (power-post beetles): Family of superfamily Bostrychoidea, suborder Polyphaga and order Coleoptera. The power-post beetles are small, elongate, smooth or pubescent, black, brown or yellow insects with deflexed head not coverd by pronotum and two segmented antennal club. The larva is characteristic in having a more retracted head and a large spiracle on the eight abdominal segments. These beetles are destructive to wooden furniture, freshly cut and old timbers, and dry wood.

Lygaeidae: Family of superfamily Lygaeoidea, suborder Heteroptera and order Hemiptera. These are small insects, conspicuous by their red white or black spots or bands, ocelli are present, four segmented antennae inserted on the sides of the head, rostrum are four segmented, wings are characterized by four or five simple veins in the membrane of the hemelytra. Legs possess rotatory coxae, three segmented tarsi and pulvilli and thoracic gland openings are evident. Most of the members are phytophagous. Examples are chinch bug, *Blissus leucopterus* (pest of wheat, rye, corn, barley and grasses in U.S.A.). *Oxycarenus hyalinipennis* (pest of cotton), *Lygaeus hospes* and *L. militarius* (pests of safflower and sorghum ears in South India). Few Lygaeids, like *Geocoris tricolor* is predaceous on insect pests like *Myllocerus viridanus, Calocoris angustatus, Ragmus importunitas*, painted bug *Coptosoma cribraria, Aphis gossypii, A. nerii* and coccids.

Lygaeoidea: Superfamily of division Gymnocerata, suborder Heteroptera and order Hemiptera. This includes three families, *viz.*, Lygaeidae (chinch bug, milk weed bug etc.), Pyrrhocoridae (red bugs, stainers) and Neididae (Berytidae).

***Lygaeus civilies* Wolff.:** Belongs to order Hemiptera and family Lygaeidae. It is an insect pest of ornamental plants, commonly known as the milk weed bug.

Lymantriidae [Liparidae, Ocneriidae] (tussock moth): Family of super family Noctuoidea, suborder Ditrysia and order Lepidoptera. The insects are moderate sized, the males having bipectinate antennae. Ocelli are absent. Tufts of hairs are seen at the anal end in the female and the eggs are covered with these anal hairs. The larvae are densely hairy and often have compact tufts of hairs on the dorsal aspect of certain segments. On the sixth and seventh abdominal segments osmeteria are often seen. The larvae of *Euproctis* possess urticating hairs on their body.

Lymexylonoidea: Superfamily of suborder Polyphaga, order Coleoptera. This includes a single family Lymexylonidae (Lymexylidea). These are soft-bodied elongate insects with long flabellate palpi in male and larva with enlarged prothorax and without paired processes on the ninth abdominal tergite. They bore into hard wood and drill cylindrical holes. In India, *Atractocerus reverses* is a pest of forest trees.

M

M: Abbreviation stands for median vein in the wings.

Machilidae: A family of order Thysanura. Possess large compound eyes or ocelli and eversible vesicles on abdominal segments 1-7 and styli on thoracic coxae and on abdominal segments 2-9. For example, *Petrobius* sp., *Machilis* sp., *Praemachilis* sp.

Macro: Large.

Macrocephalic female: The females having disproportionately large head as found in the social Halictinae.

Macropterous: Insects having long or large sized wings.

***Macrosiphum miscanthi* Takahashi:** Belongs to order Hemiptera and family Aphididae. It is an insect pest of wheat, commonly known as the wheat aphid.

***Macrosiphum rosaeiformis* Das:** Belongs to order Hemiptera and family Aphididae. It is an insect pest of ornamental plants, commonly known as the rose aphid.

Macrotrichia: These are the hollow hairy exocuticular processes. They are of great taxonomic importance as their distribution differs from species to species. They are of following types: (a) Clothing hairs: These cover the entire body and appendages as found in honey bees, (b) Glandular hairs: These occur around the opening of various epidermal glands and can be seen in a few Lepidopteran larvae, (c) Sensory hairs: These constitute various mechano and chemoreceptors and are occurring on the antennae, mouth parts, legs and various regions of the body in all groups of insects, (d) Wing scales: These are modified and varyingly pigmented microplate like structures that invade the wing surface of the adult butterflies and moths.

Maggot: A vermiform larva, legless, soft bodied, and do not have a distinct head capsule. Often they are found in decaying organic matter, *e.g.*, Diptera and Aculeata.

Maize borer: See *Chilo partellus* Swinhoe.

Major pest: In major pest, the GEP is close to the EIL and in some cases both may be at the same level. The population often crosses EIL, thereby, repeated control measures are necessary but economic damage is avoided by timely interventions, *e.g.*, cotton jassid and whitefly; brown plant hopper and leafhopper on rice; sugarcane whitefly and scale insect; rice stem borer, gall midge and leaf folder etc.

Major worker: A worker subcaste which is largest and found in the ants.

Malaise trap: A tent like trap made out of netting used to catch flying insects.

Malaria: A human disease caused by a single celled protozoan parasite called *Plasmodium vivax* and *P. falciparum* (Protozoa, Sporozoa). This disease is carried and spread by mosquitoes of the genus *Anopheles* (see vector). There are about 60 species of this genus in tropics and subtropics.

Malathion: It is an organophosphate insecticide with contact and stomach action; chemically S-1, 2 di(ethoxy carbonyl) ethyl-dimethyl phosphorothiolothionate ($C_{10}H_{19}O_6PS_2$). It is colourless liquid with unpleasant odour slightly soluble in water (145 mg l^{-1} at 20 °C) but highly soluble in organic solvents except saturated hydrocarbons. The toxicity of the compound depends on two metabolic processes, activation by oxidation and detoxification by hydrolysis. It has got different routes of metabolism in insects and vertebrates. In insects, it is oxidized to more toxic ester of thiophosphoric acid while in invertebrates, the hydrolysis of the group in the side chain takes place by the hydrolytic enzymes carboxyesterase and carboxy amidase producing non-toxic products. These enzymes are not found in insects. Therefore, hydrolysis is out stripped by oxidation. It is one of the safest insecticides and marketed under trade names, Malathion, Cythion etc.

Malathion, metabolism: It is converted to corresponding phosphatic analogue resulting in the increase in anticholinesterase potency. Malathion undergoes the following metabolic processes. (i) Oxidation: Where malathion is converted into malaoxon which is metabolized to dimethyl thiophosphoric acid and further metabolized to phosphoric acid. (ii) Dealkylation: Malathion is converted to desmethyl malathion which is further metabolized to phosphorothioate and then to phosphoric acid. (iii) Conversion of malathion to malathion mono-acid and di-acid takes place by carboxyesterase. These acids are then hydrolysed by phosphatase to give dimethyl dithiophosphate which is further metabolized to phosphorothioate and then to phosphoric acid. In soil, malathion is degraded to desmethyl malathion by *Aspergillus*. The degradation of malathion by *Rhizobium* gives dimethyl dithiophosphoric acid. This is desulphurated to dimethyl thiophosphoric and then to dimethyl phosphoric acid.

Male reproductive system: Consists of a pair of testes with sperm tube, vas-differentia, seminal vesicle, median ejaculatory duct and accessory glands.

Mallophaga: (*Mallo*= wool, *phaga*= eat). Insect order made up of the chewing lice. They are characterized by flattened, wingless bodies, chewing mouthparts and incomplete metamorphosis. The Mallophaga or bird lice are ectoparasites mostly of birds, rarely of mammals and subsisting on hairs or feathers, scales or dried blood from scales. Head flattened from top to bottom and its long axis is on the same plane as the body. Tentorium reduced, antennae 3-5 segmented being filiform, prothorax narrower than head, almost free and distinct from the mesothorax, legs longer almost alike, abdomen eight or ten segmented. The reduced appendages appears to be represented by a pair of small appendages on the eighth sternite of the female, while in the males there is a median aedeagus.

Malpighian tubules: Excretory organs that arise near the anterior end of the hind gut and extend into the body cavity. The Malpighian tubules were first discovered by Italian anatomist Malpighi and were named after him. These are absent in Collembola, Thysanura and aphids. These are long, slender and blind tubules lying in haemocoel where they freely bathed by the blood. They open at their proximal extremities into the intestine near the junction of hindgut and midgut. (i) Distally, these tubules are usually free. The number of Malpighian tubules in different insects varied greatly (4 in Thysanoptera, Hemiptera and Diptera; 4-6 in Coleoptera and Psocoptera; 2-8 in Isoptera; 4-60 in Thysanura; 30-120 in Orthoptera and 2 in Coccidae). Waste products excreted in large amount are uric acid, ammonia, mineral, salts etc. (ii) In some insects, the distal part is closely applied to hind gut, this type of condition is known as cryptonephridial condition and occurs in larvae and adult of many Coleoptera and all lepidopterous larvae. This condition probably enables the insects to conserve water by withdrawing moisture from the faeces. (iii) In Reduviid bug, there is vascular enlargement at ampula at the base of each tubule.

Maltase: An enzyme that catalyzes the breakdown of the disaccharide maltose into glucose molecules.

Mandibles: Jaws. One of the anterior pairs of mouthpart structures. Mandibles are clearly visible strongly chitinised structures in insects with chewing mouthparts, such as grasshoppers, cockroaches, beetles and ants. They always articulate in a transverse plane. Mandibles are provided with the strong adductor and abductor muscles arisen from the dorsal and lateral walls of the cranium. Their function is to carry out the biting and chewing the food material in the orthopteroid insects. They are adapted into the sickle or stylet like piercing organs in some fluid sucking insects. They are modified into defensive organs in the termites and some beetles.

Mandibular gland: A pair of gland often opens at the base of mandibles.

Mandibulate soldier: A subcaste in the social insects having mandibles which are used for fighting against the intruders in the colony.

Mandibulate: Having primitive, usually chewing mouthparts.

Mandibulosuctorial mouth parts: The chewing mouth parts of this subtype are found in some Neuropterans (aphidlions, antlions etc.). They have long fang like jaws. Mandibles are deeply grooved on their ventral surfaces but closing lobes are long blades of maxillae which fit into mandibular grooves and thus forming the tubular channel between the two appendages through which the larva sucks juices of its victim. The maxillae undergo reduction in appendicular parts. The labium contains a middle plate differentiated into the submentum and mentum.

Manejo Integrado de Plagas (MIP): This is the Spanish equivalent of Integrated Pest Management (IPM).

Mange: A contagious skin disease caused by certain mites. The skin becomes thickened and hair coat is lost.

Mango bud mite: See *Aceria mangiferae.*

Mango fruit fly: See *Dacus dorsalis.*

Mango hard scale: See *Aspidiotus destructor.*

Mango hopper: See *Amritodus atkinsoni.*

Mango leaf cutting weevil: *Deporaus marginatus* Pasc. is a minor pest of mango.

Mango leaf gall maker: See *Procontarinia matteiana* Kieff.

Mango mealy bug: See *Drosicha mangiferae*

Mango mite: See *Eriophyes* sp.

Mango nut weevil: See *Sternochaetus mangiferae.*

Mango shoot gall psyllid: *Apsylla cistellata* Buckton. is a pest of mango that makes galls on the shoots.

Mango soft scale: See *Pulvinaria psidi.*

Mango stem borer: See *Batocera rufomaculata* Dejean.

Mango stone weevil: See *Sternochaetus mangiferae.*

Mantid: Insect belonging to the family Mantidae (order Dictyoptera). Mantids are large, elongate insects with enlarged front legs that are adapted for grasping their prey (other insects). They have incomplete metamorphosis.

Mantispidae: Family of superfamily Hemerobioidea, suborder Planipennia, order Neuroptera. With elongated prothorax and raptorial legs, serving as effective prehensile organ. Larvae on hatching thysanuriform, subsequently eruciform and feed on lycosid eggs, *e.g., Mantispa* sp.

Manubrium: A narrow strut attached to the anterior end of the tegmen.

Margarodidae: A family of superfamily Coccoidea, suborder Homoptera and order Hemiptera. Characterized by females having distinct segmented body, often covered by waxy secretions. Examples, cottony cushion scale, *Icerya purchasi*; mango mealy bug, *Drosicha mangiferae* etc.

Margin: The posterior edge of an insect wing.

Marginal cell: A cell in the distal part of the wing bordering the coastal margin in order Diptera, Hymenoptera etc.

Marginal vein: The vein forming the posterior side of the marginal cell, *e.g.,* Hymenoptera.

Marine: Pertaining to ocean or sea.

Marker pheromone: A pheromone smeared on the host by the ovipositing parasite to inform others that the host was already parasitized and thus avoid superparasitism.

Market basket survey: Pesticide residue monitoring on a wide range of food items collected from consumer points of sale and in proportions approximating consumption patterns in the local populations. Samples are prepared for the analysis according to Codex guidelines, *i.e.*, minimal preparations.

Market value: The amount of money that a seller can expect to obtain for a commodity.

Marketing: The phenomenon of production, promotion, product public relations and information services as well as distribution and selling on local or international markets.

Mask: The elongate labium which can be thrust forward to seize prey, *e.g.*, in dragonfly naiad.

Mass communication: The transmission of information among groups of individuals of a kind that cannot be transmitted from a single individual to another.

Mass culture: A term usually reserved for large scale production of bacteria, viruses or other biological control agents using insect hosts. It is also synonymous with factory rearing (mass rearing in millions per unit time or a regular basis).

Mastotermitidae: A family of order Isoptera. Primitive, well represented in fossil record, with tarsi five segmented and hind wings with well developed anal lobes. Example, *Mastotermes darwiniensis*.

Mathematics: Branch of science in which we study various calculations. Mathematics is used in almost all the branches of science. For the purpose of convenience in learning, it is divided into many branches, like arithmetic, algebra, geometry, trigonometry etc.

Mating flight: The flight taken by a virgin queen while she mates in the air with drone.

Mating stimuli: Many kinds of stimuli are concerned in the mutual attraction of the sexes at mating. In many Nematocera, Trichoptera, Plecoptera etc., the females are attracted to the males by their dancing in swarms; crickets and grass hoppers by their chirping songs; male mosquitoes are attracted by the note emitted from the beating wings of female; and males of Lampyrid beetles by the luminous organs of the female. But the most important stimuli is probably the scent emitted by glandular scent producing organs in the neighbourhood of sexual opening of female Lasiocampidae and Bombycidae. The female sex pheromones in some insects act as a good stimuli for the males. See also queen substance and androconia.

Matrix: (i) A set of numbers or terms arranged in rows and columns between parentheses and double lines, (ii) The material or component sampled for pesticide residue studies.

Maxilla (pl., maxillae): Jaw or upper jaw bone, one of a pair of mouthparts characteristic of chewing insects. Segmented, secondary jaws that aid in holding and chewing foods; located between the mandibles and the labium; stylets in many insects with piercing-sucking mouthparts; variously modified in other insects. Maxillae are used mainly for grasping and tasting of food. The maxilla has lobes and bears a multi-segmented palp or palpus.

Maxillary lobe: One of the two apical lobes of the maxilla (galea or lacinia).

Maxillary palpus (pl., maxillary palpi): A small feeler like structure arising from the maxilla.

Maxillary: Pertaining to the maxilla.

Maximum residue limit (MRL): Maximum concentration or residue of an insecticide that is legally permitted or recognized as acceptable in or on a food, agricultural commodity or animal feedstuff as set by Codex or a national regulatory authority. The term tolerance used in some countries is in most instances, synonymous with MRL. Normally, it is expressed in mg kg^{-1} fresh weight or ppm of fresh weight of food. It is calculated from food factor, acceptable daily intake (ADI) and average weight of the consumer.

Maximum tolerated dose (MTD): Highest dose of pesticide in chronic toxicity testing, *i.e.*, expected dose on the basis of sub-chronic studies which result in limited toxicity when administered for duration of the test period.

Maybug: See cockchafer.

Mayfly: Any of various insects of the order Ephemeroptera. Mayfly is a fragile winged insect that develop from aquatic nymph. In the adult stage, it lives no longer than a few days. Sometimes, it is also called dayfly.

MB: See methyl bromide.

MCPA: 2-methyl-4-chlorophenoxyacetic acid. It is a hormone used as a weed killer for broad leaf weeds at the rate of 1-2 lbs of acid equivalent per acre, available as sodium salt, closely resembling 2, 4-D.

MCPP: 2-methyl-4-chloro-phenoxypropionic acid, a hormone herbicide used to kill weeds of cereal crops.

Mead: Honey wine.

Mealworm: Common name for the slender larvae of beetle family Tenebrionidae (order Coleoptera). They are generally found in grains and cereals.

Mealy bug: Any of various insects of the family Pseudococcidae (suborder Homoptera; order Hemiptera). These are small wingless insects that have the body covered with a white powdery waxy substance.

Measuring worm: See looper.

Mechanical agitation: Agitation of spray fluid mechanically with paddles or propellers.

Mechanical control: The types of insect control measures that include the manual or machinery operations, like hand picking, hand netting or bagging, creating artificial barriers (tree banding, trenching and fencing, screening etc.), insect traps (mechanical trap, suction trap, grid type trap, glass trap, pheromone trap etc.).

Mechanical transmission: Mechanical transmission of pathogens takes place by two ways: (i) Indirect: It is the exogenous infection carried out with the help of feet or body of insect. In this the insects act merely the passive agents, *e.g.*, house fly which picks up the microbes while visiting the dirty places on their legs and other body parts and contaminate human food and spreads various dangerous diseases, like typhoid (causal organism, *Salmonella typhi*), summer diarrhoea and dysentery (caused by *Entamoeba* sp.) and cholera (caused by *Vibrio cholerae*); (ii) Direct: It is the endogenous contamination made due to direct way in which the insect picks up the germs from an infected animal and directly inoculate them into the blood of host animal, *e.g.*, common eye flies, *Siphunculina funicola*, *Hippelates* sp. and *Chlorichaetus tuberculosa* (Cyclorrhaphan Diptera) is responsible in spreading conjunctivitis (viral infection) from man to man.

Mechanism of circulation: The haemolymph circulation always takes place in postero-anterior direction. The dorsal vessel is the sole circulatory organ. It is a contractile structure and peristaltic wave often originates at the posterior most end and runs to the anterior end including the pumping of the blood in similar direction. The heart undergoes uniformly expansion and contraction comprising diastole and systole phases of cardiac cycle, respectively. The diastole is initiated through myocardium and associated connective tissues but in some insects due to contraction of the alary muscles (*Agrion*, *Hydrophilus*). The systole is brought up by titanic contraction of the muscle fibres of the wall of dorsal vessel. The diastole phase is often longer than the systole in most of the insects except *Lepisma*. Sometimes, there occurs an intermediate or resting phase, the diastasis in between the diastole and systole phase. During diastasis, the heart remains expanded. In between diastasis and systole, appears a presystolic notch. The incurrent ostia opens during diastole allowing haemolymph inside the heart while remain closed during systole in order to prevent outward as well as backward flow of the haemolymph. The haemolymph, therefore, is drawn into the successive heart chambers through the ostia during systole. During its forward passage, some amount of blood is discharged out through the excurrent ostia and lateral segmental vessels.

Mechanoreception: The perception of stimuli carried out various mechanical changes, *i.e.*, perception of either tactile stimuli or vibration or both.

Mechanoreceptors: The mechanoreceptors include (i) trichoid sensilla on antennae, tarsi, tibia and cerci, (ii) campaniform sensilla at wing base, halteres, cerci, palpus and legs, (iii) chordotonal organs as intra-visceral mechanoreceptors are of two types- integumental and subintegumental. Some chordotonal organs are modified into Johnston's organ (in antennal pedicel) which acts as aerial or aquatic indicators; subgenal organs (in proximal part of tibia in Dictyoptera, Odonata and orthoptera) which respond to substratum vibrations and airborne sounds; and tympanal organs (auditory organs) of integumental (Acrididae, Cicadidae and Lepidoptera) and subintegumental type (Tettigonidae and Gryllidae) situated on forelegs (Tettigonidae and Gryllidae) and first abdominal segment (Acrididae).

Meconium: The reddish fluid ejected by a member of the Lepidoptera after emerging from the pupa or chrysalis.

Mecoptera: Gr., *meco*= long, *ptera*= wings. Insect order, made up of the scorpionflies. They are characterized by membranous wings, elongate faces with chewing mouthparts, a scorpion shaped abdomen (male only), and a complete metamorphosis, *e.g.*, scorpionflies. This order includes scorpionflies, so called because the males in many species have the terminal segments of the abdomen raised in the manner of scorpions. They are carnivorous, essentially terrestrial insects with an elongate beak like head, frequenting shady places and are often met with in herbage or rarely under stones.

Media (M): The longitudinal vein running through central region of the wing in most insects: often the 4th and abbreviated as M. It starts from the distal medial plate and mostly fuses with the radius. At the base, it is largely sclerotized and convex and after running for a short distance, it gives two branches, anterior and posterior media, respectively. The anterior media (MA) divides further into two convex branches (MA_1 and MA_2). The posterior media (MP) also known as media sector is concave and divides dichotomously into four branches (MP_1-MP_4).

Media worker: An individual belonging to the medium size subcaste. In polymorphic ant, two other worker subcastes are there, *i.e.*, major and minor worker.

Medial: See mesal. It also refers to the medial vein or medial area of the hind wing.

Medial area: The portion of the hind wing behind the medial bar (MP_{1+2}) and the anal fold. It is also called medial field (Kukalova-Peck and Lawrence, 1993).

Medial cross vein: A cross vein connecting two branches of the media.

Medial fleck of hind wing: A portion of the wing membrane bearing rasp-like structures and lying somewhere near the apex of the vein MP_3, also called the sub-cubital fleck. This structure may act as a holdfast when contacting a similar patch on the opposing wing or the underside of the elytron. It is also known as katastigma.

Median area of hind wing: Area of the hindwing between the median bar (MP_{1+2}) and the anal fold (between AA and AP).

Median: On the midline; along midline of the body.

Median bar: (of hind wing). The posterior of the two major supporting structures of the hind wing in beetles, formed from the vein MP_{1+2} and in most Archostemata, Myxophaga and Adephaga supporting the oblongum cell at or near its apex. In most Polyphaga its apex is joined to RP, forming the radio-medial loop.

Median caudal filament: A thread-like process at the posterior end or between the cerci as in Thysanura.

Median effective concentration (EC_{50}): Statistically derived concentration of a pesticide in an environmental medium expected to kill 50 per cent of the test organisms in a given population under defined conditions.

Median effective dose (ED_{50}): The dose or concentration of a chemical that will produce a response in half the test subjects. It is an indirect measure of the mean tolerance of a batch of the test subjects. Median lethal dose is the special case in which the death is the response.

Median lethal dose (LD_{50}): Statistically derived dose of a pesticide expected to kill 50 per cent of test organisms in a given population under a defined set of conditions. Normally, it is expressed as mg of test material per kg of body weight of the organism.

Median lethal time (LT_{50}): Period of exposure to a toxicological stimulus which will produce death in half the test subjects. The length of exposure is the direct measure of dosage and an increase in the period of exposure results in an increasing uptake and true dose increases in the same ratio. It is also known as median survival time, $ST_{50.}$

Median nerve cord: The median strand of nerve tissue produced from the ventral neuroblasts.

Median nerves: Unpaired nerves arising from the ganglia of ventral nerve cord between the roots of connectives.

Median neusosecretory cells: Cells in the pars intercerebralis region of hormones, including brain and eclosion hormones.

Median oviduct: In insects, the single duct formed by the merging of the paired lateral oviducts; this duct opens posteriorly into a genital chamber or vagina. The two lateral oviducts combine to form a median oviduct which in Dermaptera opens outside directly on seventh abdominal segment of the body. The opening is called gonopore. In most of the insects, however, the median oviduct does not open directly to outside but instead into a tubular genital chamber or vagina formed by an invagination of body wall from eighth abdominal segment and opening is known as vulva or gonopore which is served both for receiving sperms and discharging eggs.

Median plates: The first median plate is a small sclerite of the variable shape lying in the angle between the second axillary and the distal arm of the third axillary in function, and usually attached to it. The second median plate is a variable sclerotization at the base of mediocubital field, folding convexly on the outer edge of the first median plate along the plica basalis; often absent or represented by the united bases of medial and cuital veins.

Median survival time: See median lethal time.

Medical entomology: The branch of science that deals with insects that serves as vectors of organisms that cause disease in human being. In medical entomology, for example, research may pertain to the interactions of the arboviruses and their hosts and involve physiology, genetics, biology and ecology.

Medical pest: A pest that causes a pathological condition or transmits diseases in human being.

Medical science: Branch of science in which various diseases, causes of origin, their prevention and treatment are studied.

Medical toxicology: A branch of toxicology which deals with the effects of poisons in man.

Medio-cubital cross vein: A cross vein connecting the median and cubitus.

Mediproboscis: See sponging type mouth parts.

Medium volume spray: A spray application of 20-60 gallons per acre.

Medlure: A synthetic attractant for baiting the Mediterranean fruit fly; sec-butyl 4(or5)-chloro-2-methyl-cyclohexanecarboxylate.

Medulla: It is composed of a mass of axons of varying diameters, their terminal arborizations investing glia and tracheoles. The axonal fibres are at some instances, arranged into the bundles or tracts which may be branched or unbranched. The glial sheath is reduced or entirely lacking in some areas providing a site for synapses among adjacent axons. The presynaptic membrane bears the so called presynaptic knobs of 1.2 to 2.5 micrometre diameter, each bearing a cluster of synaptic vesicles. The medulla is compactly filled with the neuropile.

Medullary substance: The dense fibrous mass of nerve terminals forming the interior of a ganglion.

Meenoplidae: A family of superfamily Fulgoroidea of suborder Homoptera and order Hemiptera. A small family characterized by the presence of wax secreting poxes on 6th to 8th abdominal segments. Two species, *viz., Nisia atronervosa* and *Eponisia guttula* are known to infest rice in India, the former being more important.

Megachilidae (leaf cutting bees): Family of super family Apoidea, suborder Apocrita and order Hymenoptera. Megachilidae are moderate-sized, stout-bodied, densely haired, solitary bees. The presence of ventral scopa in the female (which is wanting in parasitic form) and two submarginal cells of about equal size in forewings are characteristic features of this group.

Megalodontidae: Family of super family Megalodontoidea of suborder Symphyta (Chalastogastra), order Hymenoptera. The adults possess flabellate antennae and the mouthparts are as long as the head capsule. The larvae web the leaves of plants.

Megalodontoidea: Super family of suborder Symphyta (Chalastogastra), order Hymenoptera.

Megaloptera: Suborder of order Neuroptera. Larvae with biting mouth parts, *e.g.*, alderflies and snakeflies.

Megalyridae: Family of super family Ichneumonoidea, suborder Apocrita and order Hymenoptera. These insects are known from Australia and *Megalyra* sp. is parasitic on wood boring beetles.

Megathripinae: Subfamily of family Phlaeothripidae, suborder Tubulifera and order Thysanoptera. Maxillary stylets broadened at apex, broader than labial palps.

This is a major subfamily with several mycophagous species, *e.g.*, *Tiarothrips subramanii*, *Kleothrips gigans*, *Nesothrips* etc.

Meiosis: The two successive divisions of a cell's nucleus in which the chromosomes is reduced by one half.

Meiotic drive: Ant alteration of the normal process of meiosis with the consequence that a parent heterozygous for two genetic alternatives produces an effective gametic pool with an excess of one type.

***Melanagromyza* (=*Ophiomyia*) *phaseoli* (Tryon):** Belongs to order Diptera and family Agromyzidae. It is an insect pest of peas, commonly known as the pea stem fly or bean fly.

***Melanagromyza obtusa* (Malloch):** Belongs to order Diptera and family Agromyzidae. It is an insect pest of pulse crops, commonly known as the red gram pod fly.

Melanastic: The cuticle characterized by excessive pigmentation or blackening of the tissues.

Melanins: Insect pigments of yellow, black or brown.

Melanization: Darkening.

Melanosis: A queen honey bee disease characterized by decolouration of egg cells and trophocytes which turn from yellow brown to black.

Meliantriol: A toxic alkaloid obtained from neem tree is a feeding inhibitor against insects at low concentrations.

Melittology: Study of bees.

Melittophile: An organism which spend at least a part of its life cycle with bee colonies.

Melliferious: Producing or yielding honey.

Mellittophily: Transfer of pollens by bees, a type of pollination.

Meloidae (oil beetles, blister beetles): Family of superfamily Cucujoidea, suborder Polyphaga and order Coleoptera. The beetles are medium sized, 1.25 to 2.5 cm long and conspicuously covered with bright metallic blue, green, black or brown. The head is hypognathous and is joined to the prothorax by a distinct neck. The wings are well developed and loosely cover the body; in some they are vestigial while others are apterous. The forecoxal cavities open behind and tarsi are five-segmented. The legs are long, appendiculate and usually with serrate tarsal claws. When disturbed the beetles emit a fluid containing the oil principle, cantharidine, which has irritant properties, through the opening in the apices of the femur. The beetles are usually found on flowers and leaves and cause injury by feeding on them.

***Meloidogyne incognita* Chitwood:** Belongs to order Tylenchida and family Heteroderidae. It is a pest of brinjal, commonly known as the root-knot nematode.

Melolonthidae (cock-chafers, june beetles): Family of superfamily scarabaeoidea, suborder Polyphaga and order Coleoptera. The cock-chafers are variously colored, black, brown, green, mahogany, blue or metallic colored. In the male the antennal club is often elongate and horny processes are present on its head. The mentum and ligula are fused. The metasternum laterally attains the coxae and the abdomen is with six tergites. These beetles are commomly attracted to light. The larvae feed on roots of plants and trees and are often destructive. The grub is fleshy, white with brown head, curved body and the terminal segment large and smooth.

Melon fruit fly: See *Bactrocera cucurbitae* (Coquillett).

Membracidae: A family of superfamily Cicadoidea, suborder Homoptera and order Hemiptera. Characterized by two ocelli, antennae lie in a depression beneath and in front of the eyes. They are generally smaller measuring upto 10-12 mm long. Stridulatory organs are absent. The usually jump. Pronotum extending backward over abdomen and ornamented with spines, horns, keels, bell like knobs etc. of various shapes. The wings are largely concealed by pronotum. For example, tree hoppers and cow bugs. The tree hopper, *Tricentrus bicolor* is widely distributed in India and a common pest of Lucerne, alfa-alfa, sorghum, sugarcane, wheat, pea rose, wild grasses etc.

Membrane: A thin film or tissue usually transparent; the thin apical part of a hemelytron as in the bugs.

Membranous: Like a membrane. Thin and transparent (in reference to wings); thin and pliable (in reference to integument).

Membranous wings: The forewings or hind wings or both are membranous with tubular veins, *e.g.*, wings of dragonfly and honey bee.

Mengedae: Family of order Strepsiptera. The members of this family are parasites of the Lepismetidae and are readily separated from others by the male possessing five-segmented tarsi with claws. The adult females are free living with legs, antennae and eyes.

***Menopon gallinae* (Linn.):** Belongs to order Mallophaga and family Menoponidae. It is an insect pest of poultry birds, commonly known as the poultry shaft louse.

Menoponidae: A family of suborder Amblycera of order Mallophaga. Infesting birds. Head broadly triangular, much enlarged at temples and evenly expanded behind. Example, domestic fowl louse, *Menopon gallinae*; poultry louse, *Menopon stramineum*.

Menotaxis (Light compass response): There are insects which move at a constant angle to a source of light; a method of orientation is known as menotaxis. In some cases this response seems to result from balanced turning effect of two lights- that from the source itself or that reflected from the back ground. Sometimes, included in this mechanism is the light compass orientation of the bees and ants which make use of the direction of Sun's rays in finding their way back to the nest.

Mental setae: Setae present on the mentum as in Odonata.

Mentum: Distal part of the labium which bears the labial palpi.

Meroblastic cleavage: Method of blastoderm formation characteristic associated with eggs containing a large amount of yolk; the fusion nucleus and cytoplasm proliferate mitotically and the daughter nuclei migrate to the egg and form the blastoderm.

Merocrine secretion: Release of enzymes across the cell membrane, without destruction of the entire cell.

Meroistic ovariole: Ovariole with nurse cells, *i.e.*, polytrophic and telotrophic ovarioles.

Meron: The lateral post-articular area of the base of coxa. In some insects it is greatly enlarged and extended distally in the posterior part of the coxa but always proximal to the basicostal suture.

Meropleuron: A sclerite consists of meron (basal part) of the coxa and the lower part of the epimeron.

Meropodite: The fourth segment of a generalized limb.

***Merops orientalis* Latham:** Belongs to order coraciiformes and family Meropidae. It is a predator of insect pests of farm animals and commonly known as the common green bee eater.

Merothripoidea: Superfamily of suborder Terebrantia, order Thysanoptera. Antennal segments moniliform, eight or nine segmented, style absent; segments three and four with a sensory area at apex, tympanum like, without sense cone; fore and hindfemora greatly enlarged; ovipositor weak, abdomen blunt. Only a single family the Merothripidae exists. Example, *Merothrips*.

Merus: Pertaining to the meron.

Mesad: Towards the midline of the body.

Mesal: At or near the midline of body.

Mesenchyme: Tissue formed due to loosely connected or scattered cells.

Mesenteron rudiments: Group of endodermal cells that regenerate the mesenteron, including an anterior, a posterior and sometimes an intermediate rudiment.

Mesenteron: See midgut.

Mesepimeron (pl., mesepimera): The epimeron of the mesothorax. The posterior part of the mesothoracic pleuron, the exposed portion of which is joined to the metaventrite and metepisternum and forms part of the mesocoxal articulation; another concealed portion lies beneath the elytra and extends anterodorsally to form part of the elytral articulation.

Mesepisternum (pl., mesepisterna): The episternum of the mesothorax. The anterior part of the mesothoracic pleuron, the exposed portion of which is joined mesally to (or sometimes fused with) the lateral edge of the mesoventrite and forms part of the mesocoxal articulation; another concealed portion lies beneath the elytron and extends anterodorsally to form part of the elytral articulation.

Mesh (Sieves): Standard sieves are used to separate the solid particles of different sizes. The mesh refers to the number of openings to each linear inch, finest sieve is the 325 mesh which has openings of 44 microns in diameter, one micron is equivalent to 0.001 mm. Such sieves have 10,500 openings per square inch. About 95 per cent of the particles of fine sulphur can pass through it. The particles small enough to pass a 60 mesh sieve are considered as dust.

Mesh work: A network or web.

Mesial: See Mesal.

Mesinfraepisternum: It is ventral subdivision of the mesepisternum.

Meso: Middle. Prefix referring to a structure forming part of the mesothorax, including mid legs (*e.g.*, Mesocoxa, mesepisternum, mesotarsus, mesepisternum).

Mesocoxal cavity: The counter sunk sclerotized cavity for housing the mesocoxa; formed from the mesoventrite and metaventrite, usually with the addition of one or more elements of the mesopleuron (or more rarely the metapleuron). The cavity or housing is almost completely sclerotized (except where the coxa is joined by membrane). In some cases, there is no real cavity or housing present, and the coxa is attached externally by a broad membrane.

Mesocuticle: A procuticular layer lying between the exocuticle and endocuticle.

Mesoderm: The middle germ layer.

Mesofemur: The femur of the mid leg.

Mesometaventral junction: The meeting of the mesoventral and metaventral processes; this may consist of a simple meeting of the two processes, but in some cases a complex ball and socket joint is formed and in others, the two processes are connate or completely fused.

Meson: (i) The midline of the body or an imaginary plane dividing the body into right and halves. (ii) The posterior portion of the coxa.

Mesonotum: The dorsal surface of the second thoracic segment, the mesothorax, usually the largest thoracic sclerite.

Mesophyll: Tissues or cells in the interior of leaves of plants.

Mesopleural bristles: Bristles on the mesopleuron in order Diptera.

Mesopleuron (pl., mesopleura): The lateral sclerite(s) of the mesothorax; the upper part of episternum of the mesothorax, *e.g.*, in Diptera.

Mesoproboscis: See sponging type mouth parts.

Mesopsocidae: A family of order Psocoptera and suborder Eupsocida. Tarsi three segmented. Example, *Mesopsocus.*

Mesoscutellum: Hind most of the three divisions of the mesonotum, often more or less triangular. It refers to the scutellum of the mesothorax.

Mesoscutum: The anterior part of the mesonotum in beetles, occasionally bearing a stridulatory plate. The middle division of the mesonotum.

Mesosoma: The middle of the three major divisions of insect body, *i.e.*, thorax but in higher Hymenoptera, it includes the propodeum.

Mesosternal cavity: See mesoventral cavity.

Mesosternum: The sternum of the mesothorax. See mesoventrite.

Mesotarsal: Referring to the tarsus (and pretarsus) of the mid leg.

Mesothorax: The middle, or second, segment of the thorax.

Mesovellidae: Family of superfamily Aradoidea of division Gymnocerata, suborder Heteroptera and order Hemiptera. These water treaders are small insects with large head and eyes and long delicate legs. Winged and apterous forms are common, *e.g.*, *Mesovelia* sp.

Mesoventral cavity: A mesal cuticular cavity on the mesoventrite, which houses the apex of prosternal process.

Mesoventral process: A posterior projection of the mesoventrite which may meet an anterior projection of the metaventrite.

Mesoventrite: The major ventral sclerite of the mesothorax. This consists mainly of combined precoxal elements (preepisternites) fused together or separated by a discrimen. This structure is usually called the mesosternum but the true sternum is no longer visible externally (Campau, 1940; Ferris, 1940; Lawrence, 1999; Beutel and Haas, 1999).

Meta: Prefix referring to a structure forming part of the metathorax, including hind legs (*e.g.*, Metacoxa, metepimeron, metatibia, metaventrite).

Metabola: With metamorphosis.

Metabolism: Sum total of all physical and chemical processes that take place within an organism. In a narrow sense, it is the physical and chemical changes that occur for a pesticide within the body, changes (biodegradation) and elimination of pesticides and their metabolites.

Metabolites: Any intermediate or product resulted from metabolism.

Metabolous: With metamorphosis.

Metacoxal cavity: The counter sunk sclerotized cavity for housing the metacoxa; formed mainly by the base of abdomen.

Metacoxal plate: A partial or complete flattened, ventral projection of the metacoxa, which forms an excavated area for housing the metafemur.

Metaldehyde: A chemical, $(CH_3CHO)_4$.

Metamere: (i) A primary body segment of insect, mostly used for the division of embryo. (ii) Any of a linear series of primitively similar segments into which the body of a higher invertebrate or vertebrate is divisible.

Metamerism: Repetitions of parts along the long axis of an animal.

Metamorphosis: The development of an embryo into adult takes place through a series of successive transitional, morphologically alike or variable forms; such type of post embryonic development is called metamorphosis. This may be regarded as a change in form during the development. After hatching from the egg, an insect grows by a series of moults. After shedding the old skin they expand into a new larger one. This moulting continues until the adult stage is reached. At each moult, some externally visible changes occur. This type of post embryonic development is called metamorphosis. The division of insects into Apterygota and pterygota (exopterygota and endopterygota) is mainly based on differences in the type of metamorphosis. (See also, complete metamorphosis, incomplete metamorphosis).

Metamorphosis, types: On the basis of variations in metamorphosis, the insects can be grouped into two main types: (i) Ametabola, (ii) Metabola (a) Hemimetabola or incomplete metamorphosis: Palaeometabola (Epimetabola and Prometabola), Heterometabola (Archimetabola and Paurometabola/ simple or gradual metamorphosis) and Neometabola (Homometabola, Remetabola, Parametabola and allometabola); (b) Holometabola/ complete /indirect/ Complex metamorphosis; (c) Hypermetabola.

Metanotum: The dorsal surface of the metathorax. It is often very small and its subdivisions are usually obscured.

Metaphase: The second stage in the cell division process.

Metapleural gland: Also called metasternal or metathoracic glands are secretory glands that are unique to ants and basal in the evolutionary history of ants. They are responsible for the production of an antibiotic fluid that then collects in in a reservoir on the posterior of the ant's alitrunk. These reservoirs are also referred to as the bulla and vary in size between ant species and also between castes of the same species.

Metapleuron: Lateral sclerite of the metathorax or sclerites comprising the side wall of metathorax.

Metapneustic respiration: In many larvae (*Dytiscus* sp., Culicidae and many other Diptera), the other spiracles have been become obliterated or atleast functionless and the insect breathe entirely through the terminal spiracles.

Metarbelidae (bark borers): Family of super family Psycoidea of suborder Ditrysia and order Lepidoptera. They are confined to tropical region. Larvae are wood borers. Moths are nocturnal. The bark borer *Indarbela tetraonis* is known to attack a number of trees and shrubs such as mango, guava, moringa, rose, orange, citrus etc.

***Metarrhizium anisopliae*:** A species of entomopathogenic fungi causing green muscardine disease in many insect species. It is marketed under trade names, Biomax and Pacer MA (1.15 WP) etc.

Metascutellum: Sternum or ventral sclerite of the metathorax.

Metasoma: In Apocrita (Hymenoptera), the abdominal segment first forms the propodeum and the remaining abdominal segments form the metasoma.

Metasternum: The ventral surface or sclerite of the metathorax.

Metasystox: Trade name of methyl demeton.

Metatarsus: The basal segment of the tarsus or foot, usually the largest segment.

Metathorax: The posterior or third segment of the thorax.

Metaventral process: An anterior projection of the metaventrite which usually meets a posterior projection of the mesoventrite.

Metaventrite: The major ventral sclerite of the metathorax. This consists mainly of combined precoxal elements (preepisternites) and paired metakatepisterna fused together or separated by a discrimen. This structure is usually called the metasternum, but the true sternum is no longer visible externally (Campau, 1940; Ferris, 1940; Lawrence, 1999; Beutel and Haas 1999).

Metendosternal laminae: Flattened processes located on the lateral arms of the metendosternite.

Metendosternite: A ventral, median, internal invagination of the metathorax forming a complex, fork-like structure to which muscles are attached; a remnant of the original sternum.

Metepimeron (pl., metepimera): The posterior part of the metathoracic pleuron, the exposed portion of which, if present, is relatively small, lying between the lateral edge of the transversely oriented metacoxa and the elytron and forming the main coxal articulation; another concealed portion extends anteriorly and forms part of the hind wing articulation.

Methamidophos: An organophosphate insecticide. See Monitor.

Methidathion: An organophosphate insecticide. See Supracide.

Methiocarb (Metmercapturon): A carbamate insecticide, sold under trade name Mesurol.

Methomyl: A carbamate insecticide marketed under trade name, Lannate or Dunet.

Methoprene: It is highly active juvelile hormone analogue developed by Zoecon Corporation, U.S.A.

Methoxychlor: A chlorinated hydrocarbon insecticide with low toxicity, related to DDT (DDT analogue), 1,1,1-trichloro-2, 2-bis (p-methoxyphenyl) ethane, less toxic to mammals as compared to DDT, acute oral LD_{50} for rat is 1000 mg kg^{-1}, not accumulated in fatty tissues hence preferred to use on animals, non-phytotoxic to crop plants and used against wide range of indoor and outdoor insects including pests of vegetables, fruits, shrubs, flowers and ornamental plants as it is a broad spectrum insecticide. Since, it is unstable, it has less residual effect. The technical compound is white crystalline with melting point 70-85 °C. Its dehydrochlorination takes place like DDT but slowly.

Methyl bromide (Bromomethane): A non-inflammable, non-explosive highly toxic fumigant with good penetrating power; odourless, available as liquid, chemical formula CH_3Br; used for treatment of mills, warehouses, vaults, ships, freight carriers against insect pests of stored grains. The insecticidal properties of methyl bromide were first reported by Le Goupil (1932) in France. It is colourless gas, boiling point 3.6 °C, molecular weight 94.94; specific gravity of gas is 3.27 and that of liquid is 1.732. It is an active alkylating agent and reacts with sulphydral (-SH) group of enzymes such as succinic dehydrogenase and accounts for the toxic action to insects and warm blooded animals. As the gas is odourless, 2 per cent chloropicrin is mixed with it as a warning gas. It is also used as soil fumigant against nematodes etc. It is toxic to growing plants. The gas is 3.3 times heavier than the air. It has been extensively used in quarantine for plants, vegetables, and some fruits, and stored products, pests in soil, mills, ware houses, ships and freight carriers. The dosages are 24-32 g per m^3 and 48 mg l^{-1} with exposure period of 48 hrs for stored grains and milled products, respectively. It has no effect on viability of seeds. However, because of its ozone depleting potential, there has been a global move to phase out methyl bromide as per the Montreal Protocol, 1992. In India, till July 2000, the fumigant was used to treat the soil structures (flour mills) and stored commodities. Later on, its use has been banned except for quarantine and shipment treatment.

Methyl conjugation: Pesticide conjugation by methylation. N-methylation is a natural process of some primary, secondary and tertiary amines. S-methylation of the sulfide bond of metasystox is known to occur, however, the process is not known to occur in insects.

Methyl demeton: Diethyl S-2 ethylthio ethyl phosphorothiolate $[(CH_3O)_2 PSO(CH_2)_2 SC_2H_5]$ is an organophosphate insecticide, the technical grade is a mixture of thiono and thiolo isomer. The thiono isomer is a liquid with unpleasant odour, soluble in water (330 mg l^{-1} at 20 °C temperature) and most of the organic solvents. It is a systemic insecticide effective against sucking insect pests. In plants, it is rapidly isomerised to thiolo isomer which is further metabolized to non-insecticidal products as a result of duration of insecticidal activity usually does not exceed three weeks. It is marketed under trade name Metasystox.

Methyl parathion: An organophosphate insecticide, O, O-dimethyl o-p-nitrophenyl phosphorothionate ($C_8H_{10}NO_5PS$); contact and stomach poison, white crystalline solid with melting point 35 °C, less stable than parathion, practically insoluble in water (35 mg l^{-1} at 25 °C); extremely toxic to mammals, acute oral LD_{50} for rat is 9-25 mg kg^{-1} (but less toxic than parathion as part of methyl parathion is decomposed as a result of demethylation reaction before it reaches active centres) and effective against defoliators, aphids etc. It breaks down more quickly than parathion and path of conversion is the same as in parathion. It is marketed under trade names Metacid and Methyl parathion.

Methyl trans trans 10, 11-epoxyfarsenate: Juvenile hormone mimic. This compound has all the chemical and biological activity relationship of pure cecropia extract.

Methyl trans trans 7, 11-dichlorofarsenate: Juvenile hormone mimic. Romanuk *et al.* (1967) isolated this compound and found that 1 nanogram (ng) of this mimic is sufficient to prevent normal morphogenesis in *Pyrrhocoris*.

Methyl trithion: An organophosphate insecticide and acaricide, S-[(p-chlorophenylthio) methyl] O,O-dimethylphosphorothioate; moderately toxic to mammals, acute oral LD_{50} for rat is 98-200 mg kg^{-1}, and effective against mites, mired bugs, aphids, boll weevil etc.

Metinfraepisternum: A ventral subdivision of the metaepisternum.

Metoxeny: A phenomenon of the change of host by the parasite.

Metric system: A system of measurement adopted by most of the countries of the world. The units of the metric system are metres (for length), grams (for weight) and litres (for volume).

Mevinphos ($C_7H_{13}O_6P$): Highly toxic insecticide and acaricide, introduced by Shell under trade name Phosdrin which is used against insect pests of fruits and vegetables. It is a systemic organophosphate with powerful contact and fumigant action and occurs in two forms, *cis* and *trans* isomer. The insecticidal property is due to the *cis* isomer which constitutes 60 per cent of the product. It is broken down in plants rapidly (90 per cent in 0.8-4 days). Mevinphos is chemically, 2-methoxycarbonyl-1-methylvinyl dimethyl phosphate. Its insecticidal properties were studied by Corey and his associates in 1953. It is colourless liquid, highly soluble in water, acetone and benzene. Extermely toxic to mammals, acute oral LD_{50} of alpha-isomer for male rats is 3 mg kg^{-1} and beta-isomer 46 mg kg^{-1}.

Mexacarbate: Highly toxic insecticide, acaricide and molluscicide marketed under trade name Zelt is broken down in plants.

Micoplasma like organism: These organisms are smaller than the bacteria and bigger than the viruses and responsible to cause many plant diseases. These organisms are formerly attributed to the viruses.

Micro: Small.

Micro applicator: Equipment to apply small measured amount of solution or suspension (topical application or injection).

Microbial control: Control of insects (or other organisms) by the use of micro-organisms.

Microbial insecticide: A pathogenic micro-organism or its products (toxins etc.) that is applied in the same way as a conventional pesticide to control a pest population. Similar terms: microbial pesticide, biotic insecticide and microbial control product.

Microbial pathogen: A microorganism which causes disease in the host. A term microbial pesticide or microbial insecticide is used to denote a micro-organism or byproduct of micro-organism used by man to suppress the pest or insect pest populations.

Microbial pesticides: Living micro-organisms, such as fungi, bacteria, viruses, protozoa and nematodes, as well as metabolites produced by micro-organisms that are used in pest control. For example, *Bacillus thuringiensis*, *Trichoderma* sp. etc.

Microbiology: The branch of biology that deals with the micro-organisms.

Microbiota: The combined microflora and microfauna of an organism present in an area or space in a given time.

Microclimate: The conditions of moisture, temperature and other abiotic factors as influenced by the topography, vegetation etc.

Microenvironment: A small distinctive set of environmental conditions, such as dead animal, fallen log etc. at a habitat.

Microfauna: Refers to microscopic forms of animal life in a given area or space in a given time.

Microfeeding: Refers to forced feeding of small quantity of solution or suspension to the insect or other animal.

Microflora: Microscopic plant life of a given area or space in a given time.

Microgram: A metric weight equivalent to 1000th part of one milligram.

Micron: 1000th part of one millimeter.

Microorganism: Microscopic animals or plant life.

Microphagous: Feeding on microorganisms.

Micropterygidae: Family of suborder Zeugloptera and order Lepidoptera. These are small moths having wing expanse less than 7 mm and include the most primitive forms of all Lepidoptera. They are diurnal and the larvae feed on the pollen of flowers. A species of *Micropteryx* is known to feed on liverworts. The larva of *Sabatinca* feeds on liverworts. Though Comstock considered micropterygids to be terrestrial Trichoptera, Chapman and Hinton suggested an independent order Zeugloptera.

Microtrichia: Minute hairs projecting from the integument which are formed around cellular filaments.

Micropyle. A minute opening or group of openings into the insect egg through which the spermatozoa enter for fertilization.

Microscope: Equipment to inspect the microscopic or minute objects.

Microscopic: The object which is visible under microscope.

Microspecies (Jordanon species): One of the number of true breeding, morphologically slightly different lines within a complex of largely interbreeding forms.

***Microtermes obesi* Holmgren:** A species of field termites (Termitidae, Isoptera).

Microtrichia: See seta.

Microtype egg: The minute eggs laid in large numbers by some mites. These eggs are hatched only after ingestion by the hosts of macrotype eggs.

Midgut (Mesenteron or ventriculus): Mid part of alimentary canal, endodermal in origin, concerned primarily with the secretion of enzymes and digestion and absorption of food. In some insects, the internal surface area of mid gut is increased by the development of varying number of sac like diverticulum known as gastric or enteric caeca. These are situated at the beginning of midgut or at the junction of foregut and midgut. In cockroaches and mantids, these are six in number and wanting in Lepidopterous larvae. Unlike the foregut and hindgut, the midgut lacks intimal layer but in many insects, the thin cuticular layer is secreted periodically from the entire surface of epithelium and food particles remain enclosed in this membrane. This protective lining is termed as peritrophic membrane which is permeable to enzymes and products of digestion.

Midgut goblet cells: In the Lepidopteran larvae, the goblet cells excrete K, Ca, metals and large organic molecules from haemolymph into the midgut lumen (Nijhout, 1975).

Migrant pest: Migrant pests are non-residential and appear periodically for a short period due to migration, *e.g.*, locusts, armyworms.

Migration: Movement over great distance during which locomotion is not inhibited by food, mates or oviposition sites; an adaptation to periodically transport insects beyond the boundaries of their old reproductive sites and into new ones.

Migratory beekeeping: The moving of colonies of bees from one locality to another during a single season to take advantage of two or more honey flows.

Migratory locust: It is *Locusta migratoria* L., polyphagous insect pest, yellowish brown in colour and found scattered in many states of India.

Milbemycins: The compounds having insecticidal properties which were isolated from soil bacteria, Streptomyces hygroscopius. These are effective against mites and aphids.

Mildew: Disease caused by a fungus and characterized by the appearance of a white, mycelial growth and spores on the surface of infected plant parts.

Milk glands: Specialized accessory glands in certain adenotrophically viviparous flies, *Glossina* sp. that produce secretions fed upon by larvae.

Milk weed bug: See *Lygaeus civilies.*

Milligram: A metric weight equivalent to 0.001 gram.

Millimeter: 0.001 meter or 0.03937 inch (approximately 1/25 inch).

Millipedes: Animal belonging to the class Diplopoda. Some characteristics of millipedes are, a cylindrical, two-parted body (head and many-segmented trunk), one pair of antennae and the trunk segments each bearing two pairs of legs.

Mimicry: The resemblance of an insect to its natural surroundings, providing some protection from enemies.

Mine: Deep hole or tunnel in a plant part caused by burrowing insects or their larvae (*e.g.*, stem borers, leaf miners).

Mineral: An inorganic homogeneous substance.

Mineral spirits: It can be used as a solvent for the insecticides. It is similar to kerosene.

Mineralization: Conversion of an element from an organic form. Mineralization of pesticides most commonly refers to the microbial degradation to carbon dioxide as a terminal metabolite.

Minima: A minor worker in the ants.

Minor pest: The GEP in case of minor pest lies below both EIL and DB. Under favourable environmental conditions, the population may cross EIL and DB for usually a short interval and easily amenable to available control measures and a single application of insecticides is usually enough to prevent economic damage, *e.g.*, cotton stainers, grey weevil, thrips and mites; rice hispa and root weevil; sugarcane mealy bugs, thrips and mites; and *Spodoptera litura* (Fab.) on oilseed and vegetable crops.

Minor worker: It is same as minima. It is smallest worker subcaste in the ants.

Miridae: Family of superfamily Cimicoidea, suborder Heteroptera and order Hemiptera which includes variously coloured leaf bugs and plant bugs. Delicate small to medium sized elongated insects. Antennae, tarsi and rostrum are three segmented. Ocelli are wanting, distinct cuneus and indistinct embolium. Ovipositor well developed present in female. Mirids are primarily plant sap feeders. Examples are *Calocoris angustatus, Megacoelum stramineum* on sorghum ears, *Pachypeltis politus* on betelvine, *Helopeltis antonii* on cashew, cacao, guava, neem, tea, grapevine etc.

Mirror: a membrane in the sound producing organ of cicadas (Hemiptera).

Miscible: That can be mixed.

Miscible liquid: A formulation in which the technical product is dissolved in an organic solvent which is then dissolved in the water carrier.

Miscogasteridae: Family of super family Chalcidoidea, suborder Apocrita and order Hymenoptera. These are parasites of dipterous insects. *Ecrizotomorpha taskhiri* is either a hyper-parasite or secondary parasite of the linseed gall midges *Dasineura lini. Systasis dasineurae* is also parasitic on *D.lini. Hypolixus truncatulus* is parasitized by *Dinarmus sauteri. Bruchobius colemani* and *B.laticeps* are parasites *Bruchus* sp., *Anysis alcocki* and *A.saissetiae* are parasites of scale insects. The grub of *Caryoborus gonagra* is parasitized by *Oedaule stringifrons.*

Misdiagnosis: Error or mistake in deciding the fact that which pest has caused the damage.

Missing link: A hypothetical form of animal believed to have existed in the evolutionary process intermediate between two closely related groups of animals.

Mist: Water droplets formed as a result of condensation of water vapour on the dust particles. It also refers to a pesticide formulation in which the toxicant comes out of the container in the form of fine droplets that condense on the target.

Mist blower: Sprayer producing a fine air carried spray.

Mite: Tiny eight legged animal with body divided into two parts (cephalothorax and abdomen). Mites belong to class Arachnida, these are allied to ticks and found in many habitats, like the soil, decayed organic matter and parasitic on both plants and animals.

Miticide (Acaricide): Pesticide that can kill mites, ticks and spiders, preferably called acaricide.

Mitochondrial pump: The plasma membrane of the papillae is deeply infolded and associated with mitochondria to form mitochondrial pumps which are concerned with active and selective absorption of ions.

Mitochondrion (pl., mitochondria): An organelle that is the principal site of energy in the cell.

Mitosis: The series of changes through which the nucleus passes during ordinary cell division and by which each of the daughter cells is provided with a set of chromosomes similar to that possessed by the parent cells.

Mitotic poison: A chemical that disrupt the cell division.

Mitox: Trade name of insecticide chlorbenside.

Mixed cropping: Growing of two or more crops simultaneously on the same piece of land in the same season without any row arrangement.

Mixed function oxidase (MFO): An enzyme that metabolizes many foreign substances.

Mixed nest: A nest having colonies of two or more species of social insects in which both the adults and the brood occur.

Mixtures: When two or more substances are added together, they form a homogeneous or heterogeneous system in which there is no chemical reaction.

Mnesarchaeidae: Family of super family Eriocranioidea, suborder Montrysia and order Lepidoptera. Represented in the New Zealand.

Mode of action: Manner in which an insecticide kills or prevents the insects.

Mode of transmission: The insects often transmit the pathogens at the time of sucking the blood from the animals and cell sap from the plants, primarily in three ways: (i) Mechanical transmission, (ii) Biological transmission and (iii) Myiasis. See details.

Model calibration: Testing of a model with known input and output information for adjustment or estimation of factors for which data are not available.

Model, computer: Assembly of numerical techniques (algorithms), book keeping and control language (*i.e.*, the computer programme) comprising a mathematical model and which carries out acceptance of input data and instructions through delivery of output.

Model ecosystem: Man made study system containing associated organism and abiotic components that is large enough to be representative of a natural ecosystem, yet small to be experimentally manipulated. There is some subjective differentiation between larger, out door model ecosystems (mesocosms) and smaller, generally indoor ecosystems (microcosms).

Modulus: It is the constant factor or multiplier for conversion of unites from one system to another.

Molar: Refers to rough grinding surface of the mandibles.

Molar solution: A solution containing 1 g mole of solute per cubic decimeter of the solution.

Molarity: It refers to g moles of a solute per litre of the solution.

Mold: A fungus making its growth in damp or decaying areas on living things. Same as mould.

Mole cricket: *Gryllaotalpa africana*. It is a major pest of potato. It burrows in the field, bores into tuber, usually infests damper parts of the field, both nymphs and adults cause damage.

Mole: The unit of the substance in g molecular weight of that substance.

Molecular biology: A branch of modern biology in which biological phenomena and processes are studied not only from the phenomenological point of view but by physical, chemical and biochemical investigation at the molecular level.

Molecular weight: The molecular weight of a substance is that number which represent that how many times one molecule of that substance is heavier than the weight of 1.008^{th} part of one hydrogen atom or weight of $1/12^{th}$ part of one carbon 12 atom, so molecular weight of oxygen is 32.

Molecule: Molecule is the smallest particle of element or compound which can exist independently in the nature. They are made up of atoms.

Mollusca: A phylum of animals covering snails, slugs, oysters etc.

Molluscicide: A pesticide used to kill or control snails and slugs.

Molluscs: Any animal belonging to phylum Mollusca.

Molya nematode: See *Heterodera avenae.*

Moniliform: Bead like antennae with spherical or rounded segments, *e.g.*, termites.

Monitor: See methamidophos.

Monitoring system: A regular system of checking up the pesticide residues whether they are escaping into the environment.

Monocondylic joint: It is a joint with a single point of articulation between the adjacent leg segments.

Monocot: A seed plant having a single cotyledon (monocotyledon) or leaf, *e.g.*, grasses, wheat, corn, palms etc.

Monocropping: The practice of growing the same crop year after year in the same piece of land.

Monocrotophos ($C_7H_{14}O_5NP$): Dimethyl-cis-1-methyl-2 methylcarbamoylvinyl phosphate, colourless, crystalline solid with mild smell, miscible in water, acetone and alcohol. It is a systemic insecticide with stomach and contact action. It is highly toxic to bees but at normal rate of application less toxic to fishes. It is hydrolyzed in plants and animals to non-insecticidal product but oxidative product hydroxymethyl amide is as toxic as parent compound. Commercially sold under trade name Azodrin, Nuvacron or Monocil is a highly toxic organophosphate insecticide used to control the sucking insect pests.

Monoculture: Cultivation of a single crop species over large area without provision for diversity or use of the land in any other way, *i.e.*, polyculture.

Monoecious: Having both male and female organs in the same plant.

Monogamy: Mating with a single individual.

Monogenic: Characteristic governed by a single gene such as resistance. In this case, the level of resistance may be very high. The well known example is 200 fold resistance to organophosphate insecticides in spider mite. See polygenic resistance also.

Monognatha: Forms having single jaw only.

Monohybrid: It is a cross between two forms of single trait, like a cross between tall plants and dwarf plants.

***Monomorium indicum* Forel:** Belongs to order Hymenoptera and family Formicidae. It is an insect pest of household, commonly known as the black ant.

Monophagous: Feeding upon only one kind of food, for example, one species or one genus of plants.

Monophasic allometry: It refers to the polymorphism in which the allometric regression line has a single slope. In ants, it implies that the relationship of some of the body parts measured is nonisometric.

Monophyletic: Of individuals derived in the course of evolution from a single interbreeding population or phyletic stock (as opposed to polyphyletic).

Monophyly: The formation of a taxon through one or more lineages from one immediately ancestral taxon of the same lower rank.

Montrysia: Suborder of order Lepidoptera. Adults do not posses functional mandibles; maxilla without lacinia; larva possesses not more than seven pairs of abdominal legs; female with one (rarely two) genital aperture behind sternite IX; male without a saccus on sternite IX; wing nearly always aculeate.

Morbidity: The relative incidence of disease or the state of wasting away.

Morestan: Trade name of an insecticide, oxythioquinox.

Moribund: The insects with no sign of recovery; dying or near to death.

Moriculture: The mulberry silk worm, *Bombyx mori* mainly feed on green leaves of mulberry plants belonging to genus *Morus* under the family Moraceae. In India, out of about 20 species, *M. alba, M. indica, M. serrata* and *M. latifolia* are generally cultivated as food plant of silk worms which is known as moriculture.

Morocide: Trade name of binapacryl.

Morph: Any of the individual variants that accounts for polymorphism.

Morphology: Gr., *morph*= form. The science of structure of the organism as a whole.

Mortality: Death rate or numbers dying per unit of time.

Morula: The mass of cells formed by holoblastic cleavage of the egg.

Mosaic: A plant disease caused by virus which is transmitted by aphids showing discoloured and curled plant leaves.

Mosaic evolution: The evolution involving unequal rates for different structures, organs etc. of the phenotype.

Mosaic theory: See apposition image.

Mosaic vision: The type of vision of the compound eye of insects; each ommatidium conveys a single point of light to each retinula, which points as a whole, combine to produce a single erect optical image.

Mosquito hawk: See dragonfly.

Mosquito net: A fine net or screen used to keep out mosquitoes.

Mosquito transmitted disease: The malaria, filariasis, encephalitis, yellow fever and rift valley fever are transmitted by the mosquitoes. See details.

Mosquito: Any of various two winged insects of the family Culicidae (order Diptera). In most species the female is distinguished by a long proboscis for sucking blood. Some species of mosquitoes are vectors of diseases such as malaria and yellow fever. Also see *Anopheles*, *Aedes* and *Culex*.

Moth: Any of numerous insects of the order Lepidoptera. Moths are generally distinguished from butterflies by their nocturnal activity, hair like or feathery antennae, stout bodies and the frenulum that holds the front and back wings together (See also, butterfly).

Motor neurones (Efferent neurones): The soma of these neurones are situated in the ganglia and are of monopolar, devoid of dendrites. Their axons form the motor nerve and innervate various organs of the body, while collaterals merge into the neuropile. They conduct impulses from the central nervous system to the effector organs.

Mottle: A pattern of indistinct light and dark areas on plant tissue. It is usually a symptom of virus diseases.

Mould: A popular name for any of numerous small fungi. It grows as hyphae interwoven into an extensive mycelium. Moulds usually occur on decaying matter or on plant surfaces.

Mould toxin: See mycotoxin.

Moult (ecdysis): To shed the outer covering or cuticle of the body. See moulting.

Moulting: Moulting or ecdysis is the act of casting off the outer layers of the integument (cuticle) and secreting new cuticle to accommodate the growth of body. It is a complex process involving the separation of old cuticle from the hypodermis (apolysis) followed by shedding of old cuticle (ecdysis). Immediately after ecdysis, tanning of cuticle (sclerotization) takes place.

Moulting fluid: This is a thin layer of liquid which lies between the newly developing cuticle and the old cuticle that will be cast off. This liquid contains a mixture of enzymes (chitinase and proteinase) secreted by the epidermal cells which will digest the substance of endocuticle but will not touch the cuticulin of the epicuticle or the sclerotin of the exocuticle. Thus the new cuticle is protected from its action by the epicuticular covering.The enzymes of moulting fluid appear to be secreted from the general surface 'of the new cuticle; perhaps they are discharged at the tips of the pore canals and products of digestion are likewise absorbed through the general surface which at this stage is still quite permeable.

Moulting gland: See prothoracic glands.

Moulting hormone: Ecdysone, the hormone that stimulates moulting. See ecdysone.

Mound building termite: A group of termites which have tendency to make mounds of different shapes and sizes.

Mound nest: A mound nest is a living place of termites, which is constructed of a mound of soil or certain material that project above the ground surface. The mound is often elaborate, specific in plan to the species and adapted to contribute to the control of microclimate within the nest.

Mouth: The anterior opening of the stomodaeum, primarily located in the ventral wall of the head, but in most sucking insects retracted into the head with the transformation of the cibarium into a sucking pump, in which the functional mouth is the entrance to the pump chamber.

Mouth aspirator: Device used for capturing the small insects by sucking through a mouth piece and drawing insects into a glass vial.

Mouth cavity: The area enclosed by the labrum-epipharynx, mandibles, maxillae and labium-hypopharynx.

Mouth hooks: The paired hardened portions of the mouthparts of fly larvae.

Mouth parts: A collective name including labrum, mandibles, maxillae, labium and hypopharynx, are -also called the feeding organs.

Movable frame hive: For the culture of honey bees, special types of frame hives are used. Langstroth (1851) designed first the movable frame hive for artificial colonization of *A. mellifera* and production of honey and wax on a large scale. The movable frames are suspended at equal distance in the hive boxes. The hive boxes are two in number, the lower one is brood chamber and the upper one is honey chamber. The space between the adjacent frames is kept more than 0.47 cm so that the bees can construct new combs and the bee space of 0.96 cm is maintained between the combs through which the bees can move freely between the combs.

Mucro: A non-articulated projection arising from the inner apex of the tibia.

Muga silk: A golden coloured silk obtained by spinning the cocoons produced by muga silkworms, *Antherea assama*.

Muga silkworm: *Antherea assama* producing the muga silk. It feeds on sum and suala tree leaves and its rearing is common in Bihar, West Bengal and Orissa.

Mulch: Material used to form a covering over the soil, usually to prevent evaporation of soil moisture. For example, rice straw can be used as a mulch to cover the soil between individual vegetable plants.

Mullerian mimicry: A group of species belonging to different genera, often different or even orders, having similar colours, possessing more or less distasteful species living in the same locality.

Multicolonial: The population of social insects is divided into colonies that recognize nest boundaries.

Multidose poisons: These are chronic poisons such as anticoagulants like warfarin.

Multiparasitism: It is phenomenon of simultaneous parasitization of a host individual by two or more different species of primary parasites at the same time, *e.g.*, parasitization of a lepidopteran egg by two primary parasitoids namely, *Trichogramma* sp. and *Telenomus* sp.

Multiple resistance: Multiple resistance is the coexistence of different defense mechanisms in the same strain. The main difference between the cross and multiple resistances is that in the former, the resistance against selective agent and other compounds is developed by the same mechanism whereas in the latter, the resistance to different insecticides is developed by different mechanisms. The multiple resistance is developed due to simultaneous or consecutive use of several insecticides under field conditions.

Multiresidue method: Analytical method which measures a number of pesticide residues simultaneously.

Multivoltine: Insects which completes two or more generations annually.

Mung dhora: See *Callosobruchus analis.*

Murine typhus fever: The oriental rat flea transmits bacterium like microorganisms, *Rickettsia mooseri* from rat to man which is responsible for causing murine typhus fever in man.

***Musca domestica* Linn.:** A species of house fly which occurs in the temperate climate found in the Himalayas and other places.

***Musca nebulo* Wiedemann:** Belongs to order Diptera and family Muscidae. It is an insect pest of household, commonly known as the housefly.

Muscardine: See silkworm diseases.

Muscidae: Family of suborder Cyclorrhapha, order Diptera. Small to medium sized Diptera, with fleshy proboscis. Oral vibrissae are present and the adults are characterized by bristles on the mesonotum. Apical cells of wings are not narrowed distally. Larvae are typical maggots with pseudopodia, metapneustic and hemispherical stigmata. Muscids are the houseflies and are agents in the transmission of diseases and a source of nuisance to man. The adults help to carry bacteria from infected to a non-infected person. *Musca nebula* and *Musca domestica* are common species.

Muscle: Fleshy fibres that serve to move the appendages and other organs of the body.

Muscoid type mouth parts: Same as sponging type mouth parts.

Museum: A building, room, etc. for keeping animal specimens or their parts, and other objects of natural history etc.

Mushroom bodies: The protocerebrum includes the protocerebral lobes and the optic lobes. It is bilobed ganglion, connected internally by a transverse commissure, the central body. Also situated in this region is a peculiar structure comprising numerous fibres and called the mushroom body or corpora pedunculata. It is associated primarily with the integration or sensory information.

Mustard aphid: See *Lipaphis erysimi* (Kalt.).

Mustard sawfly: See *Athalia lugens proxima* Klug.

Mutagen: A material that induces genetic changes in the form of mutation. Mutations may be brought about by chemical or physical agents (mutagens) or as a result of an accidental change in the genes.

Mutagenic: Capable of producing mutations.

Mutant: An individual having abnormal structure and behaviour in which it differs distinctively from the type and which has genetic rather than environmental origin so that there is chance of transmission to the offspring.

Mutation: Any detectable sudden heritable change in the genetic material not caused by genetic segregation or genetic recombination, which is transmitted to daughter cells and even to succeeding generations giving rise to mutant individuals provided it does not act as dominant lethal factor.

Mutation theory: Postulates a sudden variation in some inheritable characteristics of an individual animal as distinguished from variation resulting from generations of gradual changes.

Mutillidae (velvet ants): Family of super family Scolioidea, suborder Apocrita and order Hymenoptera. Mutillidae are densely pubescent and mostly brightly colored with red, orange and yellow markings. The apterous ant-like females may inflict a painful sting. The males are mostly winged and in others wings may be wanting.

Mutualism: A symbiotic relationship between two different species in which both derive benefit from each other and their interaction is usually obligatory for their survival in the nature.

Mycangium: These are pocket shaped receptacles used to carry symbiotic fungi in the scolytid beetles.

Mycelium: The vegetative structure of a fungus, made up of hyphae.

Mycetocyte: A cell containing symbiotic microorganisms, scattered throughout particularly within the fat body or aggregated in organs called mycetomes.

Mycetome: See mycetocyte.

Mycology: The study of fungi.

Mycoplasma: A micro-organism intermediate in size between viruses and bacteria. Mycoplasma possesses many virus like properties.

Mycotoxin: Refers to the toxins produced by the moulds. Of the various such toxins, the aflatoxins produced by the fungus, *Aspergillus flavus* are highly toxic, carcinogenic to mammals and wide spread contaminants of stored foods, particularly ground nut, maize, cotton seed, pistachio etc.

Myiasis: Infestation of the body by the larvae of flies. Infection by lodging the living Dipterous larvae in the open body cavities by man and domestic animals is called myiasis. The species are either parasitic or saprophytic. They cause primary myiasis when it is induced by true parasites or secondary myiasis which is induced by saprophagous larvae. The latter only follow diseased condition or wounds and usually where there is microbial infection. House fly produces facultative myiasis while bot flies, the obligatory one. Myiasis can be grouped into (a) The human myiasis which may be (i) cutaneous (caused by *Hypoderma* and *Gasterophilus*), (ii) infracranial (larvae of *Oestrus*, *Rhinooestrus*, *Gasterophilus* and *Dermatobia* invading orbital, nasal or auditory cavities; when secondary, it is due to larvae of *Sarcophaga*), (iii) intestinal (larvae of atleast 18 genera are associated; and (b) mammalian myiasis: Oestridae larvae occur in the nasal and pharyngeal cavities and beneath the skin.

Mylabris phalerata **Pallas:** Belongs to order Coleoptera and family Meloidae. It is an insect pest of ornamental plants, commonly known as the banded blister beetle.

Myllocerus discolor **Boheman:** An insect pest of order Coleoptera, family Curculionidae which infests apple.

Myllocerus lactivirens **Marshal:** The almond weevil (Coleoptera; Curculionidae), besides feeding on almond, reported to be feeding on apple, apricot, ber, citrus, falsa, loquat, mango, peach, plum, pear and pomegranate.

Myllocerus maculosus **Faust:** Belongs to order Coleoptera and family Curculionidae. It is an insect pest of cotton, ornamental plants, commonly known as the cotton grey weevil.

Mymaridae (fairyflies): Family of super family Chalcidoidea, suborder Apocrita and order Hymenoptera. Mymaridae are extremely minute (less than 1.0 mm long) black or yellowish insects which are parasitic on eggs of insects. *Alaptus magnanimus* measures 0.21 mm long. The eggs of certain aquatic Hemiptera are parasitized by some mymarids. Both sexes of *Polynema natans* swim beneath the water using their wings and parasitise, *Notonecta* sp. and *Anaphes* sp. are egg parasites of *Hypera postica*.

Myoneural junction: The site of nerve impulse transmission from a nerve to a muscle.

Myophily: Transfer of pollens by flies, a type of pollination.

Myriapoda: Arthropod class comprising the millipedes and centipedes.

Myriapods: Arthropods belonging to class Myriapoda.

Myrmecodomatia: Structures of higher plants that evolved in the course of mutualistic evolution to serve as dwelling places for ants.

Myrmecoladae: Family of order Strepsiptera. They are parasites of males of Formicidae and female orthopteroids. The males possess seven-segmented antennae (four very short, five to seven very long) and tarsi with less than five segments. Claws are absent. In the female, the cephalothorax has hood-like projections behind the spiracles.

Myrmecology: It refers to the scientific study of the ants.

Myrmecophile: An organism that must spend at least part of its life cycle with ant colonies.

Myrmecophilus: Insects other than ants which live in ant hill. It also refers to the plants that live symbiotically with the ants or pollinated by them.

Myrmecophily: Transfer of pollens by ants, a type of pollination.

Myrmecophites: Refers to higher plants that live in obligatory, mutualistic relationship with the ants.

Myrmeleontidae: Family of superfamily Myrmeleontoidea, suborder Planipennia, order Neuroptera. Adults with long, narrow, many veined, clear or spotted wings. Abdomen long and slender. Body soft, antennae clubbed. Larvae called antlions, conceal themselves at the bottom of a conical pit in sand and feed on ants that fall into it.

Myrmeleontoidea: Superfamily of suborder Planipennia, order Neuroptera. Antennae stout, thickened at apex, mandibles strongly developed in the larvae, stout and curved. Body with dolichasters.

***Mythimna separata* (Wlk.):** Belongs to order Lepidoptera and family Noctuidae. It is an insect pest of wheat, commonly known as the army worm.

***Myzus persicae*:** Scientific name of green peach aphid.

N

Nacoleia octasema **(Meyr.):** Belongs to order Lepidoptera and family Pyralidae. It is an insect pest of banana, commonly known as banana scab moth.

Naiad: Aquatic gill bearing nymph of an insect or larva of a dragonfly or damselfly (Odonata) which are aquatic.

Nalad ($C_4H_7Br_2Cl_2P$): An organophosphate insecticide, acaricide and fungicide; chemically 1,2-dibromo-2,2-dichloroethyl dimethyl phosphate, moderately toxic to mammals; acute oral LD_{50} for rat is 430 mg kg^{-1}; introduced by Chevron Chemical Company and marketed under trade name Dibrom; effective against a wide variety of fruits, vegetables, flowers and green house insect pests. The pure material is white crystalline with melting point 25 °C, insoluble in water but highly soluble in organic solvents.

Nanitic worker: The dwarf workers produced from either first or last brood of ants as a result of starvation. Nanitic workers occur both in monomorphic and polymorphic species.

Nano: A prefix denoting one thousand millionth, *i.e.*, 10^{-9}.

Naphtha: Naphtha is a hydrocarbon fraction of crude oil, especially valuable as a raw material for chemicals and gasoline.

Naphthalene ($C_{10}H_8$): It is a coaltar derivative used as a fumigant against cloth moths and carpet beetles, relatively non-toxic to mammals and formed by high temperature breaking up of large molecules of petroleum. It is a white crystalline compound vaporizes slowly to form an inflammable gas. The gas has specific gravity of 4.4. It is also used to a limited extent as a soil fumigant for wireworms and as a fumigant in green houses to control spider, mites and thrips. It is generally available as moth balls and moth flakes etc.

Narcotic: The substances producing sleep, stupour or insensibility, *e.g.*, morphine, heroin, cocaine etc.

Narrow spectrum insecticide: An insecticide effective against a narrow range of insect pests.

Nasal bot fly: See *Cephalopina titillator.*

Nasus: It is a snout like projection in soldiers of Nasutitermitinae which secretes poisonous substance on the intruders of termite colonies.

Nasute soldier: A soldier of termite possessing nasus.

Nasutus (pl., nasuti): A soldier caste in the termites that bears median frontal rostrum through which a defensive fluid is secreted; the jaws are small or vestigial.

Natality: Birth rate.

Natatorial: Swimming legs, *e.g.*, in aquatic bugs, beetles etc. The modifications, however, differ among different groups of insects. The coreids bear foliate processes on the hind femora and tibiae. The aquatic bugs can swim with the help of long hairs developed on the mid and hind legs. In the Notonectids, only the hind legs are natatory. In gerrids, the middle tarsi bear complex plumose setae enabling them to swim on water. In the beetls, the natatorial modifications are variable. In the Dytiscidae, the hind tibia is provided with a dense fringe of hairs or setae serving as an oar at swimming. In gyrinidae, the mid and hind legs are natatory and modified into short paddle like structures. The hind legs in aquatic beetles can fold back along their accessory hair and even can rotate so that the animal can propel forward.

Natural: Refers to occur in the nature; not artificial or synthesized.

Natural control: The collective action of environmental factors to maintain a pest population size at an acceptable level over a period of time.

Natural enemy: An animal or plant which kills or injures another animal or plant. For example, predators, parasites and pathogens that attack pests.

Natural group: A group of organisms having common ancestor.

Natural history: Life and behaviour of animals in their natural surroundings.

Natural host: A host in which the pathogenic micro-organism or parasite or an insect is commonly found and in which it can complete its development.

Natural pesticide: Naturally occurring substances of botanical, geological or zoological in origin, *e.g.*, rotenone, pyrethrins, neem oil, avermectin etc. which are used in the pest control.

Natural selection: A theory of the mechanism of evolution which states the survival of best adapted forms with the inheritance of fitness determining characteristics which arise as random variation. Selection among a group of animals or plants is done by the forces of nature. It allows that of the group best fitted to survive in the particular environment to live and reproduce, while those not fitted to that environment will die. By this means, the species or group gradually adapts to the environment as poorly adapted individuals are gradually eliminated over many generations.

Nausea: It is a feeling or inclination towards the vomiting.

***Nausinoe geometralis* (Guenee):** Belongs to order Lepidoptera and family Pyraustidae. It is an insect pest of ornamental plants, commonly known as the jasmine budworm.

NCDC: National Cooperative Development Corporation was set up with the Agricultural Produce Act, 1956; latter replaced by the National Cooperative Development Corporation Act, 1962 with the objective of food grain storage in India.

Neala: Jugal area of the wing.

Nebulisation: Formation of an aerosol of very small liquid particles (fog) or solid particles (smoke) from a pesticide formulation, generally for fumigation of an enclosed space such as a glass house.

Necrophoresis: It refers to transport of dead members of the colony away from the nesting site.

Necrophoric behavior: See necrophoresis.

Necrosis A localised and rapid destruction of a cell or group of cells or death of part of a plant, *e.g.*, rot and canker are necrotic symptoms.

Necrotic: Having symptoms characterized by the death or disintegration of cells or tissues.

Necrotising archnidism: Progressive ulceration and loss of skin around the site of a bite by a number of araneomorph spider species. Its cause is uncertain and for this reason no fully effective treatments are available at present time.

Nectar: A sweet liquid secreted by the nectaries of plants; the raw product of honey. It is a sugary liquid and mainly secreted by flowers.

Nectar guide: The colour marks on flowers believed to direct the insects to nectar sources.

Nectary (pl., Nectaries): The organs of plants which secrete nectar, located within the flower (floral nectaries) or on other portions of the plant (extra-floral nectaries).

Neelidae: A family of order Collembola. Antennae shorter than head. For example, *Neelus* sp.

Neem: A tree species, *Azadirachta indica* in the tropical and subtropical parts of the world. It is a source of toxic alkaloids including azadirachtin used in the pesticides.

Neem oil: It is the oil extracted from seed kernels of *Azadirachta indica;* used in controlling many insects as it possesses phagorepellent effect. It is also used in the soap industry and pesticide formulations.

Neem seed kernel: Seed kernel of neem, *Azadirachta indica* used to extract oil, azadirachtin and other alkaloids used in the pesticide formulations.

Negligible residue: A tolerance which is set for a food or feed crop which contains a very small quantity of pesticide at harvest as a result of indirect contact with the chemical.

Neguvon: It is the trade name of insecticide, trichlorfon.

Neididae (Berytidae): Family of superfamily Lygaeoidea, suborder Heteroptera and order Hemiptera. Small group of bugs with mosquito like appearance bearing long legs and slender wings, femora are apically dilated, *e.g., Gampsocoris delhiensis* infesting the growing leaf tips of bottle gourd.

Nemathelminthes: A phylum that includes roundworms.

Nematicide: Toxicant effective against nematodes (eelworms).

Nematocera: Suborder of Diptera, possesses antennae longer than the head and thorax and are usually many segmented. Arista is wanting. The palpi are usually 4 or 5 segmented and are pendulous. Larvae possess a well developed exserted head and horizontally biting mandibles.

Nematodes: Small unsegmented worms injurious to crop plants, man and other animals (phylum, Aschelminthes).

Nematology: A branch of zoology concerning study of nematodes including their relationship with plants and animals.

Nemestrinidae: Family of suborder Brachycera, order Diptera. Large flies with simple third antennal segment and mouthparts with very long proboscis, frequently many times longer than the body. Frequent flowers to suck nectar. *Hirmoneura annandalei* and *H.brunnea* are Indian species.

Nemopteridae: Family of superfamily Myrmeleontoidea, suborder Planipennia, order Neuroptera. With long ribbon like hindwings, head prolonged into a rostrum. Larvae with elongate, narrow prothorax and reduced meso and metathorax.

Nemouridae: A family of order Plecoptera. A transverse vein connecting R to Cu near middle of wings is present and distal portion usually without cross veins

***Neomaskellia bergii* (Signoret):** Belongs to order Hemiptera and family Aleyrodidae. It is an insect pest of sugarcane, commonly known as the sugarcane spotted fly.

Neometabola: A hemimetabolous development of insects in which they do not undergo any considerable inner metamorphosis except degeneration of parts. A retardation of development of external rudiment of wings gives rise to completely wingless nymphs. It can be divided into four subtypes: (i) Homometabola: Winged females of Chermisidae and Phylloxeridae (Homoptera) in which wing rudiments appear only in the last nymphal instar, (ii) Remetabola: The Thysanoptera with two apterous nymphal stages followed by a pronymph and one or two nymphal stages with wing pads, (iii) Parametabola: The adult male of Coccidae is incapable of feeding. Two successive feedings occur at larval stages and are without wing rudiments, moreover, their development is followed by a pronymph and a nymphal stage, (iv) Allometabola: There are four apterous depressed larval stages of which the first alone is a free larva, *e.g.,* Aleyrodoidea.

Neonicotinoids: Imidacloprid, a novel insecticide derived synthetically from a nitromethylene insecticidal chemical, nithiazine, is closely related to the nicotinoids in mode of action and structure activity relationship. Latter, several insecticides resembling imidacloprid appeared which are collectively called neonicotinoids.

Neoprene: Synthetic rubber used to manufacture the hand gloves and boots that provide protection during working with pesticides.

Neopseustidae: Family of super family Eriocranioidea, suborder Montrysia and order Lepidoptera.This family is represented in India and Formosa by three species.

Neoptera: Insects capable of flexing or folding the wings.

***Neotermes bosei* (Synder):** A termite species found in the North India and attacks the branch wood of several trees including mango, Ficus etc. The tunnels go right into the heart of wood.

Neotypes: A specimen is designated when the original types are known to have been destroyed or were suppressed by the commission.

***Nephantis serinopa* Meyrick:** Belongs to order Lepidoptera and family Cryptophagidae. It is an insect pest of coconut, commonly known as the black-headed caterpillar.

***Nephopteryx eugraphella* Rag.:** Belongs to order Lepidoptera and family Pyralididae. It is an insect pest of sapota, commonly known as the sapota leaf webber.

***Nephotettix virescens* (Distant) *and N. nigropictus* (Stal):** Belongs to order Hemiptera and family Cicadellidae. It is an insect pest of cereals and millets, commonly known as the green leaf hopper.

Nephridium (pl., Nephridia): These are tubular excretory structures in insects.

Nephrocytes: These are accessory organs of excretion. They form two major groups: a dorsal group of pericardial nephrocytes and a ventral group of infra-intestinal nephrocytes. The dorsal nephrocyes are arranged in long chain at either side of heart in the pericardial sinus in most of the insects. The ventral nephrocytes form a pair of chain which run along the ventral surface of intestine and finally are attached to the salivary glands which occur in the Dipteran larvae.

Nepidae: This is a family of division Cryptocerata, suborder Heteroptera and order Hemiptera. The antennae of these predaceous aquatic bugs are three segmented. The forelegs are raptorial, whereas, the hind legs are adapted for walking, tarsi one segmented and clawas are absent. The most common water scorpion belongs to genera Ranatara and Lacotrephes. The water scorpions inflict painful bite to man when handled.

Nereistoxin: Nereistoxin was isolated from a marine annelid, *Lumbrineris heteropoda* by Nitta in 1934 but its insecticidal properties were discovered by Sakai (1964). Nereistoxin reduces the amount of ACh released from the presynapses and also has inhibitory effect on ACh receptors.

Nerve: An aggregation of neurones form ganglion and collection of several neuronic strands form a nerve.

Nerve cord: Cord composed of delicate filaments of tissue, functioning to transmit stimulus to or from a ganglion or from or to any part of the body.

Nerve fibre: The axon or other branches of a neurocyte.

Nerve impulse transmission: Nerve impulses and their propagation along the axon are resulted due to changes in the conductance of the nerve membrane to Na^+ and K^+ ions. Na^+ ions should be more concentrated outside than inside (ENa= +60mV) and more K^+ ions inside than outside (Ek= -90mV). In membrane depolarization, which resulted due to opening of the ionic channels of the post synaptic receptors due to binding of the transmitter molecules at the chemical synapse, the membrane becomes more permeable to Na^+, *i.e.*, rising phase of the spike. This transient increase is followed by decrease in Na+ conductance associated with an increase in K^+ conductance. Both activities initiate repolarizing phase of the action potential. The presence of Na^+, K^+ stimulated ATPase activity and the Na^+, K^+ exchange pump in axon membrane has been reported in insects. Synaptic transmission in insects is facilitated by three systems: (i) Cholinergic system, (ii) Glutaminergic and GABAminergic system: At the neuromuscular junction, L-glutamic acid and gamma-aminobutyric acid (GABA) are known to occur as the excitatory and the inhibitory transmitters, respectively. The glutamic acid is obtained from the Kreb's cycle (alpha-ketoglutaric acid) or from glutamine while GABA from kreb's cycle but *via* succinic acid and succinic semialdehyde or from glutamic acid. (iii) Aminergic system: Biogenic amines such as dopamine (DA), noradrenaline (NA), adrenaline (A) and serotonin (5-hydroxytryptamine or 5-HT) are detected in the central stomatogastric nervous systems, neurohaemal organs, salivary glands, Hymenopteran venom apparatus and 5-HT in the haemocytes in cockroaches. Octopamine acts as modulator at the neuromuscular junction while dopamine and 5-HT as the neurotransmitters or neuromodulators in the CNS, visceral muscles and salivary glands. They stimulate adenylate cyclase activity and increase levels of cyclic AMP in the target cell.

Nerve poison: Poison which affects the nervous system and render the insect to behave abnormally leading to death is called nerve poison. The toxicant gets dissolved in the tissue lipoids and inhibits the production of acetylcholinesterase enzyme in insects and mammals, *e.g.*, organophosphate and organochlorine compounds.

Nerve tract: A strand of nerve fibres; usually applied to tracts within a nerve centre.

Nerve trunk: A bundle of nerve fibres in the peripheral system, the usual nerves.

Nervous system: The nervous system constitutes the connecting links between various sensillae responsible for receiving stimuli from the environment on the one hand and different organs on the other hand. Thus, the nervous system coordinates various activities. The nervous system constitutes large number of nerve cells called neurones. Each neurone consists of cell body called perikaryon with a nucleus and protoplasmic extensions. The largest protoplasmic extension arising from perikaryon is axon while the shorter ones are known as dendrites. The axon gives out the side branches known as collaterial near its origin from cell body. The axon collaterials and dendrites dorsally divides into fine fibrils called terminal arborizations. The dendrites are specialized for reception of stimuli which are then conducted through the axon.

Nervures: These are sometimes referred as veins, provides rigid support to the wings.

Nest odour: The distinctive odour of a nest by which the inhabitants of the same nest are able to distinguish the nest from those belonging to other colonies or at least from the surrounding environment. In some cases, the insects (ants, honey bees etc.) can orient towards the nest by means of colony odour. The nest bees of honey bees are known as the hive aura or hive odour.

Nest parasitism: The relationship found in some termites, in which colonies of one species live in the walls of nests of second species and feed on the carton material of which they are constructed.

Nest robbing: See cleptobiosis.

Neural groove: Refers to the median ventral groove of the embryo between the neural ridges.

Neural lamella: Outer connective tissue layer of a ganglion.

Neural ridges: The longitudinal ventral ridges which are two in number in which the lateral cords of neuroblasts are formed.

Neuration: It refers to plan of the nervures in insect wings, also called veins.

Neurilemma: The nucleated sheath or nerve tissue covering the ganglia, nerve trunk and terminal branches.

Neurite: See axon.

Neuroblasts: Embryonic cells that differentiate into neural tissues.

Neurogenic: Produced by a neurone; contractions of muscles stimulated by neuronal impulse.

Neurohaemal organs: Structures that serve as storage and release sites for neurosecretory material such as the corpora cardiaca.

Neurohormones: See neurohumours.

Neurohumours: Noradrenaline, 5-hydroxytryptamine, acetylcholine and other neurotransmitters may sometimes be liberated into the circulating blood and so have a widespread effect throughout the body. They are often called neurohumours or neurohormones or neuropeptides. There is no very sharp distinction between neurohumours and neurohormones. Proctolin is the first insect neuropeptide isolated from the whole body extract of *Periplaneta americana* and amino acid sequence of a molecule has been detected.

Neurological: Pertaining to the nervous system.

Neurone, apolar: The neurone without any process.

Neurone: The entire nerve cell including all its processes. It is a basic functional unit of the nervous system; an elongated, excitable cell that carries information in the form of electrical charges. It may be (i) sensory neurone (afferent neurone): which is concerned with perception of stimuli and conducting them from site of perception to the central nervous system, (ii) motor neurone (efferent neurone): which conducts impulses from central nervous system to the muscles (effector organ) and (iii) internuncial or association or interpolated or interneurone: which link the sensory neurone with motor neurone within the central nervous system through synapse. The neurones constitute a definite nervous system in insects; both the central nervous system as well as the sympathetic system is well developed. Besides these major systems, some bipolar neurones are distributed in the peripheral region of body innervating the integumentary hair. Similarly, some multipolar neurones form a complex nerve network beneath the integument above the muscles and on the surface of the gut. All these neurones constitute the peripheral nervous system.

Neurone, bipolar: The neurone with two processes, proximal axon and a distal long dendrite.

Neurone, monopolar (unipolar): The neurone with a single process, *i.e.*, axon.

Neurone, multipolar: The neurone with a proximal axon and many distal dendrite processes. The multipolar neurones are mostly confined to the frontal and the hypocerebral ganglia.

Neurone, pseudounipolar: This type of neurone is at first bipolar but as the development proceeds, the cell processes converge until meet at one side of the cell body; then the common converged process elongates which gives rise to a T shaped structure at the end of cell body. One branch of T is being the dendrite and other the axon.

Neurone, structure: Neurone or nerve cell is a structural and functional unit of nervous system having long processes. A neurone consists of a cell body or neurocyte or soma and two types of processes- axon and dendrite (dendron). The following structures are found in the cell body (soma or neurocyte): (i) nucleus, (ii) neuroplasma, (iii) granules, (iv) mitochondria, (v) Golgi apparatus and (vi) neurofibrils.

Neuropeptide: See neurohormones.

Neurophysin: See vertebrate hormones in insects.

Neuropile: The medullary substance or mass of fibrous tissues within a ganglion.

Neuroptera: Insect order (*neuro*= nerve, *ptera*= wings) containing carnivorous insects, like dobsonflies, fishflies, alderflies, owlflies, snakeflies, antlions, doodlebugs and lacewings. They are characterized by membranous wings, chewing mouthparts, and a complete metamorphosis. Principally, there are two subdivisions, the Megaloptera including the alderflies and snakeflies and the Planipennia including the antlions and the lacewings. As a group, the Neuropteroids are characterized by the nature of their wings which are membranous and almost subequal, with more or less primitive venation and wit many accessory veins, costal veinlets and Rs generally pectinately branched. When at rest these are held over the abdomen. The mouth parts are of the mandibulate type with more strongly developed mandible in the males and the maxillary and the labial palps, five and three segmented.

Neurosecretion: Certain cells within the central nervous system contain material which stains conspicuously with certain specific stains, and has a characteristic appearance under the electron microscope, being made up of electron dense spheres 100-400 nm in diameter. This material is also carried down the axons.

Such cells are called neurosecretory cells. Those in the dorsum of brain which produce the activation hormone to initiate moulting and which have an important influence on reproduction. The neurosecretory cells are widespread in other parts of the brain, in the corpus cardiacum, the suboesophageal ganglion and throughout the ganglia of nerve cord. Some of the examples are diapause hormone in the silkworm, the colour change hormone in stick insect and *Pieris* pupae, bursicon concerned with hardening and darkening of cuticle, the diuretic hormone of *Rhodnius* and other insects, the metabolic hormones such as the hyperglycaemic hormone and the adipokinetic hormone in *Periplaneta*. Neurosecretory hormones seem mostly to be small peptides. See also neurohumours.

Neurosecretory cells: Modified neurones found throughout the nervous system, producing hormones except ecdysteroids and juvenile hormones. See neurohormones.

Neurotoxin: A substance that damages nerves or the nervous system.

Neurotransmitters: The axon ends either in a synapse or multiple synapses in contact with the processes of other nerve cells, or on the cell body of a muscle or gland. The nerves convey electrical disturbances produced by progressing cycles of depolarization of the axon membrane. At the synapse including the neuromuscular junctions, however, this electrical transmission is interrupted; a chemical substance termed a neurotransmitter is produced and this serves to set up a progressing cycle of depolarization in the adjacent nerve or in the muscle. In the central nerves system, the chief chemical transmitter set free at the synapse is acetyl choline. Gamma-aminobutyric acid (GABA) is used as an inhibitory transmitter. Noradrenaline and dopamine also probably play some role as neurotransmitter. At the neuromuscular junctions of insects (unlike vertebrates) acetylenecholine appears not to be involved, glutamate probably serves as the neuromuscular transmitter.

Neutralism: No interaction or association between two species.

News group: This is an international level discussion group. This provides the facility to consult or exchange the information on any specific subject with other internet users.

Niche: The position or status of an organism within its community or ecosystem. (i) A habitat supplying the factors necessary for the existence of an organism or species. (ii) The ecological role of an organism in a community especially in regard to food consumption.

Nicotine ($C_{10}H_{14}N_2$): A botanical insecticide derived from the leaves and stems of tobacco plant (*Nicotiana tabacum* and *N. rustica*); an alkaloid, 1,1-methyl-2-(3-pyridyl)-pyrrolidine. As an insecticide, it was used first time in 1763 in France and the pure alkaloid was isolated in 1828 by Posselt and Reimann. It does not leave harmful residue on the treated surface. The oral LD_{50} for rats is 50-60 mg kg^{-1}. It was synthesized in 1904 and appeared in the market in the name of Black leaf 40 (See nicotine sulphate).

Nicotine sulphate: Nicotine sulphate, marketed under the trade name black flag-40 is a moderately toxic insecticide useful for controlling aphids and other insect pests of kitchen gardens and green houses.

Nicotine, metabolism: Nicotine is metabolized to cotinine in plants, animals and man. The route of cotinine formation is apparently through the immediate hydroxylation product, 2-hydroxynicotine. Cotinine is further metabolized to desmethylcotinine.

***Nictotiana tabacum* L.:** A plant species (tobacco) which is source of nicotine.

Nidi: Groups of regenerative cells located at the bases of midgut epithelial cells.

Nidus: Place in which the insects deposit their eggs.

***Nilaparvata lugens* (Stal):** Belongs to order Hemiptera and family Delphacidae. It is an insect pest of cereals and millets, commonly known as the brown plant hopper (BPH). It is a small cicada that can destroy rice crop by causing hopper burn.

Nimbin: This component of neem, *Azadirachta indica* A. Juss have been found to have antiviral activity. It affects potato virus X, vaccinia virus and fowl pox virus. It could perhaps open the way to control these and other viral diseases of crops and livestock. Nimbidin is a primary components of bitter principles obtained when neem seeds are extracted with alcohol. It occurs in sizeable quantities-about 2 per cent of the kernel.

NIOSH: Abbreviation stands for 'National Institute for Occupational Safety and Health'.

***Nipaecoccus viridis* (Newstead):** Belongs to order Hemiptera and family Pseudococcidae. It is an insect pest of tamarind, commonly known as the citrus mealy bug.

Nit: The egg of a louse often glued or attached to a feather or hair of the host.

Nitidulidae (sap beetles): Family of superfamily Cucujoidea, suborder Polyphaga and order Coleoptera. These are small, elongate or oval insects with tarsi 5-5-5 rarely 4-4-4. The fore-and midcoxae are transverse with exposed trochantins. The elytra are short and expose the terminal one or two abdominal segments. Functional spiracles in the abdomen are six (rarely five).

Nitrification: Refers to conversion of atmospheric nitrogen into nitrates by the bacteria.

Nitrogen cycle: The nitrogen in the protoplasm is break down from organic to inorganic form by a series of decomposer bacteria, each specifying for a particular part of the job. Some of this nitrogen ends up as nitrate, the form most readily used by green plants (although some organisms can use nitrogen in other forms) thus completing the cycle. The air which contains 78 per cent nitrogen is the greatest reservoir and safety valve of the system. Nitrogen is continuously entering the air by the action of denitrifying bacteria and continuously returning to the cycle through action of nitrogen fixing bacteria or blue green algae and through the action of lightening (*i.e.*, electrification).

Nitrophenol: A synthetic organic pesticide containing carbon, hydrogen, nitrogen and oxygen.

NMI: National Metrological Institute.

NMR: Nuclear magnetic resonance spectroscopy.

No effect level: Minimum amount of pesticide uptake, chronic or acute which will not produce harmful effect on living organism instantly or in future. These are assessed as daily intake values mostly through food ingestion.

No observable effect concentration/ level (NOEC/ NOEL): Highest concentration or amount of pesticide in the test system that causes no observable biological effect to the target organism.

No residue: Registration of the pesticide on the basis of absence of residue at the time of harvest in or on the raw agricultural product when the economic poison is used as directed.

No till: Sowing crop seeds directly into stubble or sod with no more soil disturbance than necessary to get seed into soil.

Noctuidae (Agrotidae): Family of super family Noctuoidea, suborder Ditrysia and order Lepidoptera. This is a very remarkable family which includes a large number of species of economic importance. These insects are nocturnal and are attracted to light. A proboscis is present but rarely atrophied. The labial palpi are well developed. A frenulum is present. The insects are generally cryptic and dull colored.

Noctuoidea: Super family of suborder Ditrysia and order Lepidoptera. Moths of this superfamily posses minute maxillary palpi and the tympanal organs lie in metathorax. The vein C_2 is absent in both wings. Most of these insects are economically very important as pests of crops. It includes the following families: Notodontidae, Amatidae, Arctiidae, Agaristidae, Noctuidae, Lymantriidae and Hypsidae.

Nocturnal: Active at night.

Node: A knob like or knot like swelling.

Nodiform: In the form of a knob or knot.

Nodule: A tumour like structure produced by the roots of nitrogen fixing plants. The nodules contain symbiotic nitrogen fixing microbes. Nodules can be found for example in legumes.

Nodus: The kink or notch on the costal margin of the dragonfly wing. The name is also used for the strong, short cross vein just behind the notch.

Noise pollution: Release of unwanted sound in the atmosphere.

Nomadic phase: This is the period in the activity cycle of army ant colony during which the colony forages more actively for food and moves from one bivouac site to another. At this time, the queen does not lay the eggs and the brood is in the larval stage.

Nomadism: Refers to relatively a frequent movement of a colony from one nest site to another.

Nominalistic species concept: Which governs that only the individuals exist in nature while species are man made.

Nominate: Containing the type of name of the higher taxon to which it is subordinate.

Nomogram: A curve obtained from the values of one or two variables which are known so that the value of a variable can be found without calculation.

Non accumulative: Refers to a pesticide which does not accumulate or store in an organism or in the organism.

Non-anticoagulant rodenticide: Any substance that does not inhibit, delay or suppress blood clotting. At present, there are three non-anticoagulant rodenticides, zinc phosphide, cholecalciferol (vitamin D_3) and bromethalin, registered and available for commensal rodent control. Since, none of these are anticoagulants, all can be used to control anticoagulant resistant rodent populations.

Non-biodegradable: Substance incapable of getting broken down in the biological systems.

Noncellular: Layer on the haemocoel side of the gut epithelium.

Non- corrosive: Substance which cannot corrode.

Non-dimensional concept: Species lacking dimensions of space and time.

Non-extruded granules: See granular formulations.

Non-infectious disease: A disease which cannot be transmitted from one plant to another. It is not caused by a pathogen but by an environmental factor.

Non-ionic surfactant: A surfactant that does not ionize in solution and is, therefore, compatible with both anionic and cationic surfactants.

Non-occluded virus: Viruses which do not form the inclusion bodies; also called the non-inclusion viruses.

Non-persistent pesticide: When the effect of pesticide lasts for few weeks or less. A pesticide may disappear because it is broken down by light, microorganisms or it may evaporate.

Non-persistent virus: A virus that persists in its vector for only a few hours (usually less than 4 hours).

Non-polar insecticides: The molecules of non-polar insecticides such as DDT, BHC, synthetic pyrethroids do not have any separation of charges because the electrons are uniformly distributed. Such insecticides are lipophilic, *i.e.*, have affinity for fat or fat like substances.

Non-polar solvent: A liquid which has molecules with no permanent electric dipole moment and hence has no drive for solvating polar compounds. Such a solvent is lipophilic in nature, *e.g.*, hexane, carbon tetrachloride etc.

Non-preference: An insect pests's response to a potential host (variety) which displays various characteristics that discourage its use for ovipositon, food or shelter (Painter, 1951).

Non-renewable resource: A natural, normally non-living resource such as a mineral which is present in finite supply and not renewed by natural system.

Non-resonating muscles: Muscles that require a nervous impulse for each contraction with resonating muscles, also called the synchronous muscles.

Non-selective: A chemical generally toxic to plants or animals without regard to insect species.

Non-target organisms: Those organisms which are not the intended targets of a particular use of a pesticide.

Non-toxic: Not poisonous.

Non-volatile: A chemical that does not evaporate.

Norbormide: Trade name is Shoxin. Low toxicity; a rodenticide specifically with single dose toxicity for Norway rats. Chemically, it is 5-(alpha-hydroxy-alphaphenyl- alpha-(2-pyridyl-methylene-5-nor-bornene-2, 3-dicarboxymide. It is used 4.5 per cent in the baits.

Nornicotine: The alkaloid other than nicotine and anbasine found in tobacco plant. Besides tobacco, the nornicotine is also found in the *Dubosia hopwoodii*.

Nosema disease: A disease of the adult honey bee caused by the protozoan, *Nosema apis*. The *N. apis* attacks epithelial layer of the midgut causing death of large number of bees.

Notal wing process: Point at which the notum articulates with the wing or at the base of the wings.

Notaulix (pl., notaulices): (i) Longitudinal furrows convergent posteriorly in the anterior part of the mesonotum of some insects. (ii) One of a pair of longitudinal grooves on the mesonotum of certain hymenopterans, dividing the mesonotum into a central area and two lateral areas.

Notifiable pest or disease: Pathogen or pest required by law to be reported to the officials of Ministry of Agriculture.

Notodontidae (prominents): Family of super family Noctuoidea, suborder Ditrysia and order Lepidoptera. The moths are moderately stout-bodied insects with elongate forewings and are nocturnal. When at rest the wings are folded roof like over the body. In some genera such as *Notodonta, Lophopteryx* and others the forewings have tuft of scales arising from the middle of hindmargin of forewings and hence the family name Notodontidae.

Notonectidae: This is a family of division Cryptocerata, suborder Heteroptera and order Hemiptera. These light coloured bugs swim upside down and hence the name back swimmers. The body is more convex dorsally. The head is inserted into prothorax and possesses four segmented antennae and a three or four segmented rostrum, tarsi two segmented, posterior tarsi clawless, *e.g.*, black swimmer, *Notonecta* sp. is a widely distributed species. In India, *Anisops bouvieri* is common in fish ponds and predator of fishes, larvae and nymphs of aquatic insects like mosquitoes, chironomid larvae, mayfly nymphs and different planktonic crustaceans.

Notopleural bristles: Bristles on the notopleuron in Diptera.

Notopleural suture: Suture separating the external propleuron from the pronotum; occurring in most Archostemata, Adephaga and Myxophaga.

Notopleuron: A triangular area on the thorax of certain flies, just behind the humeral callus and occupying parts of both dorsal and lateral surfaces.

Notosternal suture: Suture separating the prosternum from the pronotum.

Notum (pl., nota): The dorsal sclerite of a thoracic segment or the dorsal or upper surface of any thoracic segment: usually prefixed by pro or meso or meta to indicate the relevant segment.

Noxious weed: A weed arbitrarily defined by law as being especially undesirable, troublesome and difficult to marring.

Nozzle: Mouthpiece of a spout through which spray fluid is discharged. Different types of nozzles are (i) Air blast: nozzle using high velocity air to break up the spray liquid supplied at low pressure. (ii) Anvil: nozzle in which the spray liquid jet strikes a smooth, solid surface at a high angle of incidence. (iii) Cone (or swirl): nozzle in which the liquid emerges from the orifice with tangential velocity imparted by passage through one or more tangential or helical channels in the swirl chamber. (iv) Hollow cone nozzle: in which spray jet has a core of air breaking to give drops in an annular pattern. (v) Fan nozzle: the aperture is an elongate horizontal slit, producing a fan shaped spray pattern. (vi) Deflector nozzle: in which a fan shaped sheet of spray is formed by directing the liquid over a sharply inward curving surface.

NPD: Nitrogen-phosphorus detector or detection for gas chromatography (see AFID).

NPL: National Physical Laboratory.

NPV: Nuclear polyhedrosis virus.

Nuclear cytoplasm: Cytoplasm around the nucleus of an insect egg.

Nuclear polyhedrosis virus: A disease causing virus of insects, mainly in the larvae of certain Lepidoptera and Hymenoptera. Abbreviated as NPV, certain types of NPV are cultured commercially and sold as microbial insecticides. These viruses invade the cells of host and multiply there. The multiplied viruses spread and invade other cells. The polyhedra are formed in the cell nucleus; the size of the infected cell and its nucleus increase which ultimately get destroyed releasing the polyhedra in the body. The host dies due to liquidification of internal tissues. The polyhedra are crystal like with many faces and are made up of concentric layers. The NPV has been found very effective against Lepidoptera and to a lesser extent to Hymenoptera and Diptera.

Nucleocapsid: The fundamental nucleic acid of a virus plus its surrounding crystalline coat or shell. Some nucleocapsids are naked but a mature ultimate phase virus particle is enclosed by a limiting membrane and is also known as virion.

Nucleus (pl., nuclei): (i) A small hive of bees, usually covering from two to five frames of comb and used primarily for starting new colonies, rearing or storing queens; also called nuc. (ii) The spheroid body within a cell that has the major role in controlling and regulating the cell activities and contains the hereditary units or genes.

Nucleus brain: Demarcated mass of cell body in brain, *e.g.*, basal nuclei.

Nucleus culture: A small population from which any colony begins. Usually refers to a population obtained from a culture in another insectary but synonymous with parental colony if the individuals are collected from the wild.

Numerical response: The reproductive, migratory and survival response of a parasitoid or predator populations to changing density of its host/ prey; a positive response involves increased natality, survival and immigration with increased host/ prey density and *vice-versa*.

***Nupserha bicolor postbrunnea* Dutt:** Belongs to order Coleoptera and family Lamiidae. It is an insect pest of jute, commonly known as the jute stem girdler.

Nuptial flight: Refers to the mating flight of the winged queens and males.

Nurse bees: Young bees, three to ten days old which feed and take care of developing brood.

Nurse cells: Cells that are located in the ovarian tubes of certain insects and that furnish nutriment to the developing eggs. These nutritive cells in polytrophic and telotrophic ovarioles.

Nutrient: A substance needed by plants and animals for life and growth.

Nutrition: L., *nutrio*= feed. The use and conversion of food substances.

Nymph: Immature instars of ametabolous and hemimetabolous insects. The nymph is usually quite similar to the adult except that its wings are not fully developed. It normally feeds on the same kind of food as the adult. These do not have a distinct pupal stage; also the immature stages of Acarina that have eight legs.

Nymphalidae (brush footed butterflies): Family of super family Papilionoidea of suborder Ditrysia and order Lepidoptera. This is a large family with brightly colored butterflies flying in the sunshine. They have considerably reduced forelegs in both sexes which are functionless and generally folded on the thorax. The short tibiae are covered with long hairs and hence the name brush-footed butterflies. The tarsi of forelegs are unjointed in males and four or five segments in females. The larvae are characteristic in possessing numerous spinous processes and the last abdominal segment bears a pair of backwardly directed processes. The chrysalis has often prominent tubercles in rows on the abdomen and in some a pointed projection on either side of the head. It is usually suspended by the cremaster. The most remarkable of the members of this family is the Indo-Malayan *Kallima* which includes the leaf butterflies.

Nymphipharous: Development of nymphs inside the mother.

Nymphulinae: Subfamily of family pyraustidae. Larvae are aquatic possessing breathing gills and feed on aquatic plants. *Nymphula depunctalis* sometimes causes severe damage to rice. The larvae remain in cases made up of leaf bits and feed by scraping the leaf tissue. *N.fluctuosalis* also attacks rice in India. The other species known to infest rice are *N.nymphaeata* in Italy and Hungray, *N.enixalis* in Malaysia and *N.vittalis* in Japan.

O

Objective synonym: Each of two or more scientific names (synonyms) based on same type.

Obligate parasite: The organism which must develop parasitically and cannot complete its growth away from the host.

Obligate: The word is used to signify the parasite which is obligate in nature and sometimes restricted to those parasites which cannot be cultured in the laboratory on artificial media.

Obligatory parthenogenesis: Parthenogenesis in which a species reproduces entirely without males.

Oblique sternals: Lateral abdominal muscles in other than longitudinal or dorsoventral -orientation.

Oblique vein: It is a slanting cross vein in the wings of Odonata.

Oblongum cell of hind wing: A cell located at the end of medial bar and formed from rp-mp cross veins; occurs in most Archostemata, Adephaga and Myxophaga.

Obovate: Shape of the eggs in which the broad part is above and middle, and tapering towards the base.

Observation hive: A hive made largely of glass or clear plastic to permit observation of bees at work.

Obtect pupa: Pupa with wings, antennae, mouthparts and legs glued close to the body surface; such pupa is generally covered by a cocoon as in Lepidoptera.

OC: The abbreviation refers to the organochlorine insecticide group.

Occasional ectoparasites: These are often the winged free living insects and are fast fliers. They attack the animals repeatedly and suck the blood periodically without making permanent association with the host. These act as carriers of various pathogen causing fatal diseases in animals, *e.g.*, horseflies, dear flies, stable flies, black flies, buffalo gnats, mosquitoes etc.

Occasional pest: They appear only in certain regions and at a certain time. These pests are generally under natural control. It causes economic damage only sporadically or in localized areas as their population increases when natural balance is disturbed, *e.g.*, jassids, aphids etc.

Occipital arch: The area of the cranium between the occipital and post-occipital sutures, its dorsal part is occiput proper and lateral parts are post-genae.

Occipital carina: A transverse carina in the occipital region, usually located at the point where the top of head normally protrudes from the anterior edge of pronotum.

Occipital condyles: The processes on the margin of post-occiput to which the lateral neck plates are articulated.

Occipital foramen: An opening in the cranium through which nerve chords, aorta, oesophagus, salivary ducts and a pair of trachea enter or leave the neck.

Occipital ganglion: A single or paired post-cerebral ganglion of the stomodaeal nervous system.

Occipital groove: Median longitudinal groove on the posterior part of head, extending from the occipital foramen to some point on the vertex or frontal region; usually associated with an endocarina. See occipital line.

Occipital line: Median longitudinal line or ridge on the posterior part of head, extending from the occipital foramen to some point on the vertex or frontal region; usually associated with an endocarina. See occipital groove.

Occipital region: The posterior most area of the head capsule; not clearly distinguished from the vertex which lies just in front of it.

Occipital suture or sulcus: It is the horse shoe or U-shaped suture and well evident on the posterior part of head of the orthopteroid insects. It starts from posterior articulations of the mandibles of either side. Both the arms after running a long distance, join one another above the occipital foramen. It separates the genal area from the post genal area of the lateral region of head. It also differentiates the vertex from the occiput in posterior region of head. It forms an internal ridge strengthening the posterior epicranial wall.

Occiput: Hind part of the epicranium between the vertex and the neck region. It is U-shaped posterior or hinder cuticular band. In the higher insects, the occipital suture is obscure.

Occluded virus: These are the viruses characterized by enclosure of the mature replicated virions in the matrix of proteinaceous or crystal like bodies, such as capsules or polyhedra; also called inclusion viruses.

Ocellar bristles: Bristles arising around or between the ocelli in various flies.

Ocellar triangle: A triangular area, usually quite distinct from rest of the head on which the ocelli of true flies are carried.

Ocellus (pl., ocelli): There are three simple eyes known as ocelli located on the face and between the compound eyes. A simple eye may be present in an insect or other arthropod. Ocelli are often present on the head of spiders and insects. In adult insects, it is the compound eye which is responsible for the movement towards light, such movements do not occur when compound eyes are covered and only the ocelli remain. They are found in the nymphs of Hemimetabolous insects and in almost all adult insects.

Ocellus, structure: The ocelli are simple, light sensitive or stimulatory organs activating immediately the central nervous system and thus play a key role in maintaining the diurnal rhythm. It is well adapted for quick perception of light, changing in its intensity. In most of the insects, it is incapable of forming images because of wide angular separation of rhabdoms and as the images are focused far behind the retina. The basic difference between the compound eye and ocellus is that (i) the ocellus bears a dome shaped common lens for all visual cells while in the compound eye, each visual cell is provided with an independent lens, (ii) lens is cuticular, (iii) retina is made up of four retinula cells, (iv) axons form ocellar nerve, (v) ocellar nerve terminates into pars intercerebralis, (vi) cone is absent, (vii) pigment iris cells are absent, and (vii) ocelli are incapable of forming the image. The ocellus bears a single lens and in most of the insects, it is cuticular.

Octanol/ water partition coefficient (P_{ow}): Partition coefficient for a pesticide in the two phase system octanol/ water. The P_{ow} is a distribution coefficient reflecting the lipophilicity of a pesticide and its potential for bioconcentration. For convenience, the value is often expressed in logarithmic (base 10) form (log P_{ow}).

Ocular sclerite: These sclerites form a cuticular ring around each compound eye.

Ocular suture: These sutures are annular grooves on the compound eyes.

Ocular: Pertaining to eyes; visual; ophthalmic.

Odonata: Insect order, including the dragonflies and damselflies. These are characterized by membranous wings, bristle like antennae, long, slender bodies, chewing mouthparts and an incomplete metamorphosis.

***Odontotermes obesus* (Rambur):** Belongs to order Isoptera and family Termitidae. It is a polyphagous insect pest, commonly known as the field termite.

Odour: Odour is diagnosed by the olfactory stimuli. It may be classified as (i) sexual attractants: male gypsy moth, *Lymantria dispar* is attracted by the odour, gyptol and disparlure from over two miles by a virgin female of its own species. Other examples are Lasiocampidae, Bombycidae and Saturniidae. (ii) recognition odours: social insects can detect members of their own species, caste or colony by olfaction, example, termites, ants and honey bees. (iii) oviposition attractants: parasitic insects are attracted by odours such as ammonia, skatol etc arising from the decomposition of organic material by bacteria. Female *Habrobracon* discovers its host, *Ephestia* by smell through antennae. (iv) food attractants: phytophagous insects are attracted to their host plants by the smell of essential oils, *e.g.*, fruit sucking moth, *Othreis materna* to ripe orange fruits.

Odour trail: A pheromone trace or odourous trail substance laid down by one insect and followed by another.

Oecophoridae: Family of super family Tinaeoidea of suborder Ditrysia and order Lepidoptera. These are small moths generally brownish in color with apically rounded broad wings. The larvae generally web together or roll the leaves, or sometimes feed on seeds. In India, larvae of *Tonica zizyphi* roll or web the leaves of *Zizyphus jujube*, orange, lemon *(Citrus sp.)* and *Murraya koenigii.*

Oedema: The pathological acumulation of the fluid in tissue spaces.

Oenocytes: These cells are ectodermal in origin and situated between epidermis and basement membrane or these may be distributed throughout the fat body. These are probably intermediary in metabolism and become specialized for the production of some particular constituent of the cuticle such as the lipoprotein of the cuticular layer or the cuticular wax. There is some evidence that they are specifically concerned in the synthesis of hydrocarbons. These are excretory as well as secretory in function. These play important role in formation of cuticulin layer during moulting. These are known to secrete ecdysone in some insects and predominantly occur in the honeybee larvae and pupae. These undergo cyclic changes during postembryonic development in some insects.

Oenocytoids: These are blood cells of rounded or oval form, the cytoplasm is generally oesinophil, and it neither takes part in phagocytosis nor in formation of encapsulating membranes. In appearance, these cells are like diminutive oenocytes and on this account these are often called oenocytoids but these do not seem to bear any relation to the true oenocytes. Under phase contrast, these often show conspicuous inclusions and are therefore, sometimes called granular cells. Their function is quite unknown; these increase during moulting, like plasmatocytes and aggregate around foreign bodies in a zone peripheral to the plasmatocytes.

Oesophagus: Pharynx is followed by an undifferentiated tube of variable length known as oesophagus. It is a part of foregut and its function is to receive food from pharynx and pass it onwards to the crop, therefore, a seat of food conduction.

Oestridae (warble or botflies): Family of suborder Cyclorrhapha, order Diptera. Oestridae are large, stout-bodied, hairy, bee-like flies. The larvae of which are parasitic in various animals such as cattle, sheep, goats, deer, reindeer, camel and kangaroo. The flies have a large head with small eyes and facial grooves for antennae and atrophied mouthparts. In Australia, *Tracheomyia macropi* is found in the trachea of kangaroo.

***Oestrus ovis*:** Belongs to order Diptera and family Oestridae. It is an insect pest of farm animals, commonly known as the sheep bot fly.

Often cross pollination: Cross pollination often exceeds from 5 per cent and may reach 30 per cent.

Oil: These are viscous oils such as petroleum oil, diesel oil, fatty oils (glycerides) etc. The aromatic, paraffinic oils used as diluents in formulations of insecticides as carriers of toxicants. These are also used directly without adding other toxicants.

Oil formulation: The oil based formulations of insecticides or other chemicals.

Oil spray: A spray of poisons mixed with emulsion of oil.

Olefins: Reactive, hydrogen deficient hydrocarbons characterized by one or more double bond per molecule. Ethylene and propylene, the widely used building block basic chemicals are olefins.

Olfaction: The olfactory or sense related to smell.

Olfactory: Pertaining to the sense of smell.

Oligoaerobe: Refers to the organisms which can thrive at low oxygen content.

Oligocarpus: Having few fruits only.

Oligogyny: Occurrence of two or several functional queens in a single colony.

Oligolectic: Bees that gather pollen from only a few species of related flower. See polylectic.

***Oligonychus indicus* Hirst:** Belongs to order Acarina and family Tetranychidae. It is a pest of sugarcane, commonly known as the sugarcane mite.

Oligophagous parasite: A parasite feeding upon only closely- related hosts or various species of a family.

Oligophagous pest: A pest feeding upon only a closely- related host plants or various species of a family is referred as oligophagous pest.

Oligopneustic: Partially open tracheal system which there are one or two functional spiracle on each side of the body. This is of three types: (i) Amphipneustic: Only first thoracic and last abdominal spiracles are functional, *e.g.*, Dipteran pupae (ii) Metapneustic: Only last abdominal spiracles are functional, *e.g.*, first instar Dipteran pupae and (iii) Propneustic: Only mesothoracic spiracles are functional, *e.g.*, Dipteran pupae.

Oligopod larvae: These larvae are characterized by the presence of well developed thoracic legs, absence of abdominal appendages except cerci and with well developed head capsule. These larvae may be campodeiform type or scarabaeiform type. The campodeiform larvae resemble *Campodea* and typically possess a long fusiform compressed body, well sclerotiszed cuticle, sensory equipment, prognathous head, long thoracic legs, a pair of terminal abdominal processes and absence of dorsal ocelli, *e.g.*, Neuroptera, Adephaga (Coleoptera), Strepsiptera and Trichoptera. The scarabaeiform larvae are stout sub-cylindrical and C-shaped with short thoracic legs, fleshy body and without caudal processes, *e.g.*, Scarabaeoidea and other Polyphaga.

Oligotomidae: Family of order Embioptera. Tenth tergite in males asymmetrical; usually with R_{4-5} forked; echinulations on inner surface of basal segment of left circus. Instars of both sexes lacking median or second tarsal on the plantar surface of the hind basitarus, *e.g.*, Indian oligotomid, *Oligotoma greeniana* adults live in spider nests.

Omite: It is trade name of a synthetic acaricide, 2-(p-tert-butylphenosi) cyclogexyl 2-propynyl sulphite; slightly toxic to mammals; acute oral LD_{50} for rats is 2500 mg kg^{-1}. It is mainly used against mites, particularly on fruit plants.

Ommatidium (pl., ommatidia): A single unit or visual section of a compound eye. Each ommatidium despite its size and secondary modifications in different species of insects consists of a dioptric or optical apparatus receiving the light and a sensory or retina forming the image. The optical apparatus is composed of a cuticular lens and a cone. The sensory part is composed of elongated specialized nerve cells, called retinula cells are surrounded by the secondary pigment cells, while the cone is surrounded by primary pigment cells. The retina is separated at the bottom by a basement membrane through which the axons arise from the retinula cells and pass into the optic lobes.

Ommatidium, structure: It consists of (i) Cornea: The outer most part of ommatidium. It is transparent, colourless and biconvex modified cuticular area often termed as the facet or lens and cast off at each ecdysis and resecreted by the epidermal corneagen cells. (ii) Corneagen cells or layer: These are epidermal or hypodermal cells lying behind the cornea. A group of two corneagen cells secrete a single lens. After formation of the lens, they withdraw peripherally and modified into the primary iris cells. (iii) Cone or Semper cells: Beneath the corneagen layer or cornea, there are four distinct cells. They secrete the crystalline eucone,or liquid cone (pseudocone). The cone cells simply occur as the transparent cells without secreting any kind of cone (acone) or remain unmodified (exocone). In most cases, these cells are represented by the nuclei only and called Semper's nuclei. (iv) Pigment or primary and secondary iris cells: There are two groups of iris cells, one around the crystalline cone cells (primary) and the other around the retinula (secondary iris cells). Each group of iris cells is composed of six cells arranged in in a sort of circlet. The secondary iris cells separate one ommatidium from the neighbouring one. (v) Retinula cells: The crystalline cone is followed by a long retina forming the basal part of ommatidium. It is formed from a group of alike seven retinula cells as pigmented visual cells. The inner margin of the retinula cells lying aroung the ommatidial axis is highly differentiated from rest of the cell body and called rhabdomere. The rhabdomeres of all seven retinula cells extend the whole length of the retina and form a rhabdom (the internal optic rod having a fibrillar structure and becomes a central axis retina). The common photoreceptor pigment is called rhodopsin. (vi) Tapetum: The fine branches of trachea invest the basement membrane forming a tapetum. It is well developed in the nocturnal flying insects and reflects back the light, as a result the rays retraverse the rhabdom. (vii) Optic nerve fibres: The optic lobe of each side of brain is directly connected to the compound eye.

Ommatidium, types: (i) Eucone ommatidium: contain a true crystalline cone as a hard refractive body formed from an intercellular secretion of the cone cells or Semper's cells, *e.g.*, Orthoptera, Odonata, Ephemeroptera, Trichoptera, Lepidoptera, Hymenoptera, Thysanura, some Hemiptera and Coleoptera (ii) Pseudocone ommatidium: Donot contain crystalline cone but all cone cells are filled with gelatinous semi-liquid substance, *e.g.*, Dermaptera, Hemiptera,

Nematocera (Diptera) and Coccinellidae, Curculionidae (Coleoptera). (iii) Acone ommatidium: Donot contain either a crystalline or liquid cone but the cone cells are elongated and transparent, *e.g.*, Dermaptera, Hemiptera, Nematocera (Diptera) and Coccinellidae, Curculionidae (Coleoptera). (iv) Exocone ommatidium: Have extracellular cuticular cone in place of crystalline or liquid cone secreted by the cone cells. It is formed as an ingrowth from the inner surface of corneal facet in front of the unmodified cone cells. It is refractile in nature, *e.g.*, Lampyridae, Elateridae, Dermestidae, Cantharidae etc. On the basis of rhabdom, the eyes can be classified into (i) apposition eyes and (ii) superposition eyes.

Omnivore: Animals that feed on material of both plant and animal origin (See also, carnivore and herbivore).

Omnivorous: Both carnivorous and herbivorous, *i.e.*, eating meat and vegetation.

On line services: Internet users may avail many facilities, such as latest market values of stock market, fresh news, articles of various magazines, scientific documents etc.

Onchocerciasis: The black flies, *Simulium damnosum* and *S. neavei* cause onchocerciasis (blinding filarial disease) in tropical Africa, tropical America, from Southern Mexico to Columbia, Venezuela and Northern Brazil by transmitting filarial nematode, *Onchocerca volvulus*. Due to the infection, subcutaneous nodules under the skin, joints and muscles are formed. The lymph glands are swollen leading to elephantiasis. The microfilariae may attack lungs, liver and eyes.

Onchogenic: The property to produce tumours in the living organism which are not necessarily the cancerous.

Oncopoduridae: A family of order Collembola. Have scales and long mucro; vescular sense organs present on fourth antennal segment; post-antennal organ present. For example, *Oncopodura* sp.

Onion maggot: See *Hylemya antiqua.*

Onion thrips: *Thrips tabaci* Lind. is a pest of onion belonging to family Thripidae. The nymphs and adults of the pest suck the cell sap from the leaves, distorting and twisting them; also attack the garlic, cauliflower, cabbage, turnips, cotton etc.

Onisciform larva: A flattened platyform larvae like a woodlouse in appearance.

Onomatophore: Name given by Simpson to the type.

Ontogeny: The developmental history of an organism from its egg to adult.

Onychiuridae: A family of order Collembola. Have pseudocelli (aperture in the body wall) on thoracic and abdominal terga. For example, *Onychiurus* sp.

Oocyte: The egg cell differentiated from oogonium before maturation.

Oogonium: The first stage in the differentiation of an egg cell from a primary female germ cell.

Ootheca: The covering or case of an egg mass of certain insects in the orders Orthoptera (grasshoppers) and Dictyoptera (cockroaches, mantids).

OP: Abbreviation stands for organophosphate compounds which include organophosphate insecticides, thiophosphates, dithiophosphates and other esters and amides of phosphorus acids.

Operator: It is a stretch of bases along a DNA molecule located at the proximal end of an operon (a group of continuous structural genes showing coordinate expression and closely associated controlling sites).

Operculum (pl., Opercula): Refers to a cover.

***Ophideres ancilla* Cramer:** Belongs to order Lepidoptera and family Noctuidae. It is commonly known as fruit sucking moth and is a pest of citrus, mango, grapes and apple.

***Ophideres conjuncta* Cramer:** Belongs to order Lepidoptera and family Noctuidae. It is commonly known as fruit sucking moth and is a pest of citrus, mango, grapes and apple.

***Ophideres materna* Cramer:** Belongs to order Lepidoptera and family Noctuidae. It is commonly known as fruit sucking moth and is a pest of citrus, mango, grapes and apple.

***Ophiomyia (=Melanagromyza) phaseoli* (Tryon):** Belongs to order Diptera and family Agromyzidae. It is an insect pest of pulse crop, commonly known as the bean fly or red gram pod fly.

Opisthodelphic: Insects having a single ovary situated posterior to the vulva.

Opisthognathous: The long axis of head is vertical and the mouth parts are directed backward, *e.g.*, bugs.

Opisthogoneate: Having the genital opening situated terminally, at the posterior end of the body.

Opisthosoma: The hind body region of arachnids.

Optic: Pertaining to the vision or eyes.

Optic lobe: Lateral expansions of the brain of an insect innervating the compound eyes and ocelli.

Optic nerve fibre: See ommatidium, structure.

Optical system: Muller (1929) proposed the mosaic theory stating that the individual ommatidia of a compound eye form an image in parts limited to their visual fields which differ in intensities and juxtaposition. All these ommatidia, give a total erect image of an object. Each ommatidium has a field of vision.

Oral: Pertaining to the mouth.

Oral disc: See sponging type mouth parts.

Oral toxicity: The toxicity of a compound when ingested. The oral toxicity of a pesticide is expressed in milligrams of chemical per kilogram of body weight. It is the amount which when given orally in a single dose, will kill 50 per cent of the test animals. See also LD_{50}.

Oral vibrissae: A part of stout bristles, one on each side of the face near or just above the oral margin and larger than the other bristles on the vibrissal ridge in order Diptera.

Orange stem borer: *Stromatium barbatum* Fab., a major pest of citrus. The grubs bores into branch of tree and tunnels inside causing branch to wither and tree to dry up, also infests the other trees of citrus family.

Orbital plate: A sclerite on the head next to the compound eye in Diptera; also called genovertical plate.

Order. A subdivision of a class or subclass containing a group of related families.

Organ: A functional unit of an organism performing a specific activity.

Organelles: Organized microstructures inside the cells (mitochondria, chloroplasts etc.) in which certain biochemical functions are performed.

Organic compound: A compound which contains one or more carbon atoms except CO_2 which is not considered an organic compound.

Organic Farming: In organic farming, the food is grown in a sustainable manner. It is more than not using the pesticides and identical with IPM, except IPM allows for the rare use of synthetic pesticides.

Organic insecticide: An insecticide of organic origin containing chlorinated hydrocarbons, organophosphate compounds, carbamates, plant extracts, pyrethrins, nicotines etc.

Organic matter: The fraction of the soil composed of carbon based materials

Organism: Any living being including higher plants, animals, bacteria, fungi etc.

Organochlorine insecticide: One of the many chlorinated hydrocarbon insecticides, *e.g.*, DDT, dieldrin, chlordane, BHC, lindane etc.

Organoid: A body within the cytoplasm which carries out specific function in the cell.

Organomercurial compound: A combination of an organic compound with the mercury, such as ethyl mercury used in sprays or seed treatment.

Organometallic compound: An organic compound containing a carbon metal bond, *e.g.*, tetraethyl lead, $(C_2H_5)_4Pb$. These compounds are used as seed dressers and fungicides.

Organophosphate: Organic compounds containing phosphorus, an important group of synthetic insecticides. This class of insecticides is derived from phosphoric acid esters. Organophosphates inhibit the functioning of the enzyme cholinesterase.

Oriental fruit fly: See *Dacus dorsalis.*

Origin: Stationary attachment area of skeletal muscle.

Ormyridae: Family of super family Chalcidoidea, suborder Apocrita and order Hymenoptera. These differ from Torymidae in having indistinct parapsidal sutures and a short ovipositor. These are parasites of gall insects.

Ornamental plants: Plants used to beautify the houses, lawns and gardens, *e.g.*, trees, shrubs and flowering plants.

Orneodidae [Alucitidae] (many plume moths): Family of super family Tinaeoidea of suborder Ditrysia and order Lepidoptera. Both pairs of wings dividedinto six or more narrow plume-like structures, each division being fringed with hairs densely along both margins. Larvae of many plume moths are known to cause galls by burrowing into shoots, flower stalks and buds. Larvae pupate in silken cocoons formed on the ground. This family includes a number of species of the genus, *Orneodes.*

Ornithophily: Many tropical plants are pollinated by the birds. Ornithophilous flowers are tubular (*Nicotiana*), cup shaped (*Callistemon*) or urn shaped (some members of family Ericaceae). They are attractive, brightly coloured and contain plenty of nectar. Pollen grains are attached to the bodies of birds, insects etc. which visit the flowers due to these attractions, and reach other flowers along with them resulting in cross pollination.

***Orseolia (=Pachydiplosis) oryzae*:** Belongs to order Diptera and family Cecidomyiidae. It is an insect pest of cereals and millets, commonly known as the paddy gall fly.

Orthene: Trade name of acephate.

Ortheziidae: A family of superfamily Coccoidea, suborder Homoptera and order Hemiptera. Segmentation of the body of female is distinct and covered withwhite waxy plates. Antennae simple, eyes and legs are prominent. Example, Lantana bug, *Orthezia insignis*.

Orthogenesis: A racial tendency to a particular line of development, uncontrolled by external conditions.

Orthokinesis: A simple reaction of insects by which a particular stimulus can move the insect in a straight line but they come to rest as soon as the stimulus ceases.

Orthoptera: Order (*ortho*= straight, *ptera*= wings). These are characterized by leathery forewings, chewing mouthparts and a gradual metamorphosis. Usually the hind legs are enlarged and adapted for jumping. Medium to large sized, mostly tererrestrial insects, head hypognathous rarely prognathous bearing compound eyes, normally possessing three ocelli but absent in wingless species and reduced to two in long horned grasshoppers, mouth parts are ectognathous mandibulate, the forewings are thickened into protective tegmina. The insects included in this order are grasshoppers, crickets and locusts.

Orussoidea (parasitic wood wasps): Family of super family Megalodontoidea of suborder Symphyta (Chalastogastra), order Hymenoptera. This includes a single family Orussidae. The wing venation is reduced and hindwings are without closed submarginal cells. The insects are very small and their apodous larvae are ectoparasitic on larvae of wood boring Buprestidae.

***Oryctes rhinoceros* (Olivier):** Belongs to order Coleoptera and family Scarabaeidae. It is an insect pest of coconut and date palm, commonly known as the rhinoceros beetle.

Osmeterium: A usually Y-shaped, eversible glandular structure located behind the swallow tail butterfly larva that secretes a repugnatorial substance.

Osmoregulation: The osmoregulation is the vital physiological mechanism maintaining haemolymph ionic and water balance, *i.e.*, the hydromineral homeostasis. The mechanism differs in the terrestrial, fresh water and salt water insects. The excretory system, *i.e.*, the Malpighian tubules and rectal epithelia are actively involved in the osmoregulation. The diuretic and antidiuretic neurohormones control the activity of osmoregulatory organs in order to maintain the salt and water balance of haemolymph.

Osmosis: The transfer of materials from semipermeable membrane that separates two solutions or between a solution and a solvent that tends to equalize their concentrations. The walls of living cells are semipermeable membranes and much of the activity of the cells depends on osmosis.

Osmotic pressure: maximum pressure which can be developed in a solution which is separated from pure water by a rigid membrane permeable only to water.

Ostial valve: Valve like pouches of the heart wall containing ostia at their inner ends.

Ostiole: A small opening.

Ostium (pl., ostia): A slit like opening in the insect heart. The ostia are two types: incurrent and excurrent. The incurrent ostia occur in almost all insects and their number is always equal to the number of chambers of the heart. They permit the entry of blood from the haemocoel into the heart but at the same time prevent its outward flow. In the dragon fly nymphs, there is a pair of valves, ventricular valves infront of each incurrent ostia; while in mayfly nymphs, the valves develop internally from the ostia, called auricular valves which prevent the backward flow of haemolymph. The excurrent ostia are found in Thysanura and Orthopteroid insects which occupy ventrolateral position and are paired structures. These are devoid of valves and do not correlate with the chambers of heart.

Ostium bursae: Opening of the bursa copulatrix in Lepidoptera, equivalent to the vulva of the female insects having genital opening on the eighth segment.

Ostomatidae (Trogositidae, Ostomidae): Family of superfamily Cleroidea, suborder Polyphaga and order Coleoptera. Mostly tropical species, elongate or hemispherical, cylindrical, 5 to 20 mm long, head and pronotum large, elytra not truncate, metatarsus short and inhabit decaying trees. Mostly they are predaceous on larvae of other insects and other feed on fungi. The beetle *Tenebrioides mauritanicus* is a pest of stored flour, grain etc. Its larvae are called "cadelles" and feed on grains and prey upon other stored products pests found in association with it. These are the bark gnawing beetles.

***Ostrinia nubilalis* (Hb.):** Belongs to order Lepidoptera and family Pyralididae. It is an insect pest of maize, commonly known as the European corn borer.

Out breeding: Mating between genetically dissimilar, relatively unrelated individuals.

Out-apiary: An apiary situated away from the home of beekeeper.

Outer fork: The fork of R_{4+5} as in Ephemeroptera.

Outer vertical bristles: The large bristles on the vertex between the ocelli and the compound eyes as in Diptera.

Ovacide: See ovicide.

Oval: broadly elliptical, rounded at both ends, nearly twice as long as broad, broadest in the middle.

Ovarial ligament: A ligamentous strand attaching the terminal filaments of an ovary to the dorsal diaphragm or to the body wall.

Ovarian steroid hormone: Although, Karlson and Stamm-Menendez (1956) found 1.6 *Calliphora* units (1 CU= 16 ng) of ecdysteroids per gram in the ovary of silk moth, *Bombyx mori* and Horn (1971) speculated role of ecdysteroids in ovarian development, but the first sound evidence regarding the ovary producing steroid hormone was put forward by Hagedorn *et al.* (1975). Before 1975, the insect ovary was thought to be incapable of secreting the hormone. Some workers, moreover, demonstrated the ovary secreting oostatic hormone in the house flies (Adams *et al.*, 1975) and mosquitoes (Lea, 1975) but chemical nature of ovarian hormone was almost obscure.

Ovariole: The ovary is composed of a large number of tubules or egg tubes, commonly termed as ovarioles. There are in fact, the functional reproducing units of the ovaries, inside which the development of female gametes (oogenesis) takes place and finally the ova are discharged into the lateral oviducts. These are externally enclosed by a double layered coat consisting of an outer tunica externa and an inner tunica propria. The tunica externa is formed from the fat body while the tunica propria represents simply the elastic peritoneal membrane with some fine fibrils. There is a single ovariole in the ovary of *Termitoxenia* and Coprinae, two ovarioles in *Glossina* and *Hippobosca*, four in Lepidoptera, 360 in *Apis*, 1200 in army ant (*Eciton*) and about 2000 in termite queen (*Eutermes*).

Ovariole sheath: Cellular layer that surrounds an ovariole.

Ovariole, structure: The ovariole can be differentiated into four regions, anterior most terminal filament, second- germariun, third- vitellarium and the last pedicel which opens into the lateral oviduct. See details of these.

Ovariole, types: The ovarioles can be distinguished into two principal categories: The panoistic containing no nutritive cells (nurse cells) or trophocytes and the meroistic containing the trophocytes. The meroistic ovarioles are subdivided into two sub-types, the polytrophic and the telotrophic according to the location of trophocytes in vitellarium or germarium, respectively. In the polytrophic ovarioles, the developing oocytes within the vitellarium are alternating with the nurse cells. They participate in the vitellogenesis, *i.e.*, yolk deposition. In the telotrophic or acrotrophic ovarioles, the trophocytes are situated into the germarium and connected with the growing oocytes of vitellarium through the cytoplasmic canals, called the trophic cords.

Ovary: Paired gonads or egg producing organs (mesodermal in origin) in the female reproductive system, situated either side or above the alimentary canal. The reproductive system of female consists of a pair of sexual glands (ovaries), paired gonoducts (lateral oviducts) of mesodermal origin into which the sexual products are discharged and a common oviduct (median oviduct) lined with cuticle. The ovary of the female is likewise composed of a series of tubular follicles or ovarioles. In these, the oocytes arising in the germarium do not multiply repeatedly as do the germ cells in the male but each becomes enormously enlarged to form the egg cell or ovum. The anterior end of each ovary gives out a long thread like suspensory ligament of which the other end is attached to the dorsal body wall or dorsal diaphragm. The ovaries are covered with the fat bodies and richly supplied with the tracheae.

Ovate: Egg shaped with broad part below the middle part.

Overkill: Unsustainable hunting, collecting or harvesting of specific desirable species can lead to extinction of that species.

Ovex: A synthetic acricide; a sulfonate, p-chlorophenyl p-chlorobenzene sulfonate; acute LD_{50} for rats is 2050 mg kg^{-1}.

Ovicide: Pesticide that destroys eggs of insects and mites. Also referred as ovacide.

Oviduct: The tube leading away from the ovary through which the eggs pass.

Oviductus communis: Median ectodermal outlet duct of the female genital system usually opening into the genital chamber or vagina.

Oviductus literalis: One of the paired lateral ducts of the female system connected with the ovary, mesodermal in origin but sometimes partly or entirely replaced by an ectodermal branch of the median duct.

Ovigerous: Egg bearing. The term is applied to the external structures to which the eggs are attached after being discharged from the body.

Oviparae: Refers to oviparous females.

Oviparous: Reproduction by laying eggs.

Ovipore: The external opening of the female reproductive system through which the eggs are laid during oviposition.

Oviposit: To lay or deposit eggs.

Oviposition: The act of egg laying by the female insects. The eggs are carried down the oviducts by waves of peristalsis and deposited singly or in masses according to the habits of species. They are coated with the secretion of the accessory glands which may serve to glue them to the surface or to bind them together in a protective sheath. In the Hydrophilid water beetles, the accessory glands produce silk from which an egg cocoon is formed; in cockroach, the eggs are cemented together by fluid which hardens to form a tough ootheca. In this insect, the protein secreted in the left collaterial gland is tanned and hardened by means of the quinine formed by the oxidation of protocatechuic acid secreted in the right colleterial gland.

Ovipositor: The external genitalia of female mainly consists of egg laying apparatus or simple egg laying tube known as ovipositor. It is basally communicated with the female gonopore which is located on either the 8th or 9th segment. It is of gonapophysial origin. The paired gonapophyses which develop on the sterna of the 8th and 9th segments form this tubular structure. It is found in most of the Thysanura and Pterygota while absent in Collemola, Diplura and among Pterygota and Ephemeroptera, Odonata, Plecoptera, Mallophaga, Anoplura, tubuliferous Thysanoptera, Coleoptera, Neuroptera, Mecoptera, Trichoptera, Lepidoptera and Diptera.

Ovipositor, structure; The gonophyses or valvulae. In pterygota, the ovipositor is formed as a complex of paired appendages originated from the 8th and 9th segments; these appendages are either known as the gonophyses or the valvulae. They occur in three pairs. The paired appendages of 8th segment are called the anterior or ventral or first valvulae. They are modified from the mesal or endite processes. The coxopodites of the 9th segment from the lateral or the third valvulae ensheath the anterior or posterior pairs of gonophyses. They grow as the distal outgrowths of the coxopodites. A well developed ovipositor can be seen in Orthoptera, Hemiptera, Thysanoptera and Hymenoptera. The coxite of segment 8th is called first valvifer, bearing the anterior valvula. Similarly, the basal or coxal area of 9th segment is called the second valvifer bearing the posterior valvula. The paired second valvifers are often united ventrally by a narrow bar. The ovipositor is composed of a shaft and a basal apparatus. The shaft is made up of the first and second pair of elongated processes, the valvulae. The basal apparatus consists of two pairs of lobes or plates, the first and second valvifers which provide support to the shaft by the bases of valvulae.

Oviscapt: Modification of the terminal abdominal segments of a female to serve as an egg laying organ, *i.e.*, pseudo-ovipositor.

Ovoid: Oval shaped.

Ovoparity: Reproduction by laying eggs.

Ovoviviparous: Producing active young ones immediately following the hatching of eggs within the female. The eggs of some insects are retained within the genital tract of the mother until development is well advanced. In *Cimax,* in which impregnation takes the form of introduction of sperm into the body cavity by way of the so called 'organ of Ribaga', fertilization takes place in the ovary and the embryo is already developing by the time the egg is laid. In various tachinid flies, the eggs contain fully formed larvae and these escape from the chorion during oviposition (ovoviviparity).

Ovule: Minute structure in the ovary of the flowering plants that becomes seed after fertilization and development.

Ovum: Mature female germ cell.

Ox warble fly: *Hypoderma lineatum* and *H. bovis* are the pests of cattle, deposit eggs on the hair of legs; after hatching the larva bores into the skin and parasitize the host.

Oxidation: It is a chemical reaction in which a substance gains oxygen or loses hydrogen. Since oxygen is an electronegative element, so oxidation is defined as a reaction in which a substance gains an electronegative radical or loses an electropositive radical. Examples, (i) gain of oxygen: S+ $O_2 \rightarrow SO_2$, (ii) gain or addition of electronegative radical: $SnCl_2 + Cl_2 \rightarrow SnCl_4$, (iii) removal of hydrogen atom: H_2S+ $Cl_2 \rightarrow$ 2HCl+ S, (iv) removal or loss of electropositive radical or element: 2KI+ $H_2O_2 \rightarrow$ 2KOH + I_2.

Oximes: These are the chemical compounds characterized by the general formula, R-CH-NOH.

***Oxya nitidula* Wlk.:** Belongs to order Orthoptera and family Acrididae. It is a polyphagous insect pest, commonly known as the sporadic grasshopper.

***Oxycarenus laetus* Kirby:** Belongs to order Hemiptera and family Lygaeidae. It is an insect pest of cotton and ornamental plants, commonly known as the dusky cotton bug.

Oxydemeton methyl: The chemical is highly tocic and effective against aphids, mites, leaf hoppers etc. It is systemic organophosphate, S-[2-(ethylsulfinyl)ethyl] O, O-dimethyl phosphorothioate, moderately toxic to mammals, acute oral LD_{50} fot rat is 65-75 mg kg^{-1}, systemic insecticide used against sucking pests of field and vegetables.

Oxygen depletion: Exhaustion of oxygen by chemical or biological use.

Oxygen sag: Depletion of oxygen concentration during night due to respiration.

Oxythioquinox: A synthetic acaricide and fungicide (trade name, Morestan), 6-methyl-2,3-quinoxalinedithiol; slightly toxic to mammals; acute oral LD_{50} for rats is 3000 mg kg^{-1} used on citrus, most deciduous fruits and ornamental plants to kill mites, their eggs and to control mildew.

Ozone: Ozone is a gas that is naturally present in our atmosphere. In the atmosphere, oxygen is found in three forms, O, O_2 and O_3. An ozone molecule is O_3 which has three atoms of oxygen. The word ozone is derived from Greek, *ozein*, meaning 'to smell'. Ozone has pungent odour that allows it to be detected even in very low amounts. For every ten million air molecules, there are only three ozone molecules. Ozone is poisonous and once inhaled may be fatal. Fortunately, most ozone (about 90 per cent) is found in stratosphere, a region that begins about 10-16 km above the earth's surface and extends upto about 50 km. The large amount of ozone in stratosphere is referred to as ozone layer which protects atmosphere from harmful ultraviolet rays.

P

Pacemaker: A region from which waves of contraction in the heart beat cycle are propagated.

***Pachycrepoideus derbies* Ashm.:** A pupal parasitoid of melon fruit fly.

Pachytroctidae: A family of order Psocoptera and suborder Troctomorpha. Body mostly convex with hind femora not flattened, otherwise as in *Liposcelis*.

Package bees: Quantity of adult bees (2 to 5 pounds) with or without a queen, contained in a screened shipping cage.

Packaging: The container of pesticides with the protective wrappings which is used to carry these products from manufacturer to the ultimate user *via* the whole seller or retailer.

Paddy climbing caterpillar: *Pseudaletia (Cirphis) unipunctata* Haw and *P. albistigma* are insect pests of paddy.

Paddy eelworm: *Ditylenchus angustatus* Butler is a pest of paddy.

Paddy gall fly: See *Orseolia (=Pachydiplosis) oryzae* Wood Mason.

Paddy grasshopper: See *Hieroglyphus banian.*

Paddy green jassid: *Nephotettix bipunctatis* Fab., *N. apicalis* Mots. and *Tettigoniella spectrar*. Pests of paddy; the nymphs and adults suck sap of young paddy leaves, resulting in yellowing and drying up of plants.

Paddy root weevil: See *Echinocnemus oryzae.*

Paddy rough rice: The rice kernel with the husk.

Paedogenesis: In a few insects, there is a precocious reproduction in the larval stage which is known as paedogenesis. The classic examples occur in the Cecidomyiidae. This phenomenon was first observed by Wagner in 1861 in larvae of *Miastor*. The paedogenesis has also been observed in the *Micromalthus* *debilis* (Coleoptera). The oocytes develop parthenogenetically within the ovarioles until a number of daughter larvae are produced. These are set free in the body cavity where they rapidly destroy the tissues of their maternal host and ultimately escape through the body wall.

PAI: Pesticides Association of India.

Painted bug: See *Bagrada cruciferarum* Kirk.

Palaeo: A fossil species.

Palaeodictyoptera: Extinct insect order.

Palaeometabola: In this type of hemimetabolous metamorphosis, the nymphs differ from the adults palingenetically only, *i.e.*, lack secondary nymphal characters. It is further divided into two types: (i) Epimetabola: Apterous insects with very little pronounced metamorphosis, *e.g.*, Collembola, Thysanura and Protura represented mostly on the Ametabola, (ii) Prometabola nymphs are aquatic provided with archaic abdominal gills. Sub-imago stage is with wings. It latter on develops into the mature winged adult, *e.g.*, Ephemeroptera.

Palaeontology: Gr., *Palios*= ancient + *ontos*= being. Study of fossil animals and their distribution in time.

Palaeoptera: See palaeopteran orders.

Palaeopteran orders: Among the two exopterygote orders, *viz.*, Ephemeroptera (Plectoptera) and Odonata differ from the others as the insects of these orders cannot flex their wings over the abdomen at rest. They are characterized by the unusual wing articulation and the most primitive pattern of wing venation. Both insect orders are grouped into Palaeoptera (living fossils) because they resemble palaeozoic orders (Palaeodictyoptera, Proephemeroptera, Archodonata, Protodonata, Megasecoptera and Campylopterodea); and rest of the Pterygote orders into a section, Neoptera (modern orders) which are adapted for wing flexing mechanism.

Palaeopterous: Wing flexion mechanism absent, *e.g.*, order Odonata.

Pale headed stripped borer: See *Chilo suppressalis.*

Palmate: Like a hand.

Palpation: Refers to touching with labial or maxillary palpi. The movement can serve as a tactile signal to the other insect.

Palpifer: The small lobe or lateral sclerite located on the stipes of a maxilla and bearing the maxillary palpus.

Palpiger: The small sclerite or lobe of the mentum of labium that bears the labial palpus.

Palpomere: Palp segment.

Palpus (pl., palpi): A segmented antenna like structures arising on the maxilla or labium. Palpi have a sensory function and play a major role in tasting food. These also act as tactile and olfactory sense organs.

Pamphilidae: Family of super family Megalodontoidea of suborder Symphyta (Chalastogastra), order Hymenoptera. These insects are commonly called as web spining and leaf rolling sawflies. They are stout bodied insects with a simple antennal flagellum and a short ovipositor. The larvae live in rolled leaves or leaf webs and are devoid of abdominal limbs. *Neurotoma inconspicua* is a leaf roller on peach.

Pancreatic polypeptide: See vertebrate hormones in insects.

Panmictic: Randomly interbreeding population of individuals, *i.e.*, each individual is equally likely to mate with any individual of the opposite sex.

Panoistic egg tube: An egg tube in which the vitellarium contains eggs only.

Panoistic ovariole: Ovariole without nurse cells.

Panorpidae: Family of suborder Eumecoptera, order Mecoptera. Wings well developed and tarsi two segmented. Females without conspicuous ovipositor; legs not very long, normal, *e.g.*, *Panorpa* sp.

Panorpoid complex: See taxonomic grouping of orders.

Pantropical: Wide occurrence throughout the tropical and subtropical parts of the world.

Pantry: A room for provisions and table furnishings where plates, knives etc. are cleaned.

Paper factor: Prof. Slama (1965) brought with him a stock of *Pyrrhocoris apterus* from Prague (Czechoslavakia) to Harward University (U.S.A) and found that the fifth or last instar nymphs were all surprisingly moulted into the super-numerary sixth or even into the seventh instar nymphs or into adultoids in spite of normally developing into the adults. Further, it came to the notice that the paper towels used in rearing jars were the source of the JH mimic substances causing precocious effect. The newspaper in USA was prepared from wood pulp manufactured from balsam fir, *Abies balsamea* containing high concentration of JH activity. The substance was titled as the 'paper factor' by Slama and Williams (1966).

***Papilio demoleus* Linn.:** Belongs to order lepidoptera and family Papilionidae. It is commonly known as citrus caterpillar and is a pest of citrus and other species of family Rutaceae.

Papilionidae (swallow-tails): Family of super family Papilionoidea of suborder Ditrysia and order Lepidoptera. These are mostly tropical species. Most of them are large and conspicuously colored and possess a tail-like prolongation on hindwings. The wings are generally iridescent black with shades of green, red, blue and yellow. Body of larvae is smooth or with a series of fleshy dorsal tubercles or occasionally with a raised prominence on the fourth segment. Satae are absent on the body, exception being *Parnassius* which has secondary an eversible osmetrium. The pupa is characteristics in having two lateral cephalic projections and in hindwings being visible ventrally.

Papilionoidea: Super family of suborder Ditrysia and order Lepidoptera. This includes the butterflies which have clubbed or dilated antennae. A frenulum is absent and a well developed humeral lobe is present on the hindwing. Rarely a few others possess clubbed antennae but they are readily distinguished from the members of Papilionoidea by the presence of a frenulum. The following families, *viz.* Nymphalidae, Danaidae, Satyridae, Heliconiidae, Libytheidae, Lycaenidae, Pieredae, Riodinide and Paplionidae are included under this superfamily.

Papilla: A small nipple like elevation.

Papule: Small solid elevation on the skin.

Parabiosis: The utilization of the same nest or living place and sometimes the same odour trails by colonies of different species which cannot keep their brood separate.

Parachor: The product of molar volume of a liquid and its surface tension raise to power ¼. If two liquids have same surface tension under a given set of conditions, their parachors are proportional to their molar volumes.

Paradichlorobenzene: A low toxicity compound; PDB is very similar to the naphthalene in moth balls and flakes. It is a white crystalline material, vapourises slowly to form an inflammable gas which has been used as a fumigant. The odour is less persistent than naphthalene and more satisfactory for treating woolen clothes against cloth moths. It is also repellent for bats and soil fumigant for peach tree borer and apple wooly aphis. It is also used to protect museum specimens from the attack of dermestid beetle larvae. The empirical formula is $C_6H_4Cl_2$; doses over 300 mg kg^{-1} are harmful for human being, low acute oral toxicity for rats as survive at 1000 mg kg^{-1}, phytotoxic and seriously affect seed germination.

Paraffin wax: A solid mixture of hydrocarbons obtained from petroleum.

Paraglossa (pl., Paraglossae): Laterad to the glossae. One of a pair of lobes at the outer edges of the tip of labium; with the central glossae, the paraglossae make up the ligula.

Parallax: Differences in the direction or a shift in the apparent position of a body due to a change in the apparent position of the observer.

Parallelism: The independent acquisition of similar characters in related evolutionary lines.

Paralysis: A virus disease of adult bees which affects their ability to use legs or wings normally.

Paramere: One of a pair of processes usually articulated to the phallobase of the beetle aedeagus. Also referred to as lateral lobes. Occasionally used in the singular for the apicale (Spilman, 1952).

Parametabola: See neometabola.

Parameter: A measurable variable, quantity as distinct from a statistic or estimate.

Paraneoptera: See taxonomic grouping of orders.

Paranotal lobes: These are lateral lobes of the pronotum in certain fossil insects and theoretical lobes of the mesonotum and metanotum supposed to be the precursors.

Paranotal theory: This theory is also called as the 'flying squirrel theory', put forth by Miller (1873-75) and related to origin of wings. According to this theory, the wings are dorsal backward prolongations of pteronota. These are evolved from the rigid, double walled, lobe like lateral expansions of the thoracic terga, the paranota or the paranotal lobes. The fossil evidences reveal that the lateral lamellae on the paleodictyopteran winglets (typical paranota) are homologous with the wing pads of the modern insects and all these structures show venation alike that of the wings supporting strongly the paranotal theory. The paranotal theory is supported by Sharov (1966), Hamilton (1971-72), Wotton (1976) and Rasnitsin (1976). Snodgrass (1935) advocated strongly the wings as the lateral outgrowths of terga on the basis of the fact that the nymphal wing pads, prothoracic and abdominal lateral expansions of the modern insects are directly continuous with the terga and similarly, the haemocoel, nerves, and tracheae are extended directly from the body cavity into the wing pads. Kukaleva-Peck (1978), however, emphasized that paranotal theory received powerful but false support from the fossil record because of erroneous analysis and interpretations.

Paranotal: Pertaining to the paranota.

Paranotum (pl., paranota): The notal flanges in the fossil insects.

Parapatric: Of all patric populations whose ranges are in contact and between which gene exchange is geographically possible even without sympatry.

Paraproct: A pair of lobes bordering the anus lateroventrally.

Parapsidal furrows: Lateral grooves divergent anteriorly in the posterior part of scutum of the mesothorax of some Hymenoptera.

Parapsidal suture: A longitudinal suture of the mesonotum spearing the median area from the lateral area (hymenoptera). It is also called the notaulix.

Parapsides: Lateral areas of the mesoscutum in some Hymenoptera, laterad of the parapsidal furrows.

Paraptera: See epipleurites.

***Parasa lepida* Cramer:** An insect pest of castor and pine apple, commonly known as castor slug caterpillar (Lepidoptera, Limacodidae or Cochlidiidae).

Parasite: It is an organism that is usually much smaller than its host and a single individual usually does not kill the host. Numerous individuals may irritate, weaken or otherwise debilitate the host and occasionally cause its death *e.g.*, tapeworm, lice, fleas, mosquitoes etc. on animals including human beings. A parasite spends all or part of its life in close association with another species, taking food from it but giving nothing in return. Ectoparasites live on the outside of their hosts, while endoparasites live inside the host's body.

Parasitic food chain: The food chain starts from the herbivores and the food energy flow from larger animals to smaller animals. The larger animal is referred as host and smaller animal is parasite. Such food chain is quite significant in an ecosystem.

Parasitism: It is the relationship (temporary or permanent) between two species in which one (parasite) derives the benefit while the other (host) may or may not feel the loss or injury caused by the parasite. The parasite also gets shelter and depend on the host. For example, *Plasmodium*, tapeworm, *Ascaris* etc. are found as endoparasites of man, while bed bugs, mosquitoes, lice etc. are ectoparasites of man.

Parasitization: A quantitative term referring to the proportion of a host population attacked by parasites.

Parasitoid: It is a special kind of predator which kills its host and requires only one host (prey) for development into a free living adult.

Parasitology: (Gr., *para*= beside+ *sitos*= food). Study of animals that live or subsist on or in other animals.

Parasocial: See presocial.

Paraspecies: In palaeontology, isolated parts of animals or fragments when named equally as species or other taxa.

Parasporal body: An inclusion that develops with the endospore in the cells of certain bacteria during the process of sporulation.

Parategula: An apically rounded, more or less dorsally flattened lobe arising from posterolateral corner of mesoscutum and lying adjacent to the tegula. Present in some Eumeninae and some Masarinae (Hymenoptera)

Paratergite: See laterotergite.

***Paratetranychus citri* McGregor:** Belongs to Acarina and family Tetranychidae. It is commonly known as citrus mite and is a pest of international importance infesting sweet orange, lemon, grape fruit and sour lime.

***Paratetranychus punicae* Hirst.:** Mite pest of pomegranate and grapevine.

Parathion: An organophosphate insecticide, O,O-diethyl O-p-nitrophenyl phosphorothionate ($C_{10}H_{14}NO_5PS$); extremely toxic to mammals, oral LD_{50} to rats 2-6 mg kg^{-1}; contact and stomach poison with some fumigant action, phytotoxic in over dose, used against mites, aphids, eelworms. It is a clear oily liquid with boiling point 113 °C at 0.05 mm of Hg. The technical material is dark brown liquid with an unpleasant garlic odour, contains 90 per cent pure parathion and highly volatile. It is practically insoluble in water (24 mg l^{-1} at 25 °C) but soluble in organic solvents. In vivo, it is oxidized to paraoxon which is more toxic than parathion. The principal metabolic product in parathion in man is p-nitrophenol and other hydrolysis products. It is marketed under the trade names, Folidol, Ekatox and Thiophos.

Parathion, metabolism: All the thiophosphates are first converted *in vivo* to become toxic (activation). The activation is carried out by mixed function oxidase (MFO). In mammals, the major metabolites are formed from P-O bond cleavage. These are mono and diethylphosphoric acid and phosphoric acid. Such phosphate cleavages leave *p*-nitrophenol which is excreted in the urine. There appear three enzymes or enzyme system working at the parathion molecule independently in P-O bond cleavage. These are (i) MFO, (ii) GSH-S transferase, and (iii) phosphatase. Paraoxon on the other hand, is degraded by phosphatase action to give diethyl phosphoric acid. This phosphatase is sometimes called paraoxnase. Hitchcock and Murphy (1967) reported the formation of aminoparathion and aminoparaoxon among other metabolic products in many mammalian, avian and fish species. The break down products found in soil treated with parathion consist principally of aminoparathion and *p*-nitrophenol produced by the reducing capacity of yeast, chemical hydrolysis and hydrolytic capacity of soil bacteria, *Corynebacterium simplex* is able to convert parathion completely to nitrite in one week. *Rhizobium* converts parathion to aminoparathion and hydrolysis product diethylphosphorothioic acid.

Parecium: Air space which surrounds the fungus garden in the nest of a Macrotermitinae termite.

Parenchyma: A plant tissue composed of thin walled cells which often store food and usually retain the capacity to divide.

Parietals: The lateral areas of the cranium between frontal and occipital areas, separated by coronal suture.

Paris green: An organometal insecticide (green coarse powder), chemically copper acetoarsenate, $(CH_3COO)_2Cu.3Cu\,(AsO_2)_2$; highly toxic to mammals, tender plants and fruit trees; stomach poisons widely used in the 19th century against leaf eating insects. It was first used in 1867 for the control of Colorado beetle. Because of injurious to foliage and fruits in fruit trees, it is rejected for orchard spraying. It has now a limited use as mosquito larvaecide and for the control of snails.

Parotid: Saliva secreting gland situated near the ear.

Parrot: *Psitticula cyanocephala*, a bird pest causing damage in guava, mango, ber, loquat, apple, grain ears etc.

Pars intercerebralis: It is well differentiated median antero-dorsal cortical region of the protocerebrum of brain. It is bounded laterally by the corpora pedunculata, apically by the neurolemma and ventrally by the central body.

Parthenocarpy: Refers to the development of the fleshy part of fruit without fertilization of contained seed.

Parthenogenesis: Reproduction without fertilization; mostly through eggs but sometimes through viviparity. Reproduction without exchange of gametes is wide spread in the insects. Many species such as grass hoppers and some Lepidopterans occasionally develop from eggs which escape fertilization. Others, such as familiar stick insect, *Carausius morosus* are constantly parthenogenetic; males of this species are exceedingly rare. In honey bees, the unfertilized eggs produce drones and fertilized eggs produce workers and queens; the sperms are stored in the spermatheca during the copulation. Parthenogenesis may be (i) Constant: In some species it is of common occurrence, (ii) Sporadic: Under certain physiological conditions show tendency towards parthenogenesis, *e.g.*, both males and females develop from unfertilized eggs in *B. mori*, (iii) Cyclic: In aphids, some Hymenptera and some others, the asexual reproductive cycle may be interrupted by sexual reproduction which is often termed as alteration of generation.

Parthenogenetic: Refers to parthenogenesis.

Parthenogenic: Reproduction by the formation of egg cells capable of development without fertilization.

Parthenote: Offspring produced parthenogenetically; it may be male Hymenoptera or female Aphidoidea.

Partition coefficient: Ratio of the concentrations of a substance in solution in two phases which are in equilibrium.

Partition ratio: See partition coefficient.

Parts per billion (ppb): Ratio expressed as one part pesticide per 10^9 part sample. Strictly, the quantities should be in the same state, *i.e.*, weight by weight (solids) or volume by volume (liquids or gases), *i.e.*, 1 ppb = 1 µg/ kg. A common usage is for weight to volume but to avoid misunderstanding, it is recommended that the SI units are used rather than ppb; *e.g.*, µg/ l.

Parts per million (ppm): Ratio of amounts expressed as parts pesticide per 10^6 samples, 1 ppm = 1 mg/ kg. As with ppb, it is recommended that SI units are used rather than ppm, particularly for weight to volume.

Passalidae: Family of superfamily scarabaeoidea, suborder Polyphaga and order Coleoptera.Family of superfamily scarabaeoidea, suborder Polyphaga and order Coleoptera. These are dark brown or black somewhat flattened beetles with elytra having longitudinal grooves. The antennae are not geniculate. The mandibles are not developed specially in male and the mentum is deeply notched. On the head the characteristic horn-like structure is present. The adults are capable of producing a squeaking sound by rubbing the inner surface of its wings on the upper surface of the abdomen. The elongate grubs develop in decaying wood in forest areas. The hind pair of legs are much reduced with a short coax and a more elongate trochanter and function as a stridulatory organ producing sound by working across a striated area on the mesocoxa. *Epispheus neelgheriensis* and *Ophrygonius cantori* are some Indian species.

***Passer domesticus* (Linn.):** Belongs to order Passeriformes and family Ploceidae. It is a bird pest of crops, commonly known as the house sparrow.

Patella: A segment between the meropodite (femur) and carpopodite (tibia) in the legs of most Arachnida.

Pathogen A disease causing microorganism; a bacterium or virus.

Pathogen reservoir: A group of hosts harbouring a disease causing organism.

Pathogenic: Giving origin to disease or pertaining to pathogen.

Pathology: (Gr., *Pathos*= suffering). The nature, symptoms and cause of the disease.

Patrolling: To watch or investigate inside the nest of honey bee. The worker honey bees are especially active in patrolling and quick to respond in a group to contingency when it arises in the nest.

Paurometabola: See heterometabola.

Paussidae: A family of superfamily Caraboidea, suborder Adephaga and order Coleoptera. Tropical and subtropical beetles, small, red-brown or black in color, antennae characteristic in being very thick, broad and expended; elytra truncate and leave the terminal portion of abdomen exposed. They are found mostly in ant nests and sometimes on soil, under stones or attracted to light. Like Carabidae, these beetles also discharge a liquid which is irritant to the human skin. Ants lick the secretion eagerly from the body of the myrmecophilous beetles. The common Indian species belong to the genera *Platyrrhopalus, Merismoderus* and *Paussus.* The nests of the ant, *Pheidole latinoda* is inhabited by the beetles *Platyrrhopalus denticornis* and *Ceratoderus bifasciatus* in India.

pBR322: See plasmid vector.

PDB (Paradichlorobenzene): Crystals used to fumigate combs against wax moth.

PDI: Per cent disease intensity.

Pea beetle: *Laria affinis* Frol. is a pest attacking pea crop and in store; the grubs bore into the pods and attack green grain inside.

Pea leaf miner: *Phytomyza atricornis* Meign. is an insect pest of pea; maggots make shining white zig-zag galleries on leaves making them wither and dry up; also cause damage in the cauliflower, cabbage, turnips, radish, spinach, brinjal, potato, tomato, carrot, mustard, safflower, alfalfa, linseed and poppy.

Pea semilooper: *Plusia orichalcea* Fab. is an an insect pest of pea; cabbage, cauliflower and gram. The caterpillars by feeding on the leaves, defoliate the plants.

Pea stem fly: See *Melanagromyza phaseoli.*

Peach fruit fly: *Dacus zonatus* Saunders, a pest of peach, mango, fig, guava, sapota, ber, citrus, loquat, bottle gourd, tomato, long melon, brinjal, custard apple etc.

Peach leaf curl aphid: *Anuraphis helichrysi* Kalt. is a serious pest of peach and plum, sucks cell sap from young leaves which curl up, checks growth of the tree and causes fall of the immature fruits.

Peach scale: *Pseudaulacaspis pentagona* Targ., a pest of peach sucks juice from twigs to which it remains attached.

Peat: A type of soil consisting of partly decomposed organic matter that has accumulated under excessive moisture.

Pebrine: See silkworm diseases.

Pecten: (i) A comb like structure found at the base of antenna in some insects. (ii) In Hymenoptera, a comb of long tarsal setae, usually on the foreleg and is used to rake soil particles. In Lepidoptera, there may be a pecten of stiff bristle like scales or a dense scale tuft. In a few groups the scape is broadly flattened and concave beneath when it is said to form an eye-cap.

Pectinate: Pectinate or comb like antenna possesses long segments or processes only at one side which appears like teeth or comb, *e.g.*, in moth.

***Pectinophora gossypiella* Saund.:** Belongs to order Lepidoptera and family Gelechiidae. It is an insect pest of cotton, commonly known as pink bollworm.

PED: Plasma emission detector.

Pedicel: (i) The second segment of an insect antenna. (ii) The stem of the abdomen between the thorax and the gaster (in ants) or narrow waist of an ant. (iii) It is the last part or region, represents a stalk of the ovariole in the form of a short duct providing a passage to the ova from the vitellarium to the lateral oviduct. It is composed of tall epithelial cells and bears a large lumen but the upper end is closed till the time of release of first ovum from the ovariole into the lateral oviduct.

Pediculidae: A family of order Anoplura. All abdominal segments free, as also spiracles; forelegs similar, not more slender than others, *e.g.*, human head louse, *Pediculus humanus capitis* and human body louse, *P. humanus vestimentorum.*

***Pediculus humanus capitis*.:** Belongs to order Anoplura and family Pediculidae. It is an insect pest of human, commonly known as the human louse.

Pedipalp: One of the second pair of appendages near the mouth of a spider or other arachnid. Pedipalps can be modified for various functions such as reproductive, predatory or sensory.

Pedon: The lower lobe of a penis which is divided horizontally into dorsal and ventral lobes; usually referring only to the penis in Curculionoidea.

Pelagic: Inhabiting the sea or sea dwelling.

Pellet: (i) It is solid formulation of pesticides, the pellets are larger than granules and often used for molluscicide formulations. (ii) Seed coated with inert material, often incorporating pesticides, to ensure uniform size and shape for precision drilling.

Pelleted formulation: A formulation made up of inert carrier containing toxicant, *e.g.*, celphos (a formulation of aluminium phosphide).

***Pempheres affinis* Fst.:** Belongs to order Coleoptera and family Curculionidae. It is an insect pest of fibre crops, commonly known as the cotton stem weevil.

Penduncle: Stalk of flower cluster.

Penetrant: Oil added to a spray to enable it to penetrate the waxy insect cuticle more effectively.

Penetration: The process of entry.

Penis: The male intromittent organ; in beetles, it usually consists of an external, sclerotized tube and an internal, inflatable endophallus; also refers to the external tube.

Penta: Prefix of a compound denoting five.

Pentachlorophenol (PCP): Moderately toxic chemical used as wood preservative; effective against dry wood termites, wood boring insects and fungi. It is moderately toxic to mammals, acute oral LD_{50} for rats is 210 mg kg^{-1}, toxic to plants at contact point; also used as herbicide and extremely irritating to skin and eyes.

***Pentalonia nigronervosa* Coq.:** Belongs to order Hemiptera and family Aphididae. It is an insect pest of banana, large cardamom etc., commonly known as the banana aphid.

Pentatomidae: Family of superfamily Scutelleroidea of division Gymnocerata, suborder Heteroptera and order Hemiptera. These are moderate to large sized, often brightly coloured insects with five segmentes antennae and atriangular scutellum. They are broad, shield like in shape and produce a disagreeable odour. The bases of the antennae are concealed by the lateral margins of the head. The stink bugs belonging to order Pentatominae are mostly phytophagous. Those belonging to Acanthosomonae and Asopinae are mostly predaceous on other insects and some others feed on both plant sap and insects. The predaceous Pentatomid bugs are *Cantheconidia furcellata* and *Andrallus spinidens*, whereas, the phytophagous bugs are *Bagrada cruciferarum* and *Nezara viridula*.

Pepper flea beetle: *Longitarsus nigripennis* M. is an insect pest of pepper, small, shining yellow blue beetles.

Peptides: A combination of amino acids formed by the linkage of amino group of some of the amino acids with the carboxyl group of others.

Per capita income: Annual income of an individual in a population.

***Peregrinus maidis* Ashmd.:** Belongs to order Hemiptera and family Delphacidae. It is an insect pest of sorghum, commonly known as the sorghum shoot bug.

Perennial: Plant that continues to live for more than two years. Trees and shrubs are perennial plants.

Perfoliate antenna: Antenna in which most of the antennomeres are flattened and expanded on each side of the articulation.

Pergidae: Family of super family Tenthredinoidea, suborder Symphyta (Chalastogastra), order Hymenoptera. These are leaf feeding sawflies and the adults possess six-segmented antennae.

Pericardial: Surrounding the heart.

Pericardial cells: These cells are scattered along the heart, alary muscles and aorta. These have been considered in the past to have an excretory function but now a day they are regarded as comparable with the reticulo-endothelial system of vertebrates.

Pericardial septum: See dorsal diaphragm.

Pericardial sinus: Body cavity above the dorsal diaphragm.

Perikaryon: Cell body of a neurone, also called the soma or cell body.

Perilampidae: Family of super family Chalcidoidea, suborder Apocrita and order Hymenoptera. The Perilampidae are metallic green or black stout bodied insects with coarsely punctuate large thorax and small, triangular abdomen. They are mostly hyperparasites and a few are primary parasites. The insects lay the eggs on foliage and the planidium type of larvae get themselves attached to the larval host found on the foliage and enter its body.

***Perina nuda* Fab.:** Belongs to order Lipidoptera and family Lymantriidae. It is an insect pest of jack fruit, commonly known as the jack fruit leaf webber.

Perinaem: The region between the anus and the opening of the reproductive system (vulva or scrotum).

Perineural sinus: The body cavity surrounding the dorsal blood vessel, limited ventrally by the dorsal diaphragm.

Periodic release: It involves repeated liberations of artificially maintained high population levels of indigenous biotic agents in situations where such levels are unattainable naturally.

Periodicity: Recurrence of particular behaviour or physiological events in a regular pattern.

Periphallic organ: Peripheral genital processes of the ninth segment or also of other segments in the genital complex including the movable claspers, or harpagones.

Peripheral: Pertaining to the surface or superficial parts of the body.

Peripheral nervous system: All the nerves emanating from the ganglia of the central and visceral nervous system.

***Periplaneta americana* L.:** Belongs to order Dictyoptera and family Blattidae. It is an insect pest of household, commonly known as the cockroach.

Periplasm: Cytoplasm located around the periphery of the yolk of an insect egg.

Peripneustic respiration: See polypneustic.

Periproct: Posterior somite of a segmented animal in which the anus is located, also called the telson.

Peristaltic contractions: Rhythmic or waves of contractions of longitudinal and circular surrounding or composing a tubular organ, like the alimentary canal.

Peristome: The ventral marginal part of cranium formed by the clypeus, or epistoma, pleurostomata and hypostomata.

Perisympathetic organs: They are associated with the sympathetic unpaired fine median nerve and with the transverse nerves representing the neurohaemal organs of the neurosecretory cells of the ventral ganglia and are widely evolved among different groups of insects. Each ventral ganglion, therefore, constitutes its own independent neurosecretory complex.

Peritreme: A sclerite of the body wall containing the spiracular opening.

Peritrophic membrane: A chitinous membrane in insects secreted by the anterior or entire epithelial layer of mid gut which envelopes the food contents and protects the naked epithelial cells from abrasion by hard food granules as no mucus is secreted in insects. It forms an envelope around the food to pull it loose from the midgut, remains around the food and passes the faeces. In fluid feeding insects, it is wanting. It is permeable to the digestive juices and products of digestion and thus functions as an ultrafilter. On the basis of structure and mode of formation, the peritrophic membrane can be classified into two types; (i) lamellated peritrophic membrane, and (ii) non-lamellated peritrophic membrane.

Perivisceral sinus: The body cavity surrounding the digestive system, reproductive system (viscera or internal organs) -between the dorsal and ventral diaphragms.

Perlidae: A family of order Plecoptera having widely separated anterior coxae and vestigial mandibles. The third tarsal segment is longer than first and second together, *e.g.*, *Perla* and *Chloroperla* sp.

Permanent ectoparasites: These are wingless (apterous) parasitic insects and are attached on the body of host permanently throughout the life. They feed on the dry skin and hair but never pierce the skin or draw the blood of host. Sometimes, they form a wound which is infected by germs and cause cutaneous myiasis, *e.g.*, lice.

Permanent parasitism: See inquilinism.

Permanent social pasitism: See inquilinism.

Permethrin: A synthetic pyrethroid insecticide with low toxicity marketed under trade names Ambush and Pounce, effective against wide range of insects, especially caterpillars that feed on crops, trees etc. Chemically, it is 3-phenoxbenzyl (±) cis, trans- 2, 2 dimethyl-3- (2, 2- dichlorovinyl)-cyclopropane-1-carboxylate ($C_{21}H_{20}Cl_2O_3$). It is viscous liquid practically insoluble in water but soluble in most of the organic solvents. It acts by ingestion and by contact. Vapour activity is very low. It has powerful initial knock down and good level of persistence. It is highly effective against Lepidopterous pests, also controls Coleopterous, Hemipterous, and Dipterous pests. It is marketed under trade names, Ambush, Kafil, Perthrine and Permasect.

Permissible level: A concentration expressed in ppm of residue in or on a food when first offered for consumption. It is calculated from the acceptable daily intake (ADI), the food factor and the average weight of the consumer.

Persistence: Perseverance time of a chemical (pesticide and/ metabolite) in soil, animal or other defined environmental compartment. In chemicals, persistence is the result of low volatility and chemical stability. Certain organochlorine insecticides such as DDT etc. are highly persistent. See also dissipation.

Persistent virus: A virus that persists within the body of a vector; circulative virus.

Pest density: The population level at which a pest species causes economic damage.

Pest insect: An insect causing damage to man's crops, domestic animals or possessions. (See pest also).

Pest intensity: The total number of pests per unit of habitat or area.

Pest management strategies: The pest management strategies are meant to provide pest management which include (i) Sole reliance on natural forces: It is a slow and limited approach as the competence of natural enemies is very difficult in the agroecosystem, (ii) Prevention or eradication: It can be achieved by using insecticides which is effective method of controlling insect pests of cash crops and results into intolerable and irrepairable losses, (iii) Containment or correction: Sometimes, pests develop irregularly and are harmless at low levels which can be successfully contained or corrected simply by applying cultural or mechanical techniques.

Pest management: The careful manipulation of a pest situation, after extensive consideration of all aspects of the life system as well as ecological and economic factors.

Pest spectrum: The complete range of pests attacking a particular crop.

Pest upset: A condition in which innocuous species become man made pest as a result of insecticide use, which unintentionally destroys their population regulative natural enemies.

Pest, categories: Key pest, major pest, minor pest, sporadic or occasional pest, potential pest, migrant pests are different categories of the pest. See details.

Pesticide (economic poison): Any substance (chemical or microbial) which because of its toxicity is used to control pests.

Pesticide deposit: The initially laid down amount of pesticide on the treated surface.

Pesticide formulation: It is the pesticide product offered for sale and generally comprises active ingredient, adjuvant and other formulants combined to render the product useful and effective for the purpose claimed.

Pesticide residue enforcement: Pesticide residue monitoring programme where the intention is regulatory action against non-complying consignments.

Pesticide residue monitoring: Sampling and analysis of pesticide residues in biological and environmental samples taken according to pre-arranged schedules.

Pesticide residue: (i) The initially led down amount of pesticide (deposit) after being acted upon by weathering agencies which is known as residue. The residues may be present on or in the feed or food commodity, soil air or water. For regulatory purposes, it includes the parent compound and any specified derivatives such as degradation and conversion products, metabolites and impurities considered to be of toxicological significance. (ii) The pesticides, its metabolites, derivatives and related compounds to which the maximum residue limit (MRL) applies, as specified by the Codex or a national regulatory authority.

Pesticide resistance: Genetically selected tolerance of pest populations to pesticides. Resistance is caused by the repeated exposure of the pest population to pesticide treatment. Sensitive individuals are killed, while the resistant individuals will continue to reproduce. Resistance can develop to both chemical and microbial pesticides.

Pesticide trade name: Proprietary name assigned to a pesticide or its formulation by the company manufacturing or selling it.

Pesticides: A chemical by virtue of its toxicity (poisonous properties) is used to kill pest organisms. A term of wide application which includes all the more specific applications- insecticide, acaricide, bactericide, fungicide, herbicide, molluscicide, nematicide, rodenticide, etc.

Pests of medical importance: Mosquitoes, sand flies, black flies, house flies, tsetse flies, human flea and human lice are important pests transmitting the causal agents of diseases, thus are regarded as the pests of medical importance.

Pests of veterinary importance: Several species of insects attack domestic animals such as cattle, sheep, goat, dogs, poultry, ducks etc. which are regarded as the pests of veterinary importance. These pests are horse flies, stable flies, buffalo flies, horse sucking louse, horse bot flies, horse or cattle ked, cattle biting louse, cattle blood sucking louse, cattle warble flies, screw worm, sheep bot or nose fly, green bottle fly, black blow fly etc.

Petaluridae: A family of order Odonata. Have notched labia, *e.g.*, *Petalurus* sp.

Petiolate: Attached by a narrow stalk.

Petiole: The narrow waist of bees and wasps and some other hymenopterans: often known as the pedicel when referring to ants.

Petrochemical: A chemical derived from crude oil or natural gas.

Petroleum products: Any product containing gasoline, kerosene, oil or similar products derived from petroleum.

Petroleum: Petroleum is a complex mixture of hydrocarbons. Oxygen, nitrogen and sulphur are also present in small amount. It is found in the depth of earth. It is black coloured viscous liquid called crude or unfiltered oil. This oil is formed by the continuous decay of animals and plants suppressed in the Earth's surface for about 400 million of years.

Petrolium distillate: Kerosene.

Petrolium oils: These are the oils refined from the crude oil.

***Petromalus luzonensis*:** Egg parasitoid (Hymenoptera) associated with citrus caterpillar, *Papilio demoleus* Linn.

pH: Negative logarithm of hydrogen ion concentration. The pH of insect blood is usually very slightly acidic; 6.4 in *Celerio* larva, 6.6-6.8 in *Bombyx mori*, 6.83 in *Apis* larva; but it may differ slightly in the two sexes and may vary with age of the insect.

Phadka grasshopper: *Hieroglyphus nigrorepletus* Boisd. is an a polyphagous pest, serious on millets and maize; also attacks paddy, sugarcane etc. Nymphs and adults feed on leaves and also cut the ears.

Phagocytic organs: In some Tettigonids and Gryllids, triangular shaped sacs are observed in the anterior region of abdomen that are termed as phagocytic organs.

The dorsal side of the phagocytic organs is formed from the phagocytic cells while the ventral wall from the dorsal diaphragm. The heart opens into the phagocytic organs through the lateral excurrent ostia. The lumen of these organs is filled with the multinucleated phagocytic cells. These serve as the filters.

Phagocytic: Any cell that ingests microorganisms, other cells and substances.

Phalacridae: Family of superfamily Cucujoidea, suborder Polyphaga and order Coleoptera. The small shining convex beetles measuring 1 to 3mm long are usually brownish or black and oval in shape. The coxal cavities of forelegs are open behind; tarsi 5-5-5, the second and third bilobed and fourth small. Tarsal clothed are toothed and appendiculate. The larvae of genus *Phalacrus* develop on smuts on sugarcane, grasses etc. the species of *Olibrus* are known to be phytophagous in the heads of compositae.

Phalaenophily: Transfer of pollens by moths, a type of pollination.

Phallobase: Proximal part of the phallus, highly variable in development, sometimes a large structure supporting the aedeagus, often produced into a thecal fold or sheath about the aedeagus sometimes represented only by basal phallic sclerites in the wall of genital chamber. The basal lobe of aedeagus in beetles; often enlarged or combined with the fused parameres to form a tegmen.

Phallocrypt: A pocket of the phallobase or of the genital chamber wall containing the base of the aedeagus.

Phallomeres: Genital lobes formed at the gonopore in the ontogeny of some insects; in most cases they unite to form the phallus but in the Blattidae and Mantidae, they develop separately into complex genital organs of the adult.

Phallotheca: A fold or tubular extension of the phallobase about the aedeagus.

Phallotrme: The distal opening of the endophallus, usually at the end of aedeagus.

Phallus: The unpaired penis or median intromittent organ including the phallobase, aedeagus, endophallus and various processes of the phallobase.

Pharate instar: Hinton used the term apolysis for the separation of epidermis from the old cuticle and the pharate state for the period between apolysis and ecdysis when the new cuticle is in process of formation below the old. Thus, the old cuticle is known as exuviae and the developmental stage baring the exuviae is pharate instar. In other words, the pharate stage is the larval period between apolysis and ecdysis.

Pharynx: Anterior part of foregut (between the mouth and the oesophagus) which has additional dilator muscles which are well developed in sucking insects, like Hemiptera and adults of Lepidoptera as in these the pharyngeal tube is used to take the liquid diet from the host. In insects, with biting and chewing mouth parts pharyngeal muscle helps in pushing the food from mouth to oesophagus.

Phase-I metabolism: The processes by which the xenobiotics are eliminated, are divided into two general groups, *viz.*, phase I and phase II reactions or metabolism. Phase I reactions involve initial biotransformation of a pesticide in which a hydrophilic functional group is introduced in the molecule of xenobiotic.

Phase I reactions involve mainly oxidation through MFO (mixed function oxidase), reduction, hydrolytic processes, glutathione mediated reactions etc.

Phase-II metabolism: Biotransformation, where the pesticide or phase-I metabolites are conjugated with naturally occurring compounds, *e.g.*, sugars, glutathione, amino acids etc.

Phasmid vector: These are artificial vectors with the characteristics of plasmid and phage vectors.

Phasmida: Insect order made up of the walking sticks and leaf insects, characterized by short leathery wings (or wings absent), chewing mouthparts, slender body, and an incomplete metamorphosis.

Phasmidae: A family of order Phasmida, includes leaf insects. The leaf insects are flattened and leaf like, winged or wingless with lamellate expansions of the legs (forefemora).

Phasmodidae: A family of order Orthoptera, suborder Ensifera and super family Tettigonioidea. Wingless forms of grasshoppers with long slender bodies looking like stick insects.

***Phenacoccus hirsutus* Cr.:** Belongs to order Hemiptera and family Coccidae. It is an insect pest of jute, commonly known as the jute mealy bug.

Phenetic: The classification based on appearance of organisms rather than on evolution from a common ancestor.

Pheneticist: One who studies the phenotypic characters.

Phenolic compounds: See dinitrophenols.

Phenon (pl., phena): A sample of phenotypically similar specimens.

Phenothiazine: An organic stomach poison; a light brown or sometimes green, crystalline chemical; finding increasing use in the removal of parasites from livestock.

Phenotype: The totality of characteristics of an individual due to interaction between genotype and environment. The external appearance of the character is called phenotype, *e.g.*, dwarfness, tallness etc.

Phenylphosphorothioate: See EPN.

Pheromone: A substance secreted by an animal which when released externally in small amounts causes a specific reaction, such as stimulation to mate with or supply food to a receiving individual of the same species.

Pheromone (Ectohormone): A substance secreted by an insect to the exterior causing a specific reaction in the receiving insect of same species. For the first time, Carlson and Luscher (1959) have proposed the term pheromone to the chemical substances that enable members of the same species to communicate each other. The pheromones are of various types: trail making pheromones, sex attractants or courtship pheromones, aggregation pheromones, primer or sex maturity specific pheromones etc. The sex pheromones are found to be useful in controlling the insect pests. These are produced by one sex only and affect the other sex in majority of insects, while in some species these are produced by both sexes in order to perform the courtship or mating. They are reported in various groups of insects, *viz.*, Lepidoptera, Diptera, Coleopteran, Hymenoptera, Neuropteran, Mecoptera and Hemiptera.

Pheromone trap: An insect trap in which dispensers with insect pheromones are used as an attractant. Usually the inside surface of the traps is covered with glue to prevent escape.

Pheromone, synthetic: Recently, sex pheromones of some species have been identified and synthesized such as; (i) gypsy moth, *Porthetria dispar* Gyptol or disparlure, cis-7,8-epoxy-2-methylocadcand, (ii) Silk moth, *Bombyx mori*, Bombycol-trans-10-cis-hexadecadienol, (iii) Pink bollworm, *Pectinophora gossypiella*- 10-propyl-trans-5, 9-tridecadienyl acetate, (iv) Cabbage looper, *Trichoplusia ni*- cis-7-dodecenyl acetate, (v) Honey bee queen, *Apis mellifera*- 9 keto-trans-2-decenoic acid. All are volatile substances and act as olfactory receptors stimulating directly the central nervous system so that change in behaviour can be induced. They cause insects to aggregate for mating. The use of sex pheromones is being done in the pest control by mating disruption and mass trapping the adults.

Philopatry: The tendency of an organism to stay in, or return to its home area.

Philopteridae: A family of suborder Ischnocera of order Mallophaga.Very widely distributed family occurring on birds. Antennae five segmented. Example, *Goniocotes columbicola, Goniodes*.

Phlaeothripidae: A family of suborder Tubulifera, order Thysanoptera.

Phlaeothripinae: Subfamily of family Phlaeothripidae, suborder Tubulifera and order Thysanoptera. Maxillary stylets slender, never broadened, narrower than labial palps. Distance between hind coxae less than that of mid coxae. This is a major subfamily with several genera and species, *e.g.*, *Haplothrips* sp.

Phloem: Part of the plant's vascular system adapted to the transport of photosynthetic products from leaves to rest of the plants.

Phorate: Highly toxic, systemic organophosphate insecticide, O, O-diethyl S-ethylthiomethyl phosphorothiolothinoate ($C_7H_{17}O_2PS_3$), highly toxic to mammals, oral LD_{50} for rats 100-180 mg kg^{-1}. It is slightly soluble in water (70 mg l^{-1}). It is a clear liquid, technical grade contains 90 per cent phorate and easily oxidized to sulfoxide which is insecticidal in action and more resistant to hydrolysis than the parent compound which enables phorate to have a long residual action. The final products of metabolism are diethyl hydrogen phosphate and then phosphoric acid which are biologically inactive. It is effective against soil insect pests, corn rootworms, potato and vegetable insect pests and marketed under trade name Thimet.

Phorate, metabolism: The activation reaction of phorate gives sulfoxide and sulfone and increases the anticholinesterase activity in plants and animals. Sulfoxide and sulfone by the action of phosphatase are converted into non-toxic phosphoric acid and sulfone derivative of thioalcohol.

Phoresis: The usage by one animal of another solely as a means of transport, *i.e.* certain mites on various other insects.

Phoresy: A type of symbiotic relationship in which one organism associates with another species in order to obtain transportation.

Phoridae (hump-backed flies): Family of suborder Cyclorrhapha, order Diptera. Phoridae are small or minute, grayish-black or yellowish flies characterized by hunched thorax which gives a hump-backed appearance, heavily developed anterior veins on forewings and laterally flattened hind femora. The wings are often vestigial or wanting. These insects are abundant in decaying vegetation. The larvae live in decaying vegetation. The larvae live in decaying animal or vegetable matter. Some forms are known to live in fungi. Some species are parasitic on other insects. A few are known to be either inquilines in the nests of ants and termites or parasitic on ants and millipedes. The flattened larvae sometimes possess at their tail end a breathing tube and the pupa has two thoracic breathing tubes or processes.

Phosalone: A synthetic organophosphate insecticide and acaricide, S-[(6-chloro-2-oxobenzodazolin) methyl] O, O-diethyl phosphorodithionate ($C_{12}H_{15}CNO_4PS_2$); white crystalline substance with garlic odour, insoluble in water but soluble in organic solvents, very persistent and substitute of DDT, contact poison, moderately toxic, acute oral LD_{50} for rats is 100-180 mg kg^{-1} and controls certain mites and insect pests on apple, pears, grapes and field crops. In plants, it is oxidized to a relatively unstable compound which quickly breaks down. In animal tissue it is degraded rapidly. It is marketed as Zolone.

Phosdrin: Dimethyl 1-carbomethoxy-1-propenyl-2-phosphate. It is an organophosphate compound, effective against sucking pests like, aphids, mites etc. as it is systemic in action.

Phosmet: It is a moderately toxic insecticide, effective against wide variety of fruit and shrub pests, alfalfa weevils and livestock pests.

Phosphamidon ($C_{10}H_{19}O_5ClP$): 2-chloro-2 diethyl carbomoyl-1 methylvinyl dimethyl phosphate. It was synthesized by Beriger (1955) but its insecticidal properties were discovred by Bachmann and Meierhaus (1956). Pure form is colourless liquid highly soluble in water, alcohol and acetone with boiling point 70 °C at 0.01 mm of Hg. The technical grade is dark brown and commercial product is bright violet due to addition of a dye. It contains two isomers, *i.e.*, alpha and beta in the ratio of 3: 7. The insecticidal properties are due to beta isomer. It is a systemic and contact insecticide with low contact action, less toxic to fishes, highly toxic to bees. It is hydrolysed to desethylphosphamidon which has as toxic as phosphamidon. In plants, desethylphosphamidon is rapidly broken down to non-insecticidal compounds like monochloroacetone, ethylamine and CO_2. It is marketed under the trade name Dimecron as soluble concentrate.

Phosphine: A toxic gas. See aluminium phosphide.

Phosphorus cycle: Phosphorus is an important and necessary constituent of protoplasm, tends to circulate, the organic compounds being broken down eventually to phosphates which are again available to the plants. The great reservoir of phosphorus is not the air but the rocks and other deposits which have been formed in the past geological ages. These are gradually eroding, releasing phosphates to ecosystems but much phosphate escapes into the sea where part of it is deposited in the shallow sediments and part of it is lost to the deep sediments. The means of returning phosphate into the cycle may be presently be inadequate to compensate for the loss. In some parts of the world, there is no upliftment of the sediment at present and the action of marine birds and fishes (being brought by animals and man) is not adequate. Sea birds apparently play important role in returning phosphorus to the cycle.

Phosphorus: An element.

Phostoxin: Trade name of aluminium phosphide.

Photodecomposition: Decomposition of pesticide as a result of exposure to the light.

Photolysis: Chemical reaction caused by light in which a bond is cleaved. See photodecomposition

Photoperiodism: (i) A response by an organism to the duration or pattern of light and dark periods, (ii) seasonal responses to the changes in length of day light.

Photoreceptor: Receptor which perceives light stimulus. Adults and nymphs of insects have often two compound eyes and up to three simple eyes on their head.

Photosynthesis: Photosynthesis is the process in which green plants prepare their food. The green plants (which have chlorophyll) in presence of sun light convert water and carbondioxide (inorganic substances) into organic substances. Roots absorb water from the soil and transport it to the leaves. The leaves absorb carbon dioxide from the atmosphere through their minute pores. The green leaves contain chlorophyll. In this process, sugar is produced which gives chemical energy. $6CO_2+12H_2O \xrightarrow{\text{sun light and chlorophyll}} C_6H_{12}O_6+6H_2O+6O_2$.

Photovoltaic (PV): Photovoltaic is the technical term used for solar electricity.

Phragma: A partition. A projecting ridge to which muscles are attached.

Phragmata: See endothorax.

Phragmocyttarous: Pertaining to nests and especially wasp nests in which combs are attached laterally to the inner surface of a bag like envelope.

Phragmosis: A condition in which the head or tip of the abdomen is truncated and is used as a living plug for the nest entrance. It occurs in ants and termites, usually in the soldier caste.

Phthiriidae: A family of order Anoplura. Forelegs more slender than others; abdominal segments 3-5 fused and with three pairs of spiracles. The pubic or crab louse of man, *Phthirus pubis*.

***Phycita infusella* (Meyrick):** Belongs to order Lepidoptera and family Noctuidae. It is an insect pest of cotton, commonly known as the bud moth.

***Phycita orthoclina* Meyrick:** Belongs to order Lepidoptera and family Phycitidae. It is an insect pest of tamarind, commonly known as the tamarind fruit borer.

Phycitidae: Family of super family Pyralidoidea of suborder Ditrysia and order Lepidoptera. The moths are small, delicate insects with elongate forewings having no R_5. The hindwings are broad and bear pectin of hairs on the dorsal side. Males are characterized by swollen basal antennal segment. Larvae feed in various ways. Larva of the Indian meal moth *Plodia interpunctella* is destructive to maize, figs and seeds of various kinds. *Cadra cautella* is a serious pest of stored products feeding on all type of grains in granaries, warehouses, markets and houses.

Phyllocnistidae: Family of super family Tinaeoidea of suborder Ditrysia and order Lepidoptera. The larvae are apodous and mine into leaves. The citrus leaf miner *Phyllocnistis citrella* is a notorious pest of citrus tree in India. The other species include *P.toparcha* on grapevine and *P. chrysopthalma* on cinnamon.

***Phyllocnistis citrella* Stainton:** Belongs to order Lepidoptera and family Gracillariidae. It is commonly known as citrus leaf miner and is a pest of citrus and other species of family Rutaceae.

***Phyllotreta cruciferae* Goeze, *P. chotanica, P. birmanica, P. oncera, P. downesi*:** Belongs to order Coleoptera and family Alticidae. These are insect pests of cabbage, commonly known as the cabbage flea beetles.

***Phyllotreta*:** A genus within the Alticidae family (order Coleoptera).

Phylloxeridae: A family of superfamily Aphidoidea, suborder Homoptera and order Hemiptera. Most characters are of Adelgidae but wings held flat at rest; Cu and A arise on common stalk on forewings. Apterous parthenogenetic female; if possess wax, will be non-flocculent.

Phylogenetic tree: A graphic representation showing the descent relations of different organisms.

Phylogenetic: Evolutionary development of any animal species.

Phylogeny: The evolutionary history of an organism or taxonomic group.

Phylum (pl., phyla)**:** A major division of the animal kingdom, containing various suborders and classes etc. Some examples of animal phyla are Arthropoda, Mollusca, Nemathelminthes etc.

Physical control: The use of physical factors of environment such as temperature, photoperiod, humidity and ecological conditions of the pest or that of radiations in controlling insect pests (*e.g.*, artificial cooling, heat treatment, superheating, radiation etc.).

Physical gill: The hairs absorb and store the air for respiration forming the physical gills. They are capable of absorbing the respiration from the aqueous media and are found on the hind legs of aquatic bugs, *Noucoris, Notonecta, Corixa* and aquatic beetle, *Dytiscus*. They can store air for considerable long time. In *Notonecta*, the hairs (hydrofuge hairs) are concentrated around the spiracles. These are hydrophile on one side (on withdrawal from the water) and hydrofuge on the other (on immersion). As these touch the water surface, a fringe of hairs spreads outwards by surface tension. The air space within is put into communication with the atmosphere.

Physical poison: Poison which kills the insect by exerting a physical effect is known as physical poison. Heavy oils and tar oils kill the insects through asphyxiation, *i.e.*, exclusion of air. Inert dust cause abrasion of cuticle or absorb moisture from the body of insect. The insect dies due to loss of body moisture.

Physical properties: Solubility, volatility, inflammability, state etc. are physical properties of a substance or insecticide.

Physical sciences: The physical science pertains with the study of all the non-living objects found in nature. It is classified into physics, chemistry, geology, astronomy and mathematics.

Physical selectivity: Refers to the use of broad spectrum pesticide in such a way as to obtain selective action against certain insect pests. This may be accomplished by exploiting pest behaviour and feeding habits and selecting pesticidal formulations, dosages and application times. These pesticides are inherently more toxic to some pests than to others.

Physics: In this branch of science, various physical properties of matter and energy, and their interrelationship are studied. The various forms of energy such as electrical energy, mechanical energy, sound energy etc. are also studied in the physics.

Physiology: Gr., *Physis*= nature. The study of living processes or functions within the cells.

Physogastry: Refers to swelling of the abdomen to an unusual degree due to the hypertrophy of fat bodies, ovaries or both.

***Phytomyza atricornis* Meigen:** Belongs to order Diptera and family Agromyzidae. It is an insect pest of peas and ornamental plants, commonly known as the pea leaf miner.

Phytophagous: Herbivorous, feeding on plants.

Phytoplankton: Microscopic plant life living suspended in water.

Phytosanitation: The removal and destruction of infested or diseased plants from a crop so that they do not form a source of infestation for healthy plants.

Phytotoxic: A chemical liable to damage or kill plants (especially the higher plants) or plant parts.

Phytotoxicity: Injury caused to the plants by the pesticides and chemical metabolites.

Phytotoxin: A chemical produced by plants that is toxic to self or other plants (See allelochemical).

Pictured: A term used to describe wings, especially among the Diptera, which have dark mottling on them.

Piercing and sucking type mouth parts: These type of mouthparts take most valuable liquid known as sap of the growing plants and blood of living animals. In order to get the food, they puncture the skin of animals or epidermis of the plants making a very tiny invisible hole with their needle like stylet through which they suck the cell sap or blood. The labrum is usually modified to form the outer tube which is simply a protective structure of outer parts of the mouth parts and has nothing to do with piercing the tissue and drying the liquid food. The mandibles and maxillae are greatly elongated and modified. These serve for piercing the skin of an animal or epidermis of a plant. These also form the food channel or the tube through which the liquid sap is sucked up. These mouth parts are characteristic of order Hemiptera and mosquitoes. The piercing and sucking mouth parts are of various types described below: (i) Hemipteran PSM, (ii) Anopluran PSM, (iii) Dipteran mouth parts: mosquito PSM, Horse fly PSM, Muscoid PSM (iv) Siphonapteran PSM, and (v) Lepidopteran PSM.

Pieridae (whites): Family of super family Papilionoidea of suborder Ditrysia and order Lepidoptera. Medium sized butterflies, white or orange or yellow colored with black markings. Legs are alike and normal in both sexes. The larvae are devoid of osmeteria, fleshy filaments, cephalic and anal processes and their body segments are divided into annulets. Numerous secondary setae are present on the body of larvae. The crochets of prolegs are bi- or triordinal arranged in a mesoseries. The pupa is attached to the host plant by its anal end and a central band of silken thread suspends it in an upright position. The pupa is characteristics in having a single median projection or spine and in hindwings being not visible ventrally. Butterfly migration occurs in certain species such as *Appias* and *Catapsilia.* In North- India, butterflies of *Pieris brassicae, P. boeticus* and *Eurema blanda silhetana* move up to cooler regions in March-April with the onset of hot weather.

***Pieris brassicae* Linn.:** Belongs to order Lepidoptera and family Pieridae. It is an insect pest of cabbage, commonly known as the cabbage caterpillar.

Piesmidae: Family of superfamily Tingoidea, suborder Heteroptera and order Hemiptera which includes bugs which have less of lace like sculpturing on the body.

Pigeon pea pod fly: *Agromyza obtusa* Malloch is an insect pest of pigeon pea, whose larva enters pods and feed on seed.

Pigments of cuticle: The insect exocuticle possesses various types of pigments, such as melanin, carotenoids and pterins producing fluorescent colouration in the wing scales of the butterflies. In most of the carabid and cincindelid beetles, entire epicuticle is covered with a thick layer of secretory material of dermal glands and provides brilliant pigmentation.

Pilifer: The lateral projections on the labrum.

Piliform scales: Hair like.

Pill: A pelleted formulation in the form of minute tablets, oestrogen and progestron pills delays the ovulation in the ovary.

Pilose: Densely clothed with hair.

Pine apple mealy bug: *Pseudococcus bromeliae* is an insect pest of apple.

Pine apple thrips: See *Thrips tabaci.*

Pink bollworm: See *Pectinophora gossypiella*.

Pink borer: *Sesamia inferens* Walk. is a species of stem borer infesing wheat, rice, barley, oats, sugarcane, sorghum, pearlmillet. The ears do not form and if form, then dry up.

Pink waxy scale: See *Ceroplastes rubens.*

Pinkboll worm of cotton: See *Pectinophora gossypiella.*

Pinworm: Parasitic nematode of family Oxyuridae attacking horses causing irritation in the anal region.

***Piper nigrum* Linn.:** Belongs to order Coleoptera and family Alticidae. It is an insect pest of black pepper, commonly known as the pollu beetle.

Piperonyl butoxide: A synergist for pyrethrum and rotenone, chemically, a-[2-(2-butoxyethoxy)ethoxy]-4,5-methylenedioxy-2-propyltoluene; relatively non-toxic to mammals, acute LD_{50} for rats is 7500-12800 mg kg^{-1}.

Piping: A series of sounds made by a queen, frequently before she emerges from her cell.

Pipunculidae (big-headed flies): Family of suborder Cyclorrhapha, order Diptera. Pipunculidae are small flies with a mobile, subhemisherical, very large head composed mostly of eyes. A long dorsal arista is present on the antenna. These insects are chiefly parasitic on leaf hoppers and fulgorids and hence are of some importance. *Pipunculus* and *Verrallia* are two common genera in India. *Pipunculus annulifemur* is a nymphal parasite of the mango hoppers.

***Pipunculus annulifemur* Bruvelli:** It is a natural enemy of mango hopper, *Amritodus atkinsoni.*

Pirimicarb ($C_{11}H_{18}N_4O_2$): A carbamate insecticide, chemically, 2-dimethylamino-5,6-dimethylpyrimidinyl dimethyl carbamate. It is not miscible in water but very soluble in organic solvents; melting point 90.5 °C; colourless crystalline solid, non-corrosive to the material normally used in spraying equipment. Moderately toxic, marketed under trade name Pirimor; particularly used against aphids (contact selective aphicide at low concentrations) on green house plants and potatoes. It is also used as a fumigant with no unpleasant odour. Its acute oral LD_{50} for rat is 147 mg kg^{-1}. It is formulated as 50 per cent dispersible powder.

Pirimor: See pirimicarb.

Piroplasmosis: Infection with tick transmitted piroplasma protozoans. These attack the red blood cells of cattle, dogs and other animals.

Piscicide: A pesticide used to control fishes.

Pistil: Part of flower consisting ovary, style and stigma.

Pitfall trap: A trap for catching ground crawling insects. Usually it is made up of a cup like container buried in the ground into which insects fall. The beaker contains a liquid (water with soap; alcohol) or glue to prevent escape.

Pival: A moderately toxic anticoagulant, chronic rodenticide.

Plague (bubonic plague, black death): An infectious disease caused by bacterium, *Bacillus* (=*Pasteurella*) *pestis* and spreads *via* the Oriental rat flea, *Xenopsylla cheopis.*

The oriental rat flea spreads the bacterium from man to man through bites. The fleas act as reservoirs and culture site for this bacterium. After the death of infected rats, the fleas move towards human population. Other species recorded to be the vector of plague bacterium are: (i) *Xenopsylla astia* (rat flea), (ii) *Pulex irritans* (human flea), *Ctenocephalides canis* (dog flea), and (iv) *Ctenocephalides felis* (cat flea).

Planidium larva: The free living, active, legless first instar larva of those parasitic Diptera and Hymenoptera exhibiting hyper metamorphosis. These long lived fusiform creatures are highly adapted to host location.

Planipennia: Suborder of order Neuroptera. Larvae with suctorial mouthparts.

Plankton net: A net used to collect planktons.

***Planococcus lilacinus* Ckll.:** A species of mealy bug (Psedococcidae, Hemiptera) feeding on tamarind, pomegranate, cashewnut etc.

Plant coverage: The degree to which plants within a sprayed area receive a deposit of pesticide.

Plant derived pesticide: See botanical pesticides.

Plant diseases: Harmful condition or sickness as a result of the infection caused by fungi, bacteria and viruses which adversely affect plant life.

Plant growth regulator (PGR): Naturally occurring or synthetic substances which influence plant development or reproduction but has no nutritive value. It alters the growth of plants. The term does not include the substances intended solely for use as plant nutrients or fertilizers.

Plant host range: The species of plants on which the insect species feeds.

Plant insecticide: The insecticide extracted from plants, flowers, roots, stems, leaves and seeds.

Plant louse: See aphid.

Plant nutrient: A chemical compound used by plants for growth, development or reproduction.

Plant parasite: A parasite that commonly affects or is peculiar to plants as distinguished from animal parasites.

Plant pathology: The study of plant diseases.

Plant protection agent: Pesticide product intended for use in agriculture to protect crops.

Plant protection equipments: There are two major categories of plant protection equipments, *viz.*, spraying equipments and dusting equipments. Each of these categories is further grouped into two categories, *viz.*, manually operated and power operated. Those equipments which are used for dispersal of powdered or dust formulations of insecticides are called dusting equipments or dusters, and those which are used for dispersal of liquid formulations are called sprayers or spraying equipments. Hand operated dusters comprise (i) rotary type duster (ii) Plunger type duster and (iii) Bellow type duster; whereas, the power operated dusters are: (i) Rotary type and (ii) Air jet type. The hand operated sprayers comprise the (i) hand compression sprayer (ii) Stirup pump, (iii) Knapsack sprayer, (iv) Foot sprayer, (v) Rocker sprayer etc.

Plant quarantine laws: Restrictions on the movement of commodities from one country to another or from one part of the country to other part being suspected of infested due to new insect pests or diseases are governed by laws known as plant quarantine laws.

Plant resistance: Resistance is the ability of a plant to produce high yield of good quality as compared to a susceptible plant that produces a low yield of poor quality. Use of crop cultivars resistant to one or more of its pest is economically and environmentally sound.

Planta: A median ventral sclerite of the pre-tarsus distal to the ungutractor plate.

Plantation: Large acreage producing same crop like tea, coffee, tobacco or cotton using a large number of labour.

Plantula: A divided, flap like structure at the base of the hind tarsus, *e.g.*, Orthoptera, Tettigoniinae.

Plasma: The noncellular fluid portion of the haemolymph.

Plasmatic incompatibility: See cytoplasmic incompatibility.

Plasmatocytes: The blood cells or haemocytes of insects multiply and grow in the body cavity throughout the life of the insect, appearing first as small dark staining forms termed proleucocytes which are often seen dividing and not capable of phagocytosis. As they grow, they become pyriform or spindle shaped and will ingest dead bacteria, Indian ink, tissue, debris etc.; these are often called plsmatocytes.

Plasmid vector: Plasmid is self replicating, extra chromosomal genetic material present in a bacterial cell. It is double stranded circular DNA. Besides the genes for plasmid replication, the plasmid DNA contains some extra genes. The plasmids used in transformation carry antibiotic resistant genes required for identification and selection of recombinant DNA. The desired DNA to be cloned (DNA insert) is integrated into the plasmid DNA that facilitates its cloning. As the bacterium multiplies, the plasmid along with the DNA insert also replicates or the DNA insert is cloned. The most commonly used plasmid for molecular cloning is pBR322. Plasmids may be 1.5-1500 kb in size and may contain three to several thousand genes, DNA inserts of less than 15 kb can be cloned by these vectors.

Plasmid: An autonomous, self replicating, extrachromosomal DNA molecule.

***Plasmodium*:** The protozoan that may cause malaria.

Plasmon: The deoxyribose nucleic acid contents in the cytoplasmic organelles.

Plastron respiration: In some insects, such as beetles, *Haemonia* and *Elmis* and the Naucorid bug, *Aphelocheirus*, the hydrofuge hairs which carry the film of air are bent over at the tip so as to provide a hydrophile surface enclosing a layer of air which cannot be replaced by water. The plastron acts as a permanent physical gill.

Plastron: Fine pile of specialised microtrichia. See plastron respiration.

Platyform larva: An extremely flattened larva as in Psephenidae (also called onisciform larva).

Platygasteridae: Family of super family Proctotrupoidea, suborder Apocrita and order Hymenoptera. These are minute black insects with ten-segmented antennae and in several species polyembryony is noticed. They are parasites of aleyrodids and cecidomyiid flies.

Platyhelminthes: The phylum having the flatworm, such as tapeworms and liverflukes.

Plaumaniidae: Wings not sacly. Forefemora simple.

Play flight: Short flight taken in front of or near the hive to acquaint young bees with their immediate surroundings; sometimes mistaken for robbing or preparation for swarming.

Plecoptera: Gr., *pleco*= pleat, *ptera*= wings. Insect order made up of the stoneflies. They are characterized by membranous wings, chewing mouthparts, two short abdominal cerci, and an incomplete metamorphosis.

Plenary powers: Special powers granted to the commission.

Pleometrosis: See polygyny.

Plesiobiosis: The close proximity of two or more nests accompanied by little or no direct communication between the colonies inhabiting them.

Pleural apophysis: Arm like projection of the pleural ridge.

Pleural membrane: The unsclerotized cuticle in the pleural region of abdomen, terga and sterna.

Pleural region: The podial region or ventrolateral parts on the body on which the limbs are implanted

Pleural ridge: The endopleural ridge formed by the pleural suture, bracing the pleuron above the leg or between the coxal articulation and the wing support.

Pleural sulcus or suture: Groove running dorsally from coxa; often referred as pleural suture.

Pleural wing process: Ventral articulation of a wing.

Pleurite: A lateral or pleural sclerite. The pleuron of each side consists of three sclerites: one ventral, the sternopleurite and two dorsal, the anapleurite and the coxopleurite. The sternopleurite articulates with the coxa. It is fused with the sternum and becomes its integral part. The anapleurite and the coxopleurite are separated from one another in the Aperygota and the larvae of Plecoptera. In the adult pterygote insects, these fuse to form a common pleuron. The coxopleurite or its anterior remnants remain partially separated in lower Pterygote insects. In these insects, these form a distinct sclerite, the trochantin as an additional ventral articular sclerite for articulation with coxa. In the higher insects, the trochantin may reduce or obliterate or fuse with lower margin of the episternum. The pleuron of the prothoracic segments is divided by a pleural suture which extends from the pleural ring process onward to the pleural coxal process into two pleurites, an anterior epipleurites and a posterior epimeron. In some insects, the episterna and the epimera are divided into two parts by a transverse suture. The upper and lower parts of the lower parts of the episterna are termed as anaepisterna and the katepisterna. Similarly, the respective parts of the epimera are termed as the anaepimera and the katepimera to the upper and lower sides, respectively.

Pleuron (pl., pleura): A lateral region on the sides of body of certain arthropods.

Pleuropodium: Embryonic appendages of the first abdominal segment.

Pleurosternal sutures: Sutures dividing the sternum from the laterosternites.

Pleurostoma: The subgenal margin of the cranium bordering the mandible.

Pleurostomal suture: See subgenal suture.

Pleurotergite: A sclerite containing both pleural and tergal elements.

Pleuroventral line: Line of separation between pleural region and the venter.

Plica basalis: The basal fold of the wing or line of flection between the base of mediocubital field and the axillary region.

Plica jugalis: The jugal fold of the wing of some insects or radial line of folding setting off the jugal region from the vannal region.

Plica vannalis: The vannal fold of the wing or radial line of folding usually between the cubital field and the first vannal vein but somewhat variable in position.

Plictran: A moderately toxic acaricide and fungicide marketed under trade name Cyhexatin, selectively effective against mites, ticks and fungi.

***Plocaederus ferrugineus* Linn.:** Belongs to order Coleoptera and family Cerambycidae. It is an insect pest of cashewnut, commonly known as the cashew tree borer.

***Ploceus philippinus* (Linn.):** Belongs to order Passeriformes and family Ploceidae. It is a predator of insect pests, commonly known as common weaver bird

***Plodia interpuntella* (Hubner):** Belongs to order Lepidoptera and family Phyctidae. It is an insect pest of stored grains and other products, commonly known as the Indian meal moth.

Plume moth: See *Exelastis atomosa* Walshingham.

Plumose: Feather like or plumose antenna in which whorl of hair is coming from each segment, *e.g.*, male mosquito.

Plunger type duster: Plant protection equipment used to treat small area with the dust formulation. It consists of air pump, dust chamber and discharge assembly.

***Plusia orichalcea* Fab.:** Belongs to order Lepidoptera and family Nocpuidae. It is an insect pest of cabbage, commonly known as the cabbage semilooper.

***Plutella xylostella* (Linn.):** Belongs to order Lepidoptera and family Yponomeutidae. It is an insect pest of cabbage, commonly known as the diamond back moth.

***Plytela gloriosae* Fab.:** Belongs to order Lepidoptera and family Noctuidae. It is an insect pest of ornamental plants, commonly known as the lily moth.

PMP ($C_{14}H_{14}O_3$): Low toxicity, anticoagulant rodenticide, marketed under trade name Valone, Isoval and Incco. It is a chronic poison with good results. Chemically it is 2-(3-methyl-1-oxobutyl)-1 H-indene-1, 3(2H)-dione. It is relatively non-toxic when taken in single dose. It is normally used at 0.05 per cent concentration, in water baits at 0.012-0.015 per cent level and as a powder at 2 per cent concentration. The rodents do not develop bait shyness to it.

Pneumatic pump: An air pump used in a compressed air sprayer.

Pneumatic sprayer: Manually operated compressed sprayer used to spray the crops and can easily be carried on the shoulder.

Pneumoridae: A family of order Orthoptera, suborder Caelifera and super family Acridoidea. Males with abdomen inflated. Stridulatory ridges present on sides of second abdominal tergum. Found in South Africa.

Pod borer: *Adisura atkinsoni* is a major pest of pulses and beans. The caterpillar bores into the tender pods and eats away the seeds.

Podagrionidae: Family of super family Chalcidoidea, suborder Apocrita and order Hymenoptera. The Podagrionidae possess four to eight long teeth on hindfemur and a long straight ovipositor. They are parasites of mantid eggs. Some of the Indian species are *Pachytomus mantisiphagus, Podagrion mantoidae* and *P.pachymerum.*

Podite (Podomere): Refers to the limb segment.

Podopidae: Family of superfamily Scutelleroidea of division Gymnocerata, suborder Heteroptera and order Hemiptera. Smaller insects measuring about 3.5-10 mm long, generally brown with long scutellum and pronotum bearing a prominent tooth or lobe in front of the humeral angle, they are rare in occurrence, *e.g.* Japanese black bug, *Scotinophara lurida* feeding on rice.

Poduridae: A family of order Collembola. Pseudocelli absent; head hypognathous and furca extending beyond ventral tube.

Poduroidea: A super family of order Collembola and suborder Arthropleona. These insects have free well developed pronotum with setae.

***Poekilocerus pictus* Fab.:** Belongs to order Orthoptera and family Acridiidae. It is an insect pest of papaya, commonly known as the Ak grasshopper.

Poikilothermic: Cold blooded; the body temperature rising or falling with the changing temperature.

Point: A small triangle of stiff paper, which is used in mounting small insects

Point source pollution: A kind of water pollution resulting from discharges of sewers and industrial effluents in water bodies.

Poison: Any chemical or agent that can cause illness or death when eaten, absorbed through the skin, or inhaled by humans or animals. See pesticides.

Poison bait: Attractant foodstuff for animal pests (insects, molluscs, rodents etc.) mixed with a pesticide to kill them.

Poison control center: An agency or center which has current information regarding first aid techniques and antidotes against all poisons.

Poisson model: A mathematical model that describes a random dispersal of organisms.

Polar body: One of the daughter nuclei of primary oocyte with very little cytoplasm.

Polar chemical: A chemical compound composed of molecules with a dipole, such as water, alcohol and salt solutions.

Polarity: State of having two opposite poles owing to separation of charges caused by the non-uniform electron distribution in molecules of a substance such as water.

Pole cells: Certain cells of the posterior end of insect egg from which the primitive cells are derived.

Poliomyelitis: An acute viral disease showing symptoms such as fever, motor paralysis and atrophy of group of muscles.

***Polistes hebraeus* (Fab.):** Belongs to order Hymenoptera and family Vespidae. It is commonly known as yellow wasp which is a social insect and is a minor pest of grapevine.

Pollen: The male reproductive cell bodies produced by anthers of flowers. These are collected and used by honey bees as their source of protein.

Pollen basket: A flattened depression surrounded by curved spines or hairs, located on the outer surface of the bee's hind legs and adapted for carrying pollen gathered from flowers or propolis to the hive. See corbiculum.

Pollen brush: A brush of short stiff hair in honey bee used in collecting pollen.

Pollen cakes: Moist mixtures of either pollen supplements or substitutes fed to the bees in early spring to stimulate brood rearing.

Pollen grain: See pollen.

Pollen insert: A device inserted in the entrance of a colony into which hand collected pollen is placed. As the bees leave the hive and pass through the trap, some of the pollen adheres to their bodies and is carried to the blossom, resulting in cross pollination.

Pollen pot: A container made by stingless bees from soft cerumen and used to store pollen.

Pollen rake: A comb like row of bristles at the apex of hind tibia of a bee.

Pollen store: The bumble bee species stores pollen in the abadoned cocoons.

Pollen substitute: Any material such as soybean flour, powdered skim milk, brewer's yeast, or a mixture of these used in place of pollen to stimulate brood rearing.

Pollen supplement: A mixture of pollen and pollen substitutes used to stimulate brood rearing in periods of pollen shortage.

Pollen trap: A device for removing pollen loads from the pollen baskets of incoming bees.

Pollenizer: It is the plant that provides the pollens to the pollinators.

Pollination The transfer of pollen from the anther to the stigma of a flower. In self pollination, the pollen is carried between anther and stigma of the same flower; whereas, in cross pollination, the pollen is carried from the flower of one plant to another plant of the same species.

Pollinator: It is an agent that move pollen, *e.g.*, wind, bat, bees, moths, birds, butterflies etc. The most important pollinators are bees and butterflies.

Pollinium: A mass of pollen grains bearing a stalk with an adhesive disc that facilitates its transfer by insects.

Pollu beetle: See *Piper nigrum.*

Pollutant: Undesirable substance introduced into a solid, liquid or gaseous environmental medium totally or partially by human activities.

Pollution: The term applied to environmental state or manifestation which is harmful or unpleasant to life, resulting from men's failure to achieve or maintain control over the chemical, physical and biological consequences or side effects of his scientific, industrial or social habits. The air pollution, water pollution, land (soil pollution), noise pollution, radioactive pollution and thermal pollution are the kinds of pollution.

Polyandry: A female having numerous male mates.

Polyculture: Cultivation of different crop species in space and time to provide diversity in land use. Antonym of monoculture.

Polydomous: A single colony occupying more than one nest.

Polyembryonic parasitisim: It is the phenomenon of parasitization of an individual host by single larva of a single parasitic species, but develops into several parasitoids within the host body, *e.g.*, *Copidosoma koehleri* on potato tuber moth, *Phthorimaea operculella*.

Polyembryony: It is an unusual mode of asexual reproduction which occurs in some families like Chalcididae, Braconidae, Ichneumonidae and Platygasteridae (parasitic Hymenoptera). After maturation of the egg, the polar bodies may divide amitotically to form nutritive sheath; meanwhile, the egg nucleus undergoes complete cleavage and the separated cells each develop into embryos. In this way, two or four embryos may arise from single egg in some species; anything from a hundred to two thousand in others.

Polyethism: Division of labour in the members of a colony. In social insects, a distinction can be made between caste polyethism in which morphological castes are specialized to serve different functions and age polyethism in which the same individual passes through different forms of specialization as it grows older.

Polygamy: Phenomenon of single male having many female mates.

Polygenic: A characteristic conferred by more than one gene, such as plant resistance.

Polygenic resistance: In polygenic resistance, several genes are involved for the development of resistance.

Polygyny: Phenomenon of coexistence of two or more egg laying queens found together in single colony

Polyhedral virus: Virus characterized by presence of polyhedral inclusion bodies in the infected cells of the host and causes diseases in the insects.

Polyhedron: These are crystal like inclusion bodies characteristic of infection by a polyhedrosis virus.

Polyhedrosis: An insect viral disease caused by polyhedral virus characterized by the formation of polyhedron shaped inclusions in the infected cells. This disease is called the nuclear polyhedrosis or nucleopolyhedrosis if the polyhedra is found in the nuclei of infected cells and cytoplasmic polyhedrosis if they are formed in the cytoplasm of infected cell.

Polylectic: Bees that gather pollen from many kinds of flower. See oligolectic.

Polymeric molecule: A molecule that can be represented by more than two whorls.

Polymorphism: The simultaneous occurrence of several discontinuous phenotypes or genes in a population; the condition of having several forms in the adult. In the honey bees, the drones are haploid and develop from the eggs carrying merely half number of chromosomes as compared with those from which the female sex emerges. Their production from unfertilized eggs results through parthenogenesis. Queen and workers are basically females having equal number of chromosomes and develop from fertilized eggs having diploid sets of chromosomes.

Polyneoptera: See taxonomic grouping of orders.

Polypeptide: A polymetric covalently bonded arrangement of aminoacids joined together by peptide linkages.

Polyphaga: A suborder of order Coleoptera. The majority of beetles are included under this suborder and this is classified into 18 superfamilies.

***Polyphagotarsonemus latus* (Banks):** Belongs to order Acarina and family Tarsonemidae. It is a pest of tea, commonly known as the yellow tea mite (broad mite).

Polyphagous pest: A pest feeding on a range of hosts- belonging to several species of a group of families, *e.g.*, termites, locusts, hairy caterpillars, gram pod borer etc.

Polyphenol layer: This layer is described in the blood sucking bug, *Rhodnius prolixus* and appears often in the liquid state as a non-static layer containing various types of polyhydric phenols like homocatechol, protocatechol, dopacatechol, hydroquinone, pyrogallol etc. The polyphenols are transported from the epidermal cells through the pore canals and accumulate on outer surface of cuticulin layer. In some insects, in spite of forming an independent layer, polyphenols are used to impregnate the cuticulin. At the time of sclerotization, these polyphenols are oxidized to quinines in the presence of an enzyme, phenol oxidase and then the quinines tan first the proteins of cuticulin layer and latter proteins of outer procuticle to produce the exocuticle.

Polyphyletic: Of many tribes. Derived from more than one ancestral line.

Polyploidy: The state of having the chromosomes in somatic cells or tissue more than twice the haploid number. In human being polyploidy leads to death but we can find diploid or triploid (3n) cells in human skin.

Polypneustic: Open tracheal system with meso, metathoracic and eight abdominal pairs of spiracles. This is of three types: (i) Holopneustic: With all ten pairs of functional spiracles functional, *e.g.*, Hemimetabola (ii) Hemipneustic: Only mesothoracic and first seven abdominal spiracles are functional, *e.g.*, larvae and (iii) Peripneustic: Only nine pairs of spiracles are functional and metathoracic one is non-functional, *e.g.*, larvae of Neuroptera, Lepidoptera, Mecoptera and Cecidomyiidae etc.

Polypod larvae: These are eruciform larvae of most of the Lepidoptera, sawflies and scorpionflies which are characterized by well defined segmentation of the body, presence of abdominal limbs or prolegs, peripneustic tracheal system, presence of little developed antennae and thoracic legs, and ten pairs of abdominal legs in Hymenoptera endoparasitic larvae of Fitigidae.

Polytene chromosome: Giant chromosomes found in most of the Dipterous insects during the development.

Polytrophic egg tube: A kind of egg tube in which groups of trophocytes accompany the oocytes.

Polytypic: A taxon containing two or more taxa in the immediately subordinate category, genus with several species and species with several subspecies.

Polyvoltine: An animal completing several generations in a season; the term is applied generally to certain silkworms.

Pomegranate butterfly: See *Virachola isocrates.*

Pomegranate mite: See *Paratetranychus punicae.*

Pompilidae: Family of super family Pompiloidea, suborder Apocrita and order Hymenoptera. Pompiloidae are called as spider wasps as they mostly prey upon spiders. These wasps are mostly dark colored with smoky or yellowish wings. Some are brightly colored. They measure 1 to 2.5cm long but a few are 7.5cm long. A characteristics suture is present across the mesopleura. The hindlegs are very long and spiny and a definite pedicel in the abdomen is wanting.

Pompiloidea: Super family of Suborder Apocrita, order Hymenoptera. This superfamily includes two families, *viz.* Pomppilidae and Rhopalosomatidae. Pompilidae is the most important.

Poneroid complex: The name is based on the subfamily ponerinae, one of the major taxonomic groups of ants.

Pons cerebralis: It is a medial neuropilar body. It is transversely situated in the frontal region of the protocerebrum as a lattice neuropile behind the pars intercerebralis. The globuli cells are lying laterally. It is connected with various neuropilar masses except the corpora pedunculata.

Population: Assemblance of individual organisms of defined ages and growth stages belonging to one species within a specified location in space and time.

Population density: The number of individuals of one population per unit area or volume.

Population dynamics: The study of changes in population size of an organism in time and space and processes which created such variations.

Population regulation: The process by which the density of living organism is determined, necessarily involving at least one density dependent morality factor responsible for stabilization of population size at or about some mean characteristic density.

Porcupine: A rodent quadruped belonging to Hystricidae with body covered with erectile spines, living in colonies in burrows, bund and hill sides, sides of river beds, vegetable feeder damages potato tubers, gardens and forests.

Pore canal: There are large numbers of ducts in procuticle running from epidermis upto inner cuticular layer. Electron microscope revealed that these are helical shaped rather than straight and act as means to transport material during the formation stage of cuticle. These probably carry the moulting fluid, polyphenols and perhaps wax.

Porrect: Extending horizontally forward, applied especially to antennae. **Post abdomen:** Modified posterior segments of the abdomen which are more slender than the anterior segments, *e.g.*, crustacean, scorpion etc.

Post coxal bridge: The postcoxal part of pleuron, often united with the sternum behind the coxa.

Post cubitus: The sixth vein of the wing, associated basally with the cubitus (first anal vein in most of the cases).

Post embryonic development: See metamorphosis.

Post emergence application: The application of a chemical (weedicide) when the field crop has already come up.

Post emergence: Period after crop or pest has appeared. Herbicide usage can be referred to as post-emergence (weeds) or post-emergence (crop).

Post emergence treatment: Same as post emergence application.

Postalar bridge: Lateral extension of postnotum of a wing bearing segment behind the wing base, generally united with the epimeron.

Postcoxal line of abdomen: Raised line or carina arising from the hind edge of the metacoxal cavity and extending posteriorly or posterolaterally and sometimes recurved; rarely there may be more than one line on each side.

Postcoxal line of metaventrite: Raised line or carina arising from the hind edge of mesocoxal cavity and extending posteriorly or posterolaterally and sometimes recurved. Rarely, there may be more than one line on each side.

Postcoxal projection of hypomeron or prothoracic pleuron: A mesal projection of the hypomeron or prothoracic pleuron which extends behind the procoxa and may meet the prosternal process or in some cases, the opposing postcoxal projection.

Posterior: Facing or concerning the rear.

Posterior: Hind or rear.

Posterior angles of pronotum: The hind corners of the pronotal disc. These may be rounded, angulate or produced backwards forming acute processes.

Posterior callus: The rounded swelling on each side of mesonotum between the base of wing and the scutellum.

Posterior cell: Cell extending the hind margin of wing between the third and sixth longitudinal veins.

Posterior cross vein: Cross vein at the apex of discal cell in Diptera.

Posterior intestine or rectum: The terminal section of the proctodaeum commonly termed as rectum but usually divided into an anterior rectal sac and a posterior rectum proper.

Posterior mesenteron rudiment: The posterior group of cells of the ventral endoderm remnant that regenerates the mesenteron.

Posterior notal wing process: A posterior lobe of lateral margin of the alinotum supporting the third axillary sclerite of wing base.

Posterior pharynx: The pharyngeal chamber of stomodaeum behind the brain, present in orthoptera, Coleoptera and some other insects.

Postfrontal suture: Posterior suture present in some insects diverging from the coronal suture laterad of the antennal bases.

Postgena: Part of the head which extends downwards and back of the gena.

Post-harvest: The period between maturity of the crop and its consumption.

Posthumeral bristle: A bristle on the anterolateral surface of mesonotum near margin of the humeral callus in Diptera.

Postlabium: The basal or proximal part of the labium.

Postmentum: The basal portion of the labium proximad of the labial sutures.

Postnodal cross vein: A series of cross veins just behind costal margin of the wing.

Postnotum: An intersegmental plate of the dorsum of thorax associated with the tergum of preceding segment bearing the antecosta and usually a pair of phragmatal lobes (phragmanotum).

Postoccipital suture: A transverse suture on the head which embraces the occipital foramen magnum from the dorsal and lateral sides. Each end of the suture merges in the hypostomal suture in ventral region. This site is well defined due to the presence of posterior tentorial pits. It possesses internally a strong epodemal ridge, the post-occipital ridge. It provides a site for attachment of prothoracic muscles which move a pair of ventro-lateral condyles. The neck membrane is firmly attached to this ridge. The condyles articulate with the neck sclerites. It is the only suture having the metameric significance. It alone marks the fusion of maxillary and labial segments of the head.

Post-occiput: Behind the occipital suture, it is a narrow ring like sclerite. It lies between the occiput and the neck. The post-occiput is marked off from the occiput by a transverse post-occipital suture. In absence of this suture, the posterior tentorial pits separate the occiput from the post-occiput. The dorsal thoracic muscles are attached to the endoskeletal ridge of post-occipital suture. These muscles articulate the head freely. The anterolateral inner margin of post-occiput produces small processes. These structures are termed as occipital condyles which articulate with the anterior lateral cervical sclerite.

Post-petiole: The third abdominal segment in ants; the first abdominal segment (propodeum) is fused to the thorax.

Post-scutellum: A small tranverse piece of thoracic notum immediately behind the scutellum; in Diptera, the area immediately behind or below the mesoscutellum.

Post-sternite: The narrow post-coastal lip of a post notal thoracic plate.

Post-tergite: The narrow post-coastal lip of a postnotal thoracic plate.

Postural hairs: Rows of spines or patches of hairs that serve to indicate the degree of flexion in a joint; they are called postural hairs.

Post-vertical bristles: A pair of bristles behind the ocelli; the hind most pairs of bristles usually situated on the posterior surface of head in Diptera.

Potable: Water suitable for drinking purpose.

Potassium pump: In the caterpillars of Lepidoptera, the midgut plays an active part in ionic regulation. There is a potassium pump located in the so called globlet cells which transport potassium from the haemolymph to the gut lumen. The midgut of *Periplaneta* has a comparable function.

Potato aphid: See *Myzus persicae* (Sulz.)

Potato beetles: *Epilachna vigintioctopunctata* Fab., *E. duodecastigma* Muls., *E.ocellata* Rdt. and *E.circularis* Kobs. These beetles of potato are more serious on hills and also attack brinjal, tomato, bitter gourd etc. The grubs and adults feed on leaves, making them dry up and fall off.

Potato jassid: *Empoasca kerri* and *E. motti* Pruthi are green soft bodied insects that suck cell sap from the leaves and other tender parts of the plants.

Potato tuber moth: See *Gnorimoschema operculella*.

Potency: Refers to the strength in terms of toxicity of an insecticide.

Potential natality: The reproductive rate of individuals in an optimal environment.

Potential pest: These are presently not causing any economic damage, their GEP lies below the DB and does not cross EIL even under favourable conditions. Any change (cropping pattern, cultural practices etc.) in the ecosystem may, however, push their GEP higher and there is a danger of economic damage from these pests if control operations against the other categories of pests are undertaken indiscriminately. In North India, *Spodoptera litura* on cotton and sunflower and armyworms on wheat are potential pests.

Potentiation: The joint action of two pesticides to bring about an effect greater than the sum of their individual effects.

Pouch makers: The bumble bee species that makes special pouches adjacent to groups of larvae and fill them with pollen.

Poultry shaft louse: See *Menopon gallinae.*

Poultry stickfast flea: See *Echidnophaga gallinacea.*

Pounce: See permethrin.

Ppb: Abbreviation stands for parts per billion.

Ppm: Abbreviation stands for parts per million.

***Pratylenchus* sp.:** Belongs to order Tylenchoidea and family Tylenchidae. It is a nematode pest of ornamental plants, commonly known as the root lesion nematode.

Pre-access: The interval of time between the last application of pesticide to an area and safe access to the area for domestic livestock and man.

Pre-alar bridge: A lateral extension of the prescutal area of alinotum before the wing base, sometimes connected with the episternum.

Pre-apical: Just before the apex.

Pre-baiting: Refers to tempting rats to feed on non-poisoned bait for a day or two before the use of poisoned bait.

Precautions: Warnings or safeguards.

Precipitant: A solid substance that sinks in the bottom of a container having liquid in it.

Precipitation: The amount of rain, snow and sleet hail that falls.

Precipitin reaction: The formation of visible precipitate at the interface when an antigen is brought in contact with the corresponding antiserum.

Precision: A measure of the degree of error in making estimates of population size.

Precocene: Bowers and coworkers (1976) discovered the chromene derivatives, Precocene, 6-7 dimethoxy-2,2 dimethyl chromene, an antijuvenile hormone isolated from the plant *Ageratum houstonianum* having antiallatotropic action. These substances induce precocious metamorphosis in various insects, like *Oncopeltus, Dysdercus* etc. and, therefore, these workers recommended them as the most promising insect growth regulators. The antigonadotropic activity of precocene has been noticed in *Oncopeltus* and some other species of Heteroptera, Coleoptera and Diptera. Bowers (1976), further suggested that the precocene-II is more potent than precocene-I.

Precosta: Small first vein of the wing in certain fossil insects.

Precoxal bridge: Exoskeletal region of a wing bearing segment that unites the sternum.

Predaceous: Feeding as a predator.

Predation: One population adversely affects the other by direct attack but they do not depend on each other.

Predator: It is a free living organism throughout its life, normally larger than prey (host), requires more than one host to develop and is not very specific unlike parasitoids. For example, ladybird beetles (*Coccinella septempunctata, Menochilus sexmaculatus*) and green lace wing, *Chrysoperla carnea* etc.

Predator control: A predator prey interaction in which the predator controls population size of the prey. The predator population is the limiting factor for the prey population size.

Predictive value: Usefulness of a classification in making predictions on newly employed characters or newly discovered taxa.

Pre-emergence: Period before a specified crop or pest has emerged. Generally, it is applied to the time of herbicide application.

Prefemur: The second trochanter segment of the leg.

Preference The factor by which certain plants are more or less attractive to insects by virtue of their texture, colour, aroma or taste.

Preferential flow: Leaching phenomenon whereby water and a dissolved pesticide percolating down through the soil profile move more rapidly through soil macro pores or sand/ gravel through the network of smaller pores in the bulk soil.

Pregenital: Anterior to the genital segments of the abdomen.

Pregenital segments: Terminal abdominal segments to which the aedeagus is attached by connecting membrane; consisting of the 9th and 10th abdominal segments (the last often highly reduced and not distinguishable). See genital capsule.

Pregular area: The area in front of the gular sutures (and posterior tentorial pits) and behind the attachment of the mouthparts.

Pre-harvest interval (PHI): The interval of time between the last application of' pesticide and the safe harvesting of edible crops for immediate consumption.

Preinfective stage: Early developmental stage (egg to larva) in the life cycle of a parasite during which it is incapable of being established in the host; the stage that precedes the infective stage.

Prelabium: The distal part of the labium, comprising the prementum, the ligula and the palpi.

Prementum: Apical portion of the labium that bears the labial palpi.

Preoral cavity: Cavity formed by the mouthparts.

Preoral: Anterior or front part of the mouth.

Pre-oviposition period: The period of time between the emergence of an adult female insect and the start of its egg laying.

Prepectus: Anterior marginal sclerite of the sternopleural areas of a segment, set off by a transverse suture continuous through the sternum and episternum. This thoracic plate is significant in Hymenoptera.

Prepupa: The last larval instar after it ceases to feed; often it takes on a distinctive appearance becoming quiescent and rather shrunken, and often looks dead.

Prescutal suture: Line that delineates the prescutum.

Prescutum: Anterior piece of the dorsal part or tergum of a thoracic segment of an insect.

Presocial: Behaviour that includes one or two of the three characteristics of social behaviour.

Presternal suture: The line that separates the presternum from rest of the sterna.

Presternum: A sclerite on the anterior part of the sternum divided off by the presternal suture.

Presumptive organization: Arrangement of cells in the embryo into groups which in normal development become a particular organ or tissue.

Presutural bristle: Presutural bristles are situated on the mesonotum immediately anterior to the transverse suture and adjacent to the notopleuron in Diptera.

Pretarsal claw: See Tarsal claw.

Pretarsus (pl., pretarsi): The terminal segment of the leg representing the foot, usually bearing a pair of terminal, little curved claws (ungues). In Coccidae and some lice, the pretarsus bears a single claw. In Thysanoptera, the claws undergo enormous reduction and appear in the form of minute spines. In Notonectid bugs, they are wanting. In most of the Orthopteroid insects, median lobe lying in between the terminal claws is well distinct and termed as arolium. In the higher insects like Diptera, however, the arolium is replaced by a long prominent spine or bristle known as empodium. In Diptera, a pair of lobe like structure occurs ventral to the paired claws termed as the pulvilli. The pulvilli bear a large number of pores and hairs which secretes the adhesive secretion that helps in walking on the slippery surfaces like glass. On the ventral side, the pretarsus bears a median plate, the unguitractor. The unguitractor serves as a site for attachment to the flexor muscles of the claws.

Pretracheation theory: Comstock and Needham (1898-1899) developed this theory after establishing homology in wing venation among various orders of insects. The main principle of the pretracheation theory is that the tracheation in the nymphal wing pads recapitulates the wing venation of their adult ancestors. They further suggested that tracheation alone determines the position of veins.

Preventive: A measure applied in anticipation of pest attack.

Preventive application: Application of control measures in anticipation of an attack of an organism or outbreak of a disease.

Preventive treatment: Treatment designed to prevent a plant becoming infected.

Prey: The food animal of a predator.

Primary consumer: Heterotrophic organisms (invertebrates, such as *Daphnia* sp.) using organic substances directly from primary producers (*e.g.*, algae) as a carbon and energy source.

Primary control measures: These are the ordinary farm practices through which the outburst or migration of insect pests can be prevented in the beginning itself. These measures require basic knowledge of the agroecosystem and vigilant care of the crop. These should be taken well in advance. These differ from the chemical and biological methods as the control involves reduction of pest population without killing them, on the contrary help in maintaining the sound ecosystem of the region. The primary control measures are simple, safe and cheaper than other methods of pest control. These involve mechanical, physical, cultural, legislative and quarantine measures.

Primary decomposers: Heterotrophic organisms (*e.g.*, bacteria) using dead organic matter from all trophic levels as a carbon and energy source.

Primary infection: Initial introduction of a crop disease by insect vectors invading from elsewhere.

Primary insect: An insect capable of attacking, injuring or killing the growing parts of a living plant unaided by other insects or disease organisms.

Primary oocyte: Oocyte that is located proximally in an ovariole and that is ready to undergo vitellogenesis.

Primary orientation: Assumption and maintenance of body position in space.

Primary parasite: A parasite which establishes itself in or upon a free living host, *i.e.*, the host is not a parasite.

Primary parasitoid: An insect primary parasitoid feed on a phytophagous or free living host and ultimately lead to death of the host.

Primary producer: Autotrophic organisms (algae, higher plants) which convert inorganic compounds during the process of photosynthesis or chemosynthesis into organic compounds (cell material) of higher energy content. These organisms represent the first trophic level of the food chain.

Primary reproductive: Those members of a social group of insects whose primary role is reproduction (often the founders of colony). Compared to secondary reproductive which may produce some young but are primarily involved in some other activity. In termites, the colony founding type of queen or male derived from the winged adult.

Primary sample: Collection of one or more units or portions initially taken from a population. These portions may be combined (aggregated) or kept separate.

Primary segmentation: The arrangement of body units in such a way that intersegmental grooves form the attachment points for longitudinal muscles.

Primary spermatocytes: Diploid cells that will undergo meiotic cell division and eventually form into spermatozoans.

Primary storage pests: The insect pests of stored grains that can damage sound grains, *e.g.*, rice weevil, lesser grain borer, khapra beetle, pulse beetle, Angoumois grain moth, rice moth, Indian meal moth, almond or fig moth etc.

Prime swarm: The first swarm to leave the parent colony, usually with the old queen.

Primer pheromone: When a single substance can give a complicated variety of functions, then the pheromone is known as primer, *e.g.*, 9-keto-1-decenoic acid is an excellent example which plays an important role in the behaviour in honey bee society. It acts as a sex pheromone by a virgin queen to lure during mating, flight and after mating the same substance is used for inhibitory behaviour by the queen.

Primitive insect: The insects evolved from wingless arthropod which is intermediate between insects and millipedes. The insects such as springtails (Collembola), silverfish (Thysanura), proturans (Protura) and Campodeans and Japygids (Diplura) are nearest to the hypothetical ancestor.

Prioninae: A subfamily of Cerambycidae, Coleoptera. Adults elongate-oval, somewhat flattened, bright reddish-brown, about 2 cm long and with short antennae not extending beyond the base of the prothorax. The grubs bore into dry dead wood of logs and stumps. The genera included under this subfamily are: *Acanthophorus, Dorysthenes, Macrotoma.*

Pro: Prefix referring to a structure forming part of the prothorax, including forelegs (*e.g.*, procoxa, prosternum, protrochantin, protarsus).

Proboscis: The extended beak like mouthparts of sucking insects. The mouthparts of the bee form the sucking tube or tongue.

Procephalon: The region of the definitive head anterior to that of gnathal segments representing the primitive protocephalon formed in the embryo from the cephalic lobes and the tritocerebral segment; bearing the labrum, mouth, eyes and antennae. See gnathocephalon.

Processed food: Product resulting from the application of physical, chemical or biological processes or combinations of these (*e.g.*, canning) and intended for sale to the consumer for use as an ingredient in the manufacture of a food product or for further processing.

Processes of body wall: Unlike the true appendages, processes of body wall by no means invariably represents by embryonic counter parts. The principal types of processes are: (i) pseudopods which are characteristic of many Dipterous larvae, (ii) Scoli or thorny processes are characteristic of nymphalid and saturniid larvae, (iii) Branchiae or gills found in most larvae of aquatic insects, and (iv) Wings which are always confined to meso or metathorax.

Proclinate: Inclined forward or downward.

***Procontarinia matteina*:** A pest of mango, commonly known as mango gall midge or mango gall fly.

Procoxal cavity: The countersunk sclerotized cavity for housing the procoxa; formed from the prosternum and sometimes the propleuron (Archostemata, Adephaga, Myxophaga) or hypomeron (Polyphaga). The cavity or housing may be almost completely sclerotized (except where the coxa is joined by membrane), but in most cases, it consists of a small sternal shelf, sometimes joined to the pleuron or notum by a slender bar. In some cases, there is no real cavity or housing present and the coxa is attached externally by a broad membrane.

Procoxal rest of mesoventrite: One of a pair of concavities on the anterior edge of the mesoventrite which houses one of the procoxae at rest. Sometimes, the procoxal rest continues onto the anterior edge of the mesepisterna.

Proctodaeal feeding: Ingestion by one individual of material from the anus of another individual.

Proctodaeal invagination: Ectodermal invagination that develops into the hind gut.

Proctodaeal valve: A valve at the anterior of hindgut which serves as an occlusion mechanism.

Proctodaeum: The hindgut or the hind most of the three major divisions of the alimentary tract from the Malpighian tubules to the anus. See hindgut.

Proctodaeal valve: In insects, a valve in the anterior end of the hindgut that serves as an occlusor mechanism.

Proctolin: The insect neurohormone secreted by the neurosecretory cells of last abdominal ganglion and transported along the nerves innervating the hindgut. It is a myotropic hormone acting as an excitatory gut factor and causes contraction of longitudinal muscles of the hindgut.

Proctotrupoidea: Super family of Suborder Apocrita, order Hymenoptera. Proctotrupoidea are small, black insects which are parasitic on eggs, larvae and pupae of other insects. A few are hyperparasites and certain forms are inquilines. The insects are frequently apterous and in winged forms the venation is much reduced. The pronotum extends laterally to the tegulae. The ovipositor generally arises from the abdominal end. The larvae pupate mostly in cocoons and in aphid-infesting species the pupa is seen within the body of the host.

Procuticle: The cuticle is divisible in two layers; outer epicuticle and inner procuticle. The procuticle is composed of chitin (chief constituent forming 20-50 per cent of the dry weight of cuticle), protein and other substances. This is thicker as compared to epicuticle. It is divisible into exocuticle and endocuticle.

***Prodioctes haematicus* Chevr.:** Belongs to order Coleoptera and family Curculionidae. It is an insect pest of cardamom, commonly known as the rhizome weevil.

Proepimeron (pl., proepimera): The epimeron of the prothorax.

Proepisternum (pl., proepisterna): The episternum of the prothorax.

Prognathous: The long axis of head is horizontal and the mouthparts are directed forward.

Progoneate Having the genital opening in the anterior region of body.

Progressive provisioning: The act of providing the larva with meals at intervals during its development.

Proinsecticide: An insecticide precursor that gets converted into an insecticide *in vivo*. N-sulphonylated, N-sulphenylated or n-phosphinothiolated derivatives of an insecticide carbamate can act as proinsecticide. The carbosulfan is a proinsecticide which produces carbofuran *in vivo* and have spectrum of activity similar to the carbofuran.

Projapygidae: A family of order Aptera (Diplura). Styli present on first abdominal segment; cerci few segmented, short and stout, perforated apically. For example, *Projapyx* sp., *Anajapyx* sp., *Symphylurinus* sp.

Prolate: See phosmet.

Prolegs: Bilateral fleshy abdominal legs characteristic of lepidopteran (five pairs, four on III to VI segments and fifth on X segment) and many Hymenopteran larvae bearing a characteristic arrangements of crochets. These are unsegmented and lateral extensions of integument, however, can be differentiated into three regions; a proximal ring, middle cylindrical tube and a distal cuticular retractile lobe, a pad called the planta. It bears the curved claws or the crochets. The planta is provided with retractile muscles. The prolegs of X segment are called the anal legs or postpedes and resemble with other prolegs in structure, though differ in musculature. Some Coleoptera also bear the prolegs.

Prolific: Producing numerous fruits; marked by abundant productivity.

Prolin: An anticoagulant rodenticide having warfarin and equivalent amount of sulphaquinoxaline, marketed under trade names Banarat and Prolin. Rats sometimes do not die after eating warfarin because a bacterium develops in their stomach and the bacterium is responsible for preparing vitamin K which acts as an antidote against warfarin that is why the warfarin does not act. When warfarin is mixed with sulfaquinoxaline, it acts as good poison which is known as prolin due to which the bacterium dies in the stomach of rat and can not produce antidote.

Prolonged exposure: Long term contact with an insecticidal compound or its residue.

Promesothoracic clicking mechanism: A complex jumping mechanism involving a long prosternal process, deep mesoventral cavity and anterior procoxal rest (on the mesoventrite) which occurs in most members of the beetle families, Cerophytidae, Eucnemidae, Throscidae and Elateridae (Evans, 1972, 1973).

Prominence/ prominent: A raised or projecting portion

Promoter: The region of DNA that binds RNA polymerase to initiate transcription.

Pronotal Comb: A row of stout spines on the hind margin of the pronotum of certain fleas.

Pronotal disc: That portion of the prothorax which is visible from above, or that lies above the lateral pronotal carinae when these are present.

Pronotal hypomeron: See hypomeron.

Pronotum (pl., pronota): The dorsal body plate of the first section of thorax, which is frequently enlarged and prolonged in many insects.

Pronucleus: The nucleus of spermatozoa and ova that unite to form the nucleus of a fertilized ovum.

Propagules: Viable seeds and living plant parts capable of producing new plants under proper environmental conditions.

Propellent: An agent in self pressured pesticide products that produces the force required to dispense the active ingredient from the container.

Proper dosage: The amount of insecticide according to the label information.

Properties: The characteristics or traits which describe a pesticide.

Prophylactic control: The control of insect pests by adopting preventive measures.

Propisternum: The episternum of the prothorax.

Propleuron (pl., pleura): The pleuron or lateral wall of the prothorax. In Myxophaga and Polyphaga, the propleuron is more or less reduced and fused to the trochantin, and in the latter group it is not or only slightly visible externally.

Propneustic: See hemipneustic.

Propodeum: The first abdominal segment in the Hymenopteran suborder, Apocrita, it is completely fused with the thorax.

Propupa: In thrips, the next to the last nymphal instar in which the wing pads are present and the legs short and thick. Also found in male scale insects.

Propolis: Sap or resinous material collected from trees or plants by bees and used to strengthen the comb and to close up the cracks etc.; also called bee glue.

Propoxur: Highly toxic carbamate insecticide, O-isopropoxyphenyl methyl carbamate, effective against household, lawn, turf, flower, fruit and garden insect pests. It is effective against resistant strain of cockroach; moderately toxic to mammals; acute oral LD_{50} for rats is 97-175 mg kg^{-1}, marketed under trade name Baygon (1 per cent aerosol). Its quick action and long residual life makes Baygon a useful and general purpose insecticide.

Proprietary chemical: A chemical made and marketed by a person having exclusive right to manufacture and sell it.

Proprietary name or Trade name: Name given to a product sold by a company to distinguish it from similar products made by other companies.

Proprietory mixture: Technical mixture that is commercially available.

Proprioception/ Proprioreceptor: A mechanoreceptor that provides information either about an insect's position in its environment or the position of the insect's body parts or both.

Propylene: A colourless gas usually produced as a by product of ethylene, manufactured from propane, butane or naphtha. An important ingredient for making isoprpanol, butanol, propylene oxide, polypropylene, glycerin, acrylonitrile and phenol.

Proscopiidae: A family of order Orthoptera, suborder Caelifera and super family Acridoidea. Mostly apterous, confined to South America, body elongated rod like and legs long and thin.

Prosoma: The foremost body region of the arachnids (ticks, mites, spiders etc.) to which the walking legs are attached.

Prosophyles: The minute intracellular pores through which water enters the body of sponge.

***Prospaltella citufila* Silv.:** A major parasitoid of citrus whitefly, *Dialeurodes citri* (Ashmead).

***Prospaltella divergens* Silv. :** Belongs to order Hymenoptera and family Eulophidae. It is a parasitoid of citrus blackfly, *Aleurocanthus woglumi* Ashby.

***Prospaltella lahorensis* (How.):** A major parasitoid of citrus whitefly, *Dialeurodes citri* (Ashmead).

Prostaglandins: See vertebrate hormones in insects.

Prosternal grooves: A groove in the sternum or the ventral sclerite of the prothorax, *e.g.*, Elateridae, Reduviidae etc.

Prosternal process: The posteriorly projecting, intercoxal process of the prothoracic sternum; the process may be absent or very short, so that the coxal cavities are confluent or it may extend behind the procoxae and overlap the mesoventrite (in rare cases reaching the metaventrite).

Prosternal spine: The curved mucro in Elateridae, which extends backward into a mesodermal cavity; the cone or tubercle between forelegs in some Orthoptera.

Prosternum (pl., Prosterna): The major ventral sclerite of the prothorax, joined laterally either to the paired propleura (Archostemata, Adephaga, Myxophaga) or paired hypomera (Polyphaga).

Prosternum: Ventral surface of the first thoracic segment.

Prostheca (pl., Prosthecae): A membranous (or rarely partly sclerotized), usually setose, structure lying distal to the mola.

Prostomium: See acron.

Prot, proto: First; primary

Protease: A digestive enzyme which splits or breaks up proteins.

Protection items: This may include rubber gloves, boots, apron, respirator, face mask, etc.

Protective clothing: Clothing protecting spray operator from adverse effects of crop protection chemicals. It includes rubber gloves, rubber boots, apron, respirator, face-mask etc.

Protective colouration: Any pattern or arrangement of colours in an animal which enables it to escape or protect from natural enemies by resembling or blending into its surrounds.

Protective equipment: Materials and devices that shield against unintended exposure to pesticides.

Protective gears: Clothes and equipments that guard a person against injury or death when using poisonous pesticides. These include gloves, apron, shoes, coveralls, hat, cartridge, respirator, gas mask etc.

Protective insecticides: See stomach poisons.

Protective value: Ability of a substance (protectant) to prevent infection of the host.

Protein: A chain of amino acids that make up cell structure and controls cell function.

Protein, cuticular: The proteins of cuticle are of three types: arthropodin, resilin and sclerotin. See details of these.

Protein epicuticle: The innermost layer of the epicuticle (inner cuticle) made up of proteins and polyphenols.

Protelean parasite: An insect species in which only the immature stages are parasitic.

Protein taxonomy: An approach to classify organisms on the basis of the differences in structure of protein.

Protentomidae: A family of order Protura. Characterized by absence of tracheae and spiracles and the third abdominal appendage being one-segmented. Combs much reduced or absent on the eighth abdominal segment. Typical abdominal terga each with a transverse row of setae and without transverse grooves or laterotergites. For example, *Proturentomon* sp., *Protentomon* sp.

Proteolysis: The splitting of proteins into smaller components by breaking peptide bonds.

Proterosoma: In mites, the anterior part of the body when there is demarcation of the body between the second and third pair of leg.

Prothetely: It would seem that the gradual changes in form that occur in the successive stages of insects, as well as the more striking changes at metamorphsis are likewise controlled by a delicate balance in the timing and concentration of the secretions from the corpus allatum and thoracic gland. Sometimes, this balance is upset within the intact insect, particularly in species hybrids or in insects reared under abnormal conditions, and then monsters intermediate between larva and pupa, or between pupa and adult are produced. This phenomenon is termed as prothetely or metathetely. It may sometimes be induces by microbial or virus infections in insects.

Prothoracic cavities: Paired, broad cavities on the ventrolateral, dorsolateral or anterolateral portions of the prothorax and housing the antenna, antennal clubs, or fore legs.

Prothoracic gland: An endocrine organ in the prothorax of immature insects that secretes ecdysone, also called the ecdysial gland or thoracic gland. One of a pair of endocrine glands located in the prothorax near the prothoracic spiracles.

Prothoracic gland hormone: See ecdysone.

Prothoracic grooves: Paired, narrow grooves, usually on the ventrolateral portions of the prothorax and housing the antennae and/or fore legs.

Prothoracicotropic hormone: See brain hormone.

Prothorax: The anterior of the three thoracic segments. This segment never bears wings.

Protocephalon: Bilaterally expanded anterior region of the germ band embryo.

Protocerebral neuropile: The protocerebral neuropile contains some distinct neuropile masses: (i) pons cerebralis or protocerebral bridge, (ii) corpora pedunculata or mushroom bodies, (iii) corpus centrale or central body, (iv) corpora ventralia or accessory lobes, and (v) corpora optica or optic lobes. See details.

Protocerebrum: The dorsal lobes of the brain, representing the fused paired preantennal or archecerebral ganglia. It bears on each lateral side of the optic lobes and receives the optic nerves (nerves opticus) from the optic lobes, the ocellar nerves (nervi ocellarii) from the median and lateral ocelli and it gives out the cardiac nerves (nervi corporis cardiaci) innervating the corpora cardiaca and the aorta.

Protocooperation: Association in which both species get benefited but their relations are not obligatory.

Protomecoptera: Suborder of order Mecoptera. Wings short and broad, males without swollen genitalia; body generally depressed.

Protonymph: The first nymphal instar of mites.

Protopam chloride: An antidote for organophosphate poisoning but not for carbamate poisoning.

Protoplasmic poisons: Substances that kill the insects by precipitating or destroying cellular protoplasm of the mid gut epithelium. Mercury, copper and arsenic compounds are the examples of this group.

Protopod: These are inactive larvae with the following characteristics: The protopod larvae emerge in an early embryonic phase, lies in high nutritive medium either in egg or body of the host, devoid of abdominal segmentation, cephalic and thoracic appendages are rudimentary, respiratory and nervous system undeveloped, digestive system embryonic. These larvae show highly developed cephalothorax, powerful jaws and caudal outgrowth, *e.g.*, parasitic Hymenoptera.

Protopodite: The basal stalk of some Crustacean limbs composed of the united coxopodite and basipodite.

Protozoa: A phylum which include single celled animals.

Protura: *Prot*=first, *ura*=tail. Insect order, made up of the proturans. This is a primitive order whose members are characterized by narrow, wingless bodies, sucking mouthparts, and no metamorphosis, *e.g.*, proturans.

Proventriculus: Part of the digestive system in insects. It is an area of the foregut just anterior to the midgut, often modified for grinding. This is also called gizzard and acts as the valve between foregut and midgut. In nearly all the insects, there are special structures where the foregut and mid gut join. There is always a muscular sphincter by which the contents of two regions can be kept separate (prevent regurgitation of food from midgut tract to the foregut); and there is often a discrete organ known as the proventriculus. It is a morphological entity in different groups, sometimes composed solely of foregut, while sometimes both foregut and midgut contribute to it. In some insects, it is generally represented by a simple valve (cardiac valve) at the origin of midgut. In Acrididae, it is teeth like and arranged in six longitudinal folds. In the Blattidae, it is a powerful muscular organ, provided with rows of massive teeth which seize and triturate the solid particles in the crop. In the bee and other hymenoptera, the proventriculus is termed as honey stopper. It forms an elongated plug with an elaborate musculature. The lumen bears rows of curved spines by means of which the pollen grains are filtered out from the nectar and passed on to the stomach while the clarified nectar is retained in the crop. In the flea, it is lined with long needle like teeth directed backwards; during digestion it contracts rhythmically, driving the teeth backwards into the midgut; at the same time the midgut contents are driven up against it by antiperistaltic waves, and in this way the blood corpuscles are pounded and disintegrated.

Proximad: Towards the end or portion nearest the body.

Proximal (proximad): An adjective (or adverb) denoting position near or at or movement towards the end of an appendage nearest the body.

Proximate: Adjacent to or nearest the point of attachment or origin.

Proxol: Trade name of trichlorfon.

Pruinescent: Covered in a whitish bloom.

Pruinose: Covered with a powdery deposit, usually white or pale blue; especially applied to Odonata.

Przibram's factor: See Dyar's law.

Pselaphidae (short-winged mould beetles): Family of superfamily Taphylinoidea, suborder Polyphaga and order Coleoptera. The pselaphids are small, yellowish or brownish beetles with short elytra covering half of the abdomen. They occur under stones and logs, in rotting wood or in moss; a few are myrmecophilous or termitophilous. They are predaceous on mites and other small animals. *Mastiger abruptus* is a common species in India.

Pseudergate: A special caste found in the lower termites comprising individuals having either regressed from nymphal stages or by moults that reduced or eliminated the wing buds or derived from larvae by undergoing non-differentiating moults. These serve as the principal element of the worker caste but capable of developing in other caste through further moultings.

Pseudo: False; temporary.

Pseudoarolium or pseudarolium (pl., pseudoarolia): A pad at the apex of tarsus resembling an arolium as in Hemiptera.

Pseudocelli : See onychiuridae.

Pseudococcidae: A family of superfamily Coccoidea, suborder Homoptera and order Hemiptera.

***Pseudococcus filamentosus* Cockerell:** Commonly known as citrus mealy bug (Pseudococcidae, Hemiptera) and is a pest of *Citrus* species, ferns, begonia, poinsettia, cactus etc.

***Pseudococcus* sp. (Hemiptera; Pseudococcdae):** Mealy bug infesting grapevine.

Pseudocone ommatidium: See ommatidium, types.

Pseudomedia: A vein appearing like media but formed by the fusion of branches of Rs as in Neuroptera.

Pseudoplacemental viviparity: See viviparous.

Pseudopod: See processes of body wall.

Pseudopupa: A larva in the quiescent pupa like stage; a coarctate larva; one or two instars before the true pupal stage. For example, the sixth instar grub of a blister beetle, also called the pseudopupa (Coleoptera, Meloidae).

Pseudoresistance: Apparent, thought to be usually non-heritable pest resistance in potentially susceptible hosts resulting from some transitory ecological situation (Painter, 1951).

Pseudoscorpions: Small arachnids, seldom over 5 mm long, scorpion like in general appearance but without sting.

Pseudotrachea: A structure like a trachea; a false trachea.

Pseudovipositor: Slender tube to which the posterior part of abdomen is reduced in the females of certain insects.

PSI: Pounds per square inch is the unit of pressure.

***Psitticula krameri* (Scopoli):** Belongs to order Psittaciformes and family Psittacidae. It is a pest of crops, commonly known as the rose ringed parakeet.

Psocidae: A family of order Psocoptera and suborder Eupsocida. Tarsi two segmented. Example, *Stenopsocus* and *Psocus.*

Psocoptera: Gr., *psoco*= rub small, *ptera*= wings. Insect order, made up of the barklice and booklice. They are characterized by flattened bodies, membranous wings (or wingless), chewing mouthparts, and an incomplete metamorphosis. These are minute insects 3-6 mm long with or without wings, feeding on the glue of books, decaying animal and vegetable matter. They live in tree trunk, under bark, bird nests, straw and chaff, on fungi and lichen. These are soft bodied with integument moderately sclerotised, the large head bearing distinct epicranial suture, protruding large compound eyes and long filiform antennae (mostly 13 segmented). Three ocelli distinct in winged forms and absent in winless forms. Anteclypeus and postclypeus ae distinct sclerites of head. Mouth parts are biting and chewing. Prothorax is characteristically small in alate while larger in apterous forms. The ovipositor is small.

Psoroptes communis: Belongs to order Acarina and family Sarcoptidae. It is a pest of farm animals, commonly known as the scabies mite.

Psychic: Beyond natural or known physical processes.

Psychidae (bag-worm moths): Family of super family Psycoidea of suborder Ditrysia and order Lepidoptera. Bag-worm moths are widely distributed insects. The females are apterous whereas the males are winged. Wings are thinly clothed with imperfect scales and exceptionally large frenulum. Antennae are biceptinate. Labial palpi are short. Some females are devoid of antennae, mouthparts and legs. Larvae inhabit cases which exhibit great variety of shape and of materials used in their construction. The bag worm *Clania crameri* is a very common species found in South India on *Acacia, Casuarina equisetifolia* etc.

Psychodidae (mothflies, sandflies): Family of suborder Nematocera, order Diptera. Small moth-like insects with body and wings densely covered with hair, with long antennae composed of 15 or 16 segments. The lanceolate wings have the anterior cross vein closer to wings base. The bloodsucking sandflies (Phlebotominae) come in here and the females alone suck blood. Several species are known, but the well known ones are *Phlebotomus minutus, P. argentipes* and *P. papatasi,* the latter being the vector for kala-azar, caused by the flagellate protozoan *Leishmania donovani.*

Psychoidea: Super family of suborder Ditrysia and order Lepidoptera. Proboscis is a atrophied in moths with maxillary palpi either wanting or vestigial. Cu_2 usually represented on both pairs of wings; M present within the cell.

Psychology: Gr., *psyche*= mind. The study of animal mind.

Psychophily: Transfer of pollens by butterflies, a type of pollination.

Psyllid: Jumping plant louse of the genus *Psylla* beonging to Family Chermidae or Psyllidae (Hemiptera). These are serious pests of crops and orchards.

Psyllidae: A family of superfamily Psylloidea of suborder Homoptera and order Hemiptera. Psyllids are very small, active insects measuring 2-5 mm in length with long antennae and three segmented rostrum. The hind legs are stronger than the other legs and at the slightest disturbance they jump out quickly. They rest on the plants in a characteristic way with the body raised up and the head almost touching the surface. The nymphs are in general flat. Many species in their nymphal stages produce white waxy secretions. They are phytophagous and most species are host specific. Examples are pear psylla, *Psylla mali;* Citrus psylla etc.

Psyllipsocidae: A family of order Psocoptera and suborder Trogiomorpha. Head not transverse but long and vertical, maxillary palp without sense organ. Antennae 27-46 segmented. Example, *Psyllipsocus.*

Psylloidea: A superfamily of suborder Homoptera and order Hemiptera. Characterized by mouth parts and two pairs of wings in bot the sexes; forewings are rather harder in consistency than the hind wings; antennae 10 segmented; tarsi two segmented with paired claws, femora thickened.

Pteralia: Articular sclerites of the wing base, including the humeral plate and axillary plate or axillary sclerites.

Pteridine: It is a by-product of synthetic activities of the cells and origin of many pigments of insects. The pterydine pigments of many types occur in the wing scales of Pieridae and many Lepidoptera, or beneath the integument of wasps. Pterydines are also common pigments occurring in insect urine. They occur in small quantity. The leucopterine and isoxanthopterines and creatine are reported from the urine of some insects.

Ptero: A Greek prefix meaning 'wing'.

Pteromalidae: Family of super family Chalcidoidea, suborder Apocrita and order Hymenoptera Pteromalidae are found in large numbers parasitizing a wide variety of insects of economic importance. They are minute, black or metallic green or bronze in color with large but short thorax and small subtriangular gaster. The second or fused second and third tergites cover most of the surface of the gaster. In many species the adults female feeds on the body fluid that exudes through the puncture made on the body of the host by its ovipositor.

Pteronarcidae: A family of order Plecoptera. Characterized by forewings having cross veins but the anal lobe of hindwing without such cross veins. Coxae of forelegs are closely approximated. Mandibles vestigial. Vestiges of tracheal gills are retained by adults, *e.g., Pteronarcys* sp.

Pteropleural bristles: Bristles on the pteropleuron as in Diptera.

Pteropleuron: A sclerite on the side of thorax just below the base of wing and consisting of the upper part of mesepimeron as in Diptera.

Pterostigma: Gr., *ptero*= a wing, and *stigma*= a mark. A small coloured area near the wing-tip of dragonflies, bees, and various other clear winged insects; also called the stigma.

Pterothorax: The wing bearing segments of the thorax, *i.e.,* mesothorax and metathorax.-

Pterothysanidae: Family of super family Callidulоidea of suborder Ditrysia and order Lepidoptera. Moths resemble geometrids and are found in the Oriental and African regions. They have a slender body and large wings. In the oriental region, *Pterothysanus* is prevalent and is characterized by long double hair-fringe on the inner margin of the hindwings. In India, Eastern Himalaya mountain is inhabited by this insect.

Pterygopolymorphism: See alary polymorphism.

Pterygota: The subclass of Insecta comprising the winged insects.-

Pterygote: Winged insect, a member of the subclass Pterygota.

Pterygote insect orders: See taxonomic grouping of orders.

Ptilinal suture: A transverse fissure crossing just above the antennae and extends down laterally in the form of an inverted U towards the clypeus. It represents the closed lips of the ptilinum, an eversible sac used in eclosion, *e.g.,* Schizophora (Diptera).

Ptilinum: In Diptera an organ that can be inflated to a bladder like structure and thrust out through a frontal suture of the head at the time of emergence from the puparium. In cyclorraphous Diptera, the ptilinum facilitates escape from the puparium and also enables the insect to burrow to the surface of the debris in which it is buried. The ptilinum is a membranous sac on the front of the head which can be everted by blood pressure while the rest of cuticle of the head remains relatively soft. In the adult fly, its position is identified by the ptilinal suture.

Ptinidae (spider beetles): Family of superfamily Bostrychoidea, suborder Polyphaga and order Coleoptera. These are small, cylindrical beetles with filiform antennae and with contiguous fore and midcoxae and transverse hindcoxae. They are mainly scavenging beetles as they feed on dead insects, excrement or dry vegetable matter. Some feed on stored products and the store-house beetle *Gibbium psylloides* is well known.

Pubescence: Downy or hairy covering on certain plants and insects.

Pubescent: Downy, covered with short fine hair.

Pulsatile organ: See pulsatory organs.

Pulsating membrane: Small muscular membrane found in the head, thorax and appendages of various insects, the rhythmic contractions which probably contribute to the blood circulation.

Pulsatory organs: Sac like structures at the base of antennae, legs and wings ensuring adequate circulation of blood. In majority of insects, the blood circulation is brought about by muscular movement, respiratory movement and differential pressure of pericardial and perineural sinus.

Pulse beetle: *Callosobruchus chinensis, C. analis, C. maculates*, very serious pests of stored grains like gram, pea, cowpea, bean, lentil, soybean etc. The infestation starts from field or godown; larva bores the sound grain and feeds inside making the grain hollow. The pea beetle, *Laria affinis* grub feeds on the developing pods in the field and grain in the stores.

Pulvilliform: Lobe like or pad like a pulvillus; pulvilliform empodium.

Pulvillus (pl., pulvilli): A pad or lobe beneath each tarsal claw as in Diptera (the lower surface of tarsi as in some Orthoptera). The little pad beneath each claw on the leg of a fly.

***Pulvinaria psidi* Maskell:** Belongs to order Hemiptera and family Coccidae. It is an insect pest of mango, commonly known as mango soft scale.

Punctate: Covered with tiny pits or depressions, like the elytra of many beetles and the thorax of many Hymenopterans. Beset with punctures; marked with minute spots or depressions.

Pupa (pl., pupae): The stage between larva and adult in insect with a complete metamorphosis. It is a non-feeding and usually inactive stage (see chrysalis, cocoon). In the exarate pupae, the appendages are free from the body; while in obtect pupa, the appendages are glued to the body. When articulated mandibles are present, it is called decticous pupae; while with immobile mandibles, it is called adecticous pupae. Decticous pupae are always exarate (Neuroptera, Trichoptera and some Lepidoptera). Adecticous pupae may be exarate (Cyclorrhapha, Siphonaptera, Coleoptera, Hymenoptera) or obtect (Lepidoptera, Neuroptera, Nematocera, Brachycera etc.). If adecticous-exarate pupae are enclosed in a cocoon or puparium are called the coarctate pupa (see also pre-pupa).

Puparium (pl., puparia): A protective case formed by the hardening of next to the last larval skin in which the pupa is formed (in the order Diptera).

Pupate: To turn into and exist as a pupa.

Pupation: To transform to a pupa.

Pupiparous: Insects which give birth to fully grown larvae which pupate almost immediately are said to be pupiparous. The main examples are various blood sucking flies.

Purinose: Covered with a whitish waxy powder.

Pusa bin: It is a storage structure for grains designed by Pradhan and Mukherjee (1968) at Indian Agricultural Research Institute, New Delhi to demonstrate the utility of polythene. The bin required for storing 2000 kg grain measures about 1.40 m long, 1.0 m broad and 1.6 m high (internal dimensions). The walls consist of two layers of mud or unburnt bricks (each layer 7.6 cm thick) with a polythene film sandwiched in between them. The floor and roof comprise similar structure with each layer 5 cm thick. The roof is supported on wooden framework. The manhole is 60X60 cm in a corner of top surface. About 9 m polythene or alkathene (700 guage), 1.8 m wide (preferably ISI brand) is required. An outlet pipe is kept 9.0 cm in diameter and 24 cm long with a tight fitting lid.

Putrefaction: The decomposition of proteins by microorganisms under anaerobic conditions resulting in the production of incompletely oxidized compounds.

Pygidial glands: Exocrine glands that secrete a repugnatorial substance and that at the posterior end of the abdomen in some species of beetles.

Pygidicraniidae: Family of order Dermaptera, suborder Forficulina and superfamily Pygidicranioidea. Paired aedeagus distinct, telson and tergum of eleventh segment also distinct, forefemora compressed and keeled. Telson is as large as pygidium.

Pygidicranioidea: Superfamily of order Dermaptera and suborder Forficulina.

Pygidium: The terminal exposed and sclerotized tergite; this may be tergite 7 or 8 or last tergite of the abdomen.

Pyloric valve: The midgut opens into the hindgut through a valve known as pyloric valve.

Pylorus: An anterior part of the proctodaeum usually having the pyloric valve sometimes distinctly differentiated from the true intestinal region and forming anatomically the rear end of the stomach..

Pymetrozine: A new insecticide belonging to the group pyridine azomethines, marketed under trade names, Plenum, Chess, Fulfill, Reley and Sterling. It has no direct toxicity but blocks the stylet penetration of sucking insects and causes immediate cessation of feeding. It has both contact and systemic activity and effective against sucking insect pests.

Pyralidae: Family of super family Pyralidoidea of suborder Ditrysia and order Lepidoptera. A small group found widely distributed in the tropics. The moths are small and the larvae feed on dried vegetable matter, wax etc. Larvae of *Pyralis farinalis* feed on stored potato and that of *Hypsopygia mauritialis* feed on wax in wasp nests.

Pyralidoidea: Super family of suborder Ditrysia and order Lepidoptera.

Pyramid of biomass: Ecological pyramid showing the total dry weight or total amount of living matter in ecosystems. In grasslands and forests, there is a decrease in biomass of organisms at successive levels from the producers to carnivores. Thus pyramids of biomass are upright. But inverted shape of pyramid in aquatic ecosystem is there because biomass of producers is least.

Pyramid of energy: Ecological pyramid which gives knowledge of available energy at each trophic level in a food chain and amount of energy transferred from one trophic level to next trophic level. The total energy content of an ecosystem is the highest in the producers and least in the top consumers in any ecosystem hence the pyramids of energy are always upright.

Pyramid of numbers: The ecological pyramid showing the number of individual organisms present at each trophic level. It is usually straight, *e.g.*, crop ecosystem, grassland ecosystem, pond ecosystem etc.

Pyraustidae: Family of super family Pyralidoidea of suborder Ditrysia and order Lepidoptera. These are relatively large moths having vestigial proboscis. A large number of them are important as pests of crops. Larvae of these insects feed variously by boring into stems and pods or fruits of plants, webbing or rolling the leaves, webbing the flowers in the inflorescence or forming cases with leaves and feeding by scraping them. Some are aquatic in the larval and pupal stages. The following three subfamilies are recognized: Schoenobiinae, Nymphulinae, Pyraustinae.

Pyraustinae: Subfamily of family pyraustidae. This subfamily contains a large number of forms which are abundant in the tropics. Larvae of these feed variously by remaining in silken web amongst spun- up leaves, stems, fruits or roots. In the USA, *Ostrinia nubilalis* is the most notorious pest of corn borer. In India the family is well representation and a large number of them are pests of economic importance on a variety of crops.

Pyrethrin: Either of the two liquid esters derived from *Chrysanthemum;* the active ingredient of pyrethrum.

Pyrethroids, synthetic: The synthetic pyrethroids are based on the earlier natural pyrethroids but differed so markedly from them in properties and action so as to constitute a new class of insecticides. The first synthetic pyrethroid was described by Elliot and coworkers. The structure of the natural pyrethrins, of which pyrethrin I is the prototype, has been modified to give successive synthetic compounds with improved properties. The synthetic pyrethroids are quick in action, have high insecticidal efficiency at low dosage with low mammalian toxicity, good persistence, less risk of environmental contamination and with less chances of development of insecticidal resistance. The examples are permethrin, cypermethrin, decamethrin, fenvalerate etc.

Pyrethroids: Natural pyrethrum found in the flowers of *Chrysanthemum* plants. Chiefly, four esters are found of which two are pyrethrins.

Pyrethrum dust: A dust formulation of pyrethrum made by mixing the pyrethrum with talc or charcoal powder.

Pyrethrum: A botanical insecticide derived from the achenes of *Chrysanthemum cinerariaefolium* (family: Compositae), harmless to warm blooded animals, non-toxic to plants; effective against household insects, stored food and garden insects. It was used as insecticide for the first time in 1800. The active ingredients of pyrethrum are four esters, pyrethrin I, pyrethrin II, cinerin I and cinerin II. The esters are derived from two acids (chrysanthemic acid and pyrethric acid) and two alcohols (cinerolone and pyrethrolone). Pyrethrolone+chrysanthemic acid= pyrethrin I; Pyrethrolone+ pyrethric acid= pyrethrin II; cinerolone+ chrysanthemic acid+ cinerin I; cinerolone+ pyrethric acid= cinerin II. Pyrethrum is generally mixed with the sesamin, sesamolin and piperonyl butoxide for synergistic action.

Pyrilla or sugarcane leafhopper: See *Pyrilla perpusilla.*

***Pyrilla perpusilla* Wlk.:** Belongs to order Hemiptera and family Fulgoridae. It is an insect pest of sugarcane, commonly known as the pyrilla or sugarcane leafhopper.

Pyrophoric: Tendency of substance to ignite spontaneously.

Pyrophyllite: A mineral used as dust carrier in insecticidal formulations having chemical formula, $Si_8Al_4O_{20}(OH)_4$. It is white hydrous aluminium silicate similar to talc in physical appearance. It has low absorption power.

Pyrrhocoridae: Family of superfamily Lygaeoidea, suborder Heteroptera and order Hemiptera. Insects are generally brightly coloured with red and black markings and elongate oval, distinguished from the Lygaeids by lack of ocelli and possession of more branched veins and cells in the hemelytra, rostrum four segmented, glands open in the metathorax; legs possess rotatory coxae, three segmented tarsi and pulvilli. Phytophagous. Example is cotton stainer, *Dysdercus cingulatus.*

Q

Quacking: The calls emitted by virgin honeybee queens in their cells in response to the piping sounds of the first virgin queen to emerge in the same hive.

Quadrangle: A cell immediately beyond the arculus (Odonata, Zygoptera).

Quadrate: Four sided.

Quadrilateral cell: A cell near the base of a damselfly wing, shape of this cell is used as a feature to determine damselfly families.

Quadripectinate: Antennae that have four rami per segment.

Qualitative loss: Apart from physical loss in weight (quantitative loss), there are adverse transformations; qualitative depletions in nutrient contents, caloric value and aesthetic value due to damage of insect pests and diseases in crops and stored commodities which is known as qualitative loss.

Quality control: The application of the theory of mathematical probability to sampling the output of an industrial process with the object of detecting and controlling any variation in quality like percentage of the active ingredient in a pesticidal formulation, its shelf life etc.

Quantitative: Dealing with quantities as well as nature of the substances (*e.g.*, components in mixture) under consideration.

Quantitative analysis: Analysis for determination of concentration of one or more of the component of a sample. Classical methods include volumetric and gravimetric analysis; while recent methods use sophisticated instrumental methods, *e.g.*, atomic absorption spectrometry, polarography, ultraviolet and visible spectroscopy, gas chromatography, high performance liquid chromatography and mass spectrometry.

Quantitative loss: Physical loss is concerned with the loss in weight of a commodity due to pests or diseases.

Quarantine: (i) Voluntary or compulsory isolation or restriction on free movement or importation imposed to prevent the spread of an insect pest or disease in a territory or country, (ii) The isolation of an individual insect or group of insects in order to determine or eliminate any associated or potentially hazardous organisms.

Quarantine laws: Laws that enable enforcement of restriction on the introduction of infested material from foreign country or within the country from one place to another; the latter is known as domestic quarantine.

Quartz (SiO_2): It is one of the most common and widely distributed minerals of the earth's crust and forms an essential constituent of sedimentary rocks. The powder is white in colour and used in insecticide dusts as inert carrier.

Quasisocial: Applied to the condition or the group in which members of the composite nest cooperate in brood care.

Queen: A fully developed female, larger and longer than a worker. This is a primary reproductive female in a colony of social insects (bees, wasps, termites etc.).

Queen bee: A fully developed mated female bee; longer and larger than a worker bee; main function is laying eggs. She lays 1500-2000 eggs per day. The eggs are either fertilized which produce worker bees or unfertilized which produce drones. The egg which produces queen is a normal worker bee egg.

Queen cage: A small cage in which a queen and three or four worker bees may be confined for shipping and/ or introduction into a colony.

Queen cage candy: Candy made by kneading powdered sugar with invert sugar syrup until it forms a stiff dough; used as food in queen cages.

Queen cell: A special elongated cell, resembling a peanut shell, in which the queen is reared from egg to the adult stage. It is usually an inch or more long, has an inside diameter of about 1/3 inch and hangs down from the comb in a vertical position.

Queen control: The inhibitory influence of the queen on the reproductive activities of the workers and other queens.

Queen cup: A cup shaped cell made of bee wax or plastic which hangs vertically in a hive and which may become a queen cell if an egg or larva is placed in it and bees add wax to it.

Queen excluder: Metal or plastic device with spaces that permit the passage of workers but restrict the movement of drones and queens from one specific part of the hive to another (brood chamber to super chamber).

Queenlessness: A hive which has lost its queen; resulting in swarming.

Queen pheromone: In the social wasps, *Polybia sericea* and *Vespa orientalis*, queen pheromone induces diversified behavioural and physiological processes. It is involved in swarm formation and emigration. It is secreted by a gland bearing opening at the base of fifth sternite of gaster rather than seventh abdominal tergite found in *Apis mellifera*. It is a multifunctional pheromone. It stimulates workers for active nectar foraging. Within a colony, it is responsible for attraction of workers to the queen. It inhibits queen rearing by worker wasps and development of ovaries in workers. It directs the workers towards food sources, nest entrance and workers to assemble for a swarming.

Queen right: Referring to a colony, especially a honeybee colony, that contains a functional queen.

Queen substance: Pheromone material secreted from glands in the queen bee and transmitted throughout the colony by workers to alert other workers about the queen's presence and other purposes. This substance also acts as sexual attractant during the nuptial flight.

Quelitox: See fenthion.

Quiescence: A state of relative biological inactivity induced by physical and biological factors. Growth and development in insects may be temporarily arrested at any stage in the life cycle by adverse conditions of many kinds. Low temperature, starvation, lack of vitamins or other essential substances may lead to a state of quiescence. But many insects, particularly, in temperate latitudes, periodically enter a state of arrested growth which is not readily eliminated when conditions again become favourable. For this deep seated arrest, the term diapause is commonly used.

Quill: The hollow basal part of a feather.

Quilt: A thick, protective cover placed over the frames of a beehive.

Quinalphos: An organophosphate insecticide; chemically diethyl 2-quinoxalinyl phosphorothionate ($C_{12}H_{15}O_3H_2PS$); an odourless, white crystalline solid with melting point 30 °C, practically insoluble in water (22 mg l^{-1}) but readily soluble in organic solvents. It is wide spectrum insecticide with acaricidal property. It is marketed under trade name Ekalux.

Quinones: Organic compounds; benzene derivatives known to play a role in the hardening and darkening of arthropod cuticle.

Quiscent: Inactive, resting or non-feeding stage.

R

Rabbet: A narrow piece of folded metal fastened to the inside upper end of hive body from which the frames are suspended.

Rabon: Trade name of gardona.

Race: A variety of a species; a subspecies. Equated with subspecies in taxonomy; in general a category of variant individuals within a species and differing slightly in characteristics from typical members of the species.

Radial bar of hind wing: The anterior of the two major supporting structures of the hind wing in beetles, formed from the veins Sc and RA and in many Polyphaga supporting the Radial cell at or near its apex.

Radial cell of hind wing: In polyphagous beetles a cell formed by the diverging and subsquent converging of veins RA_{1+2} and RA_{3+4}. May also be applied to a series of cells formed mainly by radial cross veins in Archostemata, Adephaga and Myxophaga.

Radial cross vein: A cross vein connecting R_1 and branch of radius immediately behind it.

Radial sector: The posterior of the two main branches of the radius, usually abbreviated to Rs. It usually has several branches of its own.

Radiation hazard: Potential danger of ionization radiation to human being and other animals and plant life.

Radical: Group of atoms or charged atoms which function as a single unit and maintain their separate identity in the chemical reaction are known as radicals. In the aqueous solution of acid, base and salt cations and anions are present in the free state. In this process of electrolysis cations are attracted towards the cathode and anions are attracted towards the anode. In the aqueous solution of simple salt, sodium cation (Na^+) and chloride anion (Cl^-) are present and they maintain their individual identity.

Radient heat: See heat generation.

Radioactive pollution: Radioactive substances present in nature emit high energy radiations which pollute the environment and causes hazardous effects. Apart from natural resources, the man made nuclear weapons, atomic reaction, radioactive isotopes etc. also cause radioactive pollution. The radioactive pollutants cause various diseases like tumours, skin cancers, defective eye sight in animals and affects the plants.

Radioactive: Capable of giving radiant energy in the form of particles or rays as alpha, beta and gamma by the disintegration of atomic nuclei.

Radioisotope: One of a broad class of elements capable of becoming radioactive and giving off the atomic energy. Some radioisotopes are produced artificially.

Radius: It represents the third major longitudinal vein of wing and abbreviated as R. After running for a short distance as a strongly sclerotized, convex vein, it divides into two branches: the outer branch R_1 runs directly towards the outer margin. The second branch is termed as radial sector which appears to be concave and gives origin to four branches (R_2-R_5).

***Ragmus importunitas* D.:** Belongs to order Hemiptera and family Miridae. It is an insect pest of jute, commonly known as the sunnhemp capsid.

Rainfed crop: A crop or cultivated plant completely dependent on rains for their water requirement.

Ramus (pl., rami): A projection on an antennomere. See uniramose, biramose.

Randomized block design: An experimental design in which the experimental area is divided into blocks and all the treatments are randomly arranged within each block.

Range: The geographical area of the occurrence of a species or a region in which a given form occurs naturally or after introduction.

Rape: *Brassica napus* var. *biennis* belonging to mustard family and grown chiefly as pasture crop for forage.

Raphidioidea: Superfamily of suborder Megaloptera, order Neuroptera. Larvae terrestrial with well developed spiracles, wings with pterostigma and ovipositor exserted.

***Raphidopalpa foveicollis* Lucas:** Belongs to order Coleoptera and family Galerucidae. It is an insect pest of cucurbits, commonly known as the red pumpkin beetles.

***Raphimetopus ablutellus* Zeller:** Belongs to order Lepidoptera and family Pyralidae. It is an insect pest of sugarcane, commonly known as the green borer.

Raptorial: Preying or prehensile legs, *e.g.*, forelegs of mantids, aquatic bugs, hindlegs of parasitic Hymenoptera etc. These legs are elongated and spiny. In the mantids, the tibia can be accommodated in the groove of femur and tarsus can be inserted in the apical hook of the femur. The femora tibial joint is well developed. The foretibia of the bug, *Naucoria* is modified into a single arrow like pointed structure.

Rasping and sucking type mouth parts: Intermediate between the piercing and sucking type and chewing type but rasping and sucking in their action serving to lacerate the epidermis of the plants and sucking up the exuding sap. The right mandible is reduced making the head and mouth parts somewhat asymmetrical. The left mandible and maxilla according to Borden hypothesis is elongated, suggesting the stylets of piercing type and adapted to move out of circular opening at the apex of cone shaped head. The stylets are contained in a separate pouch that has invaginated from the surface to give them their internal position. The stylets apparently neither form food channel external to the wall of head nor do they enter deeply in the wound. The sap as it exudes on the surface is sucked up by the cone shaped mouth rather than stylet. According to Snodgrass, the food channel within the head capsule lies between the labrum and the hypopharynx and saliva passes to the tip of the stylet between the hypopharynx and the labium. Both pairs of palpi are present, *e.g.*, thrips.

Rat: A notorious pest of fields and stored grains in houses and godowns. This belongs to phylum chordata, subphylum vertebrata, division gnathostomata, super-class tetrapoda, class mammalia, subclass theria, infraclass eutheria, order rodentia and family muridae. Common rat species are house rat, *Rattus rattus* Linn.; brown rat or sheep rat, *Rattus norvegicus* Erxleben; house mouse, *Mus musculus* Linn.; large bandicoot, *Bandicota indica* Bechstein; Indian mole rat, *Bandicota bengalensis* Gray; short tailed bandicoot, *Nesokia indica* Gray; field mouse, *Mus booduga* (Gray); soft furred rat, *Rattus meltada* (Gray); Indian gerbil, *Tatera indica* (Hardwicke); Indian desert gerbil, *Mriones hurrianae* (Jerdon); Pacific rat, *Rattus exulans* etc.

Rat bait: Any food attractive to the rats and to which a substance poisonous to them is added.

Rat flea: See *Xenopsylla cheopis.*

Ratooning: Raising the crop from the shoots emerging from stumps of the harvested crop.

***Rattus rattus*:** A long tailed rat (family: Muridae) commonly known as house rat, bigger than a mouse, a serious pest in field and storage, also responsible for spreading plague, cosmopolitan in distribution.

Recapitulation: The theory and ontogeny recapitulates phylogeny.

Receptaculum seminis: A sac like structure in the female reproductive system for storing sperm cells required for fertilizing the eggs.

Receptive apparatus: The part of sense organ, primarily responsive to the stimulus transmitted through the peripheral parts, formed of the sense cells.

Receptor: Sense organ or specialized structure of integument responsive to external stimuli.

Recessive characteristics: A characteristic governed by the allele that does not express itself in the presence of another allele of the same gene. The recessive traits are only expressed in homozygous individuals.

***Recilia dorsalis* (Motschulsky):** Belongs to order Hemiptera and family Cicadellidae. It is an insect pest of cereals and millets, commonly known as the zig-zag leafhopper.

Reclinate: Bent downward so that the apex is below the base.

Recommended dosage: The amount of pesticide prescribed to be used for preventing damage by insect pests or diseases which is not always the maximum allowed by law.

Recovery: It refers to obtaining a simple qualitative assessment of the presence or absence of the species soon after its initial release in a new locality with little or no reference to its probable efficiency.

Recovery rate: The amount of insecticide recovered in relation to the amount applied.

Recruitment pheromone: Recruitment by chemical communication appears to be wide spread in social insects and has been demonstrated in bees, termites and ants. Trail making substances are used by animals as navigational aids in directing other members of the colony to a different location varying from hundreds of metres in terrestrial insects. Trails are laid by foraging workers as they return from a food source. These trails are then used by other foragers. Trails may be laid to recruit workers for slave raids, colony emigration, or the repair of a breach in the nest wall (Honeybee- geranoic acid or geraniol; solitary bee- nerotic acid or nerol and termite- hexanol acid).

Rectal chloride epithelia: In Odonate larvae, the basal pads of gill lamellae are modified into the excretory or osmoregulatory epithelia and are commonly referred to as the rectal chloride epithelia. The structures absorb sodium and chloride ions from exogenous water entering through the anus and transport to the haemolymph. In order to maintain osmotic hyper regulation, the structure and function are extensively reviewed by Komnick (1977).

Rectal gill: In dragonfly nymphs, anterior two third part of the rectum is modified into a barrel like chamber and the rectum wall epithelium is differentiated into thick and thin regions alternating with each other. There are six thick epithelia and each is differentiated into the basal thick pads and distal gill filaments. There are thus six rectal pads, each is composed of alike two padlets and each padlets laterally bears obliquely overlapping flattened gills. The gills are richly supplied with tracheoles arising from the efferent branches of longitudinal tracheal trunk.

Rectal glands: Pad like or papillaeform structures on inner wall of the rectal sac.

Rectal pads: Swammerdam (1669) first discovered the rectal pads in the rectum and Wigglesworth (1932) found that these reabsorb the water from rectal contents.

Rectal papillae: The epithelial pads are replaced by four large papillae in Diptera and Hymenoptera, six rectal gills in dragon fly nymphs and an anal sac in Thysanura. The rectal papillae of the adult blow fly, *Calliphora erythrocephala* has been extensively studied by Gupta and Berridge (1966). Each papilla is composed of a cortex and medulla. The cortical region is formed from a single layer of columnar epithelial cells while the medullary region is composed of connective tissue. The papilla is externally covered with a thin layer or cuticle facing the lumen. The rectal papillae absorb water from the faecal matter.

Rectal sac: The enlarged anterior part of the rectum, sometimes produced into a large rectal caecum.

Rectal valve: A circular or lobate fold of the proctodaeal wall between the anterior intestine and the rectum.

Rectum: The pear shaped posterior expanded part of hindgut. The posterior intestine including the rectal sac and the posterior, narrow rectum proper opening at the anus.

Recurrent vein: One or two transverse veins immediately posterior to the cubital vein as in Hymenoptera.

Recurved: Curved upward or backward.

Red ant: *Oecophylla smaragdina* Fab., a species of reddish brown ant which is an important pest of mango.

Red cotton bug: *Dysdercus cingulatus* Fab., an insect pest of cotton causes damage by sucking the cell sap from tender plant parts and spoils lint. It also causes damage in lady's finger, holly hock etc.

Red gram plume moth: *Exelastis atomosa* is a slender moth with wings arranged in plumes. Greenish brown caterpillar feeds on pods of red gram.

Red gram pod fly: See *Melanagromyza obtusa.*

Red hairy caterpillar: *Amsacta moorei* Butler and *A. albistriga* Moore causes damage in clusterbean, green gram, sorghum, maize, pearlmillet, sunnhemp etc. by feeding on plant foliage.

Red mite: *Tetranychus bioculatus* Wood Mason is a serious pest of jute and tea, sucks cell sap from leaves which turn yellow and fall off.

Red palm weevil: See *Rhynchophorus ferrugineus*

Red pumpkin beetle: *Raphidopalpa (=Aulacophora) foveicollis*, (blue pumpkin beetle, *R. atripennis;* grey pumpkin beetle, *A. stevenisis;* yellow pumpkin beetle, *A. cincta*). These are insect pests of cucrbits.

Red rust flour beetle: *Tribolium castaneum* Herbst, is a pest of stored products, like flour, *maida*, *suji* etc.

Red spider mite: *Tetranychus telarius.* A pest of vegetable crops.

Red squill: A low toxicity, tasteless rodenticide. Rats develop bait shyness against this poison, therefore, it is less useful. It is a biological extract of the bulb of *Urginea maritima*. This toxicant is used mostly against the brown rat at 10 per cent.

Red vegetable mite: *Tetranychus telarius.* Same as red spider mite.

Redistribution: Movement of pesticide subsequent to the initial application to other parts of the plant, usually by rain.

Reduction: It is a chemical reaction in which there is a gain of hydrogen or any electropositive radical or a loss of oxygen or an electronegative radical. Examples, (i) gain of hydrogen: $Cl_2 + H_2S \rightarrow 2HCl + S$, (ii) gain or addition of electropositive radical: $SnCl_2 + 2HgCl_2 \rightarrow Hg_2Cl_2 + SnCl_4$, (iii) removal of oxygen atom: $CuO + H_2 \rightarrow Cu + H_2O$, (iv) removal or loss of electronegative radical or element: $Fe_2(SO_4)_3 + H_2 \rightarrow 2FeSO_4 + H_2SO_4$.

Reduviidae: Family of superfamily Reduvioidea, suborder Heteroptera and order Hemiptera. These are usually brownish or black and medium sized to large insects with elongated narrow head, the portion behind the eyes being neck like. Rostrum three segmented, usually curved and its tip fits in the prosternal groove. The antennae are filiform, metathoracic scent glands are absent. Abdomen is broad at the middle and margins of the segment beyond the wings are exposed. They are mostly predaceous on other insects and a few are blood suckers. The causal agent, *Trypanosoma cruzi* of human trypanosomiasis in South America is carried out by *Triatoma* sp. and *Rhodnius prolixus*. In India, *Harpacter costalis* is predaceous on red cotton bug, *Dysdercus cingulatus*; *Reduviolus* sp. On the mired, *Calocoris angustatus*; *Acanthaspis siva* on honey bee and *Platymeris laevicollis* on coconut beetle, *Oryctes rhinoceros*.

Reduvioidea: Superfamily of division Gymnocerata, suborder Heteroptera and order Hemiptera.

Reentry level: The time period between the pesticide application and entry in the field to conduct manual operations.

Reflex: A functional unit of nervous system involving entire pathway from a receptor cell to an effector.

Refraction of light: The process of deviation of light from its original path at the surface of separating plane is called refraction of light.

Regeneration: The new growth of members or lost parts in insects.

Regenerative cells: Replacement cells at the bases of the midgut epithelium.

Regenerative crypt: A pocket of the ventricular epithelium having a group of regenerative cells.

Regression coefficient: The amount of change in one variable associated with a unit change in another variable.

Regulated pest: An organism considered by a state or country to be a serious pest requiring regulatory restrictions or control procedures in order to protect the crop, man or environment.

Regulation (Population regulation): The natural phenomenon of pa fluctuating within certain bounds for long periods of time thus keeping the population in existence.

Regulatory control: See legislative control.

Regulatory officials: People working with the state or central government to enforce rules, regulations or laws.

Reinfestation: Return of insect pests after they left over or destroyed.

Relapsing fever: A disease caused by *Borrelia* species of bacteria and transmitted by human body louse. The European epidemic fever is caused by *Borrelia recurrentis*, transmitted by human body louse and characterized by alternating periods of fever.

Relative growth: Relative growth or increase of one body part with respect to another as total body size is varied. Allometry is a special kind of relative growth.

Relative humidity: The amount of moisture in the air compared to the total amount that air can hold at that temperature. See humidity for detail.

Relay crop: Crop planted in already standing crop shortly before the latter is to be harvested.

Remetabola: See neometabola.

Remigium: The main part of wing, chiefly responsible for maintaining and propelling the insect in flight.

Remote: To move posteriorly.

Rendering wax: The process of melting combs and cappings and removing refuse from the wax.

Reniform: Kidney shaped.

Repackaging (Repacking): Transfer of pesticide from commercial package to smaller container for subsequent sale.

Repeated contact or inhalation: To touch or breathe in a pesticide several times over a period of time.

Repellent: A chemical which has the property of inducing avoidance by a particular pest. A chemical which has the property of inducing avoidance by a particular pest due to unpleasant odour, colour, taste or mechanical effect. For example, mosquito repellents. Some plants have repellent properties, when interplanted in a crop their smell will cause certain pest insects to avoid that crop.

Replacement cells: See regenerative cells.

Replacement control: A type of cultural manipulation involving the encouragement of non-alternate host plants to succeed and competitively replace the weedy alternate hosts of pest insects, thus eliminating pest breeding areas.

Replacement reproductive: See supplementary reproductive.

Replete: An ant whose crop is greatly distended with liquid food to the extent that the abdominal segments are pulled apart and the intersegmental membranes are stretched tight. Repletes usually serve as living reservoirs, regurgitating food on demand of their nestmates.

Replication: Repetition of the same treatment in the experiment so that it appears two or more times.

Repressor: Any of the specific allosteric protein molecules (product of regulatory genes) which bind to the operator of operon and prevent RNA polymerase from proceeding into the operon to transcribe messenger RNA from DNA (negative control).

Reproduction: Giving rise to offspring by the living beings.

Reproductive control: See autocidal control.

Reproductive isolation: Prevention or restriction of gene flow by genetically controlled differences in the reproductive behaviour and fertility relationship of the individuals which are members of different populations.

Reproductive potential: The maximum possible rate of increase of an organism in the absence of any limiting factor or stress etc.; also called biotic potential.

Reproductive system: See male and female reproductive system.

Reproductives: The male and female members of a social insect colony those are capable of reproducing. Males are referred to as drones (in bees and in wasps) or kings (in termites). Females are referred to as queens.

Reptiles: Animals of class Reptilia, cold blooded possessing scaly skin, *e.g.*, snakes, turtles and lizards.

Repugnatorial glands: Glands in insects which secrete malodorous or no vapours and discharged through special openings or other structures, as a defense against enemies, stink bugs.

Requeen: To introduce a new queen bee into an old colony to replace a malfunctioning old queen.

Reserve: Identical deposits known to be recoverable with current technology under present economic conditions.

Residue: Amount of pesticide remaining in or on plant tissues (or in soil) after a lapse of time, especially at harvest time.

Residual activity: Insecticidal activity resulting from the presence of an insecticide and its metabolites on or in a treated site after an insecticide is applied.

Residual deposit: Amount of insecticide remaining on a surface after application.

Residual herbicide: Herbicide showing persistent effect when applied to soil.

Residual insecticide: An insecticide with properties that make it suitable for application to surfaces which will latter be visited by insects. It remains effective after application.

Residual pesticide: A pesticidal chemical that can destroy pests or prevent them from causing disease, damage or destruction for more than a few hours after it is applied.

Residual spray: An insecticide which when applied to surface where insects crawl will remain effective for a long period of time. The examples are spray of DDT, methoxychlor, chlordane, lindane etc.

Residue analysis: The process of estimation of residual level of toxicants in animal feed, food, fibre and environment. The process involves essentially the three steps, (i) Extraction and clean up to initially eliminate the bulk of allied unrelated materials from the sample, (ii) separation of most of the pesticide derivatives in the sample by the chromatography, (iii) detection of chemical in question at the highest possible sensitivity without interference.

Residue tolerance: The amount of chemical pesticide residue which may legally remain in or on a food crop.

Resilience: The ability of any system (*e.g.*, an ecosystem) to resist or recover from stress.

Resilin: A cuticular protein with rubber like properties, and evident as a colourless elastic cuticular ligament such as the wing-hinge ligament, clypeolabral suture (spring), tergosternal ligament and various sites of articulation. It is composed of amino acids which are arranged in the linear fashion and formation of cross links is avoided to ensure elastic property of the protein chain. In the cuticle containing resilin, the pore canals are entirely lacking. It can stretch under tension and return to its normal form as soon as the tension is over.

Resistance: (i) Plants: All properties enabling them to fight and overcome, partially or completely, the pathogenic effects of a disease or pest attack. This also includes tolerance, the ability of a plant to grow and develop in spite of pest or disease attack. (ii) Pests and diseases: The ability of a pest population or disease to survive the poisonous effect of a pesticide or the natural or induced capacity to avoid or repel attack by pests (or parasites). It also refers to the ability to withstand the toxic effects of a pesticide or a group of pesticides, often by metabolic detoxification. Hoskins and Gordon (1956) defined the resistance as 'the added ability to withstand an insecticide acquired by breeding of those individuals which survive exposures to that particular toxicant insufficient to wipe out the whole colony'. The official definition of resistance is 'the development of an ability in a strain of insect to tolerate doses of toxicant which would prove lethal to the majority of individuals in a normal population of the same species'.

Resistant pests: Pests which are not susceptible to the environmental stresses and insecticides at the recommended dosage.

Resistant plant: The host plant capable of counteracting, suppressing or retarding the development of a pest or pathogen or other injurious factors including the conditions of stress.

Resistant strain: A strain of a species which is difficult to be controlled with a recommended dosage of insecticide against normal population of the same species.

Resmethrin (SBP-1382): A synthetic pyrethroid insecticide used to kill diseased honey bee colonies.

Resonating muscles: Muscles capable of undergoing several successive contractions stimulated by a single nerve impulse; usually associated with insect flight production, also called the asynchronous muscles.

Resource management: The comprehensive decision making process regarding the rational use and conservation of our natural resources and environment.

Respiration: Respiration is the process of interchange of gases between the environment and the blood or cellular tissues of the organism. This includes the chemical and the physical phases. The first phase includes the intracellular oxidative break down of carbohydrates resulting into carbon dioxide and water, whereas, the second involves the transportation of air to the tissues and elimination of carbon dioxide.

Respirator (Respiratory mask): A gas mask used to filter out the poisonous gases and dust particles from the air so that a person can breathe and work safely in polluted atmosphere.

Respiratory mechanism: The gaseous exchange along tracheal system can be divided into two parts: (i) transport along tarcheae, and (ii) transport along tracheoles (Weis-Gogh, 1964). (i) Transport along the tracheae (Air tube diffusion)**:** Among most of the terrestrial insects, oxygen enters the tracheal system either through all (holopneustic) or some of the spiracles, mostly by diffusion mechanism, called the air tube diffusion. Diffusion mechanism is simple in small insects like fleas, while in others, the spiracles are provided with the opening and closing apparatus that controls the rate of diffusion. (ii) Transport along tracheoles (Tissue diffusion)**:** Each tracheole ends blindly and the blind end of tracheole is made up of a single cell. All tracheoles are filled with a fluid at the blind end. The tracheole fluid is not an untrafiltrate or haemolymph but it resembles the amniotic fluid having similar composition as that of the moulting fluid. It is a watery fluid rich in protein contents but without chloride. It is replaced by active transport or absorption but not by osmosis.

Respiratory poison: Poison which block the cellular respiration and render the respiratory enzymes inactive, *e.g.*, HCN, CO etc.

Respiratory shield: See respiratory mask.

Respiratory system: The exchange of gases is entirely carried out by the complex tracheal system. The tracheal system is invaginated from the integument. In large number of Collembola, Protura, some Chironomids and endoparasitic Hymenoptera and Dipteran larvae, the tracheal system is lacking and the integument serves the function of gaseous exchange. In insects, the blood (haemolymph) does not play any role in transport of gases and the blood cells (haemocytes) are devoid of respiratory pigments like haemoglobin or haemocyanin etc. See respiratory mechanism and respiration also.

Respiratory toxicity: The intake of pesticides through air passage into the lungs. Pesticides may reach the lungs as a gas or as extremely fine droplets or particles. Most compounds are more toxic by this route. Safety equipments are necessary to protect the pesticide handler or applicator from the hazards of respiratory exposure as recommended on the pesticide label.

Responsible authority (statutory authority): The government or agencies responsible for regulating the manufacture, distribution or use of pesticides more generally for implementing pesticide legislation.

Restricted pesticide (Restricted use pesticide): Pesticide of which the use is restricted by an appropriate governmental authority. It can be applied only by certified operators because of high toxicity or potential hazard to the environment.

Restriction: Limited use.

Restriction endonuclease enzyme: The progress in the field of recombinant DNA technology was mainly due to specialized enzymes that can make precise cuts at specific sites in DNA resulting in the formation of small DNA fragments. These enzymes occur naturally in prokaryotes and known as restriction endonuclease enzyme.

Restriction fragment length polymorphism (RFLP): Restriction enzymes cleave the DNA segments of different strains or related species into fragments of different sizes which is known as restriction fragment length polymorphism. This technique is used in DNA finger printing, in identifying organisms with useful characters, as standard marker for identification of the varieties and to discover interrelationships between different strains and species.

Resurgence: Also called pest resurgence. It is rapid reappearance of a pest population in injurious numbers, usually brought about after the application of a broad spectrum pesticide which has killed the natural enemies that normally keep a pest population in check.

Retention: Refers to the ratio of amount of a pesticide deposit at a given time at a given surface to the initial surface to the initial deposit on that surface.

Reticulate evolution: Evolution depending on repeated intercrossing between a number of lines and thus once both convergent and divergent.

Reticulate: Like a network. To be covered with a network pattern.

Retina cells: The cells composing the retina of the eye.

Retina: Portion of a dorsal ocellus in which light is converted to a nervous impulse.

Retinaculum: The loop into which the frenulum of male is fitted in Lepidoptera; horny, movable scales serving to move the sting or to prevent it being darted far in Hymenoptera; a pair of minute appendages present in third abdominal segment on ventral side to retain the furcula for springing organ in Collembola; also called as the hamula or catch.

Retinue: Refers to a group of honey bee workers not necessarily permanent or even long lasting that closely attend the queen.

Retinula cells: See ommatidium, structure.

Retractile: Capable of being pushed out and drawn back in.

Reverse osmosis: Osmosis is the process of spontaneous flow of water through a semipermeable membrane from a dilute to a concentrated slolution. The diffusion of water takes place on account of hydrostatic pressure known as osmotic pressure which drives water molecules in search of equilibrium.

Revocation: To nullify or withdraw; to revoke an insecticide license.

Rhabdom: A rod like light sensitive structure formed of the inner surfaces of adjacent sense cells in the ommatidium of a compound eye.

Rhabdomere: The rod like distal portion of a retinular cell that contributes to the formation of rhabdom.

Rhagionidae: Family of suborder Brachycera, order Diptera. Predaceous flies with elongate weakly pilose bodies, non-annulate third antennal segment and the tibiae mostly with spurs. Wing veins unlike in stratiomyids not concentrated in the costal region and squmae almost absent. Larvae carnivorous, feeding on other insects. Example, *Rhagio* sp.

Rheotaxis: Directed response in relation to water current.

Rhinoceros beetle: See *Oryctes rhinoceros.*

Rhinotermitidae: A family of order Isoptera. Subterranean termites with pronotum of workers and soldiers flattened, wings oten reticulate and without anal lobe on hindwings. Example, *Rhinotermes, Reticulotermes, Coptotermes.*

Rhipiceroidea: Superfamily of suborder Polyphaga, order Coleoptera. Adults with strongly flabellate antennae and labrum replaced by a nose like projection. This includes two families, *viz.*, Rhipiceridae and Callirhipidae.

***Rhipiphorothrips cruentatus* Hood:** Belongs to order Thsanoptera and family Heliothripidae. It is commonly known as grapevine thrips and is a pest of grapes, rose, jamun, large cardamom etc.

Rhizome weevil: See *Prodiocates haematicus.*

Rhizome: Thickened underground stem producing roots and leaf shoots, *e.g.*, Johnson grass and horse nettle.

Rhodopsin: See ommatidium, structure.

***Rhynchophorus ferrugineus* Olivier:** Belongs to order Coleoptera and family Curculionidae. It is an insect pest of coconut and date palm, commonly known as the red palm weevil.

Rhysodidae (wrinkled bark beetles): A family of superfamily Caraboidea, suborder Adephaga, order Coleoptera. Slender insects, brownish with moniliform antennae and pronotum with three deep longitudinal grooves; mostly tropical. They inhabit rotten wood. Some species that occur in India are *Clinidium aptertum, Rhysodes atterrimus.*

***Rhyzopertha dominica* (Fab.):** Belongs to order Coleoptera and family Bostrichidae. It is an insect pest of stored grains and other products, commonly known as the lesser grain borer.

Ricaniidae: A family of superfamily Fulgoroidea of suborder Homoptera and order Hemiptera. The insects are moth, like hind trochanter is directed ventrally and possesses a short basal tarsal segment in the hind leg. Infestation in the grass and rice by *Ricania zebra* is common.

Rice blue leafhopper: See *Typhlocyba maculifrons.*

Rice bug: *Leptocorisa acuta* Thunb., popularly known as Gundhy bug, a serious pest of rice, causes damage by sucking the cell sap.

Rice caseworm: *Nymphula depunctalis* Guen. is a serious pest of paddy; the caterpillars feed on leaves, cut them to form tubular cases inside which they live and feed.

Rice grasshoppers: *Hieroglyphus banian* Fab., *O. velox* Fab. are common pests of paddy; nymphs and adults cause damage by feeding on leaves, adults feed on the ears.

Rice hispa: *Dicladispa armigera* Oliv., a pest of paddy; small blue black insect with spines all over the body; grubs burrow into the leaf tissues and produces yellow patch on leaf surface.

Rice leaf-folder (leaf roller): See *Cnaphalocrocis medinalis.*

Rice leafhopper: See *Cofta spectra.*

Rice mealy bug: See *Ripersia oryzae.*

Rice moth: *Corcyra cephalonica* Staint. is a serious pest of stored rice, sorghum, oil cakes, pearlmillet, coconut, dry fruits, milled products etc.; larva feeds on grain form webs and silken cocoons enclosing broken grains and frass material.

Rice stem borer: *Tryporyza incertulas* Walker.

Rice weevil: *Sitophilus oryzae* L. is a destructive pest of stored grain, cause damage in wheat, rice, maize, sorghum, barley, pearlmillet etc. This pest is an internal feeder of grains; damaged grains become hollow and unfit for human consumption.

Rickettsia (pl., rickettsiae): An obligate intracellular parasite of arthropods, many among them are pathogenic to human beings and animals. They are intermediate between bacteria and viruses because they have characteristics of both. The crystals produced by them have deleterious effect on the host. The best known species is *Rickettsiella melolonthae* which induces lorsch disease in the grubs of *Melolontha melolontha*, *Anomala* sp. and others. The death of host is slow and takes 2-3 months. The optimum temperature for the disease is 14-20 °C and at 24 °C and above, the host withstands the infection. Older grubs are more resistant than the younger ones. The diseased larvae lose their colour due to free rickettsiae in the haemolymph. The propagation takes place in the cytoplasm and not in the nucleus with the exception in ticks where it takes place in the nucleus. The behavioural change takes place in the grub, *i.e.*, the healthier grubs go deep in the soil, whereas, the diseased grubs come on the surface.

Rift valley fever: The mosquito, *Aedes caballus* and *A. circumluteolus* transmitted disease caused by a virus in Kenya, Uganda, Mozambique, Rhodesia and South Africa.

Riker mount: A thin glass topped showcase filled with cotton.

Ring gland complex: In the Cyclorrhaphous Diptera, formation of a ring gland complex takes place due to unique fusion of cephalic endocrine glands around the aorta anterior to the brain. It was described by Weismann (1864) in the larvae of *Calliphora vomitoria* and since then it is popularly referred as the Weismann's ring. Scharrer and Hadorn (1938), Thompson (1951) and Possompes (1953) described the structure and function of the ring gland complex in various species of Diptera. The ring gland complex is composed of (i) the corpus allatum situated dorsomedially, (ii) corpus cardiacum situated ventromedially, (iii) ecdysial glands (pericardial/ peritracheal) forming the lateral portions of the ring, and (iv) the hypocerebral ganglion closely associated with the corpus cardiacum. The ring gland complex is connected to the brain by a single pair of composite nerves, the NCC terminating into the corpus cardiacum. The recurrent nerve also communicates with the hypocerebral ganglion of the complex.

Riodinidae (Erycinidae, Nemeobiidae): Family of super family Papilionoidea of suborder Ditrysia and order Lepidoptera. These insects are distributed in Neotropical region, United States and the eastern hemisphere. The butterflies are small with short board forewings. In the male the forelegs are brush like with single segmented tarsi having no claws and not useful in walking. In the female the forelegs are perfect though they are slightly smaller than the remaining pairs. The family has two subfamilies, *viz.* Nemeobiinae and Libytheinae. Nemeobiinae includes *Nemeobius lucina,* an European form and the Indian *Dodonia* sp. Libytheinae includes the genus *Libythea.*

Ripe honey: It refers to the thickened honey due to evaporation of water.

***Ripersia oryzae* Green:** Belongs to order Hemiptera and family Coccidae. It is an insect pest of cereals and millets, commonly known as the rice mealy bug.

Risk: It is the degree of physiological, biochemical and histological changes acceptable in terms of usefulness of a chemical and its possible effect on public health.

RNA: Ribose nucleic acid.

Robber bee: A bee which robs food from the other colonies.

Robbing: The act of stealing of nectar or honey by bees from other colonies.

Rock bee: *Apis dorsata* Fab., also known as giant bee is a species of honey bee found in India. This species makes comb with a diameter of 5-7 feet across, suspended from rocks, branches of tall trees and other places, has ferocious temper, a good honey collector and sensitive to smoke. It cannot be domesticated.

Rocker sprayer: Sprayer provided with pump and pressure chamber fitted on wooden board, operated by handle connected with pump. This sprayer is suitable for spraying medium sized fruit trees, crops and garden plants in small holdings.

Rocking type sprayer: See rocker sprayer.

Rodent: Animals belonging to order Rodentia such as rats, mice, gophers, squirrels or woodchucks.

Rodenticide: Pesticide applied to kill rodents (mice, rats etc.).

***Rodolia cardinalis* (Mulsant):** It is commonly known as vedalia beetle (family, Coccinellidae) and is a predator of cottony cushion scale, *Icerya purchasi* Maskell.

***Rodolia fumida* (Mulsant):** It belongs to family Coccinellidae, order Coleoptera and is a predator of mango mealy bug.

Roguing: To remove diseased or abnormal specimens from a group of plants or a crop or to remove deviant plants.

Roost: A perch or a place for resting and sleeping of the birds.

Roosting: Resting or sleeping during the night.

Root aphids: Certain aphid species (family, Aphididae) which feed on roots of many plants, *e.g.*, corn root aphid.

Root knot eelworm: *Meloidogyne* sp. is a nematode pest of brinjal, potato, tomato, potato, tea etc. This pest attacks the roots and tubers forming knots and galls, cause plants to wither and tubers show protuberances.

Root knot nematode: *Meloidogyne incognita*, a species of nematode causing knots on the roots of solanaceous vegetables. Same as root knot eelworm.

Root-knot: Galls and swellings produced on roots as a result of nematode attack.

Root lesion nematodes: See *Pratylenchus.*

Root parasite: Any plant which is completely parasitic on the roots of another plant.

Root soak: Soaking the roots of seedlings for sufficient time in the insecticidal solution to enable the plant system to take up the insecticide and ultimately protect the plant from damage of sucking insect pests.

Rootworm: Any insect larva that infests the roots of a plant; any root infesting nematode.

Root zone application: Application of a systemic insecticides into an anaerobic layer of soil, below the soil surface where the roots are located, *e.g.*, in flooded rice.

Rose aphid: See *Macrosiphum rosaeiformis.*

Rose ringed parakeet: See *Psitticula krameri.*

Rosette: Pattern of leaves radiating from centre or growing point usually close to soil.

Rostrum: The beak or proboscis of Hemiptera. See also the sponging type mouth parts. Beak, usually applied to a piecing mouth of a bug.

Rotation: The practice of growing different crops on the same land in a regular, recurring sequence.

Rotavator: A rotary hoe. The treatment of soil with this implement provides the best means of mixing an applied substance into the surface soil.

Rotenoids: A group of related toxic compounds (rotenone, elliptone etc.) that are found in certain leguminous plants.

Rotenone ($C_{23}H_{22}O_6$): A botanical insecticide; the main toxic constituent in the roots of certain leguminous plants, such as *Derris elliptica, Lonchocarpus utilis* and *L. uruca*, moderately toxic to mammals, non-toxic to plants, suitable for the control of garden and vegetable insect pests, extremely toxic to fishes. The active chemical ingredients were isolated by Geoffroy in 1882 and named nicoulene.

Roundworm: Refers to the nematodes.

Royal cell: A special food given by worker bees to the larvae which are to be developed as queen bee. The royal jelly is secreted primarily by the hypopharyngeal glands and consists of a rich mixture of nutrients, many of them possessing complex chemical structure.

Royal chamber: The space reserved for the accommodation of queen and king termites, or the space reserved for queen of an ant colony.

Royal jelly: A highly nutritious glandular secretion of young bees, used to feed the queen and young brood.

RPAR: Rebuttable Presumption Against Registration. An EPA process whose objectives are to identify pesticide chemicals which present, unreasonable adverse effects on the environment.

RPM: Abbreviation stands for 'revolutions per minute.'

Rubber termite: *Glyptotermes dilatus* Havil, a termite species which attacks the woody part of rubber.

Rudimentary: Reduced in size; vestigial; atrophied; undeveloped.

Ruelene: Trade name of crufomate.

Rugose: Wrinkled. A surface covered in a sculptured pattern.

Runaway pest: Any pest organism which enter a new territory where they lack natural enemies and, therefore, reproduce with little interference resulting in a large population which can over run an area.

Run-off: The process of spray shedding from a plant surface during and immediately after application, when dropletss coalesce to form a continuous film and surplus liquid drops from the surface.

Rural godowns: The godowns are small sized, ranging from 100-500 tonnes capacity with the exception of few large sized godowns which may be as large as 3000 tonnes. These godowns are located essentially in rural areas. The scheme was taken up in 1970's and the subsidy is provided to the tune of 5 per cent by the central or state government.

Russetting: Rough brownish markings on the leaves, fruit or tubers that are caused by some insects, fungi, pesticides, chemicals or weather.

Rust red beetle: See *Tribolium castaneum.*

Rutelidae (shining leaf-chafers): Family of superfamily scarabaeoidea, suborder Polyphaga and order Coleoptera. These are mostly brightly colored insects which are metallic green or blue, reddish or golden yellow. They are characterized by unequal claws on hindlegs and a sclerotised labrum. Both adults and grubs are destructive, the former feed on foliage, flowers and fruits and the latter feed on roots. The Japanese beetles *Popillia japonica* is well known pest of fruit trees in the USA and got introduced into the country from Japan in about 1916.

Ryania: A botanical insecticide of ground stem wood of *Ryania speciosa* having active ingredient as an alkaloid, namely ryanodine; slightly toxic to mammals, acute oral LD_{50} for rats is 1200 mg kg^{-1}.

S

Sabadilla: A botanical insecticide obtained from the ground seeds of lily, *Schoenocaulon officinale*, and active ingredient is a crude mixture of alkaloids termed veratrine; relatively non-toxic to warm blooded animals. It is effective against plant bugs.

Sacbrood: A brood disease of honey bees caused by a virus.

***Saccharicoccus sacchari* (Cockerell):** Belongs to order Hemiptera and family Coccidae. It is an insect pest of sugarcane, commonly known as the sugarcane mealy-bug.

Safener: Material added to the pesticide formulation to reduce phytotoxic effect.

Safety: Certainty that harmful effect or injury will not result from the proper use of pesticide mixture.

Safety period: See waiting period.

Safflower bud fly: See *Acanthiophilus helianthi.*

Safflower caterpillar: *Perigea capensis* G. is an insect pest of safflower; the caterpillars are green with purple markings and feed on the leaves.

Safflower fly: *Acanthiophilus helianthi* Rossi is an insect pest of safflower; light green fly, the maggots of which feed on buds and flowers of safflower.

Sagittae: The lateral processes of Hymenopteran aedeagus.

Sagridae: Family of superfamily Chrysomeloidea, suborder Polyphaga and order Coleoptera. This family is closely related to Donaciidae but is distinguished in having the antennae separated by the entire front of head. *Sagra nigrita* infests *Lablab niger* in South India.

***Saissetia coffeae* (Wlk.):** Belongs to order Hemiptera and family Coccidae. It is an insect pest of coffee, commonly known as the helmet scale.

Salannin: A toxic alkaloid (triterpenoid) obtained from neem tree is a feeding inhibitor at low concentrations but does not influence insect moult. The migratory locust,

California red scale, striped cucumber beetle, house flies and Japanese beetle have been strongly deterred in both laboratory and field tests.

Salididae: Family of superfamily Aradoidea of division Gymnocerata, suborder Heteroptera and order Hemiptera. These small flattened bugs can be seen along the shores of streams, ponds or oceans, inhabiting mud, moss or salt-marsh plants, *e.g.*, shore bugs.

Salivarium: The pocket between the base of hypopharynx and the labium into which the salivary duct opens; in higher insects converted into a salivary pump or a spinning apparatus.

Salivary gland: Most insects possess salivary glands (labial glands) which are situated in the thoracic region or either side of the foregut. Sometimes in the Apterygota and in the bees, several kinds of glands open in the region of mouth and even the single pair of labial glands are present in most insects with varied functions: (i) The secretion may moisten the food, *e.g.*, cockroach and butterfly. (ii) Their secretion often contains digestive enzymes, *e.g.*, amylase in cockroach and invertase in butterflies and bees, inveratse and amylase in plant sucking bugs and aphids, lipase and protease in jassids and capsids which cause much necrotic injury on plants on which they feed. (iii) In the Hemiptera, Diptera and Siphonaptera which have mouth parts adapted for sucking up fluids, the saliva is discharged at the tip of proboscis and one of its functions may be to keep the tube up which the fluids are sucked, moist and clean. (iv) In those forms which suck the blood of living animals, the saliva usually contains an anticoagulin-probably antithrombin. It has been shown that if salivary glands are removed from the living tsetse fly, although it continued to take food for some time, ultimately its crop and proboscis become blocked with coagulated blood. (v) In some insects, the labial glands are not concerned with feeding and digestion but become modified for production of silk or for other purposes, *e.g.*, in the silkworm, *Bombyx mori*.

Salt: The most common salt is sodium chloride, the sodium salt of hydrochloric acid. Other metal or metal like salts in food may include phosphorus, calcium, potassium, sodium, magnesium, iron, cobalt, zinc etc. These may be present as chlorides, sulphates, lactates, citrates or as combination of proteins as in calcium casienate. The salts are represented by a class of compounds formed when the hydrogen atom of an acid radical is replaced by metal or metal like radical.

Saltatorial: Leaping or jumping legs, *e.g.*, in grass hoppers, crickets and Halticinae (Coleoptera). The femora of the hind legs are greatly enlarged to lodge strong levators which bring the leaping movement. The proximal margin of the femora bears a dorsal cuticular process.

Sampling: It is the first step in the residue determination in which representative sample is drawn randomly. Imperfect sampling leads to erroneous conclusion even with sophisticated and sensitive instruments used for residue analysis. The ensuing steps are extraction, clean up and analysis.

San Jose scale: *Quadraspidiotus perniciosus* Comstock is a flat round scaly insect attacking apple, peach, pear, gooseberry, plum, currant, quince, walnut, cherry and almond fruit trees in the hills. Nymphs and adults suck juice from fruits, stem and branches; attacked fruits show pinkish spots; badly attacked plants show grey scurvy appearance.

Sanctuary: Legal restrictions imposed for the preservation and protection of wild life in an area.

Sand fly: *Phlebotomus* sp. (Psychodidae, Diptera) is a pest of medical importance. The females are blood sucking, nocturnal and responsible for transmitting various diseases. These flies are 5 mm long, entire body is hairy and are also called owl midges.

Sand fly fever or papatasi fever: The sand fly, *Phlebotomus papatasi* and *P. sergenti* transmit the virus responsible for sand fly fever or papatasi fever.

Sand fly transmitted diseases: The sand flies transmit various diseases like leishmaniases, visceral leishmaniases (kala azar), dermal leishmaniases (oriental sore), Carrions' disease (bartonellosis) and sand fly fever or papatasi fever. See details.

Sanitation or crop sanitation: The removal and burning of infected plant parts. The decontamination of tools, equipment, hands etc.

Sap: Dissolved materials in water moving through the conductive tissues of the plants.

Sapota leaf webber: See *Nephopteryx eugraphella.*

Sapota mealy bug: *Phenacoccus iceryoides* Gr. and *P. lilacinus* Ckll. attack the shoots and stem of sapota, sucking cell sap from them.

Saprophagous: Feeding on dead or decaying animal or vegetable matter.

Saprophyte: An organism living upon dead or decaying organic matter.

Sapwood: The outer living layer of the trunk of tree by means of which water is transported from roots to branches. It is often distinguishable from the heartwood by its lighter colour. It contains carbohydrates and preferred by the wood borers.

Sapygidae: Family of super family Scolioidea, suborder Apocrita and order Hymenoptera. These are small insects which are parasitic in the nests of bees of the family Megachilidae.

Sarcophagous: Animals feeding on the flesh.

Saturated: When all the valencies of the constituent atoms of a compound are fully satisfied.

Saturniidae: Family of super family Bombycoidea of suborder Ditrysia and order Lepidoptera. The insects are large and found mainly in tropical countries, a few being present in temperate regions. Both sexes have pectinate antennae, the rami being longest in the males. Labial palpi are very small and frenulum is wanting. Almost all moths have a transparent eye-spot near the centre of each wing. The larvae are stout and smooth an possession of scoli is a characteristic feature not met with in other families. The position and number of scoli are some taxonomic importance. The larvae pupate in a dense, firm cocoon which in several species yields silk of commercial value.

Satyridae (Agapetidae): Family of super family Papilionoidea of suborder Ditrysia and order Lepidoptera. Butterflies of this family are known as meadow brown, heaths, graylings and marbled whites. They are small to medium sized largely shade-loving dusky butterflies with eye like or annual spots on wings. Certain veins at the base of forewings are greatly swollen excepting in *Melanitis.*

Sawflies: The insects belonging to family Tenthredinidae, Pamphilidae, Cimbicidae, Diprionidae (Hymenoptera) which derive their name due to insect pests having the saw like ovipositor which is used to lay the eggs in leaves, twigs, blossoms etc.

Saw toothed grain beetle: *Oryzaephilus surinamensis* L. (Cucucjidae) is a pest on stored cereals, groundnut, cashewnut, coconut, milled products etc.; larvae feed on food particles and forms cells by gluing them with its secretion.

SBP-1382: See resmethrin.

Scab: A plant disease causing symptoms characterised by rough, crusty lesions formed by excessive cork production.

Scabby appearance: The appearance of crust formation.

Scabies mite: See *Psoroptes communis.*

Scalariform: Refers to ladder shaped, such as the veins in insects.

Scale insect: Any of various small insects of the superfamily Coccoidea (suborder Homoptera, order Hemiptera). Female scale insects secrete a waxy scale. They remain under this scale while sucking the juices of plants.

Scales: Thin skin flakes or plates comprising the body covering of some animals such as fish and reptiles. In insect order Lepidoptera, the wings are covered by coloured scales.

Scape: First antennal segment.

Scaphidiidae (shining fungus beetles): Family of superfamily Taphylinoidea, suborder Polyphaga and order Coleoptera. Scaphidiidae adults and larvae live in fungi, rotting woods, dead leaves and under bark. They are small shiny convex oval beetles appearing pointed at each end. The elytra are truncated exposing the last abdominal segment. Some Indian species include *Scaphidium conjunctum, S.cyanellum, S.lunatum* etc.

Scarab: Scarab beetle or beetle in the family Scarabaeidae (order Coleoptera).

Scarabaeidae (scarabs; dung beetles): Family of superfamily scarabaeoidea, suborder Polyphaga and order Coleoptera. These are small to very large, smooth, shiny, usually round or oval, very convex, robust bbetles with widely separated midlegs, invisible or small scutellum and a spur on hindtibia. Many spines and horns on the head and prothorax. These are often nocturnal. They feed mostly on dung of mammals which is rolled into balls of convenient size and buried in underground chambers. The beetles also breeds in such dung rolls. A common Indian species noticed is *Heliocopris bucephalus*. The other genera found in India are *Carthasius, Copris, Oniticellus, Drepanocerus, Scarabaeus, Sisyphus, Gymnopleurus* etc.

Scarabaeiform larva: A grub like larva with a thickened cylindrical body, well developed head and thoracic legs, without abdominal prolegs and sluggish in behaviour.

Scarabaeiform: Applied to larva, to have a thick body, strong legs on thorax and well developed head, lamellicorn beetle larvae.

Scarabaeoidea: Superfamily of suborder Polyphaga, order Coleoptera. The beetles are characterized by the three to the seven apical segments of antennae forming a lamellate club. The tarsi are usually five-segmented and the foretibia often toothed or spinose with an apical spur. The exposed pygidium marks the eighth abdominal tergite. This superfamily includes the following families, *viz.*, Lucanidae, Passalidae, Trogidae, Scarabaeidae, Geotrupidae, Pleocombidae, Glaphyridae, Aegiliidae, Aphodiidae, Melolonthidae, Rutelidae, Dynastidae and Cetoniidae.

Scarecrow: Anything deployed to scare the birds.

Scavenger: An animal that feeds on dead plants and animals, on decaying matter or on animal faeces.

Scelionidae: Family of super family Proctotrupoidea, suborder Apocrita and order Hymenoptera. Scelionidae are small insects parasitic on eggs of Orthoptera. Dictyoptera, Lepidoptera, Hemiptera and occasionally of spiders. The phenomenon of phoresy, *i.e.* transportation by attaching itself to another, is noted in some species which are parasitic in the eggs of mantids and grasshoppers. The female scelionid attaches itself to the female of the host species and gets itself transported to the site of oviposition.

***Scelodonta strigicollis*:** See grapevine flea beetle.

Scent gland: A gland producing an odorous substance.

***Schistocerca gregaria* (Forskal):** Belongs to order Orthoptera and family Acrididae. It is a polyphagous insect pest, commonly known as the locust.

Schizodactylidae: A family of order Orthoptera, suborder Ensifera and super family Tettigonioidea. Burrowing forms, *Schizodactylus monstrosus* is a large Indian species with tegmina coiled.

Schoenobiinae: Subfamily of family pyraustidae. Characterized by proboscis being vestigial or absent and fringed hindwings. The eggs are covered with hairs from the anal tuft of the female. This subfamily includes *Tryporyza incertulas,* the yellow borer of rice, which is one of the most important pest rice in India. *Tryporyza innotata* is the white rice borer. *Scirpophaga nivella* is the top borer of sugarcane in India and Pakistan.

Schradan (OMPA): A synthetic, systemic organophosphate insecticide and acaricide, octamethyl pyrophosphoramide ($C_8H_{24}O_3N_4P_2$), extremely toxic to mammals, acute oral and dermal LD_{50} for rats are 8-10 mg and 50-100 mg kg^{-1}, not markedly toxic at insecticidal concentrations but the over concentration is injurious to some crops. The technical material is brown liquid and contains 40 per cent of schradan, boiling point is 154 °C. It is soluble in water and remarkably stable to alkaline hydrolysis. It is converted slowly in plants and rapidly in animals to the toxic monoamide oxide. It is also known as Pestox III, Systam and OMPA.

Schrader valve: A valve which is used for filling the gas in a container from a source of compressed gas.

Science: (L., *Scientia*= knowledge). Science is the exact knowledge or verified human experience. In another words, science is experimentally verified, classified knowledge gained by the sequential study of nature.

Scientific name: The internationally recognized Latin name of an animal or plant species. The scientific name consists of two parts, genus and species, followed by the describer (author) of the species. The scientific names (excluding the author's names) are always printed in Italics. The genus name is always capitalized, while the specific name is not capitalized. For example, the scientific name of the honey bee is *Apis mellifera* L. The L. is an abbreviation for Linnaeus (see also, common name).

Scion: Shoot or bud used in the vegetative propagation for grafting or budding.

***Sciothrips cardamomi* (Ramakrishna Ayyar):** Belongs to order Thysanoptera and family Thripidae. It is an insect pest of cardamom, commonly known as the cardamom thrips.

***Scirpophaga incertulas* (Wlk.):** Belongs to order Lepidoptera and family Pyralididae. It is an insect pest of cereals and millets, commonly known as the yellow stem borer or paddy borer.

***Scirpophaga nivella* (Fab.):** Belongs to order Lepidoptera and family Pyralidae. It is an insect pest of sugarcane, commonly known as the sugarcane top borer.

***Scirtothrips dorsalis* Hood:** Belongs to order Thysanoptera and family Thripidae. It is an insect pest of large chillies, commonly known as the chilli thrips.

Sclerite (pl., sclerites): A hardened body wall plate of an insect's exoskeleton, usually separated from other sclerites by a suture or membranous area.

Scleroma: The sclerotic annulus of a body segment in distinction to the membranous conjunctiva.

Sclerotin: It is also known as the tanned protein. It is brown or amber coloured protein which occurs in the exocuticle. In fact, it is a tanned arthropodin produced due to the action of quinine with the n-terminal amino groups of arthropodin molecule and finally the quinoid protein links onto the another protein molecule by the cross linkages.

Sclerotization: The deposition or formation of other substances than chitin in the cuticle in a definite area is known as sclerotization. Sclerotization or tanning is the last process of moulting which involves differentiation of the procuticle into outer hard exocuticle and inner soft endocuticle. The exocuticle represents a tanned part of the cuticle. It is simply a conversion of protein arthropodin into sclerotin by initiating formation of cross links between adjacent protein molecules. The cross links are formed by quinones. The concentration of tryosine increases highly in the haemolymph just before the ecdysis. Trosine is converted into dihydroxyphenylalanine (DOPA)-dopamine to N-acetyldopamine (diphenol) is transported via pore canals into the outer part of epicuticle where it is oxidized to quinone by an enzyme, phenol oxidase. The quinone diffuses across the epicuticle into the outer part of procuticle and after forming cross links among the protein molecules, produces the epicuticle. The darkness of cuticle is due to sclerotization as well as polymerization of excess quinones into melanin. The bursicon hormone (secreted by cerebral neurosecretory cells in *Sarcophaga*) controls the process of sclerotization, while ecdysone plays a key role in controlling various moulting events.

Scoli: See processes of body wall.

Scoliid wasp: A wasp of genus *Scolia* which lays its eggs on the larvae of certain beetles acting as bioagent in the beetle control.

Scoliidae: Family of super family Scolioidea, suborder Apocrita and order Hymenoptera. The Scoliidae are characterized by a continuous meso-and metasternum, overlying the bases of mid and hindcoxae. In the females, the wing membrane is striolate and the last sternite in the male is produced into three spines. The insects are large, hairy and black with yellow or red bands.

Scolioidea: Super family of Suborder Apocrita, order Hymenoptera. The Scolioidea have the pronotum appearing more or less triangular in lateral view and the lateral angles extends back to or nearly or, the tegulae. They are usually parasites of larvae of coleopterous insects. Sexual dimorphism is noticed in the group. This includes the following families: Plumariidae, Scoliidae, Mutillidae, Sapygidae and Tiphiidae.

Scolopale cell: The sheath cell that envelops the dendrite in a chordotonal organ.

Scolophore: A spindle shaped bundle of sensilla usually attached to the integument, believed -to be auditory in function; a bipolar nerve cell continuous with a fibre of the chordotonal nerve.

Scolopidia: In a chordotonal organ, the combination of three cells, the cap, scolopale and dendrite.

Scolopophorous organ: Typically, a bundle of sensilla consisting of one or more specialised bipolar neurons, the sclopidia; also called the scolopophores stretched between two internal integumental surfaces; often has a proprioceptive function; also called the chordotonal organ; also see campaniform sensilla.

Scolops: A sense rod or minute rod like capsule enveloping the distal end of the sense cell in certain sense organs.

Scolytoid larva: A fleshy larva resembling the larva of a scolytid beetle.

Scopa (pl., scopae): (i) A small, dense tuft of hair. (ii) The pollen collecting apparatus of a bee, whether it is the pollen basket on the leg or a brush of hairs on the abdomen. It is also located on the abdominal terga in Megachilidae.

Scopula: A small tuft of hairs.

Scorch Disease symptom: The burning of leaf margins as a result of infection or unfavourable environmental conditions.

Scorpion: Any member of the arachnid family Scorpionidae; they have an elongate body and a poison sting at the end of abdomen.

***Scotolinx quadristriata* Rao and Rama:** Larval parasite of citrus leaf miner, *Phyllocnistis citrella* Stainton.

Scotopic (eye): See superposition (eye).

Scout bees: Worker bees searching for a new source of pollen, nectar, propolis, water, or a new home for a swarm of bees.

Scraper: The sharpened anal angle of the forewing (tegmina) of a cricket or grasshopper, a part of the stridulating mechanism.

Scutellary striole: A shortened stria or puncture row lying just laterad of the scutellum but not extending very far posteriorly.

Scutelleridae: Family of superfamily Scutelleroidea of division Gymnocerata, suborder Heteroptera and order Hemiptera. The scutellum is very large and extends to the apex of the abdomen exposing the wings at its edge. The insects are generally brownish, 8-10 mm long, phytophagous and commonly known as shield backed bugs, *e.g.*, *Cantao ocellatus;* litchi bug, *Chrysocoris stollii.*

Scutelleroidea: Superfamily of division Gymnocerata, suborder Heteroptera and order Hemiptera. This includes the families Cydnidae, Corimelaenidae, Scutelleridae, Podopidae and Pentatomidae.

Scutellum: A triangular sclerite, appearing as more or less triangular segment behind the pronotum. The scutellum is clearly visible in the order Heteroptera (true bugs) and Coleoptera (beetles) and in some members of the suborder Homoptera (Cicadas).

Scutoscutellar suture: Line dividing the alinotum into the anterior scutum and posterior scutellum.

Scutum: The middle division of a thoracic notum, just anterior to the scutellum.

Scydmaenidae (ant-like stone beetles): Family of superfamily Taphylinoidea, suborder Polyphaga and order Coleoptera. These beetles are small, ant-like, brownish, hairy beetles which occur in moss, under stones and logs and in nests of ants. The hindcoxae are separated and the abdominal is completely covered by elytra. Indian species belong to the genera *Scydmaenus, Eumicrus, Syndicus* etc.

Sealed brood: See capped brood.

Search image: Behaviour that enables predators to increase ability to search the prey.

Seasonal cycle: The sequence of insect life cycles of a species that occurs in one year.

Seasonality: The phenomenon that show cyclic or repeated behaviour according to the season.

Sebaceous gland: A gland producing a greasy lubricating substance which is fatty or oily material.

Second axillary: The pivotal plate of the wing base resting on the pleural wing process, connected with the base of the radial vein.

Second median plate: A variable sclerotisation at the base of mediocubital field, folding convexly on the outer edge of first median plate along the plica basalis; often absent or represented by the united bases of the medial and cubital veins.

Second reproductive: The same as nymphoid reproductive in termites.

Second spiracle: The metathoracic spiracle located near the anterior margin of metapleuron between the mesopleuron and the metapleuron or the posterior margin of mesopleuron.

Second trochanter: the second segment of the telopodite, often not distinct from the base of femur; in insects usually fused with the first trochanter.

Secondary compound: See allelochemicals and secondary plant substances.

Secondary copulatory apparatus: In the male dragonflies, the primary external genitalia are almost rudimentary. The functional genitalia and the secondary copulatory apparatus is developed on the ventral surface of second and the third abdominal segments. It consists of anterior lamina, posterior lamina, genital lobe and genital fossa, supporting framework, hamules, ligula and the penis. The laminae are the sclerotized plates and articulate with the supporting framework. The genital lobes are flattened or elongated structures with the spherical tip bearing tubercles and setae. The hamules are elongated tubular structures lying on either side of the penis head. The ligula is the triangular, cuticular penis sheath. The penis is composed of four segments, *viz.*, vesicle, stalk, median segment and the head.

Secondary infection: The infection of healthy plants within a crop after the disease has entered the crop from outside (primary infection).

Secondary insect: An insect that infests after another has already infested the host; incapable of initiating injury by itself.

Secondary metabolites: See secondary plant substances.

Secondary orientation: Positioning of the insect body relative to various external or internal stimuli, superimposed on primary orientation.

Secondary parasite: A parasite that establishes itself in or upon a host which is a primary parasite.

Secondary parasitoid: A parasitoid that establishes itself in or upon a host which is a primary parasitoid. See also primary parasitoid and hyperparasitoid.

Secondary pest: Pest species that are usually present at low levels and are held in check by the action of natural enemies, can assume full pest status when natural enemies are destroyed by a pest management tactic, *e.g.*, insecticide application

Secondary plant substances: Chemicals found in plants that have no apparent metabolic and that often either influence the behaviour or physiology of herbivorous insects both, *i.e.*, may play a role in defense against herbivore.

Secondary segmentation: Characterized by membranous areas between adjacent segments, slightly anterior to the attachment points of the longitudinal muscles, typical of adults and many larval insects.

Secondary spermatocytes: Developing male gametes that form from the first meiotic division.

Secondary storage insect pest: These feed on broken or already damaged grains, *e.g.*, rust red flour beetle, flat grain beetle, saw toothed grain beetle etc.

Section comb honey: Honey produced in the square or rectangular frames.

Sectorial cross vein: A cross vein connecting two branches of the radial sector.

Seed bed: Land prepared and used for sowing the seeds.

Seed disinfectant: See disinfectant.

Seed dressing: A coating (either dry or wet) of protectant pesticide applied to seeds before sowing. Dry seed dressings are often physically stuck to the testa of seed by a sticker such as methyl cellulose.

Seed protectant: A chemical applied to seed before sowing to prevent disease and insect attack to the new seedlings or stored grains.

Seed tick: The larvae of the ticks.

Seed treatment: See seed dressing.

Segment: A subdivision of the body or of an appendage between joints or articulations with its own musculature.

Segmental muscles: The principal longitudinal muscles running between the intersegmental constrictions associated with primary segmentation.

Segmentation: The body of an insect consequently presents a jointed structure and is divided into a series of successive rings variously known as segments, somites or metameres.

Segmentation gene: Gene in insects that regulates the number or polarity of the body segments.

Selective: A term applied mainly to herbicides or insecticides, meaning that they are capable of killing some kind of plants or insects while not killing others.

Selective herbicide: Herbicide possessing differential toxicity to certain weeds and crop plants. For example some herbicides kill only plants belonging to the family Gramineae.

Selective insecticide (selectivity): An insecticide which kills selected insects, but does not kill most other organisms including beneficial species. When an insecticide is selective in action by creating more toxicity to a particular group of organisms belonging to particular taxa in comparison to the organisms of other taxa, it is termed as phenomenon of selective toxicity. Resistance is also a case of selectivity; however, the difference between the resistance and the selectivity is that resistance refers to such selective toxicity which is exhibited by the individuals of the same species while selectivity may be exhibited by members of any taxa.

Selective pressure: The main force that induces a genetic change through natural selection.

Selectivity, types: Selectivity may be divided into two distinct groups: (i) Ecological selectivity: When one organism is killed by a poison and the other, although equally sensitive to poison, survives because it avoids the exposure, is called the ecological selectivity or extrinsic selectivity or avoidance. The ecological selectivity is of significant importance because the hazards can be avoided or minimized by minimizing the exposure of non-target organisms. The examples of ecological selectivity are: (a) application of insecticides, particularly, carbamates at dusk when bees are not foraging, (b) use of systemic insecticides which would be toxic to insects feeding on treated plants and not to the predators and parasites, (c) use of specific attractants or baits, (d) formulations of insecticides such as controlled release formulations (granules) and microencapsulation. Penncap-M is a microencapsulated formulation of methyl parathion. The formulation is 4-5 times less toxic to mouse than the pure compound when given orally. Ecological selectivity is also related to the biodegradation. Pyrethroids, juvenile hormone analogues and some organophosphate and carbamate insecticides are degraded faster in the environment and biological system. The non-target organisms are, therefore, least affected by these chemicals. (ii) Physiological selectivity: When two organisms are exposed to a toxicant but one is killed and other is survived by some physiological or biochemical mechanism tolerating the exposure, is called physiological selectivity or intrinsic selectivity or tolerance.

Self pollination: It is the process of transfer of pollen grains from an anther to the stigma of the same or another flower of the same plant. It may be of two types, *i.e.*, autogamy and geitonogamy. In the former, the anthers are transferred to the stigma of the same flower, whereas, in the latter, pollen grains of one flower are transferred to the stigma of another flower of the same plant.

Self spacing frames: Frames constructed so that there is bee space apart when pushed together in a hive body.

Selocide: An acaricide containing about 30 per cent potassium ammonium selenosulphide.

Sematectonic communication: Stimulation of behaviour or physiological change in one animal in response to stimuli associated with the result of evidence of work by another animal.

Semen: Fluid discharged at ejaculation by the male consisting of spermatozoa and secretions of glands associated with the genital tract.

Semiaquatic: Living in wet places or partially in water.

Semi-looper: Caterpillars which look like loopers, but they do not belong to the family Geometridae. Usually they have two or three pairs of abdominal prolegs. Like real loopers, these move with a looping movement of the body.

Seminal vesicle: Dilated region of the vas deferens in which spermatozoa are stored prior to discharge.

Semiochemical: Any chemical involved in communication among organisms and change or modify behavior of insects by inducing in them the response for sex, feeding, fear or alarm etc. These are further classified into pheromones, secreted by insects for stimulating the insects of the same species (intraspecific stimulators) and allelochemicals (substances used for interspecific communication). The semiochemicals are effective at micronanogram level.

Semisocial: Applied to the condition or to the group which members of the same generation cooperate in brood care and there is also a reproductive division of labour, *i.e.*, some individuals are primarily egg layers and some are primarily workers.

Semispecies: A taxonomic group intermediate between a species and subspecies, especially as a result of geographical isolation.

Senescence: Changes in structure and function that decrease an individual's capacity for survival and lead to death.

Sense cell: The receptive cell of a sense organ with a proximal nerve process going to a nerve centre.

Sense cone or sense peg: A minute cone or peg, sensory in function.

Sense organ: These are sense organs of insects include both the exteroceptors and interoceptors along with the proprioceptors perceiving the stimuli from the external and internal environment, respectively. On the basis of functions, the sense organs of insects besides the eyes, are often classified into (i) mechanoreceptors and (ii) chemoreceptors.

Sensilla basiconica (sing., *Sensillum basiconicum*): They are peg like or cone like organs occurring on the antennae of many species, palps of *Hydrophilus piceus*, *Agriotes* larvae, antennae and palps of *Periplaneta*, lepidopteran larvae, antennae of muscidae, palps of butterflies, hypopharynx, epipharynx and on surface of labrum of many insects. These are sensory cells provided with pegs.

Sensilla campaniformia: See campaniform sensilla.

Sensilla coeloconica (sing., *sensillum coeloconicum*): The cuticular processes are sunk below the general level of integument. The organs are very deeply sunk and connected to the surface by an elongated tube and are found on the antennae of various insects of Coleoptera, Lepidoptera and Hymenoptera. These are sensory cells provided with pits.

Sensilla placodea (*sensillum placodeum*): These are sensory cells provided with plates and found in the antennae of Aphidoidea, Coleoptera, *Apis* etc.

Sensilla trichoidea olfactoria (sing., *sensillum trichoideum olfactorium*): These are thin walled hairs which have been described from the tarsi, labella of *Tabanus*, tarsi of *Pyrameis*, *Musca* and antennae of several species. These are sensory cells provided with hairs.

Sensillum (pl., sensilla): Sensory hair.

Sensillum chaeticum: A sense organ of which the external part is spine like.

Sensillum opticum: One of the ommatidia of compound eye.

Sensillum scolopophorum: A sense organ in which the sense cells contain sense rods of the scolops type.

Sensillum squamiformium: A sense organ of which the external parts are scale like in shape.

Sensitive areas: Places where chemical pesticides can create hazards if not used with special care and caution, *e.g.*, houses, barns, public parks, ponds, streams etc.

Sensitive crop: Crops which are easily injured or subjected to phytotoxicity by the pesticide chemicals.

Sensitivity: (i) High susceptibility to the low dosage of pesticides in terms of toxicity and not capable of withstanding the effect, *e.g.*, broad leaved plants to 2,4-D. (ii) The property of protoplasm that makes it responsive to stimuli which is highly developed in the nerve tissues.

Sensory hair: These are macrotrichia (cuticular modifications). These constitute various mechano and chemoreceptors and occurring on antennae, mouth parts, legs and various regions of body in all groups of insects.

Sensory nerves: Afferent nerves carrying sensory information to the central nervous system. The nerve poisons, like organophosphate and organocarbamate insecticides cut off this information system resulting in paralysis of the organism.

Sensory neurones (Afferent neurones): Neurones capable of generating an action potential in response to an external stimulus (or responsible for perception of external stimulus) such as physical displacement, temperature, humidity, chemicals etc. These are bipolar neurones (with a proximal axon and a distal dendrite) terminating into a sense organ or receptor and an axon running towards the central nervous system and, therefore, their cell body is always situated outside the ganglia. These conduct impulses from a receptor to the central nervous system.

Sensory organs: The sensory organs in insects include visual organs (eyes), mechanoreceptors and chemoreceptors.

Sensory thresholds: The sensitivity of the organs of the taste may be very high; under conditions of starvation, the butterfly, *Pyrameis* and the fly, *Calliphora* may be able to detect with their tarsi low concentration of sugar which the human tongue cannot.

Septicemia: A morbid condition caused by invasion and multiplication of microorganisms in the blood.

Septum (pl., septa): A wall or partition.

Sequelae: A condition following as a consequence of another disease.

Sequestration: Processing or synthesizing newer compounds from the compounds derived from plants; one of the mechanisms of detoxification of plant toxic chemicals and store or transfer to offspring as plant defense, *e.g.*, *Poekilocerues pictus* is able to sequester cardenolides from asclepiadaceous and apocyaneous hosts and gain protection from its predators.

Serial dilution: Preparation of a series of gradient solutions step by step from a concentrated solution.

Sericulture: The rearing of silk moths and their larvae for the production of silk.

Serosa: Dorsal blastoderm or its extension as the outer layer of the amniotic covering of embryo.

Serrate antenna: Antenna in which most antennomeres are triangular and project slightly in one direction forming a more or less acute process, giving the antenna a saw like appearance, *e.g.*, pulse beetles and buprestid beetles.

Serrate: Notched or toothed on the edge.

***Serratia marcescens*:** It is a non-spore forming bacterium commonly known as red pigmented bacterium. It is effective against a number of lepidopterous pests and gaining importance in biological control of insects.

Serum: Watery fluid that separates from blood on coagulation.

Sesame gall fly: *Asphondylia sesami* Felt., a pest of sesame which is a mosquito like fly, the larvae form galls in the flower buds.

Sesame leaf and capsule caterpillar: *Antigastra catalaunalis* Dup. is a serious pest of sesame, the caterpillars of which feed on leaves, web top leaves and bore into tender pods and shoots.

Sesamex: A synergist for pyrethrum and allethrin; chemically it is 2(2-ethoxyethoxy) ethyl-3,4-methylenedioxy)phenyl acetal of acetaldehyde; relatively non-toxic to mammals, acute oral LD_{50} for rats is 2000-2270 mg kg^{-1}.

***Sesamia inferens* (Wlk.):** Belongs to order Lepidoptera and family Noctuidae. It is an insect pest of cereals and millets, commonly known as the pink borer.

Sesamin: A mixed function oxidise inhibitor produced by the *Chrysanthemum* plant.

Sesiidae [Aegeriidae] (clear wings): Family of super family Tinaeoidea of suborder Ditrysia and order Lepidoptera. The clear wings are often brightly colored with mostly scaleless transparent wings and in appearance sometimes resemble ichneumons, wasps, bees etc. The larvae bore into the root stocks of plants and wood of trees, vines etc. They may either pupate in the larval gallery or in the soil in an earthen cocoon. The European and Americanria currant borer, *Sesia tipuliformis* and the American peach tree borer, *Aegria exitiosa* are economically important pests in these continents.

Sessile: Attached; Fastened. A sessile insect is unable to move because it is attached directly (without stem or petiole) to an object, *e.g.*, scale insects.

Seta (pl., setae): The setae or microtrichia are bristles or stout hair (sensory in function) present on cuticle and unicellular. The seta originates from the epidermal cell called trichogen cell and penetrate through another cell called tormogen cell.

Setaceous: Bristle-like antennae. The size of each antennal segment become smaller and smaller from base to apex, *e.g.*, cockroach, leaf hopper and Odonata.

Setose: Bristly.

Setula, setulose, setulae: Short blunt setae or bristle like hair.

Setulose. Bearing short, blunt bristles.

Sevidol: Granule containing 4 per cent in each of carbaryl and gamma BHC is formulated with the name of Sevidol.

Sevimol: Sevimol 40 LV is a product containing 40 per cent carbaryl with molasses.

Sevin: Trade name of carbaryl.

Sewage effluent: The liquids that flow out from the refuse carried off by channels which receive the discharge from house drains and streets.

Sex attractant: See pheromones.

Sex pheromone. An intraspecific chemical messenger emitted by one sex which attracts the opposite sex, *i.e.*, female moths release chemical substance into the air to signal their availability and thereby attracts males over long distances. Female emperor moths are able to attract males living upto 11 km away (Butenandt, 1963). Males usually perceive the female sex substances by means of their antennae. Removing the antennae or coating them eliminate all sexual responses from the males. The sex attractant produced by the female is bombykol in the *Bombyx mori*; whereas, the sex attractant produced by male butterfly, *Lycora ceres ceres* is 7-methyl-2, 3-dihydropyrrholizidin-1-one, acetyle acetate. The pheromone attracting both sexes is classified as assembling scent. Such pheromone is produced by hindgut of male bark beetle, *Ips* sp. and it is secreted with the faeces. It is non-anal and chemically geranyl acetate.

Sex ratio: Proportion of males to females within a population.

Sexual dimorphism: The presence of marked differences (in shape, size, colour etc.) between male and female individuals of the same species.

Sexual reproduction: Reproduction involving germ cell and union of gametes.

Shelf life: Period or time for which the pesticide may be stored without becoming unfit for use.

Shellac: The purified product of lac.

Shingling: Refers to placing of shingles or pieces of lumber among crop plantings for day time hiding places for slugs, squash bugs, earwigs etc. which can then be destroyed.

Shipping: The sending of insects from one place to another by land, air or sea.

Shivering: Contractions of the flight muscles against each other that result in elevation of thoracic temperature.

Short term pesticide: A pesticide which breaks down immediately after application into non-toxic by-products.

Shut off nozzle: A spray nozzle provided with a valve to restrict the flow of solution.

Sialidae: Family of superfamily Sialoidea, suborder Megaloptera, order Neuroptera. Without ocelli, seven pairs of gills.

Sialoidea: Superfamily of suborder Megaloptera, order Neuroptera. Larvae aquatic with abdominal wings without pterostigma and tracheal gills and ovipositor not exserted.

Sibling species: Reproductively isolated populations that are morphologically similar or identical and frequently sympatric (occupying the same geographical area).

Side dressing: To put fertilizer or a pesticide in granular form on or near the ground near plants after they have started to grow.

Sieve plates: Porous covers associated with spiracles that are thought to retard or prevent entry of airborne particles.

Siglure: Sec-butyl 6-methyl-3-cyclohexane-1-carboxylate is a synthetic attractant.

Sigmoid: Shaped like the letter S sign.

Sign: Outward signal of the pesticide poisoning or disease in animals and plants.

Silica (SiO_2): It is mineral found in free and combined form and is a dominant component of earth's crust. It occurs both in crystalline and amorphous forms and used as an inert carrier in the insecticide dust formulations.

Silica aerogel: It is a desiccant insecticide which kills insects by disrupting or dissolving the waxy outer coat (epicuticle) of the insect body, thus inducing excessive water and body fluid loss and leading to death from dehydration. Better results are obtained when it is mixed with the pyrethrum. Silica aerogel is non toxic to human beings.

Silk: Fibre produced by insects (Lepidoptera and Neuroptera), especially a product of the larva of silk moth, *Bombyx mori*, being used to construct cocoon in which the larva pupates. The fibroin is the chief elastic protein of cocoons, the other being sericin which is gelatinous protein.

Silk gland: In silkworm, *Bombyx mori* the silk glands are greatly developed about four times the body length and folded enormously covering hind region of the gut. They are basically the labial glands and homologous with the salivary glands of other insects. Anteriorly, each gland opens into a long salivary duct. Both salivary ducts unite apically and open at the apex of the median cylindrical structure known as the spinneret.

Silkworm: See *Bombyx mori* (Bombycidae, Lepidoptera).

Silkworm diseases: There are four major diseases of silkworm: (i) Pebrine: Protozoan, *Nosema bombycis* produces pepper like spots on body, wrinkled skin, sluggishness etc.; (ii) Flacherie: caused by bacterium, *Bacillus thuringiensis.* Body putrifies and becomes black or green; (iii) Muscardine: caused by white muscardine fungus, *Beauveria bassiana* (common occurrence in humid region); green muscardine,

Spicaria parssina; yellow muscardine, *Iscaria farinosa*. Hyphae come out of the intersegmental membrane and (iv) Grasserie: caused by *Borrelina* virus, causing swelling of segments and rupture of skin.

Silkworm fly (Uzi fly): *Tricholyga bombycis*, a fly resembling the house fly but slightly bigger with black and white stripes; a natural enemy of silkworm, maggots enter the worms and feed inside.

Silkworm pests: (i) Uzi fly, *Tricholyga bombycis* (Diptera). Live maggots bore the body of the silkworm, live in and eat fat body for about a week causing death of the silkworm; (ii) Dermestid beetle, *Dermestes cadeverinus* larvae and adults feed on cocoons. (iii) The other pests include ants, lizards, rats, squirrels and birds like sparrows.

Silphidae (carrion beetles, sexton-beetles, burying beetles): Family of superfamily Taphylinoidea, suborder Polyphaga and order Coleoptera. The beetles are relatively large, often brightly colored and possess 11-segmented short clubbed antennae inserted under the front margin. In cavericulous species eyes are wanting but in *Pteroloma* a pair of ocelli is present. The forelegs have large conical coxae which are contiguous, fossorial tibiae with spurs, five-segmented (rarely four) tarsi. These beetles and their larvae occur about the bodies of dead animals and feed on carrion. Some occur in decaying animal matter, some in fungi and a few in ant nests. A few are predaceous on maggots and other insects which are saprophagous. Some Indian species are: *Silpha rufithorax, Nicrophorus nepalensis, Diamesus osculans, Nodynus nitidus, Catops vestita, Aclypea sculpturata* etc.

Silvanidae (flat grain beetles): Family of superfamily Cucujoidea, suborder Polyphaga and order Coleoptera. Adults small, elongate, flat, usually less than 3mm in length with usually clubbed antennae. Tarsi are 5-5-5 in both sexes. Some beetles occur under bark and others feed on stored grains or meal. The saw-toothed grain beetle *Oryzaephilus surinamensis* is a pest of stored grains in India and other countries.

Silver shoot: A gall produced by an abnormal growth of the leaf sheath in response to the attack of rice gall midge, *Pachydiplosis oryzae* maggot.

Silverfish: See *Lepisma saccharina.*

Silvicide: Herbicide used for the control of undesirable brush and trees as in the wooded areas.

Silviculture: The care and cultivation of forest trees.

Simple: Unmodified, not complicated, not forked.

Simple dorsal eye: See dorsal ocellus.

Simple eye: A simple eye is a single lens that tells the difference between light and dark. An ocellus.

Simple lateral eye: See lateral ocellus.

Simple metamorphosis: Metamorphosis in which the wings (when present) develop externally during the immature stage and there is no prolonged resting stage (*i.e.*, pupa) preceding the last moult; stages included are the egg, nymph and adult. It is also called gradual or partial metamorphosis, and paurometabolous development.

Simuliidae: Family of suborder Nematocera, order Diptera. This family includes the buffalo gnats, which are thick, dark flies, 3 to 6mm long, with short, stout legs and long mandibles. The females have eyes dichoptic, while they are holoptic in males. Antennae are 9 to 11-segmented, without hairs in both sexes. These flies have a humpbacked appearance due to the convex mesonotum.

Singhara beetle: *Galerucella birmanica* Jacoby, an insect pest of waternut; grubs and adults attack leaves feeding on upper surface and causing dark brown decayed patches, badly damaged plants turn black.

Single action atomizer: Type of sprayer which only work with the stroke of pump and consists of a pump and a receptacle for holding the spray fluid.

Single dose poison: An acute poison such as zinc phosphide.

Sinuate: Having the wavy margin with strong indentations.

Sinus: A partitioned chamber of the body cavity.

***Siphona exigua*:** Belongs to order Diptera and family Tachinidae. It is an insect pest of farm animals, commonly known as the buffalo fly.

Siphonaptera: Insect order (*siphon*= a tube, *aptera*= wingless), *e.g.*, fleas. Insect order, made up of the fleas. They are small insects, characterized by wingless, sideways flattened bodies, jumping legs, sucking mouthparts, and a complete metamorphosis, blood sucking insects, mostly ectoparasitic on warm blooded animals, living among the hairs of mammals, feathers of birds and in the nests of their hosts. Their importance is due to their role as vectors of plague in man and as intermediate hosts of many parasitic helminthes.

Siphonapteran piercing and sucking type mouth parts: A pair of maxillary stylets and an epipharngeal stylet together form the fascicle, held in a channel formed by grooves in inner margin of the labial palps. The labium is rudimentary and at apex bears opening of salivary duct. The roof of food canal is formed by labrum-epipharynx and the floor by the maxillary stylets. Hypopharynx is absent or not demonstrable. The wound is made by protraction of laciniae which bear backwardly pointed teeth at distal end. The epipharyngeal stylet alone enters the host blood vessel and maxillary stylets remain outside. The blood is drawn into pharynx by the action of cibarial and pharyngeal pump. The maxillary palps serve as levers while the biting mandibles are lacking and labrum is a very small sclerite.

Siphoning type mouth parts: This type of mouth parts are much specialized in which labrum or mandibles are absent or vestigial. Labium is represented by a pair of labial palps. There is only a long tubular proboscis which is broad towards the base and gets narrower towards the distal end. The slender hollow tube is formed by the galeae of maxilla. Two galeae become jointed to form this. This tube of proboscis remains under the head just like a spring of a watch. When the insect is not feeding, the proboscis is not capable of piercing the skin of animal or epidermis of the plant. Feeding is accomplished by uncoiling the proboscis and projecting its tips into some exposed liquid or nectar and then sucking the liquid through the food channel which runs full length through proboscis.

Siphunculi: The paired tube like protruding organs near the terminal end of abdomen of Aphidoidea, also called cornicles, through which a waxy secretion is extruded.

Siricidae (wood-wasps or horn-tails): Family of super family Megalodontoidea of suborder Symphyta (Chalastogastra), order Hymenoptera. The insects are fairly large and conspicuously colored with black and yellow or metallic blue. On the last abdominal tergite a horny spear-like plate is seen in both sexes of the insect. The female has a long, strong ovipositor with which it bores into the woody plants and trees and lays the eggs singly in each hole.

Siricoidea: Family of super family Megalodontoidea of suborder Symphyta (Chalastogastra), order Hymenoptera. This contains three families, *viz.*, Xiphydriidae, Syntexidae and Siricidae of which the last named family is important.

Sisyridae: Family of superfamily Hemerobioidea, suborder Planipennia, order Neuroptera. Aquatic larvae with abdominal gills.

Sitation: Refers to the deposition of silt sized particles.

Site: It is the place, location, plant, animal or other organisms to be treated with a pesticide to protect it from insect pests.

***Sitophilus oryzae* (Linn.):** Belongs to order Coleoptera and family Curculionidae. It is an insect pest of stored grains and other products, commonly known as the rice weevil.

Sitophore: Food trough at the base of hypopharynx of insects.

***Sitotroga cerealella* Olivier:** Belongs to order Lepidoptera and family Gelechiidae. It is an insect pest of stored grains and other products, commonly known as the Angoumois grain moth.

Skeletal muscle: In insects, a muscle that stretches across the body wall and serves to move one segment on another.

Skeleton: Supportive shell.

Skeletonizer: Any insect that eats away the green fleshy part of leaves, leaving only the veins and a transparent membrane.

Skep: A bee hive made of twisted straw without movable frames.

Slaking: The addition of water or dilution with water.

Slatted rack: A wooden rack that fits between the bottom board and the hive body. Bees make better use of the lower brood chamber with increased brood rearing, less comb gnawing, and less congestion at the front entrance.

Slug: Gastropod related to the snail but without shell; feed on vegetables, banana, cuts germinating and young plants at the base. It can be controlled with 5.0 per cent calcium arsenate or 2.0 per cent metaldehyde in wheat bran moistened with jaggery solution.

Slug caterpillar: See *Parasa lepida.*

Slum gum: The refuse from melted comb and capping after the wax has been removed.

Slurry: Paste like liquid used in seed coating.

Sminthuridae: A family of order Collembola. Antennae longer than head. For example, *Sminthurus* sp., *Bourletiella* sp.

Smoke: Aerial dispersal of minute solid particles of pesti-cides through the use of combustible mixtures.

Smoker: A device for smoking out bees from hive to facilitate easy handling. It consists of a tin can which is provided with spout to direct the smoke from smoldering material inside it with the help of bellows.

Snail: A member of phylum Mollusca having a single coiled shell and a broad, flat foot.

Snake doctor: See dragonfly.

Social: Living in more or less organized communities of individuals.

Social homeostasis: The maintenance of steady state at the level of society either by the control of nest microclimate or by he regulation of population density, behaviour and physiology of the group members as a whole.

Social hormone: See pheromone.

Social insects: Insects that live together in communities, such as ants and termites.

Social parasitism: Amongst social and solitary Hymenoptera, the encroachment of the nest of one species by other species is of quite frequent occurrence and such cuckoo like behaviour is called the social parasitism. For example, cuckoo bees, *Psithyrus* (social parasite) are solitary bees without workers and are lacking the pollen collecting apparatus on their hind legs, therefore, incapable of forming a colony for themselves. They also have a thicker cuticle, stronger mandibles and powerful and longer sting than the social bumble bees, *Bombus* sp., however, both resemble in colour. The *Psithyrus* females enter a young *Bombus* colony identifying by a scent. The *Bombus* workers at first fight the intruders but could not prevent the *Psithyrus* females due to their powerful skin, mandibles and skin. The *Bombus* queen is killed or both species may live together side by side. The eggs laid by a *Bombus* queen are often broken or hatched larvae are killed by *Psithyrus* females and they feed upon the food collected by *Bombus* workers. After mating, the *Psithyrus* females enter a early hibernation (diapause).

Socialism: Some species of insects are good examples of social relationship, for example, honey bees, termites, wasp etc. Social insects are found in the form of colony and members of a colony are differentiated into queen, males, soldiers etc. They have physiological division of labour amongst them.

Socil: Lateral appendicular processes of tenth abdominal segment in Trichoptera and Lepidoptera, possibly homologous with the circus like appendages of the tenth segment in lower Hymenoptera.

Sociology: L., *socius*= companion. The study of societies of animals and that of man.

Sociotomy: Method of nest formation among the termites in which immatures and reproductives separate from the parent colony or a migrating colony dividing into daughter colonies.

Socket: Receptacle.

Sodium arsenate: Inorganic stomach poison, highly soluble form of arsenical, therefore mainly used in poison baits, extremely phytotoxic, extensively used as a weed killer. It is used with wheat bran as 2 per cent poison bait against locusts and grasshoppers.

Sodium fluoaluminate (Na_3AlF_6): An inorganic stomach poison. In naturally occurring mineral cryolite, it is found up to the extent 98 per cent but can also be manufactured synthetically. It is relatively less toxic to mammals. An inorganic insecticide, sodium fluoaluminate; useful in controlling codling moth, orange tortrix and several other chewing insects; low acute mammalian toxicity, phytotoxicity may seriously injure peach trees, corn or grapes otherwise little plant injury.

Sodium fluoride (NaF): Extremely toxic white powder mixed with blue or green powder so that it could not be confused with flour, sugar, salt etc. It is highly phytotoxic and used for controlling cockroaches and many other insects.

Sodium fluoroacetate (Sodium monofluoroacetate): Chemically, it is $C_2H_2O_2FNa$ and marketed under trade name Compound 1080 or 1080. Extremely toxic rodenticide, effective against rats and mice, very hazardous to handle and use. The rodent bait can be prepared by mixing 10 kg crushed grain, 12 g 1080, 100 g plasticized lecithin, 4 g yellow aniline dye and 100 ml water.

Sodium fluosilicate: An inorganic stomach poison. It is a white powder containing 60.6 per cent fluorine. This compound is also soluble in water, therefore, should not be sprayed on the plants. It is widely used in poison baits in controlling grasshoppers, locusts, cutworms, crickets etc. Since, more effective and better insecticides are available, therefore, it is not much used now.

Sodium pump: Mechanism whereby sodium ions are actively pumped out of cells and ions across the plasma membrane that helps to control cell volume, drives transport of sugars and amino acids and renders nerve and muscle cells electrically excitable.

Sodium selenate (Na_2SeO_4): An inorganic systemic insecticide and acaricide that is applied to soil, use is limited to ornamental plants but toxic to chrysanthemum at dose of about 250 mg per sq. ft., highly toxic to warm blooded animals.

Soft green scale: See *Coccus viridis.*

Soft pesticides: The pesticides obtained from the plants rather than from petroleum and other sources. Such pesticides represent the hope that agricultural pests can be controlled while maintaining environmental stability.

Soft scale: Unarmoured scale belonging to family Coccidae.

Soft water: Water with few minerals and other chemicals dissolved in it.

***Sogatella furcifera* (Horvath):** Belongs to order Hemiptera and family Delphacidae. It is an insect pest of cereals and millets, commonly known as the rice white backed planthopper.

Soil application: Application of a pesticide or other chemical to the soil.

Soil disinfectant: Chemical used to treat the soil to get rid of harmful insects, bacteria or microbes from soil.

Soil drench: To wet or soak the surface of the soil with a pesticide.

Soil dwelling termites: These termites construct nests partially or completely in the soil. Their nests are partially subterranean, exclusively subterranean, mounds or carton nests. Partially subterranean nests are mainly constructed under the soil and are extended upwards on trees, plants, upward structures of buildings, poles etc., *e.g.*, Common Indian species include *Coptotermes heimi*, *Heterotermes indicola*, *Reticulotermes assamensis* etc. The exclusively subterranean nests are constructed in the forest areas. These are indiscrete consisting long, narrow, subcylindrical subterranean galleries. The galleries run horizontally in all directions under the surface of a ground, from a few centimeters to even 100 metres and run down to a depth of 50-200 cm in different species of termites. The galleries consist chambers at some intervals which are filled with food like grass, leaves, leaf humus, wood debris, cow dung etc. The Hodotermitidae and Termitidae form these nests. *Acanthotermes macrocephalus* constructs nest in desert areas of Rajasthan (India), Pakistan and Afghanistan. In India, the *Odontotermes obesus*, *O. redemanni*, *O. wallonensis*, *O. brunneus* etc. construct huge mounds. The mounds are two types: (i) Unilocular: Because of a single central hollow in a mound, it is called unilocular mound. In unilocular type, the royal chamber is a discrete flat, solid earthy structure with numerous outer galleries. Internally, it bears a spindle shaped royal cell measuring 6-10 cm in length, 5-8 cm in width and 2-3 cm in height accommodating a royal pair. (ii) Multilocular type: These are dome shaped, subconical with few holes on outer surface and with small turrets. In side the mound are numerous chambers and vaults for fungus combs. The royal chamber is built at the ground level. The carton nests (made up of hard material, semi-digested wood and lignin) are built by the *Microcerotermes*, *Nasutitermes* and *Hospitalitermes* in stump of a tree or on a branch, stem or damaged part of a tree.

Soil fumigant: A pesticide which is used to fumigate the soil to kill the pests. When the fumigant is applied, the gas formed kills insect pests and microorganisms responsible for causing the diseases.

Soil incorporation: Mixing or incorporation of the pesticide in the soil.

Soil injection: Mechanical placement of the pesticide beneath the soil surface with a minimum of mixing or stirring.

Soil insecticide: Insecticide applied to the soil to prevent the soil inhabiting insect pests.

Soil organism: Organisms ordinarily residing and reproducing in the soil.

Soil pollution (Land pollution): Any physical or chemical change in soil conditions that may adversely affect the growth of plants and other organisms living in the soil is known as soil pollution.

Soil sickness: Unproductiveness of soil; often due to presence of parasitic soil organisms.

Soil sterilant: Toxicant added to or injected into soil for the purpose of killing pests and pathogens.

Soil treatment: Treatment of the soil with pesticides in order to protect the crops from soil inhabiting pests and diseases.

Solanaceous: Plants belonging to family solanaceae, *e.g.*, potato, tomato, brinjal, chillies etc.

Solar energy: The energy obtained from the sun. The Sun is the biggest source of energy. It is estimated that solar energy equivalent to over 15,000 times the world's annual commercial energy consumption reaches the earth every year. Man has been using solar energy mainly in preparing salt from the sea water, drying clothes, heating water and in form of light.

Solar wax extractor: A glass covered insulated box used to melt wax from combs and capping by heat of the Sun.

Soldier: A member of a social insect (sterile male or female) colony responsible for protecting the nest. Often the soldiers are equipped with large heads and strong mandibles. In termites, sterile males or females with large heads and mandibles; they function to protect the colony.

Solid cone: Jet with air cone reduced to give a solid cone of spray droplets.

Solid cone nozzle: A spray nozzle emitting a solid cone of spray.

Solitary parasitism: It is the phenomenon of parasitism of an individual host by more than one larvae of a single parasitic species but only one survives, *e.g.*, *Campoletis chlorideae* Uchida parasitic on the larva of *Helicoverpa armigera* (Hubner).

Solitary: Living alone, not in groups. See gregarious also.

Solubility: The ability of a chemical to dissolve in another chemical or solvent, a measure of the amount of a substance that will dissolve in a given amount of another substance.

Soluble powder: A powdered formulation that dissolves and forms a solution in water.

Soluble: A substance that can be dissolved.

Solution: A preparation made by dissolving a solid, liquid or gaseous substance into another liquid substance without a chemical change.

Solvent: Carrier solution in which the pesticide (technical product) is dissolved to form the concentrate.

Soma: Body of an animal as distinguished from the germ cell. See perikaryon.

Somatic cell: The body cells as distinguished from germ cells.

Somatic layer: The external layer of the mesoderm.

Somatostatin: See vertebrate hormones in insects.

Somites: One of the longitudinal series of segments into which the body of many animals is divided (metameres).

Sonagraph: A graph of sound analysis of an animal.

Sorghum earhead bug: *Calocoris angustatus* a pest of sorghum; the adults and nymphs suck the cell sap from the tender portions of plant and ears in milky stage, turning ears into chaffy without ripe or healthy grains.

Sorghum earhead gall midge: See *Contarinia sorghicola* (Coquillet)

Sorghum shoot bug: See *Peregrinus maidis.*

Sorghum shoot fly: *Atherigona varia soccata*, an insect pest of sorghum; the maggots bore the growing portion of the plant due to which the plant growth is arrested. If the attack is at early stage, the plant will produce tillers as the main plant is damaged.

Sorghum stem borer: Same as maize stem borer.

Sound producing organs: These include (i) Stridulatory apparatus: consisting a file with a series of ridges and a spine like scraper. Movement of scraper over the file causes the membrane (to which file is attached) to vibrate so that a sound is produced (Orthoptera, Heteroptera, Coleoptera etc.). Location of file and scraper varies among insects: (a) Elytral apparatus is present in Grylloidea and Tettigonioidea, and (b) Femoroelytral apparatus in Acridoidea; (ii) Tymbal: Situated in the dorsolateral region of the first abdominal segment in cicada, Platypleura etc., a membranous tymbal is encircled by a thick cuticular rim and striated by dorsoventral ribs, while externally protected by cuticular covering and internally with a fibrillar tymbal muscle and backed by an air sac which communicates outside *via* a metathoracic spiracle. Sound is produced with contraction of a tymbal muscle causing pulling of the tymbal; (iii) Epipharynx: In the sphingid moth, *Acherontia* sp., air is sucked up through proboscis which causes vibration of epipharynx and a sound is produced.

Sound waves, effect on insect population: Very intense sound may carry sufficient energy to kill the insect. The death occurs due to heating caused by absorption and degeneration of the energy. Sound of this intensity is difficult to be transmitted efficiently and changes insect behaviour. The infra sound leads to setting up of resonance among the internal organs causing violent agitation and hemorrhage.

Southern blotting technique: This technique was developed by E. M. Southern in 1975 for identification of the DNA fragments. In this technique, DNA fragments are separated by agarose gel electrophoresis.

Soybean grey weevil: *Myllocerus* sp. is a small insect pest that feeds on the leaves of soybean.

Soybean leaf miner: See groundnut leaf miner.

Soybean stem borer (soybean girdle beetle): *Oberea brevis* is an important insect pest of soybean. This is small, brown longicorn beetle whose grub bores into stem.

Space bomb: An aerospace spray that can be used in rooms or buildings. A container having a pesticide and a chemical under pressure which forces the pesticide out as a spray or mist.

Space fumigation: In the space fumigation, the entire space in the grain godown is considered for treatment. It helps in destruction of all insects hiding in the cracks and crevices, besides those infesting bagged seeds. However, in space fumigation, quantity of a fumigant is used as per the volume of store irrespective of its occupancy. The success of the fumigation largely depends on making room sufficiently air tight for particular length of time. For narrow cracks and small holes, stick masking tape or paste craft paper. Plaster of Paris can also be used for sealing of some holes. Similarly, large cracks should be repaired or sealed with suitable sealing material. Out side openings, like doors, windows or ventilators should be covered with high density (1000 guage) polyethylene sheet and joint should be plastered with suitable sealing material or adhesive tape.

Space spray: A pesticide forced out of an aerosol container or sprayer as tiny droplets which fill the air in a room and destroy insects and other pests.

Spadix: Thick fleshy spike of flowers.

***Spalangia philippinensis* Mill.:** The parasitoid associated with mango fruit fly, *Bactrocera dorsalis*.

Spatulate: Spoon shaped; broad apically and narrowed basally and flattened.

***Spatulicraspeda castaneiceps* Hamps:** Belongs to order Lepidoptera and family Cochlidiidae. It is an insect pest of Tea, commonly known as the humped slug caterpillar.

Special adaptation: The occurrence of genetic changes in a population or species as a result of natural selection so that it adjusts to new or altered environmental conditions.

Speciation: The evolutionary formation of species which in essence, occurs by the production of barriers to gene exchange (reproductive isolation) between genetically diverging populations.

Speciationist: One who studies speciation.

Species: A group of organisms whose members are similar in structure and physiology actually or potentially interbreed and produce fertile offspring and that are reproductively isolated from all other such groups. It is division of a genus. The basic unit of living things, consisting of a group of individuals which all look more or less alike and which can all breed with each other to produce another generation of similar creatures.

Species diversity (species biodiversity): Refers to the different kinds of species in a specific area and their relative abundance.

Species diversity index: Any of several mathematical indices which express the number of species and the relative numbers of each in an area.

Species dominant: The species encountered most frequently in a given population or area.

Species group: Includes species and subspecies.

Species inqurenda: Species of doubtful status because of inadequate descriptions and lack of preserved specimen.

Species odour: The odour found in or on the body of social insects which is peculiar to a given species.

Species, rare: Unusual species in an area.

Specific category: Any one of the pesticides use categories designated by state regulations; a special area (such as agricultural plant pest control) requiring state certification for use of restricted pesticides by commercial applicators.

Specific epithet: Second part of a species name, indicating a particular species.

Specific gravity: The specific gravity of insect blood varies from 1.03 in the larva of *Celerio* (Lepidoptera) to 1.045 in the larva of *Apis*.

Specific heat: The specific heat of a substance is the amount of heat in Joules required to raise the temperature of 1 kg of the substance by 1 °C.

Specific miticides: Chemicals that are used to control mites but relatively non toxic to insects.

Specific search image: An impression left on a predator by several chance encounters with a kind of potential prey which eventually leads to recognition of intensive search for the acceptable prey species.

Spectrograph: An instrument for producing a photographic record of a spectrum.

Spectrophotometry: Instrument to investigate the chemical nature of a substance by examining its spectrum using infra-red rays.

Spectrum of activity: The range of insect pest species against which an insecticide or pesticide is active.

Sperm cyst: The cellular capsules in the testis containing the spermatocytes.

Sperm duct: A tube connecting the bursa copulatrix of Ditrysian Lepidoptera to the vagina.

Sperm follicle: A tube like subdivision of the testis in which spermatogenesis occurs.

Sperm transfer: In some insects such as Gryllidae, spermatozoa are forced out of the spermatophore by the swelling of inner walls of sheath. In any case, whether the sperms are discharged into the female directly or by way of spermatophore, they are next transferred to the receptaculum seminis or spermatheca. Sometimes, it is ascribed to pumping or respiratory movements within the female system. In some species of lice, the sperms are directly deposited into spermatheca. In some other insects, the sperms are deposited in the haemocoel directly or to a special internal pouch, mesospermalage situated in the fourth abdominal gland.

Sperm tube: One of the secondary divisions of the testis.

Spermatheca (pl., spermathecae): (i) A small sac-like branch of the female reproductive tract of arthropods in which sperm may be stored. (ii) This gland is invagination of body wall of eighth abdominal segment in insects. This structure in the females receives sperms from male and often stored prior to fertilization and maintained them in viable condition. The spermathecal duct opens into the genital chamber or in its absence into the vagina. This special organ is also present in the queen of honeybee in which the sperms of drone are stored. The sperms can be stored for years together as found in the termite queen. Its usual number is one but two are found in *Dacus* sp. and three in *Culex* sp. The secretion of the gland epithelium provides nutrients to the sperms.

Spermathical gland: A gland associated with the spermatheca in the female reproductive system of insects.

Spermatids: Developing male gametes that form from division (second meiotic) of the secondary spermatocytes.

Spermatocytes: The cells arising from the spermatogonia; the primary spermatocyte arises till growth of one of the last generation of spermatogonia; each primary spermatocyte divides into two secondary spermatocytes, each of which gives rise to two spermatids.

Spermatogenesis: Development of male reproductive cells (spermatozoa) in the testicular follicles.

Spermatogonium: The first stage in the differentiation of a sperm cell from a primitive male germ cell.

Spermatophore: Gelatinous covered packets containing sperms or a sperm case. In most insects, the sperms are not conveyed to the female in a free fluid but are enclosed in a membranous sac or spermatophore formed by the secretions of male accessory gland. In primitive Collembola and Thysanonura, these sperm packets are just deposited at random on little stalks; on coming into contact with the female vulva these spermatophores rupture and the spermatozoa enter the vagina. In many insects such as Lepidoptera, crickets and many beetles, the spermatophore is a complex structure which is formed during copulation and transferred with its contents to the female genital tract.

Spermatozoon (pl., Spermatozoa): The sperm or male gamete. The testis of the male is made up of a series of tubular follicles of varying number and arrangement. Each follicle contains a succession of zones in which the sex cells are in different stages of development. Each spermatogonium arising in the germarium at the apex of follicle divides repeatedly to form a cyst filled with spermatocytes covered with a mantle of somatic cells. The spermatocytes divide to form prespermatids and then spermatids which finally in the lower extremity of the follicle, become transformed into the elongated and flagellated spermatozoa. The spermatozoa are usually discharged through an intromittent penis or aedeagus in adherent bundles or packets mixed with the secretion of a pair of accessory glands.

Spermiogenesis: Process of transforming spermatids into functional sperm cells.

Sphecoidea: Super family of Suborder Apocrita, order Hymenoptera. Sphecoidea are characterized by the pronotum terminating laterally in a rounded lobe that does not reach the tegula. These solitary wasps are mostly fossorial and occasionally contrruct free mud-cells. They sting and paralyse caterpillars, bugs, genera of Orthoptera and Arachnida etc. and carry them to their nests. Rarely, they are parasitic on their allies. Some which die due to stinging remain undecomposed for several days. This includes the family of Sphecidae.

Sphecology: The scientific study of wasps.

Sphecophile: An organism which have spent at least part of its life cycle in the wasp colony.

Sphecophily: Transfer of pollens by wasp, a type of pollination.

***Sphenoptera perotetti* G.:** Belongs to order Coleoptera and family Buprestidae. It is an insect pest of groundnut, commonly known as the groundnut stem borer.

Sphingidae: Family of super family sphingoidea, suborder Ditrysia and order Lepidoptera. The insects are widely distributed throughout the world and are moderate sized to very large moths. They are mostly tropical. The forewings are elongate, the outer margin being very oblique. The antennae in the males are ciliated with partial whorls. Generally they are thickened towards and beyond the middle and pointed apically. The moths are mostly crepuscular or nocturnal. A few such as *Macroglossa, Hemaris* etc. are diurnal. These insects are powerful fliers.

Sphingoidea: Super family of suborder Ditrysia and order Lepidoptera. The adults are with well developed proboscis and frenulum and without tympanal organs. Antennae are apically pointed and usually hooked. This includes a single family Sphingidae.

Spiculum gastrale: Narrow strut attached to the anterior end of sternite and serving as a muscle attachment.

Spider: Small animals having eight jointed legs, two body regions, no antennae (feelers) and no wings. These are often grouped with ticks and mites (class Arachnida).

Spider mite: Mite of family Tetranychidae feeding on leaves and tender plant parts.

Spike: Elongate indeterminate inflorescence bearing sessile flowers.

Spiked sample: A sample inoculated with a known concentration and volume of insecticide which is used as a standard for comparison in the insecticide residue estimation.

Spillage: The leaking or running over or dripping of any pesticide chemical. It should be cleaned up immediately for safety.

Spin: Median apodemal process of the spinasternum.

Spinasternum: Small, posterior sclerite on the venter of a thoracic segment found in many generalized pterygote insects.

Spindle shaped: Elongate and cylindrical, thickened in the middle and tapering at the ends.

Spine: A multicellular, thorn like process or outgrowth of the integument not separated from it by a joint.

Spines and spurs: These are cuticular appendages, connected with cuticle with no membranous articulations unlike seta, epidermal and multicellular in origin, *e.g.*, dragonfly larvae, some adult beetles possess the cuticular horn on apical region of head, particularly in males

Spinneret: A structure from which the silk is spun.

Spinning glands: The glands which produce viscous secretion forming silk.

Spinose: Spiny; covered with spines.

Spinosyns: Fermented products (insecticides) obtained from soil actinomycete, *Saccharopolyspora spinosa*. Spinosyn A and spinoyn D are two active ingredients those are isolated and showed good insecticidal activity.

Spinosad: The combination of spinosyns A and D is called spinosad. It is a neurotoxin and acts on nicotinic acetylcholine receptors, whereas, only spinosyn A alters the function of GABA gated chloride channels. It is effective against lepidopterous, Dipterous, Thysanopterous and termites. It is commercially available as Tracer, Success and Naturalyte (45 SC and 2.5 SC).

Spiracle: Breathing pore opening of the tracheal system- through which diffusion of gases takes place. These are external opening of the tracheal respiratory system. In almost all insects, except Diplura, the spiracles are wanting from the prothorax. There are only two pairs of spiracles- one on each meso and metathorax and abdominal I to VIII segments, situated laterally often primarily on the pleura of either side.

Spiracle, types: They vary widely in their structure and can be classified into the following types: (i) simple spiracles: are tubular and open directly from exterior to the lumen of trachea, *e.g.*, Thysanura; (ii) aritate spiracles: open first into the cavity or tubular chamber, the atrium containing sieve plate or hairs and with peristigmatic glands. It bears external opening called the atrial orifice which may be simple or provided with a valve. The internal opening is called the tracheal orifice, *e.g.*, various Pterygota; (iii) bichambered spiracles: in which the spiracular cavity is two chambered, anterior and posterior, *e.g.*, abdominal spiracles of the ants; (iv) biferous spiracles: bear two external openings which are secondarily developed while the primary atrial orifice is closed, *e.g.*, larvae of Coleoptera; and multiferous spiracles: The spiracles have more than two openings, found in the third instar larvae of cyclorrhaphous Diptera. In *Glossina* and oestrus, spiracles open to exterior by about 500 pores developed on the chitinous plate and remain open due to the absence of closing apparatus.

Spiracular bristle: A bristle very close to the spiracle as in the Diptera.

Spiracular flap theory: Bocharova-Messner (1971) proposed the spiracular-flap theory on the consideration that the wings are originated from integumental evaginations which initially serve as the spiracular-flaps. From the wing development of Odonata, it was suggested that the small spiracular flaps have achieved first mobility by acquiring the muscular and sensory mechanisms to perform respiratory and tactile functions. Later on large folds have been developed in order to protect the spiracles and form a chamber around each spiracle. With the time they have enlarged and become mobile, performing the function of protective covering. Lastly, they became large enough to serve a ventilating mechanism for the tracheal system and in addition started to help the forward locomotory movement. The evolutionary process finally has modified these structures in the locomotory organs by providing powerful musculature and large surface area with a high frequency of beating.

Spiracular gill: Spiracles of one or more pairs bear long filament like processes in the aquatic forms of beetle, *Psephenoides* and that of Dipteran families (Tipulidae, Blephoroceriae, Deuterophlebiidae, Simulidae and Empididae etc).

Spiracular plate: A plate like sclerite next to or surrounding a spiracle.

Spiracular trachea: The short and usually unbranched trachea arising directly from the spiracle.

Spittle: In insects, a frothy fluid produced by the nymphs of spittle bugs (family, Cercopidae).

Spittle bug: Small bug belonging to family Cercopidae, not true bugs and related to leaf hoppers. The young nymph produces white frothy material in which it lives. Many among the spittle bugs are harmless, whereas, several others are pests of crop plants.

Splanchnic layer: The inner layer of mesoderm applied to the walls of alimentary canal.

Splinter: A small molecule or a group of atoms that has broken away from a bigger molecule as a result of some chemical or biochemical reaction.

Split application: Application of total amount in more than one small lots; generally applicable to fertilizer application.

Split plot design: A statistical design applied to the field experiments in which one type of treatments are assigned to main plot and other type treatments within the subplots.

***Spodoptera exigua* (Hb.):** Belongs to order Lepidoptera and family Noctuidae. It is an insect pest of jute, commonly known as the indigo caterpillar.

***Spodoptera litura* (Fab.):** Belongs to order Lepidoptera and family Noctuidae. It is an insect pest of cabbage, commonly known as the tobacco caterpillar.

***Spodoptera mauritia* (Boisd.):** Belongs to order Lepidoptera and family Noctuidae. It is an insect pest of rice, commonly known as the paddy armyworm.

Sponging type mouth parts: In these mouth parts, the labrum is long, tapering, strongly sclerotized flap which arises from the distal margin of rostrum. The labrum unites with the epipharynx which is jointly termed as labrum-epipharynx. The labium is fully developed forming long proboscis consisting three parts: (i) basiproboscis (rostrum), (ii) mediproboscis (haustellum) and (iii) distiproboscis (oral disc or labellum). The mandibles are entirely wanting. The maxillae are absent, only single segmented maxillary palpi are found. This type of mouth parts are found in the dipterous flies; the house fly is an excellent example.

Spontaneous teratogenesis: See teratogenesis.

Sporadic grasshopper: See *Oxya nitidula.*

Sporadic pest: The population is usually negligible but in certain years under favourable environmental conditions, appear in a epidemic form crossing many times DB and EIL and the pest has to be controlled by management strategies. These pests are highly sensitive to abiotic conditions and once the favourable season is over, only a residual population survives, *e.g.*, white grubs, hairy caterpillars, cutworms and grasshoppers are polyphagous but some oligophagous pests, *e.g.*, sugarcane pyrilla may also be sporadic in nature.

Spot infestation: Insect infestation confined to a small area or infestation of local nature.

Spot treatment: A spray or other type of treatment directed to at specific plants or areas rather than a general application.

Spotted bollworm: See *Earias insulana* and *E. vittella*.

Spray: A liquid solution or suspension of material such as fungicides, insecticides etc. made available in the form of fine droplets applied to plant or other surface. These are (i) Air-carried: Spray propelled to target in a stream of air. (ii) Course: Dispersion of droplets of mass median diameter over 200 µm. (iii) Concentrate: Undiluted com-mercial pesticide preparation. (iv) Fine: Dispersion of drop-lets of mass median diameter from 50-150 µm. (v) Floor: Spray applied to the litter on the ground surface. (vi) High- volume: Over 100 gal. /acre (1200 1/ha) on bushes and trees; 450-1000 l/ ha on ground crops (vii) Low -volume: Spray of 20-50 gal. /acre (250-600 1/ha) on bushes and trees; 12-120 l/ ha on ground crops (viii) Median-volume: 50-100 gal. /acre (600-1200 l/ha) on bushes and trees; 120-450 l/ ha on ground crops. (ix) Mist: Dispersion of droplets of 50-100 µm in diameter. (x) Ultra-low volume: 0-5 1/ ha on ground crops.

Spray angle: Angle between the sides of a jet leaving the orifice.

Spray boom: An assembly of several spray nozzles on a horizontal pipe to cover a larger area at one time; a vertical pipe used for fruit trees.

Spray concentrate: A liquid formulation of pesticide which is diluted with another liquid (water or oil) before use.

Spray drift: The tendency of fine droplets of a spray produced by a low volume nozzle, to drift in the air from the field of application to another fields. Such drift damage the other crops and grazing animals and other animals like bees if the spray contains toxic chemicals. The drift is minimized if spray is done in suitable weather conditions and by using sprays of sufficient dilution. The addition of drift retardant or spray adjuvant oils in the formulation can reduce the drift.

Spray gun: A short hollow hand operated equipment with nozzle and cut off, used for directing sprays with wide fog for a nearby foliage or a closely compact stream to reach high tree tops.

Spray injury: Injury caused to plants as a result of spraying pesticidal chemicals. The damage may be severe on the leaves, blossoms, fruits, twigs and rarely on entire plants sometimes resulting in necrosis and death.

Spray lance: A lengthy rigid brass tube held and directed by the operator and fitted with a spray nozzle or spray boom.

Spray nozzle: The attachment on the end of sprayer extension or lance that breaks up spray liquid into fine drops so as to apply the material evenly on the plants.

Spray pump: Any pump used for pressuring spray material to deliver through spray nozzle with sufficient force.

Spray rig: A mobile spray unit.

Spray rod: Same as spray lance or extension tube.

Sprayer tank: It is the container in the sprayer to hold the spray mixture to be sprayed on the plants. The sprayer tank is made up of non-corrosive material like plastic or brass and the capacity varies according to the power of machine.

Sprayer: Apparatus for applying pesticide sprays; not to be confused with spray operator.

Spread: Uniformity and completeness with which a spray deposit covers a continuous surface such as a leaf or a seed.

Spreader: Material added to a spray to lower the surface tension and to improve spread over a given area (wetter).

Spreading agent: A material which causes the pesticide solution to spread or disperse in a thin film over a surface instead of depositing it as droplets by reducing surface tension.

Springing organ: The organ found in Collembola which helps in jumping the insects.

SPS: Abbreviation stands for 'scientific pesticide supervision' and is used in the emblem of Society of Pesticide Science, India.

Spur embedder: A device used for mechanically embedding wires into foundation by employing hand pressure.

Spur: An articulated spine, often on a leg segment, usually the tibia. A tibial spur is characteristic of the Delphacidae (Homoptera).

Spurious claw: A stout bristle looks like a claw in spiders; a false claw.

Spurious vein: (i) A false vein formed by a thickening of the wing membrane and usually unconnected with any of the true veins. (ii) A vein like thickening of the wing membrane between two true veins; an adventitious longitudinal vein between the radius and media, crossing the r-m cross vein as in Diptera.

Squama (pl., squamae): (i) Any of the membranous flaps that arise near the base of wing in many true flies. (ii) A dorsal lateral lobe of the phallobase in Hymenoptera.

Square: An unopened flower bud of cotton.

St. Louis encephalitis: A viral disease transmitted by mosquitoes which affect the brain producing languor, apathy, lethargy and death.

Stability: The ability of a pesticide formulation to resist chemical degradation over a period of time.

Stabilizer: A substance added to a pesticide formulation to prevent or retard a chemical change. The term is also used for compounds retarding oxidation (antioxidation).

Stabilmentium : A zig-zag ribbon of silk spun across the centre of web of some web weaving spiders.

Stable fly: The common stable fly, *Stomoxys morsitans* (Muscidae, Diptera) is a pest of veterinary importance and feeds on horses, mules, cattle, dogs, sheep, goats and rarely man. See also, *Stomoxys calcitrans.*

Stable fly transmitted diseases: The stable flies transmit polymyelitis virus causing infantile paralysis. These are vectors of filariid nematode, *Setaria cervi* and it is transmitted in the cattle and deer causing filariasis. These cause anthrax, surra and swamp fever or infectious anaemia in horses besides encelomyelitis. The animals become weak, drowsy and unfit for even standing on their feet.

Stadium (pl., stadia): The time interval between two moults in developing insects.

Stage: (i) A distinct, sharply different period in the development of insects, *i.e.*, egg, larval, pupal and adult stage. (ii) Any definite period in the development of plant, *i.e.*, seedlng stage, tillering stage, milk stage etc.

Stalk borer: See *Chilo auricilius.*

Stalked: With a stalk or stem with a narrow stem like base; fused together to form a single vein.

Stamen: The part or organ of a flower that produces pollen and consists of a sac like structure (anther) and a filament.

Standing crop: The reapable biological mass of certain or all living organisms of an area or volume at some specific time, *i.e.*, what could be harvested.

Staphylinidae (rove beetles): Family of superfamily Taphylinoidea, suborder Polyphaga and order Coleoptera. The rove beetles are minute or medium-sized, slender, elongate, black or brownish, somewhat flattened with very short elytra leaving exposed a considerable portion of the abdomen. The well-developed hindwings are kept folded under the elytra at rest. The mandibles are long, slender and sharp and cross in front of the head. Some larger species are capable of inflicting a painful bite.

Staphylinoid aedeagus: Aedeagus in which the parameres and penis are well developed but the phallobase is reduced to a narrow strip of cuticle or membrane only; in the latter case the parameres appear to be joined directly to the penis and the aedeagus resembles the adephagan type.

Starlicide: See ommatidium, structure.

Statary phase: The period in which an army ant colony is in the quiescent and does not move from one place to another.

State vector: The process of transferring control (in a programme) to an intermediate vector.

Static organs: The static organs are proprioceptive organs maintaining the equilibrium of body and in orienting the insect with respect to gravity. The larva of aquatic insect, *Limnophilic fascipennis* bears a pair of sacs in the last abdominal segment. Each sac is provided with sensory hair at its blind end and granules or statoconia in the lumen. Some nymphs contain specialized static organ in the form of hairy sensory apparatus.

Stelocytarous: Pertaining to the nests, specifically the wasp nests in which the combs are attached to pillars etc.

Stem eelworm: A nematode, *Ditylenchus dipsaci* which is a pest of oats and red clover, causing seedlings to become distorted and swollen and in oats causing the condition known as tulip root.

Stemma (pl., stemmata): The simple eye in holometabolous larvae. It is also called lateral ocellus. The lateral eyes of the Holometabolous larvae and some adult insects (Collembola, Lepismatidae, Siphonaptera, Anoplura, Sterpsiptera) are more developed than the dorsal ocelli but less than the compound eyes and commonly called as the lateral simple eyes or stemmata. The stemmata or lateral simple eyes of larvae occur in the absence of compound eyes and responsible for orientating responses to light in maggots and caterpillars. The stemmata can form images on proximal rhabdoms when object is closer within the distance of 0.08 mm and are formed on the distal rhabdom when the object is at long distance. The caterpillars form mosaic images.

Stenopelmatidae: A family of order Orthoptera, suborder Ensifera and super family Tettigonioidea. Wingless forms of grasshoppers hiding in caves. This group is poorly known in India.

Stephanidae: Family of super family Ichneumonoidea, suborder Apocrita and order Hymenoptera. A small group of insects measuring 5 to 19 mm in length. They are parasitic on larvae of wood-boring coleopterous insects and hence possess a long ovipositor. The antennae are long and have 30 or more segments. The hindfemur are with spines beneath.

Stereomicroscope: A microscope with dual eye piece and objective lens system which provides stereoscopic vision.

Sterile insect technique: A pest control method that makes use of artificially sterilized populations of the pest to mate with and thereby interfere with the normal reproductive efforts of the target species. It is abbreviated as S.I.T. and has been used successfully to eradicate insect pest species from localized areas such as islands.

Sterilization: (i) Refers to killing or destruction of microorganisms and other living beings in an area or volume. (ii) Inducing the disability in males and females to reproduce.

Sterilize: To treat a certain area or volume with certain chemical or sterilizing agent to kill the microorganisms and other living organisms.

Sternacosta: The transverse internal ridge of the sternal suture through the bases of sternal apophyses.

Sternacostal suture: The external suture of the sternacosta separating the basisternum from the sternellum.

Sternal apophyses: The lateral apodemal arms of the eusternum posterior to the bases of the sternal apophyses of sternacostal suture.

Sternal: Pertaining to the sternum.

Sternauli (pl., sternaulices): A longitudinal furrow on the lower part of the mesopleuron as in Hymenoptera.

Sternellum: The part of eusternum posterior to the sternal apophyses of sternacostal suture.

Sternite: The corresponding sclerite of sternum is known as sternite. In insects, the sterna are usually single, large sclerites, and external. However, if they are divided in two or more part, these subunits are called sternites.

***Sternochaetus mangiferae* Fab.:** Belongs to order Coleoptera and family Curculionidae. It is an insect pest of mango, commonly known as mango stone weevil.

Sternopleural bristles: Bristles on the sternopleuron as found in the Diptera.

Sternopleurite: The infracoxal sclerotization of a generalized thoracic pleuron, generally united with the primary sternum in the definitive eusternal plate.

Sternopleuron (pl., sternopleura): A sclerite in the lateral wall of the thorax just above the base of middle leg as in the Diptera.

Sternum (pl., sterna): The ventral portion of a segment of an arthropod thorax or abdomen.

Sterols: Large molecular alcohols found in plant and animal cells combined with fatty acids and soluble in fat solvents.

Stick insect: Insect in the order Phasmida. Stick insects resemble sticks. Leaf insects (in the same order) look like leaves.

Sticker (Adhesive): A sticker or adhesive is added to a formulation to increase the surface retention (persistence) of a pesticide. The material of high viscosity is also used to stick powdered seed dressings on to seeds; two commonly used stickers are paraffin and methyl cellulose. A solution of methyl- cellulose can be added to a spray to increase retention on plant foliage.

Stigma: A small coloured area near the wing tip of dragonflies, bees, and various other clear winged insects: also called the pterostigma.

Stigmal vein: It is a short vein extending posteriorly from costal margin of the wing, usually a little beyond middle of the wings.

Stigmelloidea: Super family of suborder Montrysia and order Lepidoptera. Characterized by reduced wing venation and female possessing a fleshy ovipositor and a short cloacal duct. Larvae of many species are leaf miners in India. *Nepticula hoplometalla* and *N.elachistarcha* are leaf miners on *Butea monosperma* and *Zizyphus rugosa,* respectively. This includes the families, *viz.* Stigmellidae, Opostegidae and Tischeriidae.

Stimulant: A chemical or physical factor which increases some specific activity of an organism, *e.g.*, feeding and oviposition in insects.

Stimulatory organs: These organs serve both to maintain the muscle tone and to enhance the kinetic efficiency of muscles when they are brought into action. Prominent among them are halteres in Diptera and Strepsiptera, one important function of which is to sustain the muscular vigour of the insect during flight by the stimuli which their own vibrations generate.

Stimulus (Pl., stimuli): An agent which provokes active response or reaction in a living organism.

Stimulus filtering: The process by which the multitude of stimuli that impinge on the insect nervous system are reduced to those that are pertinent to the behavior of insects.

Sting: The modified ovipositor which is adapted for injecting venom, found in aculeate Hymenoptera.

Stink gland: See repugnatorial glands.

Stipes: The second division of the maxilla borne on the cardo that bears lacinia.

Stipital: Pertaining to stipes.

Stirofose: Low toxicity insecticide marketed under trade name Gardona or Rabona, effective against wide range of insect pests such as household, yard, orchard, flower beds, green house, garden and arachnid pests.

Stirup pump: Sprayer bucket.

Stolon: Sub aerial stem; a shoot from base of a plant growing horizontally under the soil, rooting and budding at the nodes (*Colocasia* sp.). (ii) Above ground runners or slender stems that develop roots, shoots and new plants at the tip of nodes as in the strawberry plants or Bermuda grass.

Stoma: Small apertures in the surface of a leaf, opening or closing in response to light intensity, time of the day and other factors, adapted to control exchange of carbon dioxide, oxygen and water vapour between active internal leaf surface and the atmosphere.

Stomach: See ventriculus.

Stomach poison: A toxicant which operates by absorption through the intestine after having ingested by the insect, generally from plant material. Such insecticides are sprayed or dusted on the plants and when the insects ingest the parts of these plants such as leaves, tender shoots, fruits etc., they reach the stomach and kill them. Stomach poisons are very effective against insects having biting and chewing type of mouth parts. Examples, various compounds of element arsenic.

Stomach pump: An apparatus used to aspirate the contents of stomach during pesticide poisoning.

Stomatogastric system: The part of the visceral nervous system associated with the brain, aorta and foregut, also called the stomodaeal system.

Stomatostylet: A stylet or spear in many plant parasitic nematodes, presumed to have evolved from fusion of the walls of stoma.

Stomodaeal invagination: The embryonic ectodermal invagination that develops into the foregut.

Stomodaeal nervous system: The nervous system centering in the ganglia of the stomodaeum (stomatogastric, sympathetic or visceral nervous system).

Stomodaeal valve: A structure formed by an intussusception of foregut and midgut tissues. In insects, these are cylindrical or funnel shaped invagination of the foregut into the midgut.

Stomodaeum: The foregut, beginning with the mouth and endings with the proventriculus or midgut.

***Stomopteryx nertaria* Meyrick:** Belongs to order Lepidoptera and family Gelechiidae. It is an insect pest of groundnut, commonly known as the groundnut leaf miner.

***Stomoxys calcitrans* (Linn.):** Belongs to order Diptera and family Muscidae. It is an insect pest of farm animals, commonly known as the stable fly.

Storage bin: A compartment room, structure or receptacle for storing farm produce.

Storage excretion: It is the temporary or permanent accumulation of excess of waste products, not dealt by the usual excretory organs, in special cells called the urate cells in the fat body.

Storage life: Refers to as shelf life.

Storage pots: Containers made of soft cerumen for storage of food in the nests of social bees. Some of the pots constructed by meliponine bees contain only honey in honey pots and others only pollen in pollen pots.

Stover: Corn stalks used as fodder for animals.

STP (NTP): Abbreviation stands for standard temperature and pressure or normal temperature and pressure. This is internationally accepted conditions for stating quantities like density or volume of a gas. The values are approximately 101325 pascals (corresponding to a barometric height of 0.76 m of mercury at 0° C) for pressure and 273.15° Kelvin (for 0° C, the melting point of ice) for temperature.

Strain: A group within the species that is in some way physiologically different from other members of the species, *e.g.*, insecticide resistant strain of insects.

Strap like: Like a narrow strip.

Strationyidae: Family of suborder Brachycera, order Diptera. Small to large, almost flattened, bristleless flies with annulate third antennal segment; a conspicuous scutellum is present. They occur in damp or moist areas and wings with veins more crowded near costa. Larvae are terrestrial or aquatic and carnivorous or saprophagous. Pupae remain enclosed in larval skin. Some of the Indian species are *Sargus metallicus, Stratiomyia barca, Cyphomyia indica* and *Clitellaria heminopla.*

Strepsiptera: Order (*strepsi*= turning or twisting, *ptera*= wings), *e.g.*, stylopids (includes endoparasites of bees, wasps, leaf hoppers etc). The order comprises remarkably small insects whose larvae are known to be internal parasites of members of the families Delphacidae, Cicadellidae and Membraccidae of the auchenorrhynchan Homoptera, the Gryllidae of Orthoptera and the superfamilies Vespoidea, Sphecoidea and Apoidea of Hymenoptera. Economically they are little or no importance to man as their role in biological control of pest species is very much limited. In most of the cases the females live throughout their life in a puparium which projects like a lump or tumour from the body of the host. In view of the similarities in the larvae and metamorphosis the stylopids were included among Coleoptera, placing them near the families Meloidae and Rhipiphoridae. However, the peculiar characters of the stylopids justified the classification into an order, the Strepsiptera.

***Streptococcus pluton*:** Bacterium that causes European foulbrood.

Stress: The result of a consequent state of physical or chemical or social stimulus on an organism or a system; a state of a strain resulting from stress.

Stretch receptors: Many insects possess stretch receptors which consist of connective tissue strands with associated nerve cells; they are probably involved in the rhythmical movements of the abdomen and thorax as in respiration. Over stimulation of stretch receptors in foregut and body wall leads to the arrest of feeding; if the nerve cord is cut, feeding will continue until the body wall is ruptured.

Stria (pl., stripe): A groove or depressed line.

Striae: Grooves running across or along the body applied especially to the grooves on beetle elytra.

Striate: With grooves or depressed lines.

Striated border: The inner cytoplasmic layer of the ventricular epithelial cells with fine lines perpendicular to the surface.

Striated muscle: In general, the voluntary muscle. These muscles are composed of fibers with alternate light and dark bands.

Stridulate: To make sound by rubbing two body parts together.

Stridulation: The production of sounds by rubbing two parts of the body together; best known in grasshoppers and other orthopterans.

Stridulatory file of mesoscutum: A shiny area on the mesoscutum of Megalopodidae, Disteniidae, Vesperidae, many Cerambycidae and some Nemonychidae which contains many very fine transverse ridges (visible only under very high magnification) and acts with a plectrum on the hind edge of pronotum to produce a sound.

Strigil: A tibial comb; an antenna cleaner in Hymenoptera

Strigilation: The licking of body secretions of another animal.

Stripe: A longitudinal colour marking.

Stripped mealy bug: See *Ferrisia virgata*.

Strobane ($C_{10}H_{11}Cl_7$): A chlorinated hydrocarbon insecticide (bicyclic structure), mixture of camphene and pinene and containing 66 per cent chlorine. The technical material is straw coloured liquid. It is practically insoluble in water but soluble in most of the organic solvents. It is not more stable to alkali, have strong insecticidal and acaricidal properties. Its oral and dermal LD_{50} values are 200 and 250 mg kg^{-1}, respectively for rat and 500 mg kg^{-1} for rabbit.

Structural pests: Pests which attack and destroy buildings and other structures, clothing, stored foods and processed goods, *e.g.*, termites, cockroaches, cloth moths, rats, dry rot fungi etc.

Structure-activity relationship: A biological method to analyze the action of pharmacologically active chemicals such as pesticides by utilizing their physicochemical characteristics.

Strut: A rod like structure to which muscles are attached.

Strychnine alkaloid ($C_{21}H_{22}N_2O_2$): Highly toxic substance used against rodents, mice, birds, snakes, moles, gophers, squirrels and other higher animals. It is a biological product from plant, *Strychnos nux-vomica*. The poison bait is prepared in the following manner: (a) 27 g strychnine is dissolved in 56 cc of water, (b) 2 g jaggery is mixed in 455 cc warm water. Now mixture a and b are mixed. Now, 13 kg over night soaked seeds are mixed in the above mixture. Now, 24 g of this poison bait can be placed at every active burrow.

Strychnine hydrochloride: A rodenticide, chemically, it is $C_{21}H_{22}N_2O_2.HCl.\ 2H_2O$. This salt is more bitter than the strychnine alkaloid and replaced by it. The dosages for white rats and dogs are 5 and 75-300 mg kg^{-1} body weight, respectively.

Strychnine sulphate: A rodenticide, chemically, it is $(C_{21}H_{22}N_2O_2)_2.H_2SO_4.\ 5H_2O$. This salt is more bitter than the strychnine alkaloid and replaced by it. The doses for white rats and dogs are 5 and 75-300 mg kg^{-1} body weight, respectively. It is soluble in water and the bait is prepared by mixing 5 g strychnine sulphate to 250 cc boiling water and then adding 1.0 kg of wheat to it.

Stubbles: The stumps of plants left after harvesting.

Stylate: A type of antenna in which the last segment is bristle like known as style. This type of antenna is found in the robber flies and horse flies.

Style: A bristle like process at the apex of an antenna.

Stylet: Sword like or needle like piercing structure in piercing and sucking mouthparts.

Stylet-borne virus: A non-persistent virus carried on the mouth parts of its vector.

Stylet sheath: The plant sucking bugs, *Oncopeltus* and *Dysdercus* sp. produce two types of saliva coming from different lobes of the salivary glands: watery saliva containing the digestive enzymes and a viscid material which lines the path made by the stylets through tissue of the plant to form a stylet sheath. This appeared to be composed of self tanned material resembling the cuticulin of epicuticle. It serves to filter off the particles in the sap which might obstruct the sucking canal and to prevent the sap from exuding to the exterior around the stylets.

Stylophora: The part of head (capitulum) of mites before stylets.

Stylopidae: Family of order Strepsiptera. Insects belonging to this family are parasites of Aculeate Hymenoptera. The adult males possess four-to six-segmented antennae, the third segment only laterally flabellate. The cephalothorax in the female is with one pair of spiracles and there are two to five genital tubes.

Stylus (pl., styli): A short, slender and finger like process on the ventral surface of abdomen of Thysanura.

Stylus theory: Wigglesworth (1973, 1976) proposed a novel hypothesis that the wings might have been originated from coxal styli of the Apterygota. He suggested the homology between the abdominal and coxal thoracic styli of Apterygotes with the abdominal gill plates of mayfly nymphs and wings of Pterygotes. Moreover, the development pattern of the gill plates and that of wings is very much alike and both the structures exhibit true venation, articulation between the subcoxa and tergum and their innervation by the subcoxal muscles. The homology in between the gill plates and the wings has been strongly supported by Matsuda (1976).

Subacute toxicity: Poisonous symptoms produced in test animals by long term exposure to repeated doses or concentrations of a substance.

Subalar muscle: A single, usually large, muscle lying against the epimeral wall of the pleuron on each side of each wing bearing segment, attached ventrally on the meron of the coxa.

Subalar sclerite (subalare): Small sclerite posterior to the pleural wing process.

Subantennal groove: Groove or concavity usually lying below the antennal insertion and sometimes extending posteriorly along the ventral portion of head.

Subantennal sulcus or suture: A groove on the face extending ventrally from the base of the antenna.

Subapical: Located just proximal of the apex or situated just before the tip or apex.

Subbasal: Located just distal of the base.

Subclass: A major subdivision of a class containing a group of related orders.

Subclimax: A stage in the development of a community, *i.e.*, succession before its final stage (climax).

Subcosta: The longitudinal vein between the costa and the radius, although it is often missing or very faint; abbreviated to Sc.

Subcoxa: Leg segment of primitive arthropods, basad of the coxa.

Subcubital fleck: See medial fleck.

Subcuticle: Newly secreted endocuticle in which the microfibres are not yet oriented.

Subdiscal vein (subdiscoidal vein): The vein forming the posterior margin of the third discoidal cell as in Hymenoptera.

Subequal: Approximately or almost equal in size or length.

Subfamily: A taxonomic subdivision of a family containing a group of closely allied genera; different from other allied groups, yet not sufficiently so as to make a family series; named after the type genus and ending in inae.

Subgenal areas: The narrow lateral marginal areas of the cranium set off by the subgenal sutures above the gnathal appendages, including the pleurostomata and hypostomata.

Subgenal ridges: A pair of sharp longitudinal ridges extending from the maxillary articulations to the posterior region of head and usually forming the lateral edges of a concavity. The subgenal ridges usually occur in conjunction with a strongly declined head and fit against the procoxae when the head is at rest.

Subgenal sutures: These sutures occur on the lateral sides of head. Each suture joins the posterior tentorial pit with anterior tentorial pit of the same side. Each suture is divisible into two parts. One part of the subgenal suture is called the hypostomal suture. The hypostomal suture runs from the posterior tentorial pit upto the posterior articulation of the mandible. The other part of the subgenal suture is called the pleurostomal suture. The pleurostomal suture arches over the mandible and finally joins with the frontoclypeal and the frontogenal sutures just above the anterior articulation of the mandibles. Internally, the subgenal sutures form the submarginal, subgenal ridges. Each ridge extends from the posterior tentorial pit to a point just above the anterior articulation of the mandible. The subgenal sutures strengthen the cranial wall along the lines of attachment of the gnathal appendages.

Subgenital plate: A plate like sternite that underlies the genitalia.

Subgenual organ: Group of chordotonal sensilla located in the basal portion of tibiae. The organ is made up of 10-40 scolopidia and are provided with the attachment and accessory cells. The subgenual organs are sensitive to the vibrations of the substratum and airborne sounds of high intensity and highly developed in Odonata, Diptera and Orthoptera.

Subgenus (pl., subgenera): A major sub division of a genus containing a group of related species. In scientific names, subgeneric names are following the genus name.

Subglabrous: Bearing very short, fine and/or scattered setae only, and so appearing glabrous under lower magnification.

Sub-imago: The sub-imago or dun is the winged sub-adult (immature adult) which emerges from the nymphal skin. It is rather dull in colour but very soon moults again- the only example of a winged insect undergoing a moult to reveal the imago and found only among the mayflies (Ephemeroptera).

Subjective synonym: Each of two or more synonyms based on different types but of the same taxon.

Sub-marginal cells: Cells lying just behind the stigma in the Hymenopteran forewing; important in the identification of bees and sphecid wasps.

Submarginal vein: A vein immediately behind and paralleling the coastal margin of wing.

Submedian cell: The cell behind the median cell in the basal posterior portion of wing.

Submentum: Proximal portion of the postlabium.

Submersed plant: An aquatic plant that grows with most of the parts below the water surface.

Subocular sutures: Vertical lines that may run beneath the compound eyes. These are well evident in the orthopteroid insects.

Suboesophageal commissure: The two nerves that pass around the oesophagus and connect the brain to the suboesophageal ganglion.

Suboesophageal ganglion: The ganglion, ventral to the foregut that innervates sense organs and muscles associated with mouthparts, salivary glands and the neck region.

Suborder: A major subdivision of an order containing a group of related superfamilies or families.

Subphylum: A major subdivision of phylum containing group of related classes.

Subquadrangle: A cell immediately behind the quadrangle as in Odonata and Zygoptera.

Sub-sample: A part or fraction of a sample.

Subscutellum: A well developed thoracic plate especially in Dipera, defines the family Tachinidae.

Subspecies: A subdivision of a species. Usually a geographic race that varies in size, colour or other characteristics. Within one species, the different subspecies are usually not sharply differentiated. They are capable of interbreeding.

Substance-P: See vertebrate hormones in insects.

Subterranean: Living in the ground.

Subterranean termite: A group of termites whose nests are partly underground; build nests with small chambers connected by narrow passages; makes superficial tunnels on ground to attack growing plants, timber, dry leaves etc.

Subtriangle: A cell or group of cells proximal to the triangle as in Odonata.

Succession: A group of species which successively occupy a given habitat as the conditions of the habitat change.

Successional cline: Cline that succeeds parental cline during ecological succession process.

Sucking cattle louse: See *Haematopinus eurysternus.*

Sucking insect: An insect having the piercing and sucking type of mouth parts, *e.g.*, bugs, scale insects, aphids etc.

Sucking pump: The highly modified cibarium; in insects which ingest liquid food (sap, blood etc.), the cibarial pump or the pharyngeal pump.

Sucrose: Principal sugar found in nectar.

Suction: The production of a vacuum in a container so that the external atmospheric pressure forces the surrounding organisms into it (aspirator).

Suction hose: A hose through which water is pulled from a pond or stream or spray from the spray tank to the pump.

Suctorial: Adapted for sucking.

Sugarcane black bug: *Cavelerius excavatus* Dist. is an insect pest of sugarcane; nymphs and adult bugs suck the cell sap.

Sugarcane leaf hopper: *Pyrilla perpusilla* Walk. is an insect pest of sugarcane; nymphs and adults suck the cell sap from tender parts of the plants.

Sugarcane mite: *Oligonychus indicus* Hirst is a red mite which feeds in colonies on under surfane of sugarcane. The nymphs and adults suck juice; in severe attack, plant turns sickly, leaves dry up; also attacks sorghum and pearlmillet.

Sugarcane pink mealy bug: *Saccharicoccus sacchari* Ckll. is an insect pest of sugarcane; nymphs and adults suck the cell sap, plants turn sickly, leaves dry up; infested plants are rendered useless for sowing.

Sugarcane root borer: *Emmalocera depressella* Swinh. is an insect pest of sugarcane causing severe damage to the crop in early stage; central shoot dries up which is greyish white in colour and known as dead heart.

Sugarcane scale insect: See *Aulacaspis tegalensis.*

Sugarcane shoot borer: *Chilo infuscatellus* Snell. is an insect pest of sugarcane infesting young shoots; caterpillars bore into stem making holes in more than one node, tunneling inside and producing dead hearts.

Sugarcane spotted fly: See *Neomaskellia bergii.*

Sugarcane stalk borer: *Chilo auricilia* Ddgn. is an insect pest of sugarcane infesting unopened leaves and stem causing dead hearts.

Sugarcane stem borer (Gurdaspur borer): *Acigona steniella* Hmps.is an insect pest of sugarcane infesting cane 2-4 ft. above the ground level, bores upwards in a spiral manner, making cane weak, dry up or fall down in wind or rain.

Sugarcane termites: *Odontotermes obesus* Ram., *O. assmuthi* Hlmgr., *Trinevitermes helmi* Wasm., *Ereinotermis araspoxalis* Hlmgr. are insect pests of sugarcane infesting buds of planted sugarcane sets and roots.

Sugarcane top borer: See *Scirpophaga nivella.*

Sugarcane top shoot borer: *Scirpophaga nivella* Fab. is an insect pest of sugarcane infesting midrib of top leaves, central shoot producing dead hearts, as a result the side shoots are developed in bunches at later stage (bunchy top condition).

Sugarcane whitefly: *Aleurolobus barodensis* Mask. is an insect pest of sugarcane infesting tender plant parts, especially of ratoon crop; adult fly and nymph sucks cell sap from leaves and other tender parts turning them yellow; in severe attack, crop shows sickly appearance.

Sulcate: With a groove or furrow.

Sulcus (pl., sulci): (i) A groove or furrow formed by an infolding of the body wall. (ii) Furrow or groove. A shallow furrow of the surface of brain separating adjacent convolutions.

Sulfotepp: A synthetic organophosphatic insecticide and acaricide, tetraethyl dithiopyrophosphate, highly toxic to mammals, acute oral LD_{50} for rats is 5 mg kg^{-1}.

Sulfoxide: A synergist for pyrethrum, chemically, 1,2-methylenedioxy-4-[2-(octylsulfinyl) propyl] benzene; relatively non-toxic to warm blooded animals, acute oral LD_{50} for rats is about 2000 mg kg^{-1}.

Sulfuryl fluoride (profume): It is a chemical possessing many positive fumigant qualities including moderate toxicity, odourless, efficacy in controlling a broad spectrum of insect pests, low boiling point, excellent penetration power, low reactivity potential, limited sorption, rapid aeration, inorganic, non-combustible, non-corrosive, has no flash point and highly stable under most fumigant environments. It is marketed under trade names Vikane and Profume and effective against a number of insect pests including stored grain pests, particularly useful against dry wood termites because of high penetrating power carries it deep into infested wood. It can also be used for short and long fumigation (4-12 hrs.) to mitigate the time constraint. The gas is packed in cylinders, has high vapour pressure and quickly penetrate the fumigated area.

Sulphenone: A synthetic sulfone acaricide, *p*-chlorophenyl sulfone and related sulfones; relatively non-toxic to mammals, acute oral LD_{50} for rats is approximately 3650 mg kg^{-1}. It is toxic to grapes, most varieties of pear and apple, green house plants, cucurbits etc.

Sulphur: It has strong, acaricidal and fungicidal action and is formulated as dust and wettable powder. It may also serve as diluent for some insecticidal dusts. It is non-toxic to mammals. Lime sulphur is the aqueous of calcium polysulphide which was used for the control of San Jose scale.

Sulphuring: Treating fruits and vegetables with sulphur dioxide. This bleaches the product, minimizes the enzymatic and oxidative changes, acts as preservative and helps against insect infestation. The fruits and vegetables sulphured by dipping in solutions of sulphurous acid or in salts like sodium bisulphate and potassium metabisulphite or by exposing prepared material to fumes of sulphur dioxide by burning sulphur.

Sunflower lacewing bug: See *Cadmilos retiarius.*

Sunnhemp capsid: See *Ragmus importunitas.*

Sunnhemp hairy caterpillar: *Utethesia pulchella* L. is an insect pest of sunnhemp. The caterpillars are hairy with red, black and white markings, feeding the leaves and boring the capsules.

Super: Any hive body used for the storage of surplus honey. Normally, it is placed over or above the brood chamber.

Superclass: In classification, an aggregate of two or more classes which have in common some characters of major importance.

Supercooling: Resistance to freezing by lowering the temperature at which freezing of body fluid begins.

Superfamily: A group of closely related families; superfamily names end in oidea.

Superior appendage: One of the two upper appendages at the end of the abdomen, *e.g.*, in Odonata.

Superior oil: A dormant spray oil of high paraffinic and low aromatic content and possesses increased plant safety and satisfactory insecticidal action.

Superlingual segmental theory: Chaudonneret (1950) confirmed the presence of superlingual segment in between the intercalary and mandibular segments in apterygote, *Thermobia domestica*. Accordingly, the insect head is made up of seven segnments in all: (i) preantennal, (ii) antennal, (iii) intercalary, and (iv) superlingual segments constituting the procephalon, (v) mandibular, (vi) maxillary and (vii) labial segments representing the gnathocephalic part of head.

Superparasitism: It is the phenomenon of parasitization of an individual host by more larvae of single primary parasitic species that can mature in the host, *e.g.*, *Trichogramma chilonis*.

Superposition eye: A type of compound eye in which a superposition image is formed. The superposition eyes bear comparatively a short rhabdom. It lies far behind from the lens to which it is connected by an axial flilament, *e.g.*, Coleoptera and nocturnal Lepidoptera.

Superposition image: In Coleoptera, nocturnal Lepidoptera and skipper butterflies (family Hesperidae), the light guides are absent and light illuminating a single receptor element will enter the eye through quite an extensive group of facets. The resulting image was termed by Exner a superposition image; it will be brighter but less well defined than the apposition image.

Superposition mechanism: It is a mechanism of image formation in poor light generally during night, mostly in the nocturnal insects such as Lampyridae, moths. The secondary pigment cells are adapted to migrate themselves or their pigments move in both proximal and distal directions. Particularly, in poor light, the pigment contracts so that the light from the wide visual field can enter obliquely without being absorbed. Each rhabdom, therefore, receives the light from large number of dioptric apparatus of neighbouring ommatidia besides its own facet. The ommatidia are comparatively longer than that of apposition eye and retinulae lie far behind the cones connected by the non-refractile filament. The optical system of such ommatidia has, therefore, the focal length about half the length of lens cylinder. The image thus formed from a group of rays refracted from neighbouring cones, called the superposition image. It is erect in position and represents merely a part of an image. The compound image is formed after an amalgamation of all such images and retains a final form of a complete superposition image. See apposition and diffraction mechanism.

Supersedure: A natural replacement of an established queen by a daughter in the same hive. Shortly after the young queen commences to lay eggs, the old queen disappears.

Supervised control: The control of pest as soon as it becomes necessary (at economic threshold level) so that its population density does not touch the economic injury level. Such a control necessitates keeping a regular record of pest populations.

Supinate: To turn the trailing edge of the wing downward.

Supplement (spray): Adjuvant, like sticker, adhesive, spreader, wetter and emulsifier.

Supplementary reproductives: In termites, the caste of males and females with short wings, light pigmentation, and small compound eyes. The females lay eggs in the colony supplementing the work of queen.

Suppress: To curb the activity of the pest so that its population cannot build up and cannot cause injury or destruction.

Supra-alar bristles: A longitudinal row of bristles on the lateral portion of mesonotum immediately above the wing base as in the Diptera.

Supra-anal plate: The last abdominal tergite covering the anus from above.

Supracide: Trade name of methidathion.

Supraoesophageal ganglion: A nerve mass that lies in the head above or -the insect brain.

Surface active agent: A substance that reduces the interfacial tension of two boundary lines. In most of the insecticides, the adjuvant may be considered as the surface active agent. It is also known as surfactant.

Surface caterpillar: Same as the cutworms.

Surface feeder: The surface feeder insect pests include the leaf eating insect pests (defoliators), cut worms, army worms, leaf rollers and plant sap suckers. These pests may further be categorized as defoliators, leaf rollers, plant sap suckers and fruit sucking moths etc.

Surface grasshopper: See *Chrotogonus trachypterus.*

Surface tension: Due to molecular forces at the surface, drop of liquid forms an apparent membrane that causes it to ball up rather than to spread as a film.

Surface water: Refers to the rivers, lakes, ponds, streams etc. which are located above ground.

Surface/ volume ratio: A ratio permitting comparisons to be made on effects of size.

Surfactant: Spreader; wetter.

Surplus honey: Honey removed from the hive which exceeds that needed by bees for their own use.

Survivorship curve: A graphic representation of deaths over time for a given population.

Susceptibility: The degree to which an organism is affected by a pesticide at a particular level of exposure.

Susceptible: Capable of being injured, diseased or poisoned by a pesticide; not immune.

Susceptible species: A plant or animal that is affected by moderate amounts of a pesticides.

Suspended: (i) A pesticide, the use of which is no longer legal and the remaining stock cannot be used. (ii) It also refers to the particles that are dispersed or held in a liquid.

Suspension: A fluid containing very finely divided solid particles dispersed in the liquid.

Suspensory muscles: See dilator muscles.

Sustainable development: Development that meets the needs and aspirations of the current generation without compromising the ability to meet those of future generations.

Sutural stria: The elytral stria closest to the suture.

Suture: An external line like groove in the exoskeleton or a narrow membranous area between sclerites.

Swallow tail butterfly: The larvae that secretes a repugnatorial substance.

Swarm: A large number of insects (especially when in motion). For example, a swarm of mosquitoes or a group of bees with a queen bee while migrating to establish a new colony.

Swarm cell: Queen cells usually found on the bottom of combs before swarming.

Swarming: The natural method of propagation of the honey bee colony.

Swarming caterpillar: *Spodoptera mauritia* Boisd. is an insect pest of paddy, sorghum, maize, wheat, barley, sugarcane etc.; the caterpillars feed on the foliage at night.

Swath: Width of target area sprayed at one pass.

SWC: State Warehousing Corporation.

Sweeping: To cover a wide area with insect net by moving it to and fro.

Sweet potato leaf beetles: *Aspidomorpha miliaris* F., *Metriona circumdata* H., *Chiridia sernotata* B., *Oncocephala tuberculate* O. are the beetles which feeds on the foliage of sweet potato.

Sweet potato sphinx: *Herse convolvuli* L. is a long horned caterpillar completely defoliates sweet potato vines.

Sweet potato weevil: *Cylas formicarius* Fab. is small steel blue bodied weevils, the grubs of which bores into the stem and tubers making them rotten.

Swimmeret: An abdominal organ in Crustacea that functions in the swimming.

Swimming hairs: Relatively long and dense hairs located on the mid and hindlegs of some aquatic beetles which aid in swimming.

Swing: To sway or wave to and fro.

***Sylepta derogata* (Fab.):** Belongs to order Lepidoptera and family Pyraustidae. It is an insect pest of cotton, commonly known as the cotton leaf roller.

***Sylepta lunalis* Guenee:** Belongs to order Lepidoptera and family Pyralidae. It is commonly known as grapevine leaf roller and is a pest of grapes and other plants of family Vitaceae.

Symbiont: An organism living in symbiosis with another organism. The wood boring termites have a rich fauna of bacteria or protozoa in the gut. Here, there is an amazing population of flagellates, ciliates and spirochaetes which perform for their host the invaluable function of digesting cellulose. Termites with these fauna can thrive indefinitely upon pure cellulose, but if the insect is deprived of these organisms (by exposing it to a high tension of oxygen, for example), the power of digesting cellulose is completely lost, and unless re-infected it soon dies. Cellulase can be extracted from the flagellates in the gut of these termites and certain cockroaches (*Cryptocerus*), where this enzyme is absent from insects. See fermentation chamber also.

Symbiosis: A close association between two or more different species that benefits each other. For example, herbivorous ruminants contain vast number of cellulose digesting bacteria. Other example is the formation of root nodules by *Rhizobium* sp. in plants for nitrogen fixation.

Symmetry: A definite pattern of body organization; bilateral symmetry, a type organization in which various parts are arranged more or less symmetrically either side of a median vertical plane, where the right and left sides of body are essentially similar.

Sympathetic nervous system: The sympathetic nervous system of insects can be divided into three parts; (a) stomatogastric or stomodaeal or anterior sympathetic nervous system (innervates the stomodaeum and dorsal vessel; and also the midgut in some Orthopteroid insects and silkworm larvae), (ii) posterior or ventral sympathetic nervous system: From each ganglion of ventral verve cord arises posteriorly a fine unpaired or median nerve running in antero-posterior direction. The median nerve arising from the ganglia runs for a short distance and terminally gives off a pair of transverse nerves which innervates the spiracles, segmental and alary muscles of the region. The transverse nerves in some insects also bear fine ganlionic swellings called perisympathetic organs, and (iii) caudal sympathetic nervous system: The last abdominal ganglion gives off the caudal sympatheic nervous system. The nerves innervate the reproductive system, hindgut, Malpighian tubules and a caudal part of the dorsal vessel, (iv) peripheral nervous system consists of some bipolar and multipolar neurons, form a complex network beneath the integument above the segmental muscles and upon the surface of gut. The distal processes divide enormously over the innervated structure and their centripetal processes are terminated into the ventral ganglia, (v) automic nerves (coelopulse) system or cholinergic nervous system or coelopulse system in the insects was recently disclosed by Slama (1990). It comprises cholinergic nerve fibres in the thoracic ganglia, peripheral nerves with the metameric perisympathetic neurohaemal organs and special group of somatic or visceral muscles. The coelopulse system regulates the haemolymph circulation through appendages, water balance and iso-osmosis, and inspiration of air and expiration of CO_2 through the spiracles.

Sympatric: Two or more populations occupying the same geographical area or of a population existing in a breeding condition within the cruising range of individuals of another population.

Symphile: (i) A species which is kept as a guest in the termitarium. (ii) A symbiont, in particular a solitary insect or other arthropod, which is accepted by the insect of the colony and communicates with it amicably. Most synphiles are licked, fed or transported to the host brood chambers or treated to a combination of these actions.

Symphypleona: A suborder of order Collembola. Head hypognathous; body subspherical or globose; indistinct or obsolete segmentation due to fusion of thoracic and first four abdominal segments.

Symphypleona: A suborder of order Collembola. Head typically hyphognathous; body sub-spherical or globose; indistinct or obsolete segmentation due to fusion of thoracic and first four abdominal segments.

Symphyta (Chalastogastra): Suborder of Hymenoptera. The primitive forms of Hymenoptera are included under this suborder. These insects are characterized by the abdomen being broadly joined to the thorax with no marked construction between the first and second abdominal segments.

Symptom: Any sign of disease or poisoning in a plant or animal. These signs or indications are used to find out which insect, fungus or other pest or pesticide is causing the disease, damage or destruction.

Symptomatology: The study of the symptoms and signs of the disease.

Synanthropy: A close association with the human being.

Synapse: In general, it is the site at which the axonic end of a neurone comes close approximation with the dendritic end of an adjacent or following neurone. It is a junctional region where one neurone ends and the other begins. According to the nature of connections, the synapses are specifically classified into three groups: (i) Axosomatic synapse: between axon and cell body, (ii) Axodendritic synapse: between axon and dendrite, (iii) Axonic synape: between axon and axon.

Synaptic cleft: Space between the terminal arborizations of one neurone and the dendrites of the next.

Synaptic poison: A chemical that interrupts normal transmission of impulses across synapses.

Synaptic transmission: The transmission of information across a synaptic junction. These transmissions are brought about either by chemical or electrical or by both processes.

Syncephalon: A secondary composite head, formed of the prostomium and one or more succeeding somites.

Synchronic: Contemporary; existing at the same time.

Synchronous muscle: A muscle in which each contraction is initiated by the reception of impulse, also see nonresonating muscles.

Syndrome: Symptoms characterizing a particular abnormality.

Synechthran: A symbiont, scavenger, predator or a parasite which is treated with hostility by the host colony.

Synecology: The branch of ecology which deals with the interrelationship of living communities of organisms to each other and to their environment. See autecology.

Synergism: Increased pesticidal activity of a mixture of pesticides above that of the sum of values of individual components.

Synergist: A chemical when mixed with a pesticide, increases its toxicity. The synergist may or may not have the pesticidal properties of its own but have properties of increasing the toxicity of the pesticide in the resultant mixture.

Synockete: The guest species in the termitarium or the ant nests that are treated with indifference.

Synomone: A chemical substance produced or acquired by an organism that when perceived by an individual of another species in the natural context, evokes in the receiver or cause physiological reaction adaptively favourable to both emitter and receiver. For example, a flower scent attracting bees for pollination.

Synonym: (i) Each of the two or more different names for the same taxon. (ii) One of two or more names which have been given to a single species. The earliest name usually takes precedence.

Systemic insecticide: An insecticide capable of being absorbed into plant sap or animal blood and lethal to insects feeding on or within the treated host.

Synoxylon anale **Lesne:** Belongs to order Coleoptera and family Bostrychidae. It is commonly known as grapevine beetle and is a pest of grapes, sal (*Shorea robusta*), teak (*Ouglinia dalbergioides*), shisham (*Dalbergia sissoo*) and wooden packing cases.

Synthesis: Formation of new material from two or more substances.

Synthesis ideal: A synthetic method which gives maximum yield of the product starting from cheap and abundant materials in minimum possible reaction steps and time.

Synthetic chemical: A chemical which cannot be obtained naturally and prepared from component elements or simpler intermediate substances.

Synthetic pyrethroides: Synthetic compounds similar to pyrethrins in structure and action. These insecticides are highly potent in low dosages. Allethrin, cypermethrin, permethrin, decamethrin, fenvalerate and fluvalinate are the most important insecticides of this group.

Synxenic: The rearing of one or more individual of a single species in association with one or more known species of organisms. Conditions are referred to as monoxenic, dixenic, trixenic, or polyxenic depending on whether the number of associated species is one, two, three or more, respectively of axenic.

Syrphidae (hoverflies): Family of suborder Cyclorrhapha, order Diptera. It is a large family comprising of moderate to large-sized flies which are often brightly colored and some resembling bees or wasps. The presence of a spurious vein between the radius and the media of the wings in most cases is characteristic of the family. In their habits the larvae exhibit extreme variations.

System: An interrelated set of elements that can be identified and around which a boundary can be drawn.

System ecology: The branch of ecology which can incorporates the view points and techniques of system analysis and engineering, especially those techniques having to do with the simulation of systems using computers and mathematical models.

Systematics: The science of biological classification.

Systemic insecticide: An insecticide capable of absorption into the plant sap or animal blood and lethal to the insect feeding on or within the treated host.

Systemic pesticide: A pesticide absorbed through the plant surfaces, translocated in different parts through the plant vascular system and accumulated in plant tissues.

Systems analysis: A technique of comprehensive analysis of complex biological problems by viewing them as system of interlocking cause effect pathways.

Systole: Contraction phase of heart beat cycle. When the alary muscles contract, the paired ostial valves at the anterior end, and the auricular valve at the posterior end of the ventricle are closed and therefore the only course left for the blood is to go forward into the next ventricle through the anterior auricular valve which is opened.

Systox: Also known as demeton.It is a mixture of two parts of O, O—diethyl 0-2 (ethylthio ethyl) phosphorothionate and one part of O, O-diethyl S-(2-ethylthio) ethyl phosphorothiolate; a systemic organophosphate insecticide and acaricide, colourless, viscous liquid with very slight but characteristic odour; can be mixed with water, organic solvents and aromatic oils, but not with paraffin oil.

T

Tabanidae (horse flies): Family of suborder Brachycera, order Diptera. This is a very large family of Diptera, including the gadflies or horseflies. They have very large eyes, being contiguous or holoptic in males and ditchoptic in females. They have large heads and stout bodies, the eyes usually having a golden green and purple marks when alive. Callosities are present on the frontal portion of the head. Antennae is three segmented, segment 3 annulate, arista absent. The family includes over 2000 species.

***Tabanus* spp:** Belongs to order Diptera and family Tabanidae. It is an insect pest of farm animals, commonly known as the gad fly or horse fly.

Tachinidae (tachinid flies): Family of suborder Cyclorrhapha, order Diptera. Tachinidae are small or medium sized, black or brown flies characterized by well-developed hypopleural bristles and prominent post-scutellum and small and long bristles covering the abdomen. They are endoparasites of larvae or pupae of Lepidoptera, Coleoptera, Hymenoptera, Diptera, Hemiptera, Dermaptera and Orthoptera.

Tachinid fly: Fly belonging to family Tachinidae (Diptera), *e.g.*, blow fly and screwworm fly; the larvae are endoparasitic burrowing the tissues below the skin of animals causing festering and foul smelling wounds.

Tactile: Pertaining to the sense of touch. In insects, most of the sense organs that respond to touch are small hairs with a nerve at their base. They sense the movement of this hair if it touches another object or if it moves by the wind.

Tactile organs (Tactile receptors) or trichoid sensilla: Referring to the sense of touch. They are the simple articulated sensory hairs and distributed on the entire body surface and are commonly called as the sensilla. These are made up of cuticle and articulated within the socket with the body wall. The trichoid sensillum is formed by two cells; the hair by the trichogen cell and the socket by the tormogen cell. Both the cells are modified epidermal cells. Each sensillum is innervated by a bipolar sensory neurone. The dendrite of the neurone is enclosed at base of the hair by a cuticular tubular sheath, called the scolopale.

Taenidium (pl., taenidia): A spiral or circular thickening of the inner surface of tracheae and tracheoles.

Tagmata *(sing.,* tagma*):* The three major regions or divisions of insect body; head, thorax and abdomen. Basically, the insect body is composed of six segmented head, three segmented thorax and twelve segmented abdomen.

Tagmosis: Co-ordination or amalgamation of segments to perform some particular group of functions. It is vitally important in the evolution of insects.

Tagoreceptor: An organ of touch having a single sense cell.

Taint: Undesirable smell in the fresh or processed food ultimately results from the pesticides used on growing crops or in storage.

Talc: Powdered soapstone; anhydrous magnesium metasilicate, $Si_8Mg_6O_{20}(OH)_4$. It differs from pyrophyllite in having tetrahedral silica sheet attached to each side of an octahedral hydroxyl sheet. The colour of the powder is white and it is used in insecticide dusts as inert carrier.

Tamarind fruit borer: See *Phycita orthoclina.*

Tandem: One behind the other, the two connected or attached together.

Tangoreceptor: A sense organ responsive to touch.

Tank mix: The mixture of two or more compatible pesticides in spray tank of a sprayer in order to apply them simultaneously.

Tanning: See sclerotization.

Tannins: Vegetable originated naturally occurring compounds which are mixture of derivatives of polyhydroxy benzoic acid.

***Tanymecus indicus* Faust:** Belongs to order Coleoptera and family Curculionidae. It is an insect pest of wheat, commonly known as the ghujhia weevil.

Tap root: The main root of mocotyledonous plants; it reaches deep in the soil to provide support to the plant system and to take up the water and nutrients.

Tapetum: A reflecting surface in the eye of night flying insects, formed either of pigment or of densely massed tracheae; which reflects the light that enters the eyes and causes them to shine with a ruby or golden light.

Taphylinoidea: Superfamily of suborder Polyphaga, order Coleoptera. This is a very large group with usually very short elytra leaving a few last abdominal segments exposed. The venation is very characteristic in that M and Cu are not connected. This superfamily comprise 12 families, *viz.*, Sphaeriidae, Clambidae, Limulodidae, Hydroscaphidae, Ptiliidae, Leptinidae, Anisotomidae, Silphidae, Scydmaenidae, Scaphidiidae, Pselaphidae and Staphylinidae.

Tar oil: The heavier portion of black oil. The tar oil includes the anthracene, cresols and naphthalene. The oils for spray purpose are refined by redistillation of tar distillates to remove the lower and higher fractions and most of the tar acids or phenols. The tar oils are mostly used in dormant sprays to kill scale insects and aphids on temperate fruit trees like apple etc.

***Tarache notabilis* Wlk.:** Belongs to order Lepidoptera and family Noctuidae. It is an insect pest of fibre crops, commonly known as the cotton semi-looper.

Target, cover: The degree to which the target plant part or foliage receives a deposit of pesticide.

Target pest: The pest at which the particular pesticide or other control measure is directed.

Target surface: The surface or area intended to be sprayed or dusted with a pesticide.

Tarsal claw: A claw at the apex of the tarsus, derived from the pretarsal segment of leg.

Tarsal formula: The number of tarsal segments on the front, middle, and hind tarsi, respectively. For example, 5 - 5 - 4.

Tarsation: Touching the other insect with tarsi as a tactile signal.

Tarsomere: A subdivision or segment of the tarsus.

Tarsus (pl., tarsi): The leg segment beyond the tibia or the distal leg segment. It is composed of 1 to 6 small segments, therefore, of great taxonomic significance. The subsegment is quite large known as metatarsus. Sometimes, each subsegment of the tarsus distally bears a lobulated structure, the plantula, as seen in the cockroach, *Blatta orientalis*. The tarsal segment receives sound attention at the time of taxonomic study of insects due to its species specific organization and modifications.

Tartar emetic: Potassium antimonyl tartrate. An insecticide used in bait sprays to control fruit flies and thrips and in ant poison baits. It was one of the first chemotheraptic agents used against *Trypanosomas* in cattle. It is also an appetite stimulator for bovines.

Tasar sericulture: The tasar silkworms, *Antheraea mylitta* and *A. paphia* (Saturniidae, Lepidoptera) are wild and feed on the leaves of host plants like asan, *Terminalia tomentosa*; arjun, *Terminalia arjuna*; sal, *Shorea robusta* and ber, *Zizyphus jujuba*. The tasar sericulture in India was earlier a job of tribals bt now scientifically flourished in Bihar, Orissa, M.P., A.P., Karnataka of tropical as well as Jammu and Kashmir, Arunachal Pradesh and Mizoram of temperate region. The production of seeds (eggs) and hatching of first instar larvae are the indoor process while rearing of the worms thereafter is an outdoor process.

Tassel: The staminate inflorescence of corn.

Taxis (pl., taxes): A directed response involving the movement of an animal towards or away from the stimulus. Where a more complex system of sense organs is present, the insects can make use of directed reactions or taxes. See kineses, klinotaxis, tropotaxis and anemotaxis also.

Taxon (pl., taxa): A taxonomic group which is sufficiently distinct and can be given a name and ranked in a definite category.

Taxonomic grouping of orders: (a) Subclass Apterygota: consists of four orders, *viz.*, Thysanura, Diplura, Protura and Collembola. They are diversified from each other representing heterogenous assemblage having polyphyletic origin and some abnormal insect characters. Thysanoptera show close relationship with main evolutionary stalk of Pterygota. (b) Subclass Perygota: consists two divisions, *viz.*, exopterygota and endopterygota. Martynov (1925, 1938) distinguished exopterygota into two groups (infraclasses), Palaeoptera and Neoptera. Palaeoptera includes two orders, *viz.*, Ephemeroptera and odonata which cannot flex their wings over the abdomen; whereas, the Neoptera are adapted for wing flexing mechanism. The neoptera are divided into two groups: (i) Polyneoptera (Orthopteroid group) which includes orders, like Plecoptera, Grylloblattodea, Orthoptera, Phasmida, Dermaptera, Embioptera, Dictyoptera, Isoptera and Zoraptera which are commonly characterized by mandibulate mouth parts, hindwings bearing a large anal lobe, presence of cerci and large number of Malpighian tubules and large number of ganglia in ventral nerve cord, and (ii) Paraneoptera (Hemipteroid group) which includes Psocoptera, Mallophaga, Siphunculata, Hemiptera and Thysanoptera which are characterized by piercing and sucking or suctorial mouth parts, hindwings without large anal lobe, absence of cerci, extremely condensed nerve cord, fluid sucking or raptorial or suctorial or saprophagous mode of feeding and fewer Malpighian tubules. The endopterygota form a single group, (iii) Oligoneoptera (Panorpoid complex) which includes nine insect orders (Tillyard, 1918-35). The six orders, Neuroptera, Mecoptera, Lepidoptera, Trichoptera, Diptera and Siphonaptera constitute the panarpoid complex. Remaining three orders, Hymenoptra, Coleoptera and Strepsiptera do not show marked interrelationship either with the orders of panorpoid complex or with each other. They are supposed to be evolved independently and do not form a monophyletic group.

Taxonomy: *Taxis*= arrangement+ *nomos*= law. Classification of animals and principles there of. The theory and principles of classifying organisms.

Tea mosquito bugs: *Helopeltis theivora* Walk. and *H. antonii* Sign., are major pests of tea causing severe damage; nymphs and adults suck cell sap from young leaves and buds developing small spots which turn black giving out a brown fluid. These pests also causes damage in cashewnut.

Tea red mite: *Tetranychus bioculatus* Wood Mason and *Oligonychus coffeae* Niet. are pests of tea; the nymphs and adults suck the plant cell sap from tender parts.

Tea red slug: *Heterusia cingala* Moore and *H. magnifica* Butl. cause the damage by defoliating the tea plantations in caterpillar stage.

Tea tortrix: In South India, *Homona coffearia* is well known as a pest of tea.

Technical material (Technical product): It refers to the active ingredient in the insecticides or pesticides which is manufactured in the chemical companies. The technical product is usually of high level purity (95-98 per cent pure) but not completely pure.

Technology: The application of science for achieving the commercial objective is termed as technology.

Tectocuticle: See cement layer.

Tectum: (i) A bodily structure resembling or serving as a roof, (ii) the upper lobe of penis which is horizontally divided into dorsal and ventral lobes; usually referring only to the penis in Curculionoidea.

Tedion: Trade name of tetradifon.

Tegmina: The thickened, leathery forewing of insects belonging to the orders Orthoptera (grasshoppers, crickets) and Dictyoptera (mantids, cockroaches).

Tegula: (pl., tegulae) One of the articulated plates at the attachment point of the fore wing. A scale like sclerite basal to the humeral plate. It protects the wing base from physical damage.

Tel net: With the help of tel net, working on a computer, one can work on any other computer in any part of the world.

***Telenomus* sp.:** Egg parasitoid (Scelionidae, Hymenoptera) associated with citrus caterpillar, *Papilio demoleus* Linn.

Telopodite: The primary shaft of the limb distal to the coxopodite.

Telotarsus: The distal of the two principal segments of the tarsus in Arachnida and Chilopoda.

Telotaxis: In the case of vision, an object or a point of light is perceived from a distance. The image of such an object may be fixed upon the central point of retina of eye and the insect thus enables to go straight to it. This response is termed as telotaxis or movement towards a definite goal.

Telson: The twelfth abdominal segment present in the embryos of some insects and the adults of Protura.

Temephos: A low toxicity insecticide to mammals, birds and fishes, marketed under trade name abate is used against mosquito larvae in water and other water breeding insects such as midges and flies.

Temik: Trade name of aldicarb.

Temperature: Temperature is the measure of hotness or coldness of a body. It decides the direction of flow of heat. As the water always flows from higher level to a lower level, the heat always flows from a body at higher temperature to a body at lower temperature. Heat and temperature are two different physical quantities which are related each other. The heat is the measure of thermal energy, content of body temperature is decided in hotness. Heat is measured in unit of joule or kilojoule while temperature is measured in units of degree Celsius or Kelvin.

Temple: The lateral portion of the head between the posterior edge of the eye and an abrupt narrowing of the head to form a posterior neck.

Temporal: Limited by time: lasting only for a time.

Temporary acceptable daily intake (TADI): An acceptable daily intake established for a specified limited period to enable additional biochemical, toxiciological or other data to be obtained as may be required for establishing an acceptable daily intake.

Tenacity: Property of pesticide which resists the action of weathering agencies on its deposits.

Tenacity index: Ratio of quantity of residue per unit area at the end of given period of weathering to that present at the beginning.

Tenaculum: In Collembolans, a minute structure present on the sternum of third abdominal segment which serves as a clasp for furcula. See also, abdominal appendages, Collembola.

Tendon: It is a slender, chitinous plate, band or strap to which muscles are attached for moving appendages.

Tenebrionidae: Family of superfamily Cucujoidea, suborder Polyphaga and order Coleoptera. This is one of the largest families of Coleoptera. They are small or large, elongate somewhat flattened, usually black or sometimes reddish brown, hard, often sculptured beetles resembling superficially the carabidae. Very often they are wingless or with vestigial wings. The elytra are frequently fused. Tarsi 5-5-4, heteromerous and claws are simple. These are usually nocturnal beetles who feed on decaying vegetable and animal matter, dung meal, flour, dried seeds, cereals and other stored products. A few occur in ant nests. The larvae are elongate and cylindrical and bear often two dorsal hooks on hinded and a ventral retractile process.

Tenebrionoid aedeagus: Aedeagus in which the phallobase (or basale) and partly to completely fused parameres together form an incomplete sheath on one side (usually dorsal) of the penis; the parameres are not individually articulated but may form a combined apical piece (apicale) which is articulated to the phallobase. In some groups, there are articulated accessory lobes attached to the apicale and these may be confused with parameres.

Teneral: Refers to recently moulted individuals with a soft, pale-coloured skin. A tsetse fly is called a teneral fly from the time it emerges until the taking the first blood meal.

Tent caterpillar: *Malacosoma indica* Walk., a pest of apple; the caterpillars make thick webs on the twigs, lives inside and latter feed on leaves.

Tent or cover fumigation: Tent, sheet or cover fumigation is commonly practiced in sheds. It provides better gas tight condition and control of insects than space fumigation. The stacks of bagged seeds are covered with gas proof sheet made up of high density polyethylene sheet of 1000 guage thickness with perfect sealing at joints or rubber sheet having provision for dispensing phosphene tablets. The size of cover should be sufficient enough to cover stack properly with margins at the ground level for proper sealing. The dose should be worked out according to the quantity of seed than size of cover. It should also be protected from rodents which damage seal sometimes causing loss of gas.

Tentacle: Long slender processes of animals or feelers used for exploration, prehension or locomotion; sensitive hair or filament of plants.

Tenthredinidae (typical sawflies): Family of super family Tenthredinoidea, suborder Symphyta (Chalastogastra), order Hymenoptera. This is very large group of wasplike, medium sized to small brightly colored variations occur in both sexes. Many are found on flowers or foliage and some are carnivores feeding on small insects. Parthenogenesis occurs in the family. From unfertilized eggs either males or females, and in few cases both sexes may developed and toothed in various ways.

Tenthredinoidea: Super family of suborder Symphyta (Chalastogastra), order Hymenoptera.

Tentorial pits: The external depressions in the cranial wall at the roots of the tentorial arms; the anterior tentorial pits located in the subgenal sutures or usually in the epistomal suture, the posterior tentorial pits in the lower ends of the post occipital suture.

Tentorium: It is an endoskeletal X- shaped structure composed of three pairs of arms, anterior, posterior and dorsal tentorial arms. The anterior tentorial arms develop from the anterior pits in the subgenal or epistomal suture. The posterior tentorial arms arise from the posterior tentorial pits in the lower ends of the post-occipital suture. The dorsal tentorial arm is originated in the form of secondary outgrowths of the anterior arms. All these tentorial arms unite in the middle forming an expanded central skeletal plateform. The tentorium is bracing the epicranial walls. It serves as a site for the attachment to the ventral abductor muscles of the gnathal appendages, the retractor muscles of the hypopharynx and also to the ventral dilators of the stomodaeum. In most of the insects, the antennal muscles are attached to the dorsal arms of the tentorium. In the apterygote insects, no homologous structure is seen; therefore, it is an important evolutionary advancement in the pterygotes. The broad tentorial plate occurs in the pterygote insects which is known as corporotentorium.

Tentorium: The internal skeleton of the insect head.

Tephritidae [Trypetidae] (fruitflies): Family of suborder Cyclorrhapha, order Diptera. The Tephritidae are small to medium-sized insects usually with spots or bands on wings. The wings venation is characteristic in that the subcosta bends apically forward at almost a right angle and then fades out. The Mediterranean fruitfly, *Ceratitis capitata* on citrus and fruit trees and the apple maggot, *Rhagoletis pomonella* on apple are some notorious pests in this family.

TEPP: A synthetic organophosphate insecticide and acaricide, tetraethyl pyrophosphate ($C_8H_{20}O_7P_2$); colourless, odourless, hygroscopic liquid, miscible with water and most of the organic solvents, extremely toxic to mammals; acute oral LD_{50} for rats is 2 mg kg^{-1}, toxic to some plants and fruit trees when applied through aircraft as ULV spray. This was first synthesized by Clearmont in 1854 but insecticidal properties were discovered by Schrader (1939). It is rapidly hydrolyzed by water so it should be used immediately after mixing with water. It is rapidly hydrolysed to inactive diethyl-o-phosphoric acid and, therefore, does not leave any harmful residue on treated plants. It is effective against aphids, thrips, mealy bugs and mites and available as dust and emulsifiable concentrate under trade name Bladan.

Teratogen: A substance that can cause birth defects.

Teratogenesis: It is embryonic malformation or abnormal growth noticed in beetles, bugs, crickets, leaf hoppers and honeybees. Intense cell proliferation and mass movement of cells are susceptible leading to specific failure with teratogenesis occurring mostly during germ band formation, katatrepsis (leading to averted embryo), impair dorsal closure (leading to spiral segmentation) and head malformation (causing formation of supernumerary eyes and double heads). Spontaneous teratogenesis (malformation resulting from unknown causes) is frequent in the walking sticks. Among the known factors, ambient temperatures and some chemicals like juvenile hormone analogues cause failure of katatrepsis.

Terbufos: Insecticide marketed under trade name Counter which is highly toxic and used effectively against the corn rootworms, wireworms on corn and sugarbeet root maggots.

Terebrantia: Suborder of Thysanoptera. Female with a distinct ovipositor. Terminal abdominal segment seldom tubular, that of the female longitudinally divided beneath and usually conical; that of the male usually bluntly rounded, never tubular. Wings with microtrichia and forewing with marginal veins and at least one longitudinal vein with regularly arranged setae on veins.

Tergite: A sclerite of the tergum.

Tergum: The dorsal surface of any body segment.

Termatophylidae (Termitaphilidae): Family of superfamily Aradoidea of division Gymnocerata, suborder Heteroptera and order Hemiptera. These bugs look like woodlice, broadly oval, apterous, flat, blind bugs and inhabit termite galleries, *e.g.*, Termitophilid bugs.

Termen: Outer margin of wing.

Terminal: At the end or tip; the last of a series.

Terminal filament: It is the anterior most part of ovariole. It is separated from germarium by a transverse septa. It is merely an apical prolongation of peritoneal layer and contains a mass of nuclei with scanty cytoplasm.

Terminal growth: The tips or end of growing plants.

Termitarium: A termite nest or an artificial nest in laboratory to house the termites.

Termite: Any of numerous social insects of the order Isoptera. Termites are pale-colored, usually soft bodied insects that live mostly in warm regions. Many species feed on wood, often destroying trees and wooden structures. Sometimes, they are referred to as white ants. See *Odontotermes*, *Microtermes*.

Termitidae: A family of order Isoptera. Includes most of the living termites which are ground dwelling; pronotum of workers and soldiers narrow, with a raised median anterior lobe. Example, *Odontotermes*, *Microtermes*, *Nasutitermes*, *Capritermes*.

Termitology: Refers to the scientific study of the termites.

Termitophile: Inquilines that live in the termitaria scavenging or sharing the foods of termites and sometimes even predating them.

Terramycin: An antibiotic used to prevent American and European foulbrood.

Terrestrial: Living on land. See also, aquatic.

Tertiary reproductive: These are same as ergatoid reproductive in the termites.

Tested queen: It refers to a queen whose progeny shows she has mated with a drone of her own race and possesses other qualities that would make her a good colony mother.

Testis (pl., testes): The reproductive system of male consists of a pair of sexual glands (testes), number of sperm tubes or follicles, paired gonoducts (vas deferens) of mesodermal origin into which the sexual products are discharged, a median duct (ejaculatory duct) lined with cuticle and with or without vesicular seminalis epididymis or sperm sac etc. The testis of the male is made up of a series of tubular follicles of varying number and arrangement. Each follicle contains a succession of zones in which the sex cells are in different stages of development. Both testes of a pair are often free from each other except in some insects in which they are closed together in a common coat or scrotum. The testes are held in position by attachment of trachea or by fat bodies.

Testis steroid hormone: The testis of insects secretes 20-hydroxyecdysterone as seen in *Calliphora vicina*, *Dysdercus intermedius* (Koolman *et al.* 1979, Loeb *et al.* 1982), *Heliothis virescens* and *Gryllus bimaculatus* (Hoffman *et al.* 1981). The 20-hydroxyecdysterone promotes spermatogenesis. It increases rate of cell division in the spermatogonial cells of *Rhodnius* (Dumser and Davey, 1975). In Lepidoptera, Spermatogonial cell division normally occurs in the last larval instar and requires ecdysteroids in the absence of JH because the JH exerts negative effect on the spermatogenesis (Davey, 1974). The 20-hydroxyecdysone promoting spermatogenesis in the late larval life or during the pupal stage has been well documented in the *Bombyx mori* (Takeuchi, 1961), *Chilo suppressalis* (Yagi *et al.* 1969), *Rhodnius* (Dumser and Davey, 1975) etc.

Tetrachlorvinphos ($C_{10}H_9O_4Cl_4P$): 2-chloro-1- (2,4,5 trichlorophenyl) vinyl dimethyl phosphate is an organophosphate compound. It is a white crystalline solid with melting point 97-98 °C having low solubility in common organic solvents and practically insoluble in water. It is a contact insecticide. On foliage and in soil, it is hydrolysed at P-O-C bond to give initially trichlorophenacyl chloride and desmethyl gardona. The former then breaks down to several eight carbon non-insecticidal compound. It is marketed under the trade name Gardona.

Tetradifon: Marketed under the trade name Tedion is a synthetic acaricide, a sulfone; chemically, *p*-chlorophenyl 2,4,5-trichlorophenyl sulfone; relatively non-toxic to mammals, 14,700 mg kg^{-1} administered in rats without symptoms of systemic toxicity. It is non-phytotoxic.

Tetraethyl pyrophosphate: See TEPP.

***Tetranychus cinnabarinus* (Boisd.):** Belongs to order Acarina and family Tetranychidae. It is a pest of brinjal, some other vegetable crops and papaya, commonly known as the red vegetable mite.

***Tetrastichus radiatus* Waterston:** An important parasitoid of nymphs of citrus psylla.

***Tetrastichus sokolowskii*:** It is a pupal parasitoid (Eulophidae, Hymenoptera) of diamondback moth.

Tetrigidae: A family of order Orthoptera, suborder Caelifera and super family Acridoidea. Pronotum extended backwards to cover abdomen and tegmina reduced to small scales.

Tettigometridae: A family of superfamily Fulgoroidea of suborder Homoptera and order Hemiptera. About 70 species are known from Europe, Ethiopian region and Peru, *e.g.*, *Hilda bengalensis* infests shoots of *Ficus religiosa*.

Tettigoniidae: A family of order Orthoptera, suborder Ensifera and super family Tettigonioidea. Long setaceous antennae, four segmented tarsi.

Tettigonioidea: A super family of order Orthoptera and suborder Ensifera. Tarsi four segmented, *e.g.*, long horned grasshoppers.

Thallium sulphate (Thallous sulphate, Tl_2SO_4): It is a fiels rat poison used at the rate of 1.0-1.5 per cent in the poison baits.

Thanite: It is a contact synthetic insecticide (thiocyanate or thio-cyano compound) containing 82 per cent isobornyl thiocyanatoacetate and 18 per cent other related terpenes; low mammalian toxicity, acute oral LD_{50} for rats is 1000 mg kg^{-1}; its use is restricted due to high phytotoxicity.

Thaumastotheriidae (Thaumastocoridae): Family of superfamily Aradoidea of division Gymnocerata, suborder Heteroptera and order Hemiptera.These small insects with broad head are represented in North America and Australia. These infest the royal palm and known as royal palm bugs.

Thelyotoky: Some species are obligatorily parthenogenetic and each generation consists entirely of females. Such individuals are termed as impaternate or uniparental, *e.g.*, *Chelonus blackburni* an egg-larval parasitoid of cotton bollworms.

***Theretra alecto* Linnaeus:** Belongs to order Lepidoptera and family Sphingidae. It is commonly known as horn worm and is a minor pest of grapevine.

Thermal: Pertaining to the heat.

Thermal energy (Internal energy): In a solid, the molecules are very close to each other and a large force of attraction acts between them. So molecules maintain their position but are capable of vibrating about their respective positions. Each molecule possesses a potential energy and vibrational kinetic energy. The sum of the potential energy and kinetic energy of molecules is called thermal energy or internal energy.

Thermal equilibrium: The condition in which the two objects attain the same temperature is called the state of thermal equilibrium.

Thermal fog applicator: A spraying equipment in which oil solutions of insecticides are finely atomized by a blast of heated air or gas.

Thermal pollution: Thermal pollution is defined as the undesirable changes in the natural environment due to presence of waste heat in the water.

Thermometer: Thermometer is an instrument that measures temperature. Any property of a substance that changes when it is heated or cooled may be used as the basis of a thermometer. The length of a liquid column, electrical resistance of a wire etc. changes linearly with temperature and can be used in making thermometer. For example, mercury in glass-thermometer is based on the property of mercury that the change in the length of a column of mercury in a capillary tube is proportional to the change in temperature.

Thiamethoxam: This insecticide belongs to thionicotinyl group of neo-nicotinyl compounds. It is very effective against whitefly and other sucking insects. It is marketed under the trade names, Cruiser 25 WG and Actara 25 WG.

Thigmotaxis: A taxis based on lateral contact.

Thigmotrophic: Response of an organism to the mechanical or tough stimulation.

Thimet: Trade name of phorate, an organophosphate insecticide.

Thimeton: It is a systemic organophosphate insecticide; chemically S-2 ethylthioethyl dimethyl phosphorothiolothionate. It is colourless, oily with a strong unpleasant odour; soluble in water (200 mg l^{-1}) and highly soluble in most of the organic solvents. It is metabolized to dimethyl hydrogen phosphate and then to phosphoric acid. It is marketed under trade name Ekatin.

Thin super foundation: A comb foundation used for comb honey or chunk honey production which is thinner than that used for brood rearing.

Thio: Prefix in the names of chemicals denoting the presence of sulphur.

Thiodan: Trade name of endosulfan, an organochlorine compound.

Thiodemeton: See disulfoton.

Thiodicarb: A carbamate insecticide marketed under trade name Larvin (75 WP).

Thionate dust: A pesticide dust containing a compound of sulphur.

Third axillary: A flexor sclerite of the wing base.

Third form reproductive: See tertiary reproductive.

Third generation pesticides: A term proposed by Williams (1967) for juvenoids and other natural or synthetic chemicals used in pest suppression by virtue of their effective interference with hormonally directed natural processes in insects. Most of the synthetic insecticides which are commonly used these days suffer from two drawbacks; they lack selectivity and the target insect develops resistance against them. In third generation pesticides these two defects so far are not there as they are specific and selective and the target insect does not develop resistance against them. The juvenile hormone which is responsible for keeping the insect in larval stage shows a great potentiality as an insecticide.

Thixotropy: The property of the liquid substances to change their viscosity with the passing of time period, *e.g.*, turpentine oil, paints etc.

Thoracic appendages: There are two types of thoracic appendages, the legs and the wings.

Thoracic skeleton: The wall of each thoracic segment is typically composed of a dorsal, a ventral and two lateral skeletal plates forming a sort of compact box like structure. A dorsal skeletal plate is known as the tergum or notum. A tergum or notum in accordance with the thoracic segments is termed as the pro, meso or metatergum or notum. The ventral skeletal plate is designated as the sternum and as per segment to which it belongs is called the pro, meso or metasternum accordingly. Similarly, the lateral plates are known as the pleura and can specifically referred to as the pro, meso or metapleura with reference to a particular thoracic segment to which they represent. The sclerites of tergum, sternum and pleura are called as the tergites, the sternites and the pleurites, respectively.

Thoracic tergite: The prothoracic tergum is an undivided skeletal plate. The protergum occurs as a large plate in the Orthopteroid insects or may reduce into a narrow skeletal band as found in the higher insects. The notum of the wing bearing segments on the other hand, is typically divided by transverse sutures, *viz.*, the prescutal and scuto-scuteller into three sclerites. They are the prescutum, scutum and scutellum. The intersegmental groove, the antecostal suture gives a narrow band at the front of notum and it is known as acrotergite. The intersegmental sclerites are closely associated with the posterior sclerite called the postnotum. On the lateral side of pronotum in several Lepidopteran insects, the lobe like structures are well developed and are known as petagia

Thorax: The body region behind the head which bears the legs and the wings.

Thorough coverage: Coverage of all plant parts by dusting or spraying of a pesticide to obtain the effective control of pest.

***Thosea aperiens* (Wlk.):** Belongs to order Lepidoptera and family Cochlidiidae. It is an insect pest of pulse crops, commonly known as the green nettle slug caterpillar.

Threatened species: Species which are either endangered or rare.

Threshold: The lowest value of any kind of stimulus, signal or agency which will produce a specified effect. See threshold limit value.

Threshold limit value: The concentration of a chemical which will not produce any harmful effect on a worker, working eight hours a day throughout his entire working life term.

Thripidae: A family of superfamily Thripoidea of suborder Terebrantia order Thysanoptera. Antennal segments three and four with slender sense cones which are simple or forked. Antennae six to nine segmented. Foretarsus sometimes with a claw like appendages. The bulk of the *Terebrantian* species comes under this family.

Thripinae: A subfamily of Thripidae, family of superfamily Thripoidea of suborder Terebrantia order Thysanoptera. Terminal antennal segments are long and thin, dorsum of body not polygonally reticulate, *e.g., Scirtothrips cardamomi, Anaphothrips sudanensis.*

***Thripocentus maculates* Wat.:** It is a parasite of grapevine thrips, *Rhipiphorothrips cruentatus* Hood.

Thripoidea: Superfamily of suborder Terebrantia order Thysanoptera. Ovipositor curved downwards. Wings more or less pointed at apex. Antenna six to nine segmented; segments not moniliform; style one to three segmented. Fore-margin of wings with the fringe of hairs present. The families Heterothripidae and Thripidae are included in this superfamily. The members of the Thripidae include several economically important forms which are well known crop pests.

***Thrips orientalis* Bangnall:** Belongs to order Thysnoptera and family Thripidae. It is an insect pest of ornamental plants, commonly known as the jasmine thrips.

***Thrips tabaci* Lind.:** Belongs to order Thysanoptera and family Thripidae. It is an insect pest of onion, cotton, pineapple etc., commonly known as the onion thrips.

Thrips: Belongs to order Thysanoptera. These are characterized by fringed wings, rasping mouthparts and an incomplete metamorphosis.

Thylate: Trade name of thiram, a fungicide.

***Thyraeella collaris* Grav.:** It is a larval-pupal parasitoid of diamondback moth.

Thyrididae: Family of super family Pyralidoidea of suborder Ditrysia and order Lepidoptera. Characterized by absence of tympanal organs and hindwings Sc+R not fused with Rs and Cu_2 absent. These are different form Pterophoridae in that the wings are not being divided into plumes. In South India, the common species found are the leaf webber of *Sesbania bispinosa, Striglina scitaria* and a species of *Dysodia,* the larvae of which burrows into tender twigs and causes characteristics rolling of the affected tip in *Cocculus* sp.

Thysanidae: Family of super family Chalcidoidea, suborder Apocrita and order Hymenoptera. Members of this family are parasites of coccids, aleyrodids, other homopterous insects and rarely Diptera. Some are hyperparasites of chalcids.

***Thysanoplusia nigrisigna*:** Cabbage semilooper which feeds on foliage of cole crops.

Thysanoptera: Insect order containing thrips. The insects are characterised by minute to small size, slender usually winged insects with asymmetrical mouth parts known as rasping and sucking type, antennae 6-10 segmented, wings very narrow strap like with few veins and and fringed with long marginal setae, hemimetabolous development with one or two inactive pupal instars.

Thysanura: Insect order (*thysan*= fringe, *ura*= tail). This order covers silver fish, firebrats and relatives.

Tibia: The fourth segment of the leg between the femur and the tarsus. It is generally long, slender and tubular and provided with the spines in some insects. It articulates with the femur. It bears a single or many tibial spurs at the distal end. The femoro-tibial joint is commonly known as the knee of insect legs. The knee joint is often dicondylic.

Tibial: Relating to the tibia.

Tibial articular area: The apical surface of the tibia to which the first tarsomere and tibial spurs are attached; sometimes greatly expanded on the hind leg.

Tibial spur: An articulated, multicellular, spur like process located at the apex of the tibia; usually paired but occasionally single, and sometimes absent.

Tick paralysis: A flaccid, afebrile or without fever, motor paralysis produced due to a neurotoxin secreted by the salivary glands of feeding female ticks of certain species.

Ticks: These are tiny animals (class Arachnida, phylum Arthropoda) closely related to insects having four pairs of legs, two body regions, no antennae or feelers and no wings. These are blood suckers of animals like dog, cow, buffalo, cow, wild animals etc. and grouped with mites and spiders.

Til hawk moth: See *Acheronia styx.*

Til leaf and pod caterpillar: See *Antigastra catalaunalis.*

Tillage: The ploughing and preparation of soil for sowing the crops.

Tiller: An erect shoot arising from crown of plants belonging to family Gramineae.

Tillering stage: Growth stage of grain crops when a plant produces additional shoots.

Tillering: Producing shoots from the lower part of plant (sometimes caused by insect damage).

Timbal (pl., timbals): Sound producing organ in cicadas.

Tinaeidae (clothes moths): Family of super family Tinaeoidea of suborder Ditrysia and order Lepidoptera. The family has world wide distribution. The moth has rough hairs on its head and rather long maxillary palpi. The tibiae are hairy. The larvae are destructive to clothes, furs, carpets, dried fruits etc.

Tinaeoidea: Super family of suborder Ditrysia of order Lepidoptera. They are characterized by well developed maxillary and labial palpi; the third segment of labial palpi being usually slender and pointed. Both wings possess Cu_2 and hindwings with $Sc+R_1$ free and M_1 and R_s separate. In some the wings are divided into plumes.

Tingidae: Family of superfamily Tingoidea, suborder Heteroptera and order Hemiptera which includes bugs which have densely reticulate body and wings. Ocelli are absent. The pronotum covers the scutellum. The elevated ridges and sunken membranous oval areas present a lace like appearance on the head, lateral expansions of the thorax and wings. The nymphs are usually spiny. These bugs infest leaves of shrubs and trees and suck sap. Examples are Buddleia tinged, *Tingis* buddleiae; compositae tinged, *Cadmilos retiarius;* hollyhock tinged, *Urentius euonymus;* banana lacewing bug, *Stephanitis typicus* and brinjal lacewing bug, *Urentius hystricellus.*

Tingling: A characteristic pricking sensation.

Tingoidea: Superfamily of division Gymnocerata, suborder Heteroptera and order Hemiptera. This includes two families, *viz.*, Tingidae (lace bugs) and Piesmidae (ash grey leaf bugs).

Tiphiidae: Family of super family Scolioidea, suborder Apocrita and order Hymenoptera. The Tiphiidae are hairy, black, fair sized wasps with short legs and are parasites of scarabaied and cicindelid beetles, bees and wasps. They possess non-striolate wing membrane and the meso and metaasternum do not form a plate.

Tipulidae: Family of suborder Nematocera, order Diptera. This family includes some of the large sized forms commonly known as the craneflies or daddy-long legs. They have antennae bearing six to many segments and the front region of the head is extended forwards and in some forms an elongated proboscis is present. The mesonotum usually bears a V shaped suture and the legs are very long.

Tissue culture: This is a modern technique for developing field crop variety. This process is based on the totipotent behaviour of cell. Each plant cell has the capacity to develop into a complete plant and this capacity is called totipotency. Haberland favoured the concept of totipotency but practically proved by Steward.

Tissue: An aggregate of similar cells forming a definite and continuous fabric, and usually having a comparable function.

Tobacco beetle (cigarette beetle): *Lasioderma serricorne* Fab. is an insect pest of stored tobacco, cigarettes, cigars, coriander, cumin, turmeric and ginger. It is a dark brown beetle whose grub causes damage which is hairy.

Tobacco caterpillar: *Spodoptera litura* Fab., is an insect pest of tobacco, tomato, mustard, chillies, cabbage, cauliflower, knol-khol soybean etc., the caterpillars which are dark brown, feed on leaves and defoliates the plants.

Tobacco decoction: The tobacco decoction has an insecticidal property which is prepared by boiling or soaking overnight tobacco leaves, stalks and wastes (1.0 kg) in water (10.0 l) and adding soap (1.0 kg) to the finally prepared mixture. The stock solution so prepared is diluted 10-15 times before use as spray.

Tobacco soap decoction: See tobacco decoction.

Tolerance: (i) The maximum amount of pesticide chemical allowable on the raw agricultural commodity as set by the regulation at a specified stage of the harvesting, storage, transport, marketing or preparation of food up to the final point of consumption. (ii) It is the ability of a plant variety to withstand the attack of a pest (Painter, 1951) as a basis of resistance in which the host variety shows ability to grow and reproduce itself or repair injury despite supporting a pest population equal to that damaging a more susceptible host.

Tolerance level, median: The amount of chemical in ppm in or on the food stuff (oral tolerance) or in the environment (contact tolerance) which a sensitive species of animal can tolerate for a short duration (4-7 days) without causing any noticeable harmful effect or injury.

Tolerance level, pesticide: The maximum concentration of pesticide residue that is permitted in or on the food stuff at particular stages like harvesting, storage, transport, marketing or preparation of food upto the final point of consumption expressed in ppm. These residue levels must be the result of good agricultural practices, toxicologically acceptable and be shown to present no hazards throughout the life of human being or animal.

Tolerance limit: The maximum amount of pesticide residue that is permitted by regulation in or on a food commodity, expressed in ppm.

Tolerance, permitted: Maximum amount of toxicant allowed in or on foodstuffs for human consumption.

Tolerant: A variety of a crop possessing tolerance against an insect pest. The tolerance is one of the phases in the mechanism of resistance (Painter, 1951).

Tomato fruit borer: See *Helicoverpa armigera.*

Tomoceridae: A family of order Collembola. Vescular sense organs on fourth antennal segment absent; post-antennal organ lacking. For example, *Tomocerus* sp.

Tomosvary organ: A small post-antennal organ in the Arthropoda.

Tonic contraction: In addition to the alternate contractions and relaxation, the muscles may enter a state of prolonged steady contracture or tonus. In this state, they may support the insect in some characteristic attitude and may have to bear a considerable weight; but the rate of metabolism is very low.

Tormogen: The apical cell associated with the seta which secretes the setal membrane or socket.

Tornus: Anal angle of wing.

Tortricidae: Family of super family Tortricoidea of suborder Ditrysia and order Lepidoptera. Moths small, brownish or grey and when at rest have the wing held roof-like over the body the long fringe of hairs on the basal part of cubitus in hindwing is wanting. Similarly the first branch of vein M is absent in hindwings.

Tortricoidea: Super family of suborder Ditrysia and order Lepidoptera.

Torulus: The socket on the head into which the antennae is inserted.

Torymidae: Family of super family Chalcidoidea, suborder Apocrita and order Hymenoptera. These are somewhat elongate insects with large hindcoxae, simple or single toothed hindfemur and an exerted long ovipositor. The parasidal sutures are distinct. Both parasitic and phytophagous species occur in this family. Species of Toryminae, Erimerinae and Monodontomerinae are parasitic on caterpillars and gall insects. *Monodontomerus trichiophthalmus* is parasitic on larvae of leaf-cutting bee in India. In South India, the grubs of the seed infesting chalcid *Bruchophagus mellipes* are parasitized by *Eridontomeroidella gibboni.*

Total costs: The sum of the fixed and variable costs for any given level of production.

Total herbicide: Herbicide affecting all plants.

Totipotency: The potential throughout life to express the full behavioral repertoire of the population (even if never actually expressed) and the ability to produce offspring like oneself, exhibiting the full behavioral repertoire of the population without help.

Toxaphene ($C_{10}H_{10}Cl_8$): An insecticide which is prepared by chlorination of bicyclic terpenes. Moderately toxic, synthetic organochlorine compound (bicyclic structure or chlorinated terpenes), technical product yellowish waxy solid with melting point 65-90 °C, stomach and contact poison, effective against cutworms, grasshoppers, armyworms etc. on the crops, livestock pests and ticks. Moderately toxic to mammals having 67- 69 per cent chlorine; oral LD_{50} for rats is 60 mg kg^{-1} and toxic to cucurbits. It is a persistent contact and stomach poison having less residual action than DDT and less toxic to bees. Its formulations are 5, 10, 20 per cent dust, 25 and 40 per cent water dispersible powder and 25 per cent emulsifiable concentrate.

Toxic: Injurious to animals and plants through contact or systemic action.

Toxic residue: A poisonous residue left on the animals or plants or in the soil after spraying or dusting.

Toxicant: A poisonous substance or chemical.

Toxicity: The natural capacity of a substance or chemical that can cause injury or lethal effect. The toxicity is measured by oral, dermal and inhalation studies on the test animals.

Toxicity index: The ratio of quantity of resistance per unit area at the end of a given period of weathering to that present at the beginning.

Toxicity, classification: The chemicals which are evaluated for determination of toxicity are generally classified according to their LD_{50} values. There are six categories: (i) Extremely toxic: LD_{50} less than 1. (ii) Highly toxic: LD_{50} 1-50. (iii) Moderately toxic: LD_{50} 50-500. (iv) Slightly toxic: LD_{50} 500-5,000. (v) Practically non-toxic: LD_{50} 5,000- 15,000. (vi) Relatively harmless: LD_{50} more than 15,000 mg kg^{-1} body weight or µg kg^{-1} body weight.

Toxicogenic: Capable of producing a toxin; insects that are capable of injecting the toxin in the plant system during their feeding process are said to be toxicogenic.

Toxicognath: A poison jaw in the centipede.

Toxicological: Pertaining to poisons and their antidotes.

Toxicology: The branch of science dealing with the study of poisons in relation to living beings. It includes the study of resultant levels of toxicants after consumption of treated food by the animals, their long term effect on human beings and the environment including the pets, wild life and the ecosystems. Du Bois and Geiling (1959) defined toxicology as 'the branch of medical science that deals with the nature, properties, effects and the detection of poison. It is, therefore, the science of poisons.'

Toxin: Poisonous substance.

Trabecular layer: A row of cells bridging a cavity, a band or plate of fibrous tissue forming part of the internal supporting framework of an organ.

Tracer element: A radioactive element used in biological and other research to trace the fate of a substance or follow in the chemical reaction, such as the metabolic pathway or a nutrient or growth formation in the plants or the animals. These radioactive elements include carbon 14, calcium 45, cobalt 60, strontium 90 and phosphorus 32.

Trachea: A tube of the respiratory system, ending externally at the spiracle and terminating internally at the tracheoles. Morphologically, the tracheae are invaginations of the ectoderm. During its formation, the epicuticle of the trachea is thrown into regular spiral folds and the cells then secrete cuticular substance into these folds to form the spiral filaments or taenidia.

Tracheal commissures: Transverse tracheal trunks continuous from one side of the body to the other.

Tracheal end cell: End of a trachea that typically branches into several tracheoles.

Tracheal gills: A much more common mechanism by which the insect has become completely adapted for the respiration of dissolved oxygen, is the obliteration of all the spiracles and the development of tracheal gills.

Tracheal system: It is a system (respiratory system) of open tubes called trachea permitting direct contact with air and tissues through a number of openings known as spiracles. The tracheae are elastic tubes and invaginations of body wall which when filled with air present a silvery appearance. On the basis of functional spiracles (Kellin, 1944 and Hinton, 1947), it may be polypneustic (holopneustic, hemipneustic and peripneustc), oligopneustic (amphipneustic, metapneustic and propneusic), apneustic, hypopneustic and hyperpneustic type. See details.

Tracheoles: Fine terminal branches of the respiratory tubes or trachea. When the tracheal branches have reduced to about 2 μm in diameter, they break up abruptly to give rise to much finer tubes of 0.6-0.8 μm in thickness, termed as tracheoles.

Traction duster: A dusting machine which is mounted on the wheels to transport from one place to other and also receives its power from the wheels.

Traction sprayer: A spraying machine which is mounted on the wheels to transport from one place to other and also receives its power from the wheels.

Trade mark: It is a registered trade mark of a company manufacturing the product. See trade name.

Trade name (Trade mark name): Name given to a product sold by a company to distinguish it from similar products made by other companies (also trade mark name, proprietary name, brand name).

Trail parasitism: See trophic parasitism.

Trail pheromone: A substance laid down in the form of a trail by one animal and followed by another member of the same species.

Trail substance: Same as trail pheromone.

Transferring: The process of changing bees and combs from common boxes to movable frame hives.

Transient: A distinct individual mode in a developmental series, such as a line of descent corresponding in the time–character concept to a species in a taxonomic character.

Translaminar: Pesticide which passes from one surface of the leaf to the other, *i.e.*, lamina to lamina.

Translocated pesticide: The pesticide that is moved within the plant or animal system from the site of entry.

Translocation: Movement of pesticides or micronutrients within the plant system. It may be from roots towards the tip and margin of plants (acripetal or apoplastic), or downwards from leaves to other parts of the plants (basopetal or besipetal or symplastic. Translocation also occurs in animals and other living beings. The uptake of a pesticide into part of a plant and its subsequent dispersal to other parts of the plant body.

Translucent: Allows the light to pass through but not transparent.

Transovarial: Transmission of the pathogenic microbe from the mother to the offspring while they are still in the eggs in ovary.

Transovum: Refers to the transmission of the pathogen from one generation to the next by way of external contamination on or about the egg.

Transport host: The animal or plant which transports a pathogen to which it itself is not susceptible (a mechanical vector).

Transquilizer: Stupefying chemicals act to depress on animal's (rodent's) physical responses so that it can be readily caught and after a short period of recovery can be released unharmed. While it can be used in food, it has been used successfully in a dart gun that shoots a hypodermic needle into the muscle tissue.

Transverse: Crosswise.

Transverse costal vein: A cross vein in the coastal cell as in the Hymenoptera.

Transverse cubital vein: A transverse vein connecting the marginal and cubital veins as in the Hymenoptera.

Transverse folds of hind wing: Those more or less obliquely transverse folds which are involved in shortening the wing length.

Transverse groove of metaventrite: A groove partly or completely separating the anterior portion of metaventrite (paired pre-episternites) with the posterior portion (metakatepisternum). It is more or less complete in Archostemata but is usually more or less reduced to a short groove crossing the discrimen or completely absent.

Transverse marginal vein: A cross vein in the marginal cell as in the Hymenoptera.

Transverse median vein: A cross vein between median or discoidal vein and the anal vein as in Hymenoptera.

Transverse radial vein: Same as transverse marginal vein.

Transverse suture: A suture across the mesonotum as in the Diptera.

Trap crop: A crop planted around the main crop or in rows at a definite distance in between the main crop which serves to attract the insect pests. To save the main crop, the trap crop is latter either destroyed or sprayed with insecticides to control the insect pests.

Travel stain: The dark discoloration on the surface of comb honey left on the hive for some time, caused by bees tracking propolis over the surface.

Treated area: A building, forest, garden, field or other place where the pesticide has been applied.

Treatment: A factor varied in an experiment such as the rate of insecticide; any stimulus supplied artificially or forced upon an organism or place.

Trial use: The notation indicates that this material is worthy of trial use on a portion of grower's acreage.

***Trialeurodes ricini* Misra:** Belongs to order Hemiptera and family Aleyrodidae. It is an insect pest of castor, commonly known as the castor whitefly.

Triangle: Triangular region near the base of a dragonfly wing.

Tribe: A subdivision of subfamily containing a group of closely related genera. The names of tribe ends with 'ini'. A taxonomic division between genus and subfamily.

***Tribolium castaneum* Herbst:** Belongs to order Coleoptera and family Tenebrionidae. It is an insect pest of stored grains and other products, commonly known as the rust red beetle.

Trichlorfon: An insecticide marketed under trade name Dipterex is an organophosphatic compound, white substance soluble in aromatic hydrocarbons. Chemically, it is dimethyl (2,2,2-trichloro-l-hydroxyethyl) phosphonate ($C_4H_8Cl_3PO_4$); moderately toxic to mammals, acute oral LD_{50} for rats is 450-500 mg kg^{-1}; shows phytotoxicity in degrees to some apple varieties. This insecticide is especially effective against Lepidoptera, Diptera, Hemiptera, cockroaches etc. and in sugar bait form to the flies. It is also used in the baits; available as 5 per cent dust and 80 per cent wettable powder and marketed as Dipterex, Tugon and Dylox.

Trichobothria: In Hemiptera, specialised hairs on ventral abdominal segments (Miridae).

Trichocard: It is a paper card of 15 cm x 7.5 cm size perforated into 36 equal parts (2.5 cm x 1.25 cm) and has capacity to withhold 1 cc eggs of rice meal moth, normally sprinkled evenly over it for rearing the *Trichogramma* spp. The details of parasitoid handling, producer, supplier date and target species *etc.* are optional to print overleaf or reverse side.

Trichodectidae: A family of suborder Ischnocera of order Mallophaga. Parasitic on domestic animals. Antennae three segmented. Example, *Trichodectes bovis* on cattle, *T. ovis* on sheep and *T. canis* on dogs.

Trichodonthria: Minute sensory hairs on the tarsi of spiders.

Trichogen: An epidermal cell that generates seta.

***Trichogramma evanescens* Riley:** Egg parasitoid (Trichogrammatidae, Hymenoptera) associated with citrus caterpillar, *Papilio demoleus* Linn.

Trichogrammatidae: Family of super family Chalcidoidea, suborder Apocrita and order Hymenoptera. These are very minute insect measuring 0.3 to 1.0 mm long and characterized by three segmented tarsi and broad forewings with rows of microscopic hairs on them. They are egg parasites of a large number of insects especially Lepidoptera and *Trichogramma* sp. have been extensively utilized in biological control of crop pests.

Trichoid sensilla: See tactile organs.

Trichome: A tuft of long hairs associated on glandular areas on the body surfaces of many myrmecophilous beetles. The hairs are believed to aid in the dissemination of attractants.

Trichopore: The opening of the cuticula beneath the seta giving passage to the hair forming process of the trichogen.

Trichoptera: The insect order. The trichoptera includes the caseworm or the caddisflies, which are moth like, mostly nocturnal insects with weak powers of flight. The adults frequent the vicinity of water, and case bearing larvae and the pupa are aquatic. The head is usually small, with moderately large widely separate compound eyes and three ocelli.

Tridactylidae: A family of super family Tridactyloidea of order Orthoptera and suborder Caelifera. Live in water and burrowing into sandy areas.

Tridactyloidea: A super family of order Orthoptera and suborder Caelifera. Tarsi one or two segmented.

Trifoliate: Leaf that has three distinct leaflets.

Trigger factor: A factor, change or alteration in which sets off a chain reaction in an ecosystem.

Trigonal brace: A strong forked vein in the basal part of the wing, formed by fork of Cu_1 in front wing and of M_{3+4} in hindwing.

Trigonaloidea: Super family of suborder Apocrita, order Hymenoptera. This includes a single family Trigonalidae. These insects are characterized by a trochantellum on hindfemur, multiarticulate antennae with more than 20 segments, absence of axillary or anal lobe on hindwing, non-petiolate abdomen and vestigial ovipositor. Some of the Indian forms are: *Ischnogonalos dubia, Lycogaster rufiventris, Poecilogonalos pulchella, P. kerala* and *Pseudogonalos harmandi.*

Trilobate aedeagus: An aedeagus consisting of a well developed phallobase, paired articulated parameres and a sclerotized penis. This type of aedeagus occurs in many primitive Polyphaga.

Trimedlure: A synthetic attractant for the Mediterranean fruit fly; tert-butyl 4(or 5)-chloro-2-methylcyclohexane-carboxylate.

***Trinervitermes biformis* (Wasmann):** A species of termites found in India and Ceylon. It is generally subterranean.

Triphasic allometry: Polymorphism in which the allometric regression line, when plotted on a double logarithmic scale, breaks at two points and consists of three segments. In ants, the two terminal segments usually have slight to moderately high slopes and a middle segment has a very high slope.

Tripolite ($SiO_2.4H_2O$): It is a mineral consists mainly of amorphous opaline silica. The powder is pale yellow in colour. It is used as inert carrier in the insecticide dust formulations.

Trithion: See carbophenothion.

Tritocerebral neuropile: Each lateral tritocerebral neuropile body or lobe is composed of mixed fibres, the motor and the sensory. Both the lobes are connected with each other by the tritocerebral transverse commissural tract which runs internally or in some insects externally. It is a site of the mixed nerves, the labrofrontal nerves.

Tritocerebrum: It is the hind or ventral part of brain representing the ganglion of the intercalary segment. It gives out a pair of labrofrontal nerves (nervilabrofrontalis) which soon divide into two branches, the labral nerve innervating the labrum and the frontal connective terminating into the frontal ganglion. The tritocerebral lobes are connected with the suboesophageal ganglion (first ganglia of the ventral nervecord which lies ventral to the brain in the cephalic region) by the circumoesophageal connectives.

Triungulin: Name given to the active first instar larva of meloid beetles and some of their relatives: they appear to have three claws on each foot.

Trochanter: The second segment of leg, between the coxa and the femur. Except Odonata (two trochanters present), it is a single, short triangular or quadrangular segment lying in between the coxa and the femur in all insects. In parasitic species of Hymenoptera, the anterior part of femur is divided and this part sometimes is termed as trochentellus.

Trochanteral: Relating to the trochanter.

Trochantin: It is a triangular sclerite well evident in leg of cockroach. It is very close to the ventral articulation of the coxa. In the higher insects, the trochantin is fused with the antecoxal piece.

Trochantinal: Relating to the trochantin.

Troctomorpha: A suborder of Psocoptera. Tarsi three segmented, antennal segments with annulations and with 12-17 segments and with two segmented labial palpi.

Trogidae: Family of superfamily scarabaeoidea, suborder Polyphaga and order Coleoptera. These beetles occur in dry places feeding on carrion or dung. The larva lacks stridulatory files. *Trox indicus* and *T.omacanthus* are some Indian species.

Trogiidae: A family of order Psocoptera and suborder Trogiomorpha. Claws not toothed, body and wings without scales. Head transverse, maxillary palp with sense organ. Example, *Lepinotus* and *Trogium*.

Trogiomorpha: A suborder of Psocoptera.Tarsi three segmented, antennal segments without annulations and with more than 20 segments and with two segmented labial palpi.

***Trogoderma granarium* Everts:** Belongs to order Coleoptera and family Dermestidae. It is an insect pest of stored grains and other products, commonly known as the khapra beetle.

Trophallaxis: The phenomenon of exchanging the alimentary liquid among colony members and guest organisms, either mutually or unilaterally. In stomodaeal trophallaxis, the material originates from the mouth; whereas, in proctodaeal trophallaxis, it originates from the anus.

Trophic: Pertaining to the nourishment or feeding.

Trophic egg: An egg, usually degenerate in form and inviable, which is fed to other members of the colony.

Trophic level: Placement or grouping of all organisms which secure their food at a common step away from the first level.

Trophic nerves: In all the nerve cells, the products of the cell body including the mitochondria, move slowly down the axon towards its termination. It is generally supposed that this movement serves trophic or nutritional functions and is needed to maintain the axon in a healthy state.

Trophic parasitism: The intrusion of one species into the social system of another, *e.g.*, utilization of trail system in order to steal food.

Trophocytes: The cells of the fat bodies having a trophic in distinction from the urate cells.

Trophogenesis: The origin of different caste traits from differential feeding of the immature stages.

Tropism: The orientation of an animal with respect to a stimulus, either positive or negative.

Tropotaxis: Sometimes, the insect is able to compare the intensity of stimulus on two sides simultaneously, without movement of the sense organs or of the body; this less common mechanism is termed as tropotaxis. See also, klinotaxis, anemotaxis and telotaxis.

Truly social: See eusocial.

Truncate: Ending abruptly; squared off.

Trunk: The entire series of body segments of an arthropod including the cephalic, thoracic and abdominal.

Tsetse fly: *Glossina* sp. (Muscidae, Diptera) is a pest of medical importance in African conntries. The *Glossina gachinoidsis* is found in the Southern Arabia. These blood sucking flies are often vectors of pathogenic *Trypanosomas* that cause sleeping sickness in human beings and trypanosomiasis in livestock. Tsetse flies are viviparous.

T-test: A statistical treatment for determining the significance of different values about the effectiveness, toxicity etc. of various chemicals against one or more strains of any insect or test species.

Tuber: Swollen underground stem of a plant often used as food, *e.g.*, potato.

Tubercle: A small knob like or rounded protuberance.

Tubulufera: Suborder of Thysanoptera. Female without ovipositor, last abdominal segment in both sexes mostly tubular. Wings without pubescence and venation.

Tur pod bug: See *Clavigralla gibbosa.*

Turbidity: Suspended matter in the water which prevents light penetration.

Turgidity: Firm and tense condition by distension with water.

Turgor muscles: The muscles situated in the body wall which maintain the tense form by tonic contraction in the soft bodied insects like caterpillars (See tonic contraction also).

Turgor pressure: Pressure exerted by the cell sap on the cell wall.

Turpentine: The oleoresin obtained from various pine species; source of turpentine oil; active ingredient of some insecticides.

Tussar silk: A copperish silk obtained from tussar silkworm cocoons.

Tussar silkworm: *Antheraea mylitta*, a silkworm species producing tussar silk; wild in nature, feeding on leaves of sal, arjun and other forest trees.

Tvo: Tractor vapourizing oil (kerosene).

***Tylenchulus semipenetrans* (Cobb, 1913):** Belongs to Tylenchida and family Tylenchulidae. It is commonly known as citrus nematode and is a pest of international importance as it is associated with citrus decline.

Tymbal: A sclerotized plate in the sound producing organ of cicada.

Tympanal hood: A pair of tubercles or round prominences on the dorsal surface of first abdominal segment of Lepidoptera.

Tympanal organ: Auditory organ on the abdomen or legs of certain insects. See tympanum.

Tympanum (pl., tympana): A vibrating membrane (for sound production) or a membrane for use as an auditory organ (eardrum). The most highly developed auditory organs which occur in Tettigoniids (Orthoptera), Lepidoptera etc., consist of batteries of specially modified scolopidia stimulated by the movements of a membranous ear drum or tympanum. In these insects, the auditory organs become much more sensitive to sounds of higher frequencies and their range of perception extends into frequencies far beyond those which the human ear can detect.

Types: Specimen a species or subspecies designated when a species or group is described to serve as the reference if there is any question about what that species or group includes. The type of a species or subspecies (holotype) is a specimen; the type of a genus or subgenus is a species, and the type of a tribe, subfamily, family or superfamily is a genus.

Typhlocyba maculifrons **Motsch:** Belongs to order Hemiptera and family Cicadellidae. It is an insect pest of cereals and millets, commonly known as the rice blue leafhopper.

***Tyrophagus asiatius*:** It is a species of mite infesting the mango.

U

Ubiquitous: (i) Able to thrive in different conditions or divergent places, (ii) Widespread.

Ultra low volume spray (ULV spray): Signifies that the total volume rate of spray application is very low (5 l ha^{-1} or less). ULV pesticide formulations are specially developed for the purpose and are applied either undiluted or with very low volume of water.

Ultra-hazardous: A very dangerous activity.

Ultraviolet rays: Electromagnetic radiations in the wavelength range $4x10^{-5}$ cm to $5x10^{-7}$ cm approximately, covering the gap between the X-ray and the visible light rays. The rays of the solar spectrum beyond the violet end.

ULV spray: See ultra-low volume spray.

Uncapping knife: A knife used to shave or remove the capping from combs of sealed honey prior to extraction; then usually heated by steam or electricity.

Uncertainty factors: Factors in toxicological assessment for extrapolation of data from experimental animals to man (assuming that man may be more sensitive) or from selected individuals of the general population. For example, an uncertainty factor is generally applied to the no observed effect level to derive an acceptable daily intake.

Uncus: A solid process arising from the outer edge of the tibial apex and either hooked, curving inwardly or sometimes extending across the apex and projecting inward.

UNDP: United Nations Development Programme.

Ungues: Lateral claws of the foot secondarily developed from the base of dactylopodite, characteristic of insects and several other arthropods.

Unguifer: Median dorsal process on the end of tarsus to which the pretarsal claws are articulated.

Unguitractor plate: A lateral scerite at the base of pretarsus upon which the depressor muscle of pretarsus or retractor of the claws is inserted by a long tendinous apodeme.

Unicolonial: Pertaining to population of social insects in which there are no behavioural colony boundaries. Opposed to multicolonial.

Uniform coverage: Application of insecticide evenly over a whole area, the plant or animal.

Uniramose antenna: The presence of a single ramus on each antennomere.

Uniramous: Having simple undivided legs.

Unisexual: Of one sex only, *e.g.*, aphids where asexual females (parthenogenetic) occur.

Uniting: Combining two or more colonies to form a larger colony.

Universal antidote: Incase of poisoning by insecticide in man, it is always advisable to give universal antidote (activated charcoal 2 parts, magnesium oxide 1 part and tannic acid 1 part; 15 g of mixture in ½ glass of water) through ingestion immediately followed by gastric lavage. To induce vomiting 15 g sodium chloride with warm water can be given. To control tremors, 10 ml of calcium gluconate 10 per cent solution is injected intravenously.

Univoltine: Having one generation per year

Unsclerotized: Not hardened by the deposition of sclerotin.

Uraniidae: Family of super family Geometroidea, suborder Ditrysia and order Lepidoptera. Mainly tropical species which are brightly coloured. Species belonging to the genera *Chrysiridia, Nyctalacmon* and *Urania* resemble papilionid butterflies and the others resemble geometrid moths.

Urate cells: Some fat body cells store the granules of uric acid. In the absence of Malpighian tubules, these cells perform the role of excretory structures as seen in Collembola. In *Lepisma* sp. and higher insects, such as the nymphs of cockroaches, locusts, larvae of honey bees and ants, these urate cells act as store houses for the excretory substances, particularly, the uric acid granules. They become evident since embryonic condition. During post-embryonic growth, these gradually increase in size and in the uric acid contents. In *Apis* sp., uric acid is absorbed by the urate cells during larval and pupal stage while Malpighian tubules absorb the uric acid only in the adult stage.

Urates: See uric acid.

Urea: The insect urine often contains a small quantity of urea 0.3-0.4 per cent of dried excreta in *Melanoplus* and about 10 per cent of total nitrogenous waste products in the mosquitoes.

Urentis euonymus **Fab.:** Belongs to order Hemiptera and family Tingidae. It is an insect pest of ornamental plants, commonly known as the hollyhock tingid bug.

Urentius sentis **Dist.:** Belongs to order Hemiptera and family Tingidae. It is an insect pest of brinjal, commonly known as the brinjal lace-wing bug.

Uric acid: It is the nitrogenous end product of excretion in terrestrial insects in which for its discharge, water conservation is very essential. It requires little quantity of water from the excretory system. The uric acid in scarcity of water is excreted in the form of crystalline spherical granules. The urates may be associated with the free uric acid which represents ammonium urate, calcium urate or sodium urate and potassium urate. The uric acid is derived from metabolism of purines (adenine, guanine) and/ or from the proteins by utilizing glycine, glutamine and aspartate as substrates. Adenine→ Hypoxanthine → Xanthine→ Guanine → Uric acid.

Urothripinae: Subfamily of family Phlaeothripidae, suborder Tubulifera and order Thysanoptera. Distance between hind coxae greater than between fore and mid coxae. Anal setae exceptionally long, *e.g.*, *Stephanothrips* sp.

Urticating setae: Hair or setae connected to poisonous glands through which the poison comes out and causes mechanical irritation in the victim on contact.

Uterus: A compartment of the genital chamber or vagina in which the embryonic and a part of the post embryonic development of the young insect may take place.

***Utethesia pulchella* Linn.:** Belongs to order Lepidoptera and family Arctiidae. It is an insect pest of jute, commonly known as the sunnhemp hairy caterpillar.

Uzi flies: See silkworm pests.

V

Vacor: Chemically, it is N-3-pyridylmethyl N′-P-nitrophenyl urea. It is an acute rodenticide, produces quick and dramatic physiological disorders in the target species and hence the bait intake is relatively poor. It affects the central nervous system leading to paralysis of the hind quarters and pulmonary arrest and death is preceded sometimes by convulsions. It is light yellow, odourless powder with a poor solubility in water. It is specific to some extent in action. For the gerbils, the poison bait containing 0.5 per cent poison, 1.0 per cent oil and 98.5 per cent cereal grain is effective.

Vacuum fumigation: The primary object of vacuum fumigation is to hasten and improve the penetration of the fumigant into the material undergoing treatment. It was originally developed for this purpose when hydrogen cyanide was principal fumigant. In cacuum fumigation, most of the air in the chamber is removed before the fumigant is introduced. It is therefore, necessary to have a specially designed chamber, usually made up of steel which is capable of withstanding external pressure of atmosphere. The installation also includes a vacuum pump able to evacuate the chamber in not more than 10-15 minutes and valves and pipelines for introduction and exhaust of the fumigant.

Vagina: It is also called genital chamber, the site of coitus and fertilization. It receives the spermatheca, median oviduct and bursa copulatrix and opens to exterior through a well defined opening, the female gonopore or vulva. In the viviparous Dipteran flies, the anterior part of vagina is modified into a uterus for the larval development. The vagina, bursa copulatrix and spermathecae are the ectodermal structures, developed together from the epidermal invagination of the midventral region of the eighth abdominal sternum. See median oviduct.

Validation: The process for establishing that an analytical method or equipment will provide reliable and reproducible results in pesticide analysis.

Valone: See PMP.

Valve: One of the paired components of ovipositor.

Valvifers: The basal plates of the ovipositor, probably derived from the coxopodites of gonopods carrying the valvulae, including the first valvifers of the eighth abdominal segment and second valvifers of the ninth segment.

Valvula (pl., Valvulae): (i) A small valve or fold. (ii) The paired processes forming the blades and ensheating lobes of the ovipositor. The first and second valvulae are gonapophyses of the gonopods, the third valvulae are distal out growths of the coxopodites of ninth abdominal segment.

Vamidothion: A systemic organophosphate insecticide and acaricide marketed under trade name Kilval; chemically, dimethyl S-2-1 methyl carbamoylethylthilo ethyl phosphorothiolate ($C_8H_{18}O_4NPS_2$); white crystalline solid material. The technical grade having melting point 33-36 °C; highly soluble in water and acetone but less soluble in hexane, insecticidal properties are like methyl demeton but having long residual effect. In plants, it is metabolized to sulfoxide. Hydrolysis also occurs within the plants with the formation of dimethyl phosphoric and phosphoric acid.

Vannal region or vannus: The wing area containing the vannal veins, or veins directly associated with the third axillary; when large, usually separated from the remigium by the plica vannalis; often forming an expanded fan like area of the wing.

Vannal vein: The veins associated at their bases with the third axillary sclerite and occupying the vannal region of the wing.

Vannus, vannal area: The expanded anal area in the hind wing of many insects.

Vapona: Trade name of dichlorvos (DDVP).

Vapour: Gas steam, mist, fog or fume.

Vapour drift: The movement of vapours of a volatile chemical from the area of application.

Vapour pressure: The property which causes a chemical compound to evaporate; the lower the vapour pressure, the more volatile the compound is.

Variety: Heterogenous group of a taxonomic taxon having varied phenotype; it is now not recognized by the commission.

Vas deferens (pl., vasa defferentia): The follicles of each testis open independently through short vas deferens. It receives the sperms or sperm bundles from the testes and facilitates their transport to the ejaculatory duct. The two vasa differentia join together to form median ejaculatory duct. In some insects, like grasshopper, the posterior part of this duct enlarges to form seminal vesicle for storage of sperms. In Diptera and Dermaptera, there is single common seminal vesicle followed by union of two ducts.

Vas efferens (Vasa efferentia): From each testicular follicle, a tiny duct, the vas efferens (pl., vasa efferentia) leads to a common duct, vas deferens.

Vascular tissues: The fluid conducting tissues of a plant both xylem (water) and phloem (food).

Vasopressin: A hormone produced by the posterior lobe of the pituitary which is antidiuretic and also causes contraction of the muscles, especially arterioles. See also the vertebrate hormones in insects.

Vector control: Method of controlling a disease transmitting organism.

Vectors (sing., Vector): (i) An insect which is able to transmit viruses or other pathogens. The mosquitoes, sand flies, black flies, house fly, tsetse fly, human flea, human louse etc. are transmitting pathogens and causing various diseases in man; whereas, the whiteflies, thrips, aphids etc. transmit disease causing pathogens in plants. Female mosquito, *Anopheles stephensi* is a carrier of malarial parasite, *Plasmodium vivax*. (ii) Vectors are the DNA molecules with the ability to replicate independently in a host cell. Desired DNA fragments (also known as DNA inserts) can be integrated in these DNA molecules. The important examples of vectors are plasmids, phage and virus.

Vedalia beetle: See *Rodolia cardinalis*.

Vegatal pole: The part or region of the egg where metabolic activity is minimum.

Vegetative parts: Different parts of the plants, like roots, stem, leaves, branches etc.

Vegetoil: Mixture of vegetable oils; an emulsifier used to improve tank mixing in pesticidal sprays.

Veil: Bee veil. It is a light metal screen or diaphanous cloth material draped on the head and face when working with the bees to prevent being stung on the neck or face.

Veins: In insects, the ribs like tubes that strengthen the wings or thickened lines in the wings of insects. These are two types, the longitudinal and cross veins. Longitudinal veins: Costa (C), Subcosta (Sc), Radius (R), Media (M), Cubitus (Cu), Postcubitus (PCu), Anal/ Venal (A/ V), Jugal (J); Cross veins: Humeral (h), Costal (c), Radial (r), Sectorial (s), Radio-medial (rm), Intermedial (m), Medio-cubital (m-cu), Intercubital (cu), Inter-anal (a).

Veina arcuata: Refers to first jugal vein.

Vellidae: Family of superfamily Gerroidea, division Amphicorisae of suborder Heteroptera. Body elliptical, stout, rostrum three segmented, middle and hind legs similar to forelegs, hemelytra membranous, gregarious in habit, water crickets, *Velia*, *Rhagovelia* etc.

Vena cardinalis: Second jugal vein, usually appearing as the basal branch of vena arcuata.

Vena dividens: Secondary vein present in some orthoptera lying in the fold between the remigium and vannus.

Vena spuria: A vein like fold or thickening in the wing of Syrphidae, Diptera.

Vena triplica: Of Neuroptera, where Sc, R and Rs run parallel for a considerable length and are connected by short cross veins.

Venation: The arrangement of veins in the wings of insects.

Venom allergy: A condition in which a person, when stung by honey bee, may experience a variety of symptoms ranging from a mild rash or itchiness to anaphylactic shock. A person who is stung and experiences abnormal symptoms should consult a physician before working bees again.

Venom hypersensitivity: A condition in which a person, if stung, is likely to experience an aphylactic shock. A person with this condition should carry an emergency insect sting kit at all times during warm weather.

Venter: Lower part or lower surface of the abdomen.

Ventillation tracheae: Tracheae with collapsible walls, responding to varying surrounding pressure.

Ventrad: Towards the ventral or under side of the body.

Ventral: Referring to underside of the body.

Ventral diaphragm: A muscular membrane present in the ventral part of the abdomen of some insects.

Ventral lobe: The lower lobe of a penis divided horizontally into dorsal and ventral lobes. See Pedon.

Ventral longitudinal trunk: See longitudinal trunk.

Ventral nerve cord: The chain or connected ventral ganglia morphologically beginning with tritocerebral ganglia of the brain; in insects applied to the thoracic and abdominal ganglia.

Ventral sinus: The space of definitive body cavity below the ventral diaphragm, containing the nerve cord.

Ventral trachea: The ventral segmental trachea originating at a spiracle.

Ventral tube: A tube proceeding from the ventral side of first abdominal segment as in Collembola.

Ventriculus: Refers to the midgut or mesenteron.

Ventrite: See Abdominal ventrite.

Ventrolateral process: One of a pair of processes extending basally or posteriorly from the centre of the metendosternite bear the bases of the lateral arms.

Vermicide: Pesticide to kill worms.

Vermicomposting: Vermicomposting is the composting of organic waste into biofertilizers by earthworms. Earthworms are natural bioreactors, having specific body organs that act as a mill and grinds the waste ingested by the worms. The gut of the earthworms is such that it provides favourable physiochemical conditions for efficient growth of microorganisms that carry out biodegradation of the waste. The end product biomanure is much better with regards to its fertility as compared to all the composting processes.

Vermiculate: Worm eaten; with irregular sinuous markings.

Vermicule: Worm like.

Vermiform larva: A legless (apodous), headless (acephalic) worm like larva typical of some Diptera.

Vermin: Pests; usually rats, mice and insects.

Vernacular name: Native name of an insect species in a country or place for a scientific name and there may be many such names for the same taxon and so causes confusion.

Verson's gland: It has been observed that when the silkworm casts its skin, its body may be powdered with crystals of oxalate and uric acid derived from the moulting fluid; and this led to the suggestion that at the time of moulting, the Malpighian tubules are thrown temporarily out of action, and their excretory function is taken over by the dermal gland (Verson's gland). The cement layer is secreted by dermal glands or so called Verson's gland in the Lepidoptera.

Versonian cell: See apical cell.

Vertebrate hormones in insects: The vertebrate hormones detected in insects are insulin, glucagons, somatostatin, gastrin, pancreatic polypeptide, substance-P, vasopressin, neurophysin, enkephalin and prostaglandins. Somatostatin is a tetradecapeptide found in the brain and digestive tract of vertebrates exhibiting negative control of gastrointestinal hormone release, blockage of water and chloride secretion in colon and increased absorption of sodium chloride across ileal border. It causes inhibition of c-AMP production or activation of phosphoprotein phosphatase. Gastrin is a heptadeca or larger peptide in vertebrates that causes increase in secretion of acid containing gastric juices. Brain and prothoracic ganglionic neurones are the site of these peptides in insects. Pancreatic polypeptide in vertebrates occurs in pancreas and brain and inhibits insulin release. It is detected in four species of insects from the brain, mid gut and ventral ganglia. Substance-P is an undecapeptide extracted from brain, intestine, neural retinal tissue in vertebrate tissue in vertebrate species. It may function as hypotensive, sialogogic and contractile factor. It is detected from two species only, *i.e.*, hover fly and hornworm with unknown function. Vasopressin is a nonpeptide present in the vertebrate neurophysical extracts. It acts as an antidiuretic hormone by altering plasma osmotic pressure and also facilitates long term memory. Neurophysin is a large neuroseccretory protein found in the hypothalamus of vertebrates and is known to be involved in the transport and storage of certain hormones. It is detected in the nervous tissue, haemolymph and rectal tissue of locust and cricket. Enkephalin is pain killing neurotransmitter peptide found in the brain and intestinal nerves of contractile muscles in mammals. They are in pars intercerebralis, corpora pedunculata, ventral ganglia, optic lobes, Malpighian tubules and oviducts of locust and hornworm. The prostaglandins are lipids, fatty acids, 20- car bon unsaturated carboxylic acid with a cyclopentane ring. They have been detected in almost every tissue and fluid of the vertebrates. They mostly bind hormones to the membranes. Various types of prostaglandins, PGE1, PGE2, PGF1, PGF2 are detected from testis, seminal vesicle, spermatophores and gonads in house cricket, *Acheta domesticus*.

Vertebrates: Animals having spinal column or backbone, such as Amphibians, fishes, birds, mammals etc.

Vertimec: See Abamectin.

Vertex: The area on the dorsal part of the head between and behind the compound eyes.

Vertical infestation: A number of coleopteran beetles and weevils are capable of migrating upward or downward among the grain bulk irrespective of the pressure. Thus, infestation commences on the top of the bulk grains may migrate to the bottom and *vice-versa* which is known as vertical infestation.

Vertical resistance (Gene-for-gene resistance): Resistance governed by one or more genes in a host, each of which corresponds to a matching gene for parasitic ability in the pest species.

***Verticillium lecanii*:** A species of entomopathogenic fungi causing disease in many insect species. It is marketed under trade name, Vertilec for the management of insect pests.

Vesica: A terminal membranous part of the aedeagus.

Vesicle: A bladder, sac or cyst.

Vesicula seminalis: A dilation of the vas deferens in which the spermatozoa are retained.

***Vespa orientalis* Linn.:** Belongs to order Hymenoptera and family Vespidae. It is commonly known as yellow wasp which is a social insect and is a minor pest of grapevine.

Vespidae: Family of super family Vespoidea, suborder Apocrita and order Hymenoptera, *e.g.*, social wasps with red or yellow body with black markings. These are yellow or red insects with black markings and compromise social as well as solitary wasps. They usually have a long slender petiole and the forewings possess a very long first discoid cell. At rest the wings are usually folded longitudinally. Anal lobes are absent in hindwings. The conical sessile abdomen in the female has at its apex the ovipositor. Generally, these insects are predaceous on caterpillars which are paralyzed by stinging, collected and stored in the cells of their nests. In each cell a number of larvae are stored and an egg is suspended by a filament from the top of the cell, which is sealed subsequently.

Vespoidea: Super family of Suborder Apocrita, order Hymenoptera.

Vestibulum: An external genital cavity formed above the seventh abdominal sternum when the latter extends beyond the eighth.

Vestigial: Small, poorly developed or non functional.

Vestiture: A general term applying to the body covering, including all forms of setae (*e.g.*, hairs, scales etc.).

Veterinary toxicology: It is a branch of toxicology which deals with such poisons which deals with the effect of poisons on domesticated animals.

Viability: The seed which have ability of sprouting or germinating.

Vibrissae: The pair of large bristles just above the mouth in certain flies; usually called vibrissae. These are also stiff hairs located about the face, especially the nostrils of many mammals and often serve as tactile organs.

Victim: Who is injured; poisoned or hurt in any way.

Vinculum: The coxosternal plate of the ninth abdominal segment in Lepidoptera.

***Virachola isocrates* (Fab.):** Belongs to order Lepidoptera and family Lycaenidae. It is an insect pest of pomegranate, tamarind etc., commonly known as pomegranate butterfly.

Virga: A terminal phallic spine arising from the endophallus.

Virgin queen: An unmated queen.

Virion: A virus particle.

Virulence: Refers to the disease producing power of the microorganism, *i.e.*, relative capacity of microorganism to invade and injure the tissues of its host (Steinhaus and Martignoni, 1967).

Virus: A submicroscopic pathogen which requires living cells for growth and have ability to cause disease in plants and animals. Plant viruses are spread by insects.

Viscera: The internal organs in the body cavity of insects, particularly in abdomen.

Visceral: Pertaining to the internal organs, especially one of the large abdominal organs.

Visceral muscle: A muscle which invests an internal organ.

Visceral nervous system: See sympathetic nervous system.

Visceral trachea: The medina segmental trachea originating at the spiracle, branching to the alimentary canal, the fat tissues and the reproductive organs.

Viscus (pl., viscera): Any of the internal organs of the body.

Visual organs: The true eyes or true visual organs are found in almost all insects except few parasitic forms and are of two types, the simple and compound. The simple eyes include the ocelli which are generally three in the upper part of head in nymphs and adult insects and stemmata which are lateral groups of ommatidia in holometabolous larvae. The compound eyes are composed of large number of ommatidia in large number of insects and principal visual organs of the adult insects except Siphunculata, Siphonaptera, female coccids and some cave dwelling insects.

Vitamin A: Fat soluble accessory substances of the vitamin A type are not required by the insects in large amount but retinene in the visual purple (rhodopsin) of the eye is a derivative of vitamin A. If mosquito, *Aedes* is reared for a generation on the diet containing neither vitamin A nor its precursor beta-carotene, the structure and function of the retina is impaired.

Vitamin B group: The requirement of vitamin B in insects is strikingly similar with that of mammlals. Although, there are some specific variations, thiamin (B_1), riboflavin (B_2), nicotinic acid, pyridoxine (B_6) and pantothenic acid are indispensable.

Vitamin C: See ascorbic acid.

Vitamin D: See calciferol.

Vitamin E: It is essential for the production of viable offspring as in the Sarcophagid fly, *Agria*.

Vitellarium: It is the third and a large part of the ovariole and is characterized by a large number of growing oocytes arranged in a linear fashion. Each developing oocyte is enclosed in a follicle of the epithelial cells and is separated from each other by the interfollicular tissue. In some insects, each oocyte is provided with a group of nutritive cells or trophocytes (polytrophic ovarioles). The oocyte development occurs in anteroposterior direction inside the vitellarium and finally the fully grown ova are provided with two membranes, outer chorion and inner vitelline. It is the site of vitellogenesis.

Vitelline: It is the major yolk protein. Precursors of this oocyte protein are found in the blood stream, called vitellogenins.

Vitelline membrane: Besides the chorion laid down by the follicular cells, the egg is enclosed by inner membranes which are deposited by the oocyte itself. These may amount to no more or less delicate vitelline membrane beneath or the blastoderm of the developing embryo may lay down a tough chitinous layer, similar to the endocuticle of the free living insects which often form the most substantial layer of the egg.

Vitellogenesis: Synthesis and deposition of yolk in the oocyte is called vitellogenesis. It is a complex process utilizing substances from the haemolymph and is governed by the endocrine system.

Vitellogenins: Precursors of oocyte proteins are found in the blood stream, called vitellogenins.

Vitellophages: In some insects, all energids do not migrate to the periphery to form the blastoderm but some of them lie behind within the yolk and are called the yolk cells, merocytes or vitellophages. They initially multiply mitotically and after six or seven divisions, they increase in size through endomitosis. They reported in various species of Diptera and other insects. The vitellophages carry out the breakdown of yolk and are incorporated in the midgut epithelium.

Vitreous body: See crystalline body.

Viviparae: Female aphids which bear living young ones.

Viviparity, viviporous: The production of young ones rather than laying eggs, *e.g.*, in aphids. In psuedoplacental viparity, the egg is deficient in yolk, and usually lacks a chorion. It develops in an enlarged part of the vagina, being nourished partly by close contact and embryonic and maternal tissue. Some Blattodea, Dermaptera, Psocoptera, and Hemiptera reproduce this way.

Viviparous: Giving birth to living young ones, either larvae or nymphs, not eggs; for example, Aphididae, Strepsiptera, Diptera and parasitic Hymenoptera. (i) Ovoviviparity: The eggs are retained in the genital tract so that when they are laid, the embryo have reached advanced stage of development with the result that larvae hatch immediately after oviposition, *e.g.*, *Cimex*. (ii) Adenotrophic viviparity: In the tsetse fly, *Glossina* sp. and in the Pupipara, the larva which hatches in the uterus from an egg of normal size, is nourished until fully grown by special milk glands- modified accessory glands. These larvae are matured singly, breathing takes place by the extrusion of their posterior spiracles through the female opening; they are not deposited until they are ready to pupate. (iii) Pseudoplacemental viviparity: In the initial stage, the eggs are nourished by a single nurse cell, as the embryo reached in advanced stage, the follicular chamber also enlarges. At this stage, the pseudoplacentae also formed which remain in close contact with the developing embryo and nourished by this structure, *e.g.*, family Aphididae. These are the most familiar of the viviparous insects.

Volatile: Capable of rapid evaporation at ordinary temperature.

Volatilisation: Evaporation of a pesticide into the atmosphere from a solid or liquid form.

Volatility: The ability of a solid or liquid to evaporate quickly to ordinary temperature when exposed to the air.

Volsellae: The ventral lateral processes (parameres) of the phallobase in Hymenoptera.

Voltinism: The number of generations per year.

Volume rate: Amount of liquid formulation generally applied per unit area for effective coverage, expressed in $l\ ha^{-1}$. In general the water recommended for ultra low volume, low volume and high volume sprays are 0-5, 12-120 and 450-1000 $l\ ha^{-1}$ in field crops.

Volume: The space occupied by the solid, liquid or gas.

***Voria ruratis* Fab.:** Larval parasitoid of diamond back moth.

Vulva: See median oviduct.

Vulvar lamina: The posterior margin of eighth abdominal sternite.

Vydate: Trade name of oxamyl.

W

Waggle dance: The dance made by honeybee (*Apis* sp.) workers indicating the location of food and new nest sites. The dance is made through an eight line figure with the middle transverse line of the eight containing information about the direction and distance of the food.

Waiting period: Also regarded as the safety period; it is the interval between the treatment of pesticide and residue reaches to tolerance limit.

Wald tracht disease: The disease related to poisoning of adult honey bee foraging on honey dew from conifers, usually spruce.

Warble flies: See *Oestrus wis.*

Warehouse: Warehouse is the scientific storage structure specifically constructed for protection of the quantity and quality of stored products against rodents, insect pests and ill effects of moisture and dampness. It may be called the protector of national wealth. Warehousing in India is integrated scheme of scientific storage, rural credit, price stabilization and market intelligence.

Warfacide: The liquid bait form of warfarin.

Warfarin ($C_{19}H_{16}O_4$): An anticoagulant, first commercial rodenticide with low toxicity, requires several feedings over a period of days for effective control (chronic poison). Chemically, it is 4-hydroxy-3(3-oxo-1 phenylbutyl)-2H-1-benzopyran-2-one. It has limited use due to more potent anticoagulants. The bait is prepared by mixing 5 parts warfarin in 90 parts ground grain, 3 parts sugar and 2 parts edible oil.

Warning agent: Addition of high pungent smell substances in the space fumigants or soil fumigants as warning agent. For example, chloropicrin is added as warning agent to methyl bromide where the latter is used as a space fumigant.

Warning colouration: Often a combination of contrasting colors that warns that an animal is dangerous. For example, stinging insects often show bands of black and yellow as warning colouration.

Wasp: Any of numerous insects of the superfamilies Vespoidea and Sphecoidea (order Hymenoptera). Wasps are social or solitary living insects, having a slender body with a constricted abdomen. They have two pairs of membranous wings and the mouthparts are adapted for biting or sucking. In females, the ovipositor is often modified as a sting. See *Polistes hebraeus.*

Water dispersible concentrate: The wettable powder or water dispersible slurry formulation.

Water dispersible granule (WG): Formulation containing granules which readily disperse in water to form a suspension.

Water dispersible powder (WP): Pesticide in a dry form with surfactant often mixed with or coated on a fine solid carrier for dispersion in water to form a suspension.

Water dispersible product: Any formulation that is finally diluted with water for spray. It includes wettable powder (WP), emulsifiable concentrate (EC), water dispersible concentrate (WDC), water suspendible concentrate (WSC) or water dispersible slurry (WDS).

Water dispersible slurry: A biphase liquid formulation containing solid particles in an emulsified liquid capable of forming final sprayable water suspension.

Water pollution: The presence of some foreign substances or impurities in water in such quantity so as to constitute health hazards and making it unfit for use.

Water soluble concentrate: An insecticidal formulation which gives ready to use solution on dilution with water. These formulations are generally prepared from sodium, potassium or ammonium salts of acidic pesticidal molecules suspended in methanol.

Wave length: Distance between two successive similar points on an alternating wave.

Wax glands: The glands that secrete bees wax; located in pairs on the last four visible ventral abdominal segments.

Wax layer: See lipid epicuticle.

Wax moth: Larvae of the waxmoth *Galleria mellonella* Linn. seriously damage brood and empty combs.

Weathering: The process of wearing away of pesticides from the treated surfaces due to factors of weathering, like rainfall, snow, ice and heat.

Wedge cell: A cell in the posterior portion of the wing formed by two posterior forks of Cu_a and an anterior fork of AA. It may also be referred to as the second cubito-anal cell (the first being formed between the base of Cu_a and AA and the first posterior fork of Cua). This has been inappropriately called the anal cell, but it is formed mainly by elements of the cubitus.

Weed: A plant growing at a place where it is not desirable.

Weed control: The process of inhibiting weed growth and limiting weed infestations to grow the crops profitably and to conduct the other operations efficiently.

Weed eradication: The elimination of all live weeds, plant parts or seeds of a weed from weed infested area.

Weedicide: Chemical that kills or controls the weeds.

Weevils: The beetles characterized by beak like prolongation of the head or rostrum and elbowed antennae, which cause damage in grub and adult stage to fruits, nuts, grain etc. For example, grain weevil, *Sitophilus granarius;* rice weevil, *Sitophilus oryzae* etc.

Weighting: A way for finding out phyletic information content of a character for sound classification.

Weismann's ring: See ring gland complex.

Wettable powder: See water dispersible powder.

Wetting agent: Surfactant for use in spray formulations to assist dispersion of a powder in the diluents or spreading of spray droplets on surface. It may also incorporate the functions of a sticker.

Wheat aphid: See *Macrosiphum miscanthi.*

Wheat ear cockle nematode (Wheat gall nematode): Nematode, *Anguina tritci* causes ear cockle in wheat.

Wheat thrips: See *Anapothrips flavicinctus* (Karny) (Thripidae, Thysanoptera). Cosmopolitan, serious pest of wheat in Maharastra and Gujarat. Low temperatures favour pest incidence at early stages of the crop.

White ant: Common name of termite.

White backed planthopper: See *Sogatella furcifera.*

Whitefly: Any of various insects of the family Aleyrodidae (suborder Homoptera, order Hemiptera). Whiteflies are small insects with four long wings and a white waxy body. They are often injurious to plants by sucking the plant juices or as vector of plant diseases. See *Bemisia tabaci.*

White grub: See cockchafer and *Holotrichia insularis* and *H. consanguinea.*

White head: White empty panicles are produced due to the attack of stem borers which cut the inner portion of culm.

White spirit: A fraction of petroleum used in paints, varnishes and pesticides.

White tailed mealy bug: See *Ferrisia virgata.*

Whorl: The arrangement of leaves in a circle around the stem of a plant.

Whorled antenna: A type of antenna with whorls of hair present at every knob, *e.g.*, mango mealy bug.

Wick effect: The movement of insecticide towards the surface of soil along with water is called wick effect.

Wilt: The loss of turgidity and drooping of leaves of plants due to inadequate water supply or due to excessive transpiration or due to vascular disease which interferes with the utilization of water or due to production of certain toxin byanorganism.

Wilting: Drooping of leaves.

Wind energy: Wind is a non-conventional source of energy, completely free from pollution and always remain available. Due to the flow of air, kinetic energy contained in it can be used for the production of electricity. The wind energy is directly used in taking out the water from the well, grinding of wheat and expelling oil.

Window: A small hole in the leaf with one epidermis left intact.

Wing: The insects are only vertebrates possessing the wings (Pterygota) and because of this virtue, they are adapted for the aerial life. Except apterygotes and some insects like, lice, fleas, termites, ants etc., all other insects possess two pairs of wings- one attached to mesothorax (forewings) and another to metathorax (hind wings). Diptera and male coccids possess only a single pair of wing (forewings) as hindwings are modified into thread like structures called halteres. Forewings of some insects are modified into thick leathery structures- the tagmina (Orthoptera, Dictyoptera and phasmida) or into hardy horny covers,the elytra (Coleoptera) while in some insects only the basic region of forewings are modified hemelytra (Heteroptera).

Wing base: In Ephemeroptra and Odonata, the chitinous wing base is continued with the lateral wall of thorax while in rest of the insects several sclerites are uniformly grouped at the wing base of each wing. They form a composite joint by which the wing articulates with the thorax. They include the tegula, axillary cord, axillary sclerites and variable forms of skeletal plates. The tegulae are confined mostly to the forewings, particularly in Lepidoptera, Hymenoptera and Diptera. Each lies at the base of costal vein as a small hairy chitinous pad or a scale like sclerite. It may occur in some insects in the form of chitinous plate overlapping the wing base. The axillary code is a lobe like structure lying closely with the velar margin on the posterior region. It often appears as a ligament and known as articular membrane. The first axillary sclerite lies at the base of subcostal vein. It articulates at the anterior notal wing process and sometimes divided into two parts. The second axillary accommodates at the base of radius vein. It articulates with the first axillary proximally and with the wing process of pleuron distally. The third axillary lies between the base of anal veins and the fourth axillary. It articulates directly with the posterior notal process. The median plates occur in pair, one among them is associated with the bases of media, cubitus and the first anal vein, while the other is often fused or closely apposed with the third axillary.

Wing coupling apparatus: In several insects, coupling of fore and hindwings of either side takes place in such a way that they move together as a single unit mostly with the lobes or spines lying at the wing base. Various types of wing coupling apparatus are: (i) Jugum, (ii) Humeral lobe, (iii) Frenulum and (iv) Hamuli.

Wing margins: The wings are somewhat triangular in shape and possess three distinct margins, the outer or apical margin and the inner or anal margin. The costal margin contains a pigmented spot, the pterostigma or stigma in large number of insects.

Wing pads: The partially developed wings of the nymphs of insects with an incomplete metamorphosis. The undeveloped wings of nymphs and naiads which appear as two flat structures on each side.

Wing regions or areas: The wing is differentiated broadly into the basal articular or axillary region and rest of the alar region. The alar region can be divided into three subregions which are distinctly separated by the folding lines in several insects. The vannal fold separates the anterior region, remigium from the posterior region, vannus; similarly the jugal fold differentiates the jugum lobe like area from the vannus.

Wing scales: These are macrotrichia (cuticular modifications) modified and varyingly pigmented microplate like structures that occur on the wing surface of the adult butterflies and moths.

Wing span: The measurement across the wings when they are extended.

Wing tracheation: According to Whitten (1962) each instar (including pupal stage) possesses its own tracheal system which may more or less differ from the next instar and from the adult stage. The tracheal pattern is not necessarily homologous between instars, because new additional tracheal branches are formed and the old ones are discarded. The wing tracheation in some insects is stable enough and helpful in study of venation pattern.

Wing venation: The wings are provided by longitudinal and cross veins forming a complex venation and it is species specific, therefore, of a great economic importance. The common longitudinal and cross veins are as follows: Longitudinal veins: Costa (C), Subcosta (Sc), Radius (R), Media (M), Cubitus (Cu), Postcubitus (PCu), Anal/ Venal (A/ V), Jugal (J); Cross veins: Humeral (h), Costal (c), Radial (r), Sectorial (s), Radio-medial (rm), Intermedial (m), Medio-cubital (m-cu), Intercubital (cu), Inter-anal (a).

Winter annual: Plants which germinate in the autumn and survive during winter completing their life cycle in the spring season.

Winter cluster: The arrangement of adult bees within the hive during winter.

Wireworms: The smooth, thin, yellow larvae of the click beetles which inhabit and attack various crops, particularly, cereals. Wireworms attack the plants in spring and autumn when they eat away the plants just below the soil surface causing foliage to turn yellow before the plants die. Wireworms take 4-5 years to mature after they pupate in the soil.

Withholding period: Minimum permissible time between the last application of a pesticide on a crop (including pasture) and harvesting for human consumption or for grazing of livestock. The minimum permissible time between the final application of a pesticide on an animal and the collection of the eggs or milk or slaughter for human consumption.

Wood dwelling termites: Wood dwelling termites are two types: (i) Damp wood termites: Which live in the dead decaying logs or damaged wood of standing trees, *e.g.*, Termopsidae, Kalotermitidae; and (ii) Dry wood termites: Which live in dead and dry wood, household timber and furniture, *e.g.*, *Cryptotermes*.

Wood preservative: A substance to preserve timber, stored wood or wood products from attack of termites or fungi. These are toxic oils that evaporate slowly and relatively, and water insoluble. The water soluble salts are used as sprays or water solutions and are absorbed inside the wood and highly toxic chemicals in solvent or solvent mixtures other than water.

Woolly bear: The tiger moth or hairy caterpillar, *Pericallia ricinii* belonging to family Arctiidae.

Worker bee: A female bee whose reproductive organs are undeveloped. Worker bees do all the work in the colony except for laying fertile eggs.

Worker comb: Comb measuring about five cells to the inch, in which workers are reared and honey and pollen are stored.

Worker jelly: The food material supplied by the workers in the regular brood cells that causes larvae to develop into workers.

Workers: In termites, the sterile males and females that perform most of the work of colony; they are pale, wingless, and usually lack compound eyes; in social Hymenoptera, females with undeveloped reproductive organs that perform the work of colony. These members are responsible for nest maintenance, food gathering and other works in the nest.

World wide web (www): This is the most common source of receiving and distributing the required information through internet.

WP: See wettable powder.

Wriggler: One that moves in a fashion of a wavy track, *e.g.*, the larvae of mosquitoes.

X

***Xanthium*:** An obnoxious weed genera.

Xanthone ($C_{13}H_8O_2$)**:** An insecticide and miticide having ovicidal effect, used in the control of insects and mites.

Xenobiosis: The relation in which colonies of one species live in the nests of another species and move free among the hosts, obtaining food from them by regurgitation or other means but still keeping their brood separate.

Xenobiotic substances or xenobiotics: Natural or synthetic compounds other than carbohydrates, proteins and vitamins which are considered as foreign compounds are called xenobiotics and are eliminated through various defense mechanisms of living organisms. The term is commonly used for man made environmental contaminants. When an insecticide is applied on any living organism, it undergoes some chemical changes resulting in the formation of some new products.

Xenophagous: Feeding on non-specific host plants.

***Xenopsylla cheopis* (Rothschild):** Belongs to order Siphonaptera and family Pulicidae. It is an insect pest of household, commonly known as the rat flea.

Xerophyte: A plant species adapted for growth under dry conditions.

Xyeloidea: Super family of suborder Symphyta (Chalastogastra), order Hymenoptera. This includes a single family Xyelidae. The insect are medium-sized to small, with the most generalized type of venation and three marginal cells in the wings. The antennae are characteristics in having greatly elongated third antennal segment followed by a flagellum. The larvae have limbs on all abdominal segments. A well developed ovipositor is present in adults. This family includes the genera *Xyela* and *Macroxyela*. The larvae of *Xyela julii* feed in the staminate flowers of pine and the adults visit the flowers of *Betula*.

Xylem: It is a system of more or less continuous channels in plants which is formed by fused dead cells. This part of plant's vascular system is adapted for transport of water and solutes from the roots to aerial parts or leaves.

Xylene: Any of three isomeric hyarocarbons of the benzene series used as solvents.

***Xylocopa*:** The genus, *Xylocopa* (Hurd and Moure, 1963) or carpenter bees (family: Xylocopidae, order: Hymenoptera) are large, almost humble bee like, though, the gaster is usually bare dorsally. They excavate long galleries in timber or make nests in large plant stems.

Xylocopidae: A family of order Hymenoptera. Pygidial plate is absent and rarely represented by an inconspicuous flattened spine. Clypeus not protuberant, lateral portions transverse rather than longitudinal. Fore coxae transverse. Solitary species, *e.g.*, *Xylocopa* sp.

Xylophagous: Insects which feed on wood.

***Xylosandrus compactus* Eichh.:** Belongs to order Coleoptera and family Scolytidae. It is an insect pest of coffee, commonly known as the coffee shot hole borer.

***Xylotrechus quadripes* Chevrolat:** Belongs to order Coleoptera and family Cerambycidae. It is an insect pest of coffee, commonly known as the coffee stem borer.

Y

Yaw: Movement of a body in a lateral plane where the front region moves in one direction with the hind region moving in the corresponding opposite direction, occurs in flying and jumping insects

Yeast: Unicellular fungus that produces zymase and is capable of fermenting carbohydrates to ethyl alcohol and carbon dioxide; used in baking, brewing etc.

Yellow fever: An acute infectious disease due to a virus transmitted by mosquito, *Aedes aegypti* in humid tropics of Africa and Central and Southern America. The disease is marked by fever, jaundice, and albumin and globulin in the urine.

Yellow jacket: In the United States, a number of ground nesting, light coloured wasps of the genus, *Vespula*.

Yellow mealworm: *Tenebrioides mauritanicus* L. is a pale white grub of a flat shiny black beetle that attacks broken grains and flour of wheat and oats. It forms hard masses in maida in which the grub lives and feeds. Grub stage lasts for 10 to 20 months.

Yellow phosphorus: The toxic com-ponent of some commercial poisons.

Yellow stem borer or paddy borer: See *Scirpophaga incertulas.*

Yellow tea mite (broad mite): See *Polyphagotarsonemus latus* (Banks).

Yellow wasp: See *Polistes hebraeus.*

Yellows: One of the various plant diseases, such as aster yellows whose prominent symptom is a loss of green pigment in the leaves.

Y-vein: Two adjacent veins fused distally to form a Y-shaped figure.

Yield forming organ: That part of plant which gives rise to the fruit or the harvest.

Yield potential: Full production -or yield capability of a crop or soil.

Yolk: Deutoplasm. When the yolk is small in quantity or evenly distributed through the cytoplasm, the egg is said to be alecithal or hololecithal; when the yolk has a central position, the egg is centrolecithal; if the yolk lies at one end of the egg, the latter is telolecithal. Insect eggs are centrolecithal.

Yolk cleavage: The division of the yolk into masses containing one to several cleavage nuclei.

Yponomeutidae: It is a family of super family Tinaeoidea of suborder Ditrysia and order Lepidoptera. The moths are usually small and brightly colored. Example, diamondback moth, *Plutella xylostella* which is major insect pest of cole crops and cruciferous vegetables.

Z

Zectran: It is a pesticide highly toxic to mites and snails (see Mexacarbate).

***Zenillia heterusiae*:** It is a parasite of bark eating caterpillar, *Indarbela quadrinotata* (Wlk.).

Zero tolerance: It is the limit for a pesticide residue in food or feed which is assumed to be zero, therefore, any detectable residue on the raw agricultural commodity when it is offered for shipment is deemed illegal. Zero tolerance is used by some regulatory systems, for example in USA, where no maximum residue limits have been established for particular pesticide or crop combination.

Zeugloptera: Suborder of Lepidoptera with functional mandibles, maxillae with lacinia, eight pairs of abdominal legs, exarate pupa, wings with aculeate venation alike in fore and hind wings, forewings with hind wings without frenulum.

Zeuzeridae: Family of super family Cossoidea of suborder Ditrysia and order Lepidoptera. Adults larger than Cossidae. Hindwings with R_5 and M_1 widely separated. Larvae bore into woods of trees. *Zeuzera coffeae* is one of the stem boring species of insects which are important as a serious pest of coffee in India. The larve of *Azygophleps scalaris* bores into the stems of green manure plants, *viz., Sesbania bispinosa, S.speciosa* and *S.grandiflora.* in the Russia. The *Zeuzera pyrina* is a serious pest of apple orchards.

Zig-zag leafhopper: See *Recilia dorsalis.*

Zinc phosphide (Zn_3P_2): Zinc phosphide is a grey powdered non-anticoagulant rodenticide having garlic like odour (which attracts rodents), insoluble in water that has been used extensively (2 per cent poison bait for field rat and 1.0 per cent for commensal rats) against all species of rats and mice and any other animals as well. When zinc phosphide comes in contact with dilute acids in the stomach, phosphine gas (PH_3) is released ($Zn_3P_2 + 6HCl \rightarrow 2PH_3 + 3ZnCl_2$), which is a substance that is responsible for death of the rodent. The toxicity is not from the zinc molecule. It must be handled and applied with great caution because of the acute danger it presents to people and animals due to extreme toxicity. Rats and mice that ingest lethal amount of bait usually succumb over night with terminal symptoms of convulsions, paralysis, coma and death from asphyxia. They typically die in a prone position with their legs and tails outstretched. Because, zinc phosphide is not stored in muscles or other tissues of poisoned animals, there is no secondary poisoning with this rodenticide. The bait, however, remains toxic up to several days in the gut of dead rodent. Other animals can be poisoned if they feed enough on the gut content of rodents recently killed with zinc phosphide. Bait shyness can be a problem, therefore, pre-baiting is necessary for achieving good bait acceptance. It is also used as poison bait for mole crickets.

Zineb: A fungicide used for foliage application; chemically it is zinc ethylene bisdithiocarbamate possessing low mammalian toxicity but phytotoxic to zinc sensitive plants. Its acute oral LD_{50} for rats is greater than 5200 mg kg^{-1}.

Zinophos: It is trade name of a synthetic insecticide and nematocide organophosphate, *O*, *O*-diethyl *O*-2—pyrazinyl phosphorothioate which is highly toxic to mammals. The acute oral LD_{50} for rats is 9 to 16 mg kg^{-1} and applied in soil to control garden nematodes on vegetables and nurs-eries.

Zolone: Trade name of phosalone (see phoasalone).

Zoocides: Chemicals which kill or controls the rodents and other animals.

Zoogeographical regions: The zoogeographical regions of the world are based on geographical distribution of animals. Nearctic: New world, part of Holaretic region, Greenland, Canada and USA. Neotropical: Tropical new world, South America and West Indies. Palaearctic: Old world, part of Holaretic region, Iceland, Canary Island, North Africa, Europe, USSR, Middle East, China and Japan. Ethiopian: Parts of Arbia and Africa (below Tropic of Cancer) including Madagascar. Oriental: The East Indies, Indian sub-continent, South East Asia, Philippines and Bali Islands. Australasian: Lands that lie South-East of Asia-Australia, New Zealand, New Guinea and Lombok Island.

Zoogeography: (Gr., *Zoon*= animal + geography). The science dealing with distribution of animals in space and relationship between specific animal form and the region in which they live.

Zoology: The study of animals. The different branches are: (i) Protozoology: Study of one celled animals, (ii) Entomology: Study of insects, (iii) Helminthology: Study of helminthes, (iv) Ichthyology: Morphological study of fishes, (v) Ornithology: Study of birds, (vi) Herpetalogy: Study of reptiles, and (vii) Mammalogy: Study of mammals.

Zooneuston: Animals those are associated with the water surface.

Zoonosis: Disease that can be transmitted to human from animals.

Zoophilous: A term applied to animal loving arthropods which feed on other animals in preference to man.

Zoophily: Transfer of pollens by means of animals (*e.g.*, bait), a type of pollination.

Zooplankton: Microscopic animal forms living suspended in water.

Zoraptera: It is an insect orthopteroid order which include the minute, dimorphic, termite like insects called zorapterans. Zorapterans are characterized by small and slender bodies, membranous wings or wings absent, chewing mouthparts, incomplete metamorphosis and with and without eyes and ocelli.

Zorotypidae: The smallest order Zoraptera comprises only a single family Zorotypidae, *e.g.*, *Zorotypus* sp.

***Zorotypus*:** A single genus *Zorotypus* (having 22 species) has been included in order Zoraptera and family Zorotypidae. *Zorotypus guineensis* was discovered first by Silvestri, 1913 and fairly studied.

Zygoptera: Suborder of Palaeopteran order Odonata. Wings similar, basally stalked fore and hind wings and usually held vertically above the abdomen; compound eyes are widely separated; male with superior or inferior anal appendages, penis unjointed; female with superior appendages only; ovipositor well developed and nymphs have three caudal gills as respiratory organs.

Zygote: A fertilized egg.

Zymogenous: Organisms those are transient to particular habitat.

Bibliography

Ambrose, Dunston P. (2004). *The insects: structure, function and biodiversity*, Kalyani Publishers, New Delhi.

Ananthakrishnan, T.R. (1976). *Insect and host specificity*, Macmillan Co., New Delhi.

Anonymous (1992). *Neem: A tree for solving global problems*, National Academy Press, Washington, D.C.

Apple, J.L. and Smith, R.F. (1976). *Integrated pest management*, Plenum Press, New York.

Askew, R.R. (1971). *Parasitic Insects*, Heinemann, London.

Atkins, M.D. (1980). *Introduction to insect behaviour*, Macmillan Publishing Co., New York.

Atwal, A.S. and Dhaliwal, G.S. (2002). *Agricultural pests of South Asia and their management*, Kalyani Publishers, New Delhi.

Ayyar, T.V.R. (1963). *Handbook of Economic Entomology for South India*, Government of Madras publication.

Banerjee, B. (1988). *An introduction to agricultural acarology*, Associated Publishing Co., New Delhi.

Bhargava, M.C. and Kumawat, K.C. (2010). *Pests of stored grains and their management*, New India Publishing Agency, New Delhi.

Boarsch, R.V.D.; Messenger, P.S. and Gutierrex, A.P. (1982). *An introduction to biological control*, Plenum Publishing Corporation, New York.

Borror, D.; DeLong, D.M. and Triplehorn, C.A. (1976). *An introduction to the study of insects*, Rinechart and Co., New York.

Burges, H.W. and Hussey, N.W. (1971). *Microbial control of insects and mites*, Academic Press, London.

Brussel, E. (1970). *An introduction to insect physiology*, Academic Press, New York.

Candy, D.J. and Kilby, B.A. Edited (1975). *Insect biochemistry and function*, Chapman and Hall, London.

Chapman, R.F. (1982). *The insects: Structure and function*, Eng. Uni. Press, London.

Comstock, J.H. (1040). *An introduction to entomology*, Ithaca, New York.

Corbett, J.R. (1974). The biological mode of action of pesticides, Academic Press, London.

Cotton, R.T. 1960. The insect pests of stored grain and milled cereal products. *Pests of stored grain and grain products*. Burges Comp. Minniapolis Minn.

David, B.V. and Kumaraswami, T. (2000). *Elements of economic entomology*, Popular Book Depot, Chennai.

Davidsen, G. (1974). *Genetic control of insect pests*, Academic Press, London.

Davies, R.G. (1988). *Outline of entomology*, Chapman and Hall, New York.

DeBach, P. (1974). *Biological control of natural enemies*, Cambridge University Press.

Dent, D. (1991). *Insect pest management*, Redwood Press Ltd., Wiltshire, U.K.

Dethier, V.G. (1947). *Chemical insect attractants and repellents*, Lewis and Co., London.

Dhaliwal, G.S. and Arora, R. (1994) edited. *Trends in agricultural insect pest management*, Commonwealth Publishers, New Delhi.

Dhaliwal, G.S. and Arora, R. (2001). *Integrated pest management*, Kalyani Publishers, Ludhiana.

Dhaliwal, G.S. and Dhaliwal, V.K (1993). *Advances in host plant resistance to insects*, Kalyani Publishers, Ludhiana.

Dwivedi, A.K.; Pandey, V., Nair, Rekha and Jain, Sama (2010). *Environmental engineering and disaster management*, Neelkanth Publishers (P) Ltd., Jaipur.

Elzinga, R.J. (1978). *Fundamental of Entomology*, Prentice Hall of India Pvt. Ltd., New Delhi.

Engelmann, F. (1970). *The Physiology of insect reproduction*, Pergamon Press, Oxford.

Essig, E.O. (1982). *College entomology*, Asia Playing Cards Co., Agra.

Feuemore, P.G. (1984). *Plant pests and their control*, Butterworths, London.

Fitzwater, W.D. and Prakash, Ishwar (1989). *Hand book of vertebrate pest control*, Publications and Information Division, Indian Council of Agricultural Research, New Delhi- 110012.

Gautam, R.D. (1994). *Biological pest suppression*, Westvill Publishing House, New Delhi.

Gilmour, D. (1961). *The biochemistry of insects*, Academic Press, New York.

Gillot, C. (1982). *Entomology*, Plenum Press, New York.

Gupta, A.P. (1979). *Insect haemocytes*, Cambridge Univrsity Press.

Kunkel, J.G. (1979). *Annual Review of Entomology*, **24**: 475- 505.

Hati, A.K. (1984). *Medical entomology*, Jagmandar Book Agency, New delhi.

Hill, D. (1975). *Agricultural insect pests of the tropics and their control*, Cambridge University Press, Cambridge, Great Britain.

Huffaker, C.B. (1980) edited. *New technology of pest control*, Wiley, New York.

Huffaker, C.B. and Messenger, P.S. (1975). *Theory and practice of biological control*, Academic Press, New York.

Hutson, D.H. and Roberts, T.R. (1985). *Insecticides*, John Willey and Sons, New York.

Karlson, P. and Butenandt, A. (1959). Pheromones (ectohormones) in insects. *Annual Review of Entomology*, **4:** 39-58.

Khare, B.P. (1993). *Stored grain pests and their management*, Kalyani publishers, Ludhiana.

Kilgore, W.W. and Richards, L.D. (1967). *Pest control,* Academic Press, New York.

Kotler, P.; Keller, K.L.; Koshy, A. and Jha, M. (2009). *Marketing Management: A South Asian Perspective*, 13th Edition; Published by Dorling Kindersley (India) Pvt. Ltd., New Delhi-110017

Kumawat, K.C. (2011). *Rhyzopertha dominica* (Fab.) on wheat: Bioecology and management, Lap Lambert Academic Publishing GmbH & Co., Saarbrucken, Germany.

Lal, O.P. (1996). Recent advances in Indian entomology, APC Publications Pvt. Ltd., New Delhi.

Lefroy, H.M. (1984). *Indian insect life*, Jagmandar Book agency, New Delhi.

Mani, M.S. (1968). *General Entomology*, Oxford and IBH Publishing Co., New Delhi.

Mani, M.S. (1974). *Modern classification of insects*. Satish Book Enterprises, Agra.

Matsuda, R. (1965). Morphology and evolution of insect head, *Mem. Amer. Entomol. Inst.* **4:** 1-334

Matsuda, R. (1970). Morphology and evolution of the insect thorax, *Mem. Entomol. Soc. Can.*, **76:** 1-431.

Matsuda, R. (1976). *Morphology and evolution of the insect abdomen*, Pergamon Press, New York.

Metcalf, C.L. and Flint, W.P. (1951). *Destructive and useful insects*, McGraw Hill, New York.

Nair, M Publication.R.G.K. (1986). *Insects and mites of crops in India*, ICAR, New Delhi.

National Research Council (1992). *Neem: A tree for solving global problems*, National Academy Press, Washington D.C.

Nayar, K.K.; Ananthakrishnan, T.N. and David, B.V. (1976). *General and applied entomology*, Tata McGraw Hill Publishing Co. Ltd., New Delhi.

Patten, R.L. (1963). *Introductory insect physiology*, W.B. Saunders Co., London.

Pedigo, L.P. (1996). *Entomology and pest management*, 2nd ed., Prentice Hall, Anglewood Cliffs, New Jersey.

Pradhan, S. (1969). *Insect pests of crops*, National Book Trust of India, New Delhi.

Pradhan, S. (1983). *Agricultural entomology and pest control*, ICAR Publication, New Delhi.

Pruthi, H.S. (1969). *A text book of agricultural entomology*, ICAR Publication, New Delhi.

Richards, A.G. and Brooks, M.A. (1968). Internal symbiosis in insects, *Annual Review of Entomology*, 3: 37-56.

Richards, O.W. and Davies, R.G. (1977). *Imm's general text book of entomology*, Xth edition, Vol. I and II, Chapman and Hall, London.

Roeder, K.D. (1953) edited. *Insect physiology*, Willey and Sons, New York.

Ross, H.H. (1956). *A text book of entomology*, Willey and Sons, New York.

Roy, D.N. and Brown, A.W.A. (1970). *Entomology: Medical and veterinary*, The Banglore Printing and Publishing Co. Ltd., Banglore.

Saxena, R.C. and Srivastava, R.C. (1997). *Entomology at a glance*, Agrotech Publishing Academy, Udaipur.

Singh, S. (1975). *Bee keeping in India*, ICAR Publication, New Delhi.

Snodgrass, R.E. (1970). *Principles of insect morphology*, Tata McGraw Hill Publishing Co. Ltd., New Delhi.

Snodgrass, R.E. (1960). *Facts and theories concerning insect head*, Smithson Misc. Collns., 142: 1-61.

Somani, L.l. (1996). *Dictionary of applied entomology*, Agrotech Publishing Academy, Udaipur.

Srivastava, K.P. (1993). *Text book of applied entomology*, Vol. I and II, Kalyani Publishers, Ludhiana.

Srivastava, R.P. and Saxena, R.C. (1989). *A text book of toxicology*, Himanshu Publications, Udaipur.

Tembhare, D.B. (1984). *A text book of insect morphology, physiology and endocrinology*, S. Chand and Co. Ltd., New Delhi.

Tembhare, D.B. (1997). *Modern entomology*, Himalaya Publishing House, Delhi.

Upadhyay, R.K.; Mukerjee, K.G. and Rajak, R.L. (1996) edited. *IPM system in agriculture*, Vol.I, Principles and practices, Aditya Books Pvt. Ltd., New Delhi.

Van Der Bosch, R. and Messenger, P.S. (1973). *Biological control*, Intertext Books, Aylesbury.

Wadhwani, A.M. and Chhabra, N.N. (1972). *Bee keeping in India*, ICAR Publication, New Delhi.

Wigglesworth, V.B. (1972). *Principles of insect physiology*, Methuen and Co., London.

www.ingramcontent.com/pod-product-compliance
Ingram Content Group UK Ltd.
Pitfield, Milton Keynes, MK11 3LW, UK
UKHW041133071026
15587UKWH00009B/451

9 788170 359623